T0343460

THE MYOCARDIUM

SECOND EDITION

THE
MYOCARDIUM

SECOND EDITION

GLENN A. LANGER

Departments of Physiology and Medicine
Cardiovascular Research Laboratory
University of California at Los Angeles
School of Medicine
Los Angeles, California

ACADEMIC PRESS

San Diego   London   Boston   New York   Sydney   Tokyo   Toronto

Cover photograph: Electron micrograph from thin-sectioned, conventionally prepared rabbit papillary muscle showing the ultrastructure of the diadic cleft. See Fig. 1A of Chapter 1 by Joy S. Frank and Alan Garfinkel.

This book is printed on acid-free paper.

Academic Press
a division of Harcourt Brace & Company
525 B Street, Suite 1900, San Diego, California 92101-4495, USA
http://www.apnet.com

Academic Press Limited
24-28 Oval Road, London NW1 7DX, UK
http://www.hbuk.co.uk/ap/

Library of Congress Cataloging-in-Publication Data

The myocardium / edited by Glenn A. Langer. -- 2nd ed.
      p.   cm.
  Rev. ed. of: The mammalian myocardium / edited by Glenn A. Langer and Allan J. Brady. 1974.
  Includes bibliographical references and index.
  ISBN 0-12-436570-1 (alk. paper)
  1. Myocardium. I. Langer, Glenn A., date. II. Mammalian myocardium.
  [DNLM: 1. Myocardium--cytology. 2. Myocardium--metabolism. WG 280 M9978 1997]
  QP113.2.M96 1997
  612. 1'7--dc21
  DNLM/DLC
  for Library of Congress   97-15596
                                         CIP

Printed and bound in the United Kingdom
Transferred to Digital Printing, 2011

To All Members of

the UCLA "Heart Lab"—Past,

Present, and Future

CONTENTS

1

IMMUNOLOCALIZATION AND STRUCTURAL CONFIGURATION OF MEMBRANE AND CYTOSKELETAL PROTEINS INVOLVED IN EXCITATION–CONTRACTION COUPLING OF CARDIAC MUSCLE

JOY S. FRANK AND ALAN GARFINKEL

2

MYOCARDIAL CELLULAR DEVELOPMENT AND MORPHOGENESIS

HONG ZHU

3

ION CHANNELS IN CARDIAC MUSCLE

JAMES N. WEISS

4

MYOCARDIAL ION TRANSPORTERS

KENNETH D. PHILIPSON

5

EXCITATION–CONTRACTION COUPLING AND CALCIUM COMPARTMENTATION

GLENN A. LANGER

6

MECHANICS AND FORCE PRODUCTION

KENNETH P. ROOS

7

METABOLISM IN NORMAL AND ISCHEMIC MYOCARDIUM

JOSHUA I. GOLDHABER

CONTRIBUTORS

Numbers in parentheses indicate the pages on which the authors' contributions begin.

Joy S. Frank (1) Departments of Physiology and Medicine, Cardiovascular Research Laboratory, University of California at Los Angeles, Los Angeles, California 90095

Alan Garfinkel (1) Departments of Physiology and Medicine, Cardiovascular Research Laboratory, University of California at Los Angeles, Los Angeles, California 90095

Joshua I. Goldhaber (325) Cardiovascular Research Laboratory and Division of Cardiology, University of California at Los Angeles School of Medicine, Los Angeles, California 90095

Glenn A. Langer (181) Departments of Physiology and Medicine, Cardiovascular Research Laboratory, University of California at Los Angeles School of Medicine, Los Angeles, California 90095

Kenneth D. Philipson (143) Departments of Physiology and Medicine, Cardiovascular Research Laboratory, University of California at Los Angeles School of Medicine, Los Angeles, California 90095

Kenneth P. Roos (235) Cardiovascular Research Laboratory and Department of Physiology, University of California at Los Angeles School of Medicine, Los Angeles, California 90095

James N. Weiss (81) Cardiovascular Research Laboratory and Division of Cardiology, University of California at Los Angeles School of Medicine, Los Angeles, California 90095

Hong Zhu (33) Cardiovascular Research Laboratory, Department of Physiology, University of California at Los Angeles School of Medicine, Los Angeles, California 90095

PREFACE TO THE SECOND EDITION

The first edition of *The Mammalian Myocardium* was published in 1974. It presented the "state of the science" for the heart as it was known at that time. Over the past 22 years, knowledge in the biological sciences has grown at a phenomenal rate and the field of cardiac function is no exception. New knowledge and insight have been gained particularly at the cellular, subcellular, and molecular levels. New techniques in molecular biology, structural analysis, ion measurement, electrophysiology, biochemistry, and force analysis have allowed probing and measurement of heart cell function at its most fundamental level.

The goal of this second edition is to update the state of the science for heart muscle. The first chapter (Drs. Joy Frank and Alan Garfinkel) is ultrastructural, with emphasis on membrane transporters, channels, and cytoskeleton based on immunofluorescence, confocal, freeze–fracture, and three-dimensional reconstructions. This is followed by a comprehensive chapter (Dr. Hong Zhu) on the molecular biology of cell development and control of the hypertrophic process. The third chapter (Dr. James Weiss) presents the molecular control of the cell's ionic channels (sodium, potassium, and calcium) and is followed by the chapter (Dr. Kenneth Philipson) on molecular control of the cell's transport systems (Na–K pump, Na–Ca exchangers, sarcolemmal Ca pump, sarcoreticular Ca pump, and the Na–H exchanger). The fifth chapter (Dr. Glenn Langer) focuses on the excitation–contraction process and subcellular Ca compartmentation as they affect contractile control. Next, the molecular basis of force production at the level of the myofilaments is presented (Dr. Kenneth Roos). Finally, a chapter on the subcellular and molecular processes involved in the control of energy production in the normal and ischemic myocardium (Dr. Joshua Goldhaber) completes the text.

Included in each chapter is a succinct summary following each major subsection within the chapter. These summaries stress the most important points of the section. This provides the reader with a review or, if read first, a preview of the section's content. In addition, whenever possible, molecular and subcellular processes are related to more global cardiac function to provide an integrated picture.

As with the first edition, this edition is directed to an audience that includes the graduate student in basic medical sciences, the postdoctoral M.D. pursuing a career in academic cardiology, the medical student interested in probing beyond the standard content of the medical curriculum, the academically oriented cardiologist, and the basic scientist interested in cardiac function and desiring a reference in areas outside his particular specialty.

It is noteworthy that virtually the entire content of this second edition represents material that was either completely unknown or known only in a very limited way when the first edition was published. As we stated in the preface to the first edition, it is somewhat risky to present a text devoted to a field in which knowledge is developing at such a rapid rate. The specter of rapid obsolescence is a threat. We, however, take the risk because we feel it is worthwhile to present in one volume an in-depth and comprehensive treatment of all the major aspects of myocardial cell function at the subcellular and molecular level as they are known at the end of the 20th century.

Finally, it should be noted that all the contributors to this edition are currently members of the Cardiovascular Research Laboratory at UCLA. Having all the requisite expertise under one roof has simplified the job of the editor in making the text not only comprehensive but also cohesive.

Glenn A. Langer
Los Angeles, California

PREFACE TO THE FIRST EDITION

It is not easy to prepare and edit a text devoted to a field in which knowledge is developing at a particularly rapid rate, for the threat of immediate obsolescence is ever present. Despite this threat, we feel that the need for a comprehensive presentation of cardiac cellular structure and function is worth the risk. The student who finds himself in a rapidly developing field has difficulty in finding the basic information in the variety of disciplines invariably directed toward the subject. It is important that he have a source that summarizes the present state in each of the disciplines and directs his attention to the problems that remain. That is the goal of this book. It is directed to a diverse group of students—the graduate student in the basic medical sciences, the postdoctoral M.D. who is pursuing a career in academic cardiology, the medical student who is interested in probing beyond the standard content of the medical curriculum, the academically oriented cardiologist, and the basic scientist interested in cardiac function who desires a reference in areas outside his particular specialty.

Chapter selections have been chosen to provide overlapping and integrative discussion of normal cellular myocardial function, with the inclusion of pathology and pharmacology only insofar as these disciplines lend insight into the fundamental processes of the muscle. No attempt has been made to include the myriad facets of organ or systemic function in the limited scope of this work.

Also, this book is not intended to present a series of reviews, each with an exhaustive bibliography. The contributors have been asked to limit their references as they would in text presentation and the editors take the responsibility for this limitation. The references are intended to be a starting point for the reader who desires to pursue any of the subjects.

It is an attempt, then, to provide in a single text current information and future problems with respect to the mammalian heart as they exist at the cellular and subcellular level. Hopefully the discussions and opinions expressed here will help to motivate or provoke our readers to further thought and experimentation toward the solution of the problems of myocardial function.

Glenn A. Langer
Allan J. Brady

1

IMMUNOLOCALIZATION AND STRUCTURAL CONFIGURATION OF MEMBRANE AND CYTOSKELETAL PROTEINS INVOLVED IN EXCITATION–CONTRACTION COUPLING OF CARDIAC MUSCLE

JOY S. FRANK AND ALAN GARFINKEL

I. INTRODUCTION

The focus of this chapter is on the structures involved in excitation–contraction (E–C) coupling. This is not intended to be a comprehensive review but does cover areas that have been of particular interest to the Cardiovascular Research Laboratory at UCLA School of Medicine and that are particularly related to the other chapters in this book. The structure of key functional units in E–C coupling is covered, especially the transverse (T) tubules, junctional and corbular sarcoplasmic reticulum (SR), and the diadic space formed by the close apposition of the sarcolemma (SL) and the junctional sarcoplasmic reticulum (jSR). In addition, the subcellular localization of proteins involved in the regulation of intracellular Ca^{2+} are discussed. This includes the distribution of the Na^+-Ca^{2+} exchanger, $Na^+K^+ATPase$, the Ca^{2+} release channel/ryanodine receptor (CRC/RR), and the Ca^{2+} channel (DHP receptor). The organization of proteins within regions of the SL or SR results in domains of functional significance. Thus a discussion on the role of cytoskeletal proteins and their relationship to Ca regulatory proteins completes the chapter.

II. Ca RECEPTORS, CHANNELS, RELEASERS, AND STORAGE

A. DIADIC CLEFT

The region where the SL and the SR appose each other contains the most important functional units for the coupling of excitation to contraction. The close apposition of the lateral cistern of the jSR to the SL defines a space (~15 nm) containing channels, receptors, and exchangers whose function revolves around the regulation of Ca flux into and out of the myocyte. For all its importance, this space and the membranes that restrict it have lacked an all-encompassing name. Lederer *et al.* (1990) coined the term *fuzzy space* for this region. Langer and Peskoff (1996) have designated this space the *diadic cleft* and have produced a working model of its structural and functional characteristics (see Chapter 5).

1. Ca Release Channel/Ryanodine Receptor

The membrane of the jSR facing the SL has regularly arranged projections or "foot" processes (Franzini-Armstrong, 1980) that extend ~12 nm into the diadic cleft to contact the membrane of the T tubule forming an interior coupling or to contact the surface SL to form a peripheral coupling. The structure and function of these junctional spanning proteins or "feet" have been studied in great detail (Franzini-Armstrong, 1980; Kelly and Kuda, 1979; Lai *et al.*, 1988; Anderson *et al.*, 1989). It is well accepted that the feet that are periodically arrayed in the diadic cleft contain the ryanodine receptor (RR), a high-conductance calcium channel that provides for the release of calcium ions from the SR during contraction (Ogawa, 1994; Anderson *et al.*, 1989).

Figure 1 contains two useful images of the diadic cleft. Figure 1A is a thin-section electron micrograph from rabbit papillary muscle. It is from this conventional type of preparation that the feet were first visible, due partly to the fact that they are equally spaced along the jSR. Figure 1B presents a more three-dimensional (3-D) perspective of the structure of the diadic cleft. The tissue is prepared here with freeze-fracture/deep-etching on cells that are not chemically preserved but ultrarapidly frozen.

An "end-on view" of the feet/calcium release channel (CRC) can easily be seen to span the cleft bridging the jSR and the T tubular membrane. Immunolabeling studies with antibodies against the RR have localized the CRC/RR in adult cardiac muscle predominantly at the level of the T tubules, in clearly defined banding pattern at the Z lines [Figs. 2A and B (see color insert)].

The large tetrameric CRC/RR ion channel complex has been isolated from skeletal muscle. With the use of the isolated channels, Radermacher *et al.* (1994) were able to obtain frozen hydrated samples that were used for cryo-electron microscopy to produce excellent preservation of the macromolecular ultrastructure of the channels. With 3-D reconstruction techniques, the most detailed view to date of the CRC structure was produced. The "foot" portion of the CRC is a large assembly (29 × 29 × 12 nm) linked to a smaller transmembrane component. The resolution afforded by the 3-D reconstruction is such that a cylindrical, low-density region extending down the center of the foot assembly could be discerned and most likely corresponds to the Ca^{2+} conducting pathway (Fig. 3).

In skeletal muscle, at least, there is good structural evidence for a direct interaction between the CRC/RR and the Ca^{2+} channels (DHP receptors) located within the T tubular membrane. This is an important point because the mechanism of excitation–contraction (E–C) coupling in skeletal muscle is believed to occur by sarcolemmal depolarization acting on the DHP receptors, which in turn act as voltage sensors (Leung *et al.*, 1988; Rios and Brum, 1987) that undergo a confirmation change. It is this direct molecular interaction that is believed to induce the opening of the CRC/RR and thus the release of calcium from the jSR. Block *et al.* (1988) identified large intramembrane particles (IMPs) in the P face (membrane face adjacent to the cytoplasm) of the T tubular membrane in skeletal muscle. These are the only intramembraneous particles seen in the T tubular membrane, and they have the characteristics of the DHP receptors. In addition, skeletal muscle T tubular membranes are the richest source of DHP receptors. This adds additional support to the idea that these particles represent the DHP receptors. Interestingly, the particles are arrayed in clusters of four particles forming a tetrad. The axis of fourfold symmetry is rotated and, as a result, each tetrad of IMPs in the tubular membrane is opposite alternating foot structures of the CRC/RR. This structural arrangement between the CRC/RR in the jSR and the DHP receptors in the T tubular membrane provides direct support for the mechanical coupling hypothesis of E–C coupling in the triad of skeletal muscle. Data have been accumulating to support the idea that in skeletal muscle the peripheral SL and the T tubular membrane have distinct ultrastructural features and contain several unique proteins. The DHP receptor localization in the T tubular membrane is the most important and obvious example but several other unique proteins such as TS28 present only in the T tubules of skeletal muscle and the protein

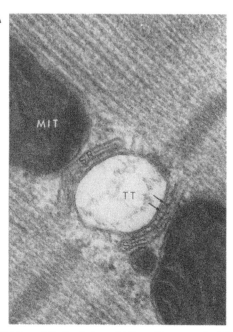

FIGURE 1   Ultrastructure of diadic cleft—rabbit papillary muscle. (A) An electron micrograph from thin-sectioned, conventionally prepared tissue. The specialized junction between the transverse tubule (TT) and the junctional side of the sarcoplasmic reticulum (SR) is clearly seen. Spanning the diadic cleft are regularly arrayed "feet" (arrows), which contain the Ca^{2+} release channels/ryanodine receptors (CRC/RRs). MIT, mitochondria. Original magnification x126,360. (B) An electron micrograph from tissue that was ultrarapidly frozen, fractured, and deep etched. The same structures seen in A are now visible in a 3-D perspective. The arrowhead points to the fractured lumen of the TT. The arrows clearly show the "feet" structures spanning the gap between the TT membrane and the junctional SR membrane, and contain the CRC/RRs. MIT, mitochondria. Original magnification x126,360. (Reprinted with permission from Frank, 1990.)

SL50 present only in the peripheral SL also show this distinct localization (Jorgensen *et al.*, 1990). This identification of membrane proteins unique to either surface SL or T tubules suggests that these distinct regions of the SL carry out distinct functions related to E–C coupling.

In cardiac muscle the situation is more complex. In contrast to skeletal muscle, the precise localization of the DHP receptor is still uncertain. The CRC/RR, while predominately located in the jSR in the diadic clefts, are also present extending from SR vesicles that are not in close proximity to the SL and thus are not associated with channels such as the DHP receptors. These vesicles are called the *corbular sarcoplasmic reticulum*.

2. Dihydropyridine Receptors

a. Localization of DHP Receptors in Cardiac Muscle

The morphology of freeze-fractured SL from cardiac muscle is studded with many IMPs ($\sim$2800/μm^2) (Frank *et al.*, 1987). Most of these particles

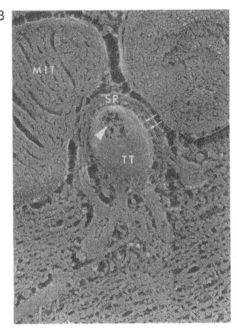

FIGURE 1 (*Continued*)

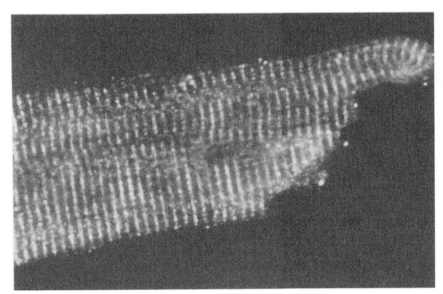

FIGURE 2A   Immunolocalization of the ryanodine receptor. Isolated rat myocyte labeled with antibodies against the CRC/RR. Immunofluorescence is clearly visible in this confocal micrograph as regularly spaced transverse bands. This localizes the CRC/RR at the level of the Z-disc and in association with the T tubular portion of the SL. Original magnification x 1500. (See color insert for Fig. 2B.)

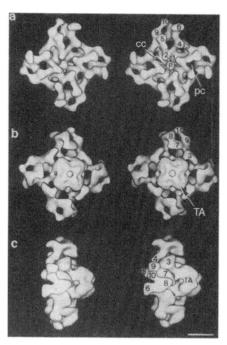

FIGURE 3   The 3-D architecture of the CRC/RR depicted from surface representations given here as stereo pairs. (a) This view shows the surface that would face the cytoplasm and the apposing T tubule in a triad junction. (b) This view shows the face that would interact with the jSR. (c) This is a side view. The protein that forms the cytoplasmic assembly appears to be arranged as domains that are loosely packed together and have been given numerical labels depicted on the right-sided pair. The other labeling is cc, central cavity, and p, plug feature that appears to be a globular mass in the center of the channel that the authors refer to as "channel plug." The plug is surrounded by four small cavities that lead to the exterior of the transmembrane assembly, labeled as pc for peripheral cavity. Bar = 10 nm. (Reproduced from *The Journal of Cell Biology*, 1994, vol. 127, p. 419, by copyright permission of The Rockefeller University Press.)

represent the numerous integral proteins involved in E–C coupling such as the Na^+–Ca^{2+} exchanger, $Na^+K^+ATPase$, and the DHP receptor. In contrast to the situation in skeletal muscle SL, in the adult cardiac SL there are no obvious differences between the number of particles on the peripheral and the T tubular SL. Within the T tubular membrane the IMPs vary in size between 7 and 12 nm (Fig. 4), but most of the particles are 8 nm in diameter. However, there are groups of large (>12 nm) particles that have not yet been identified with a particular channel or exchanger. Recent studies by Carl *et al.* (1995) determined the subcellular distribution of the DHP receptor, the SR protein triadin, and the CRC/RR in adult rabbit ventricular and atrial cells at the light microscope level of resolution with immunofluorescence labeling. In the adult rabbit ventricular cells,

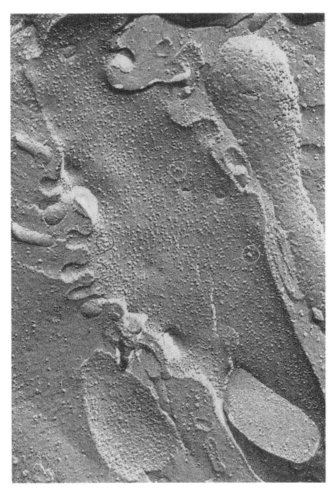

FIGURE 4 An electron micrograph from freeze-fractured rabbit papillary muscle exposing the P face of the TT. This face of the membrane contains numerous particles that represent the integral membrane proteins. Circles enclose groups of large ~12-nm particles, which may correspond to the DHP receptor. Original magnification x 101,952.

their data demonstrated that the DHP receptor is abundant in the T tubules and occurs in the peripheral SL only in discrete regions. In addition, the ventricular cells demonstrated complete overlap of the DHP receptor with the CRC/RR and the protein triadin, thus supporting the localization of the DHP receptor in the diads, closely associated with the CRC/RR. In atrial cells that lack T tubules, the DHP receptor localization was restricted to the SL and occurred in clusters closely associated with the CRC/RR. The coupling of the DHP receptor and CRCs in atrial cells appeared to be positioned primarily in the area of the SL that overlay the Z lines.

Another study by Sun *et al.* (1995) working with chick hearts, which lack T tubules, showed that the DHP receptors and the CRC/RRs are clustered and colocalized in foci along the SL in these developing cells. With freeze-fracture electron microscopy, they were able to identify large IMPs grouped in junctional domains in the SL. They proposed that these large particles are the DHP receptors and that although they are not organized into a tetrad pattern, as in the skeletal muscle T tubular membrane, they are located opposite the feet/CRC at the peripheral couplings.

Both studies provide the strongest evidence yet that the DHP receptors and the CRC/RR are closely associated in the diadic junctions in cardiac cells, and data from the adult rabbit cells suggest that although the DHP receptors are present in the peripheral SL they are more abundant in the T tubular membrane. The IMPs presumed to represent the DHP receptor in cardiac cells do not form tetrads with an obvious geometric relationship to the underlying CRC/RRs. Thus, in spite of the DHP receptor clusters in proximity to the CRC/RR, they are not specifically aligned. The DHP receptor in heart muscle is, in fact, believed to represent the L-type calcium channel (see Chapter 3). The lack of direct physical interaction between the DHP receptors and the CRC/RRs is not surprising given the importance of Ca^{2+}-induced Ca^{2+} release as a mechanism of the signal transduction mechanism in cardiac cells.

B. CORBULAR SARCOPLASMIC RETICULUM

The corbular SR is the mammalian homologue of the extended jSR first described in birds (Jewett *et al.*, 1971). It is now well recognized that cardiac myocytes contain, in addition to jSR, vesicular SR with all the anatomic features of jSR including the feet/CRCs, but their location in the cell is removed from either the peripheral or the T tubular SL (Fig. 5). Typically the corbular SR has a full array of feet that have been demonstrated in rabbit atrial cells to contain the CRC/RR and the protein triadin (Jorgensen *et al.*, 1993). In 3–4-day-old neonatal rabbit myocytes, which do not develop T tubules until ~12 days of age, the CRC/RRs are localized within the cytoplasm in transverse bands at the Z lines (Fig. 6). In the developing mammalian myocytes, while some of the CRC/RRs are in contact with the SL in the form of peripheral couplings, most of the CRC/RRs appear to be without a structural association to the SL and are functioning as extended jSR. As the cell ages and the T tubules develop, this relationship changes, and most of the CRC/RR will form junctions with the T tubules. The lumen of the corbular SR stores Ca^{2+}, which if released could be a significant percentage of the contractile calcium (see Chapter 5). Evaluation of the relative amount of calsequestrin within the lumen of jSR and corbular SR suggested that there could be as much as 40% of the SR calcium in the corbular SR in papillary muscle (Jorgensen *et al.*, 1985). This represents a

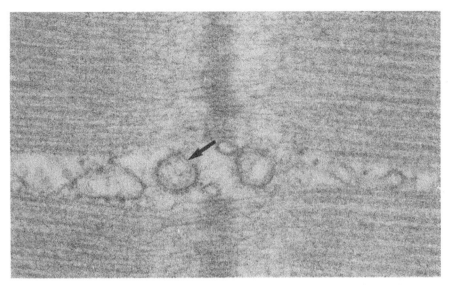

FIGURE 5 A thin-section electron micrograph from rabbit papillary muscle that illustrates the typical structure of corbular SR. Arrow indicates a corbular SR vesicle. Note the "feet" that project from the vesicular membrane into the cytoplasm and electron dense material within the lumen of the vesicle. Original magnification x 74,000.

significant amount of the total Ca^{2+} storage and release sites in mammalian cardiac cells. Sommer (Jewett and Sommer, 1971; Sommer and Waugh, 1976) first proposed that the corbular/extended jSR played a significant role in E–C coupling, and the new structural and functional data suggest that any model of E–C coupling in mammalian hearts must factor in a role for corbular SR. This is especially true in the developing cells before the presence of the T tubular system.

C. CALSEQUESTRIN

It is generally acknowledged that during relaxation some Ca^{2+} must be stored in the lumen of the SR, while some must be transported out of the cell (Langer, 1980). Calsequestrin is the major protein involved in Ca^{2+} storage in the SR (Campbell *et al.,* 1983). It has been shown that a protein with similar biochemical characteristics to calsequestrin was present in isolated SR vesicles that were ryanodine sensitive and absent from nonryanodine-sensitive SR vesicles (Jones and Cala, 1981). This study suggested that calsequestrin was present near the Ca^{2+} release sites in the SR. With the use of indirect immunocolloidal gold labeling on cryosections, Jorgensen *et al.* (1985) were able to localize calsequestrin within cardiac SR at the ultrastructural level. They found that calsequestrin was confined to the

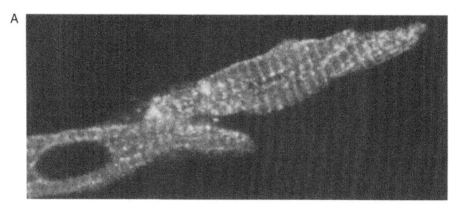

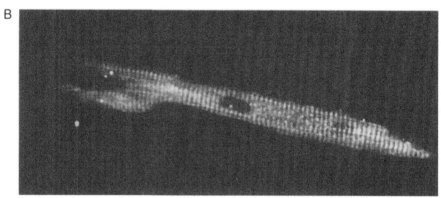

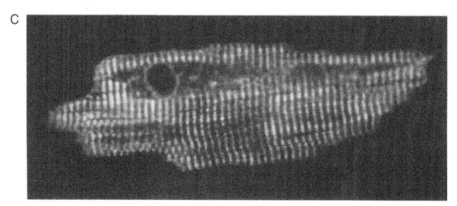

FIGURE 6   Immunolocalization of the CRC/RR in developing cardiac rabbit myocytes. These confocal micrographs are from (A) a 4-day-old myocyte that lacks T tubules, original magnification x 1680; (B) a 1-week-old myocyte, which is just beginning to develop T tubules, original magnification x 750; and (C) a 1-month-old myocyte where T tubules are almost equivalent in development to the adult, original magnification x 1095. It is clear that at 4 days and 1 week the rabbit myocyte has CRC/RRs present in organized arrays in the cytoplasm removed from sarcolemmal contact. Labeling along the cell surface in these developing cells indicates numerous CRC/RR sites in contact with the peripheral SL.

lumen of the peripheral and interior junctional SR as well as in the lumen of the corbular SR. Calsequestrin was absent from the lumen of the nonjunctional or network SR. These immunolabeling studies clearly indicate that Ca^{2+}ATPase and calsequestrin are localized within separate and distinct regions of the continuous SR membrane.

In summary, the SR of mammalian myocytes is composed of at least three distinct regions: network or nonjunctional SR, jSR, and the specialized nonjunctional corbular SR. SR morphology, especially at the ultrastructural level, reflects its different transport functions. The close apposition of the jSR to the peripheral SL and to the T tubular SL defines a 15-nm space containing channels, receptors, and exchanger molecules whose function revolves around regulating Ca^{2+} flux into and out of the myocyte. This space contains the functional units of E–C coupling including the CRC/RR, or "feet." Within the T tubular membrane in near proximity to the CRC/RR are the DHP receptors (the "L" Ca^{2+} channels), the Na^+–Ca^{2+} exchangers, and possibly the Na^+–K^+ pumps. In addition, associated proteins such as triadin are present. Although the functional roles of the E–C coupling units are fairly well defined, the ultrastructural organization of the channels, receptors, and exchangers within the 15-nm confines of the diad remains uncertain.

III. SUBCELLULAR LOCALIZATION OF CARDIAC MEMBRANE EXCHANGERS

The protein involved in Na^+–Ca^{2+} exchange and in Na^+–K^+ pumping are intramembrane proteins that are present in membranes of many cell types. The Na^+–Ca^{2+} exchanger (see Chapter 4) driven by a transmembrane Na gradient plays a key role in regulating Ca^{2+} concentration in many cells. In the heart, sodium pumps not only generate and maintain the Na^+–K^+ gradients across the SL but also regulate cardiac contractility by providing the driving force for Ca^{2+} extrusion via the Na^+–Ca^{2+} exchanger (see Chapters 4 and 5). The exchanger is the dominant mechanism of Ca^{2+} efflux from cardiac myocytes. Thus its role in E–C coupling is significant. Both of these proteins, the Na^+–K^+ pump and the Na^+–Ca^{2+} exchanger, affect Ca^{2+} regulation in the myocyte. Given that Ca^{2+} storage and release occurs at the diadic cleft and involves the jSR and the SL, especially the T tubular membrane, it is not unreasonable to hypothesize that the Na^+–Ca^{2+} exchanger and perhaps Na^+K^+ATPase might be positioned close to these E–C coupling units. This section focuses on the information currently available on the distribution of these important membrane proteins in the cardiac myocyte.

A. LOCALIZATION OF THE Na^+-Ca^{2+} EXCHANGER

Studies in other tissues indicate that the Na^+-Ca^{2+} exchanger is not distributed uniformly within membranes. In amphibian stomach smooth muscle cell membranes, the highest density of exchanger molecules are found in the plasmalemma adjacent to the jSR membrane. In an elegant immunolabeling study, Moore et al. (1993) were able to produce 3-D views of smooth muscle cells labeled with antibodies to the Na^+-Ca^{2+} exchanger, to the Na^+-K^+ pump, and to calsequestrin (a marker for the SR). They found that the Na^+-Ca^{2+} exchanger and the Na^+-K^+ pump appeared highly colocalized in the membrane. Their data suggested that there were few Na^+-Ca^{2+} exchangers that were not within at least 50 nm of a Na^+-K^+ pump. In addition, the clusters of Na^+-Ca^{2+} exchangers and Na^+-K^+ pumps occurred in regions of the membrane that were overlying the calsequestrin portion of the SR. The molecular organization seen in smooth muscle provides a mechanism for the linkage between the Na^+-K^+ pump and the Na^+-Ca^{2+} exchanger and between the Na^+-Ca^{2+} exchanger and Ca^{2+} release from the SR (see Chapter 5). Studies performed on cultured aortic smooth muscle cells also suggest a close association between the exchanger and the SR (Juhaszova et al., 1994). Immunofluorescent studies on cultured arterial smooth muscle demonstrate a reticular labeling pattern over the cell surface for the exchanger. A similar reticular pattern was observed for the intracellular labeling of the SR Ca^{2+}ATPase. In neurons, the Na^+-Ca^{2+} exchanger molecules were also localized in domains at the presynaptic terminals (Luther et al., 1992).

In cardiac myocytes both the Na^+-Ca^{2+} exchanger and the Na^+-K^+ pump are localized to the peripheral and T tubular membranes (Fig. 7; also see Fig. 14). Localizing the Na^+-Ca^{2+} exchanger to specific domains within the SL has proven to be a complex task. In studies on guinea pig cardiac myocytes, we reported a high density of exchanger sites in the T tubular membrane (Frank et al., 1992). The immunofluorescent images obtained using a monoclonal antibody against the exchanger ($R_3 F_1$) gave intense labeling of the T tubules with less label in the peripheral SL. Immunolabeling studies for the Na^+-Ca^{2+} exchanger in rabbit myocytes have suggested a somewhat more homogeneous distribution of the Na^+-Ca^{2+} in the SL (Frank et al., 1992; Chen et al., 1995). Studies by Kieval et al. (1992) suggested that the Na^+-Ca^{2+} exchanger was homogeneously distributed based on their immunolabeling experiments with the use of a polyclonal antibody against the exchanger in both guinea pig and rat myocytes. The spacial distribution of the exchanger is clearly an important issue because of the implications for E-C coupling.

A powerful approach to the visualization of the spacial distribution of proteins in membranes involves volume-rendering and gradient-shading techniques to produce 3-D images acquired from confocal microscopy of

A

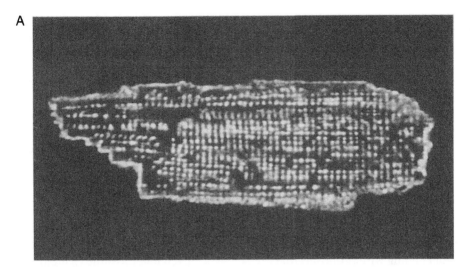

B

FIGURE 7 Confocal micrographs (A, low magnitude, original magnification x 930; B, higher magnitude, original magnification x 1590) of adult rabbit myocyte immunolabeled with a monoclonal antibody (R_3F_1) against the Na^+–Ca^{2+} exchanger. Note intense labeling of the T tubules, intercalated disc, and fairly homogenous staining of the peripheral SL.

immunolabeled cells. With this combination of techniques, the immuno-labeling of guinea pig and rat cardiac myocytes for the exchanger produced 3-D images of the T tubules and adjacent peripheral SL.

Figure 8 (see color insert) clearly shows labeled "hot spots" along the T tubules that represent high density of exchanger protein sites. What this type of analysis suggests is that, at least in guinea pig and rat myocytes, there exists domains of exchanger sites at higher concentration within the T tubular SL. The issue of possible domain structure for the exchanger is an important one in terms of defining models of E–C coupling. Clearly the exchanger is located in close proximity to the E–C coupling units involved in Ca^{2+} release. Figure 9 (see color insert) shows 3-D images of a rat myocyte labeled with antibodies to the Na^+–Ca^{2+} exchanger and the CRC/RR. These images were obtained by scanning the myocyte at 0.25-μm intervals in a digital imaging microscope and then subjecting the data to the constrained deconvolution algorithm based on regulation theory as used by Moore *et al.* (1993). Figure 9 illustrates the close association of the Na^+–Ca^{2+} exchanger to the CRC/RR within the T tubules, and shows that exchanger sites are in the region of the diadic junction (see Chapter 5).

In immature myocytes that lack T tubules, studies indicate that there are ~2.5 times more exchanger sites in the SL (per unit area) of the immature myocyte than in the mature myocyte SL (Artman, 1992). This is in keeping with proposals that the Na^+–Ca^{2+} exchanger assumes a greater importance in regulation of intracellular Ca^{2+} concentration in the developing myocyte than in the mature cardiac cell. Immunolabeling for the exchanger in these immature cells also shows an intense labeling along the peripheral SL. What is also striking is that the exchanger is inserted into the T tubule membrane as soon as it is formed (Fig. 10). Overexpression of the exchanger is also seen in transgenic mice models that exhibit a

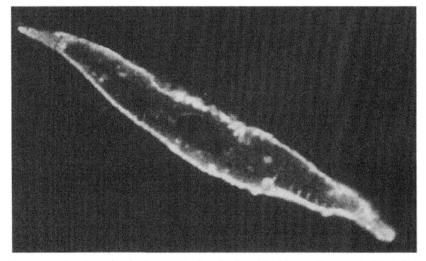

FIGURE 10   A confocal micrograph of an 11-day-old rabbit cardiomyocyte labeled with antibodies (R_3F_1) against the Na^+–Ca^{2+} exchanger. At this stage of development, T tubules are just beginning to form. Label is intense in the peripheral SL. However, the rudimentary T tubules are also labeled for the exchanger and are seen as small, evenly spaced projections from the SL (lower right-hand corner of the figure). Original magnification x 1590.

two- to threefold increase in exchanger activity and exhibit very intense immunolabeling for the exchanger in the peripheral and T tubular membrane. Intracellular labeling around the nucleus in the region of the Golgi is striking in the transgenic myocytes and reflects the increased production of exchanger protein (Fig. 11).

Attempts to study the Na^+–Ca^{2+} exchanger distribution at a higher resolution than the immunolabeling seen in the confocal light microscope have been difficult because the exchanger is a relatively sparse protein occupying less than 0.1% of the sarcolemmal protein. However, immuno-gold labeling of papillary muscle cryosections from rabbit and rat have produced some useful electron microscopic data. Given the geometry of the T tubular membrane, the likelihood of sectioning across the tubular membrane at a level to always encounter an exchanger site is problematic. In spite of this, the labeling of T tubular membrane profiles was frequent (Fig. 12), while the relative amount of label along the peripheral SL was above background but not abundant. What was surprising is the amount

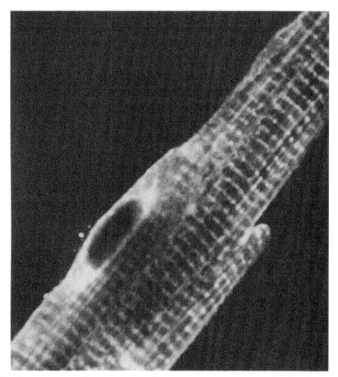

FIGURE 11   A confocal micrograph from a transgenic mouse myocyte labeled for the Na^+–Ca^{2+} exchanger. The transgene results in an overexpression of the Na^+–Ca^{2+} exchanger in these cells. Note the intense labeling in the T tubules, peripheral SL, and intracellularly in the Golgi surrounding the nucleus (seen here as a black oval). Original magnification x 2160.

FIGURE 12 A thin cryosection of rat papillary muscle that was labeled with R_3F_1 antibodies against the Na^+-Ca^{2+} exchanger and followed with goat antimouse secondary antibodies tagged with 5-nm gold. This low-power micrograph provides an overview of the cell, including a portion of the peripheral SL and several cross-profiles of T tubules. However, due to the low magnification, the 5-nm gold particles are difficult to visualize and have been enhanced so that their distribution is obvious. TT, transverse tubule. Original magnification x 32,472.

of gold label in the vicinity of the intercalated disc. At every disc encountered in a cryosection, label was more abundant than on either the T tubules or the peripheral membrane (Fig. 13). Immunofluorescent studies have reported (Frank *et al.*, 1992; Kieval *et al.*, 1992) labeling of the Na^+-Ca^{2+} exchanger in the intercalated disc membrane. Kieval *et al.* (1992) commented on the fact that the intercalated disc is a highly folded part of the

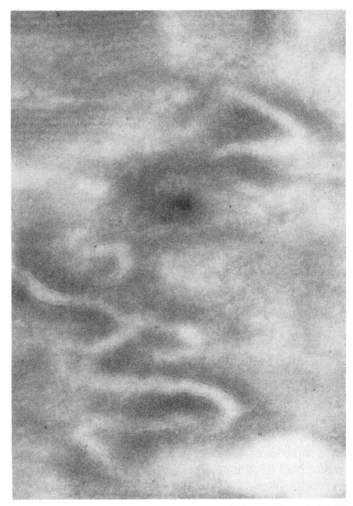

FIGURE 13   A thin cryosection of the intercalated disc portion of the SL from rat papillary muscle labeled as in Fig. 11 for the Na^+–Ca^{2+} exchanger. Gold labeling of exchanger sites is clearly seen in close association with the disc membranes. At this magnification, enhancement of the gold particles was not necessary. Original magnification x 92,000.

membrane, and this infolding could cause the augmented fluorescence that was observed. This is a reasonable explanation, but the immunogold labeling that provides superior spatial resolution does suggest a specialized role for the exchanger in this region of the cell. This is surprising since both the fascia adherens and the desmosome portions of the SL have been believed to have only a mechanical role in intercellular adhesion. However, a recent study on the intracellular localization of the inositol 1,4,5-triphospate (IP_3) receptor determined that IP_3 receptors in cardiac cells are localized to the fascia adherens portion of the intercalated disc. The presence of Na^+-Ca^{2+} exchanger sites at the intercalated disc places the exchanger in close association with another Ca^{2+} release channel. The idea that the intercalated discs are involved in intracellular Ca^{2+} regulation was first suggested by Kijma *et al.* (1993), based on the localization of IP_3 receptors at the disc. The localization of Na^+-Ca^{2+} exchangers in this region of the SL certainly adds to this intriguing possibility.

B. LOCALIZATION OF Na^+-K^+ATPase

Studies on the subcellular distribution of the sodium pump in mammalian cardiac cells have focused on whether there are isoform specific patterns of expression in the T tubules versus the peripheral SL, and whether there is a spatial coupling of Na^+ pumps and Na^+-Ca^{2+} exchangers within the SL. Na^+K^+ATPase is a heterodimer of an alpha catalytic subunit and a smaller beta glycoprotein subunit (Lingrel, 1992). The human heart expresses all three subunits. The rat heart expresses both $alpha_1$ and $alpha_2$ isoforms of the Na^+ pump, while the guinea pig expresses just the $alpha_1$ isoform (Orlowski and Lingrel, 1988; Sweadner *et al.*, 1994). In the normal adult rat, both isoforms can affect intracellular Na^+ gradients that control the activity of the Na^+-Ca^{2+} exchanger and influence intracellular Ca^{2+} stores. The localization of the various isoforms within the three membrane domains, including peripheral SL, T tubules, and intercalated discs, have been ambiguous. For example, studies on cryosections of rat hearts detected both $alpha_1$ and $alpha_2$ isoforms of the Na^+ pump in the peripheral SL, but very little labeling in the T tubules, while labeling in the intercalated discs was present in some myocytes and not others (Sweadner *et al.*, 1994). Other studies that used cell fractionation techniques also found $alpha_1$ and $alpha_2$ isoforms in rat SL subfractions but not in T tubular membrane fragments (Noel *et al.*, 1991). Another study that utilized freshly isolated rat heart myocytes, a distinct advantage for immunolabeling studies, clearly demonstrated that Na^+K^+ATPase $alpha_1$ subunits are enriched in T tubular membranes of rat myocytes (McDonough *et al.*, 1996), whereas the $alpha_2$ and $beta_1$ subunits were uniformly distributed in peripheral SL and T tubular membranes. In guinea pig myocytes that only express the $alpha_1$ isoform the Na^+ pump was uniformly distributed in the membrane. In both species, there was strong labeling of the intercalated discs. These results demonstrated that Na^+K^+ATPase is located in the same sites as the Na^+-Ca^{2+} exchanger (i.e., the peripheral SL, the T tubular, and intercalated disc portions of the SL). However, a functional colocalization between the Na^+ pump and the Na^+-Ca^{2+}, as was seen in smooth muscle (Moore *et al.*, 1991), was not detected in cardiac myocytes. The intense immunolabeling seen with several different antibodies against the Na^+ pump isoforms clearly localizes the Na^+ pump at the intercalated disc along with $Na^+–Ca^{2+}$ exchanger and IP_3 receptors (Fig. 14).

C. SARCOPLASMIC RETICULUM Ca ATPase

It is clear from structural and functional studies that the jSR contains the transport proteins concerned with Ca^{2+} release. Most of the nonjunctional or network SR is concerned with Ca^{2+} sequestration. In cardiac muscle, an important step in relaxation is removal of Ca^{2+} from the myofi-

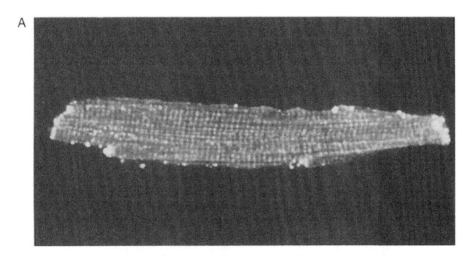

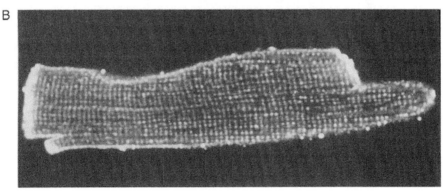

FIGURE 14   Immunolocalization of the Na^+K^+ATPase in rat cardiomyocytes. (A) Confocal image of the distribution of $\alpha1$-subunit labeled with 464.6 monoclonal antibody. Labeling is quite intense at the T tubules and the intercalated disc but less intense along the peripheral SL. Original magnification x 930. (B) Confocal image of the distribution of $\beta1$-subunit labeled with FP-B, polyclonal antiserum. Label is distributed along the peripheral SL, T tubules, and intercalated discs. Original magnification x 885. (Reprinted with permission from McDonough *et al.*, 1996.)

brils. Ca^{2+}ATPase actively transports Ca^{2+} from the myofibrils into the lumen of the SR (Tada *et al.,* 1978). The biochemical characteristics of SR Ca^{2+}ATPase are similar to those of the protein of skeletal muscle except that the Ca^{2+} uptake by cardiac Ca^{2+}ATPase is slower than the Ca^{2+} uptake in skeletal muscle. In addition, the Ca^{2+}ATPase of the SR is distinct from the Ca^{2+}ATPase within the cardiac SL (Caroni and Carifoli, 1981). Immuno-localization studies have shown that the SR Ca^{2+}ATPase is confined to the SR and is not present in the SL. In addition, within the SR, Ca^{2+}ATPase is uniformly distributed within the nonjunctional SR membrane but is lacking in the junctional SR membrane (Jorgensen *et al.,* 1995). With freeze-fracture electron microscopy techniques, nonjunctional SR membrane can be split to expose its interior architecture, and thus reveal a homogenous population of 8-nm intramembrane particles distributed uniformly through-out the membrane (Fig. 15). These intramembrane particles represent the Ca^{2+}ATPase pump proteins involved in Ca^{2+} uptake into the SR (Steward and MacLennan, 1974).

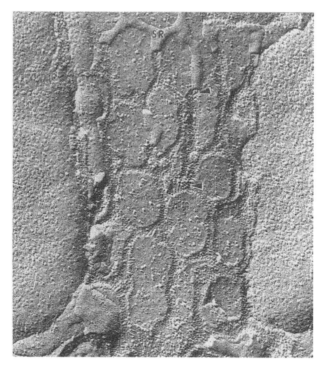

FIGURE 15   An electron micrograph from the freeze-fractured rat papillary muscle. The fracture plane has exposed the P face (arrowheads) of the network or longitudinal SR. This face of the nonjunctional SR membrane is studded with IMPs of a similar size and configuration and represent the Ca^{2+}ATPase protein homogenously distributed within this portion of the SR. Original magnification x 68,448.

In summary, although there are ample data localizing the Na^+–Ca^{2+} exchanger and the $Na^+K^+ATPase$ to the SL, a key question is whether there is preferential localization of these molecules to the diadic region, especially in the T tubules. The data from other tissues indicate that the Na^+–Ca^{2+} exchanger is localized in domains. In cardiac myocytes there are data to suggest that there are abundant exchanger sites in the T tubules, thus putting the exchanger in close proximity to the CRC/RR units. However, the degree of this regional specialization may vary with cardiac species. $Na^+K^+ATPase$ distribution appears to be isoform specific. $Alpha_1$ subunits of the Na pump are enriched in the T tubule membranes in rats, whereas $alpha_2$ and $beta_1$ subunits are uniformly distributed throughout the SL. The most intriguing regional specialization was found at the intercalated discs where both the exchanger and the Na pumps, as well as IP_3 receptors, are abundant. $Ca^{2+}ATPase$ actively transports Ca^{2+} from the myofibrils into the lumen of the SR and is uniformly distributed within the nonjunctional SR membrane.

IV. CYTOSKELETON

A. GENERAL OVERVIEW

In cardiac myocytes the cytoskeleton system usually refers to all of the filamentous structural proteins other than the contractile proteins. Titin, the intermediate filaments such as desmin and the microtubules, are well-studied cytoskeletal proteins that form an intracellular framework within the myocyte. All along the sarcomere the cytoskeletal proteins are involved in linkages with the contractile proteins, with other intracellular organelles, or with the cell membrane.

Titin (see Chapter 6), a large molecular weight protein, runs parallel to and between the thick filaments. Titin most likely functions to bind adjacent thick filaments together, thus maintaining the A-band organization (Wang et al., 1979; Furst et al., 1989). Titin may also serve to store and release elastic energy during the contractile cycle.

The intermediate filament cytoskeleton of adult heart cells is composed of desmin polymers and is oriented in the transverse direction with respect to the long axis of the cell (Lazarides, 1980, 1982). The majority of these filaments range in size from 7 to 10 nm and are arranged in bundles aligned with the myofibrillar Z-lines (Eriksson and Thornell, 1979). These filaments appear to connect the Z-bands in series with one another to reinforce sarcomere integrity. Desmin filaments are also aligned along the lateral borders of the Z-bands connecting adjacent Z-bands in parallel. All of these connections tend to stabilize the lateral registry of the myofibrillar apparatus prominent transverse network of intermediate filaments that project to the cytoplasmic surface of the nucleus. The potential function for this type of interconnection is for a propagation of mechanical force to the nuclear machinery (Georgatos and Blobel, 1988). This suggests that the intermediate filaments, in addition to acting as scaffolding, may also be involved in forming functional compartments within the cell.

Heart cells also have a prominent mictrotubular system that is closely associated with the nucleus (Rappaport and Samuel, 1988). Within the cytoplasm, mictrotubules are distributed between the myofibrils in parallel with the long axis of the cell. They come in close association with the mitochondria and may be involved in maintaining mitochondria in position between the myofibrils. Because microtubules are dynamic structures often subject to rearrangements and collapse during times of cardiac distress, they have been implicated in a possible transport function (Gelfand and Bershadasky, 1991).

B. CYTOSKELETAL MEMBRANE INTERACTIONS

The cytoskeleton is involved in many fundamental processes including cell adhesion, cell-to-cell interactions, maintenance of regional cell specializations, and the transfer of information from the cell surface into the cytoplasm. Thus, in addition to its role as a structured support system, the cytoskeletal system and its associated proteins may play a role in cell regulation. To exert these functions, the cytoskeleton must be anchored in some fashion to the SL. There is a wide variability in the structure of these cytoskeletal–membrane linkages and, in most cells, the exact molecular mechanisms are not known.

The most obvious linkage of the cytoskeletal system to the SL within the myocyte occurs between the intermediate filaments within the desmosome of the intercalated disc. The desmomal plaque contains a complex of proteins that are believed to link the intermediate filaments to the SL. The Z-bands are connected to the SL along the lateral aspects of the cell in a series of special domains termed *costameres* (Pardo et al., 1983). This linkage provides anchorage of the cell to the extracellular matrix. Vinculin, talin, and α-actinin are all localized along the inner leaflet of the SL at the Z-disc and, via transmembrane connections, link the cardiac integrin complex within the glycocalyx and the extracellular matrix, thus anchoring the myocyte to its surrounding environment (Terracio et al., 1990, 1991).

Two additional cytoskeletal proteins that may have important roles in regulating sarcolemmal stability and permeability are ankyrin and dystrophin. The rest of this section focuses specifically on the distribution of ankyrin and dystrophin in the cardiac myocyte.

C. ANKYRIN

Ankyrins are a multigene family of proteins that link the spectrin cytoskeleton to integral membrane proteins (Bennett, 1992). Such interactions between the cytoskeleton and the cytoplasmic domains of transmembrane proteins have played a major role in the regional specializations of membranes (Bennett, 1990; Luna and Hitt, 1992).

While first isolated from erythrocyte membranes, where it binds to the anionic exchanger, ankyrin also immobilizes proteins involved in transport functions in nonerythroid cells (Bennett *et al.*, 1985; Nelson and Lazarides, 1984). Ankyrin is reported to bind and colocalize with Na^+ channels in the kidneys and in the brain (Smith *et al.*, 1991). Ankyrin appears to be responsible for immobilizing Na^+ channels in the nodes of Ranvier and the axon hillock (Srinivasan *et al.*, 1988). The alpha$_1$ subunit of Na^+K^+ATPase directly binds to ankyrin as well (Devarajan *et al.*, 1994).

In skeletal muscle, ankyrin has been shown to colocalize with Na^+ channels in the postsynaptic folds of the neuromuscular junction. Ankyrin is also concentrated in the triad of skeletal muscle in close association with the DHP receptor (Flucher and Daniels, 1989; Flucher *et al.*, 1990). This was determined in immunocytochemical studies where ankyrin and the DHP receptor were localized in the triad junction formed by the T tubules and SR. The location of ankyrin in the triad suggests that ankyrin might be involved in organizing the triad and in immobilizing integral membrane proteins in T tubules and thus play a role in E–C coupling. Bourguignon *et al.* (1993) and Joseph and Samanta (1993) found that ankyrin is involved in regulation of inositol 1,4,5-triphospate (IP_3) receptor-mediated internal Ca^{2+} release from Ca^{2+} storage vesicles in mouse T-lymphoma cells and rat brain membrane, suggesting that ankyrin plays a pivotal role in the regulation of Ca^{2+} release during activation in these cells.

In cardiac muscle ankyrin is present in the SL as demonstrated by immunoblot and immunofluorescent techniques using anti-red blood cell ankyrin antibodies (Li *et al.*, 1993). While ankyrin is distributed in discrete areas all along the SL, labeling in association with the T tubules appears more abundant (Fig. 16). Ankyrin is also present within the cytoplasm in mature myocytes as faint periodic loci between the Z-lines. The function of ankyrin at these sites removed from the SL is unclear.

The Na^+–Ca^{2+} exchanger protein in cardiac SL binds to ankyrin with a high affinity. This was shown by direct binding and by immunoprecipitation experiments (Li *et al.*, 1993). In addition, the distribution of the exchanger and ankyrin demonstrated in dual immunolabeling experiments indicates that, within the T tubular SL and in discrete areas along the peripheral SL, ankyrin and the Na^+–Ca^{2+} exchanger colocalize (Chen *et al.*, in press). This interaction of the exchanger with ankyrin could restrict the movement of the exchanger and account for its localization in the T tubules. It is interesting that in immature cardiac cells that lack T tubules ankyrin was present at the Z-disc (Fig. 17). Ankyrin's presence at the Z-disc before sarcolemmal invagination occurs suggests that ankyrin is ideally situated to link to the channels and exchangers as they insert into the newly formed T tubular SL. Clearly, the possibility of physical association between ankyrin and

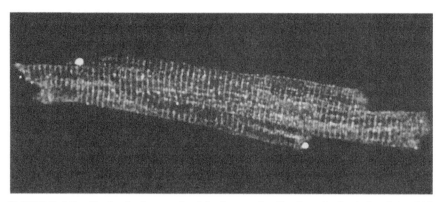

FIGURE 16   Confocal micrograph of the immunolocalization of ankyrin in adult rabbit myocyte. Ankyrin is distributed in transverse bands spaced about 1.8 μm apart. This labeling corresponds to the location of the T tubules. Label is also seen in discrete punctate regions along the peripheral SL. Original magnification x 750.

integral membrane proteins of the E–C coupling system in cardiomyocytes deserves investigation.

D. DYSTROPHIN

Dystrophin is a well-studied cytoskeletal protein that is most prominently expressed in skeletal and cardiac muscle (Hoffman *et al.,* 1987; Ahn and Kunkel, 1993). A large 400-kD protein, dystrophin is encoded by the Duchenne muscular dystrophy gene (Hoffman and Kunkel, 1989). Absence or a vast reduction of dystrophin in muscle causes severe progressive skeletal

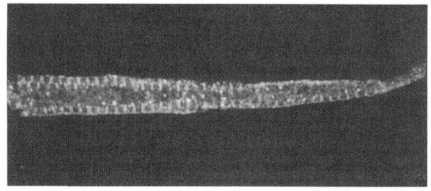

FIGURE 17   Confocal micrograph of the immunolocalization of ankyrin in a 5-day-old rabbit myocyte. Despite the lack of T tubules at this stage of development, ankyrin is present in transverse bands in the cytoplasm at spacing that coincides with the Z-lines. Label is also present in the peripheral SL. Original magnification x 1410.

muscle weakness and may result in dilated cardiomyopathy. Lack of dystrophin expression also causes myopathy in mdx mouse, a useful animal model for studying the function of dystrophin.

Recent analysis of the complete amino acid sequence of the dystrophin molecule suggests a strong structural homology with the central rod domain of spectrin (Koenig *et al.*, 1988) and an even more extensive sequence similarity to an isoform of α-actinin (Koenig *et al.*, 1988). Subcellular fractionation experiments and immunolabeling studies clearly localize dystrophin as a network on the cytoplasmic side of the SL. Dystrophin exists in the SL as a component of a large oligometric complex containing dystrophin-associated glycoproteins (DAGs) and dystrophin-associated proteins (DAPs) (Campbell and Kahl, 1989). Characterization of the dystrophin–glycoprotein complex has produced a model (Fig. 18) in which the aminoterminal domain of dystrophin is the cytoskeletal link between the subsarcolemmal actin cytoskeleton and the glycoprotein complex that spans

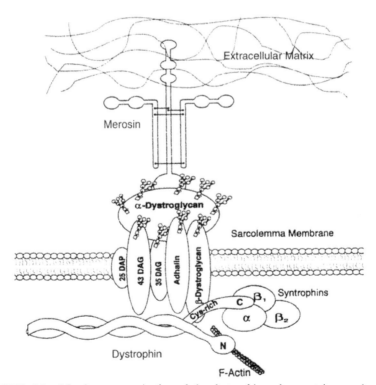

FIGURE 18   Membrane organization of the dystrophin—glycoprotein complex. This model shows all the known components of the dystrophin–glycoprotein complex in their structural associations with the SL and the extracellular matrix. Five glycoproteins span the membrane to link α-dystroglycan to dystrophin. The linkage between α-dystroglycan to the extracellular matrix is via merosin. (Reprinted with permission from Campbell, 1995.)

the SL (Hemmings *et al.*, 1992; Suzuki *et al.*, 1994). The glycoprotein complex contains five glycoproteins that are linked to the cysteine-rich domain of dystrophin by β-dystroglycan. Located on the sarcolemmal surface, α-dystroglycan is linked to dystrophin via the transmembrane glycoprotein complex and to the extracellular matrix via binding to merosin (muscle isoform of laminin) with high affinity in a calcium-dependent manner (Ervasti and Campbell, 1993). The linkage between the extracellular matrix and the cytoskeleton through dystroglycan forms a bridge across the SL from the cytoplasmic side to the extracellular matrix and is the basis for the proposed function of dystrophin in providing membrane stability.

Dystrophin is located in both skeletal and cardiac muscle along the peripheral SL (Frank *et al.*, 1994; Arahata *et al.*, 1988; Byers *et al.*, 1991; Klietch *et al.*, 1993), consistent with its function as part of a transsarcolemmal structure bridge. In skeletal muscle, most biochemical and immunocytochemical, reports agree that dystrophin is not present in the T tubular portion of the SL. This difference in the distribution of dystrophin between the peripheral SL and the T tubular SL is consistent with previous studies demonstrating that in skeletal muscle the peripheral SL and the T tubules each have distinctive ultrastructural features, contain several unique proteins and appear to carry out different functions at least as related to E–C coupling.

In most cardiac myocytes dystrophin is present and plentiful in both the peripheral SL and in the T tubular membrane (Fig. 19A) (Frank *et al.*, 1994; Klietsch *et al.*, 1993). In experiments conducted to determine dystrophin localization during rabbit cardiomyocyte development, it appears that dystrophin is present in association with the T tubular membrane as soon as it develops (Fig. 19B and C) (Frank *et al.*, 1994). Klietsch *et al.* (1993) demonstrated that in cardiac muscle dystrophin associated proteins, dystrophin, and laminin (merosin) codistribute in both the peripheral SL and the T tubular membranes. However, there may be tissue and species differences in dystrophin localization. For example, studies on rat and mouse ventricular myocytes report intense immunolabeling for dystrophin in the peripheral SL but no labeling in the T tubules (Byers *et al.*, 1991). More recent studies support the absence of detectable dystrophin in the T tubules of the mouse but do find a low level of dystrophin in the T tubules of the rat (Miller *et al.*, in press). The complexities over the subcellular localization of dystrophin are not well understood because the function of dystrophin in cardiomyocytes is still unclear. Dystrophin may serve diverse roles through its various associations. Recent work in cardiac cells show that a significant portion (~35%) of dystrophin is located in the myofibrils at the Z-disc (Meng *et al.*, 1996). The function of this unique nonmembrane location of dystrophin is unknown, although myofibrillar dystrophin is hypothesized to be critical for the maintenance of sarcomeric structure. The variations in dystrophin distribution along the SL of different cardiac species

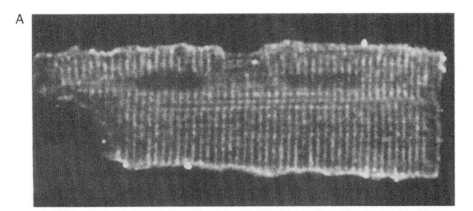

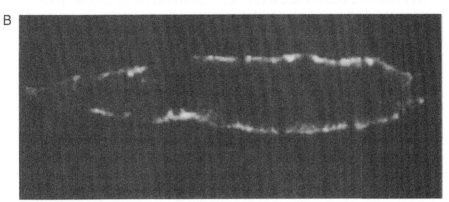

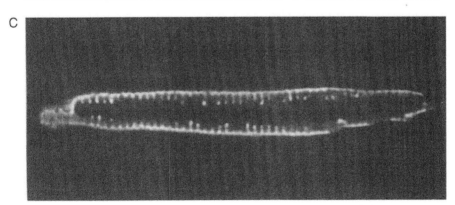

FIGURE 19   Immunolocalization of dystrophin in rabbit myocytes. (A) Confocal micrograph of an adult myocyte labeled with antidystrophin antibodies. Note uniform labeling of the peripheral SL and T tubules and lack of labeling at the intercalated discs (ends of the cell). Original magnification x1140. (B) Confocal micrographs of 4-day-old rabbit myocyte labeled with antidystrophin antibodies. At this stage of development, labeling for dystrophin is intense but somewhat discontinuous along the cell periphery. No labeling is present in the cell interior, since T tubules are absent. Original magnification x2760. (C) Confocal micrograph of a 1-week-old rabbit myocyte. Labeling with antidystrophin antibodies is present as intense fluorescence at the peripheral cell membrane and at the developing T tubules, seen here as short projections from the cell surface into the cytoplasm. Original magnification x1065. (Reprinted with permission from Frank *et al.*, 1994.)

suggest that dystrophin's roles may also vary with species. While the dystrophin–glycoprotein complex is believed to have an important role in stabilizing the SL and protecting it from the mechanical stresses during muscle contraction, it also appears that dystrophin serves a unique role in signal transduction by directly associating with and localizing nitric oxide synthatase to the SL (Brenman et al., 1995). Although the function of dystrophin along the T tubules of cardiomyocytes is still uncertain, it is possible that dystrophin has a role in maintaining a heterogeneous domain arrangement for some of the membrane glycoproteins. In this capacity, dystrophin could have a role in maintaining a nonuniform distribution of ion channels or cell surface receptors. This would be analogous to the role ankyrins play in the membrane. Indeed the relationship of these cytoskeletal proteins, both of which are located beneath the SL, requires future study.

In summary, cytoskeletal proteins have many functions in cardiac muscle, including providing a structural framework for intracellular organelles, especially the sarcomeric units in muscle; stabilizing the SL at sites of intercellular linkages, such as the intercalated disk or linking the sarcomeres to the extracellular matrix at the Z-lines; and, most important, forming cytoskeletal–membrane attachments to integral membrane proteins. This linkage of the cytoskeleton either directly or via associated cytoskeletal proteins such as ankyrin can tether channels, exchangers, and receptors to form regional specializations within the membrane.

REFERENCES

Ahn, A.H., and Kunkel, L.M. (1993). The structural and functional diversity of dystrophin. *Nature Genet.* 3:283–291.

Anderson, K., Lai, F.A., Lui, Q.-Y., Rousseau, E., Erickson, H.P., and Meissner, G. (1989). Structural and functional characterization of the purified cardiac ryanodine receptor–Ca^{2+} release channel complex. *J. Biol. Chem.* 264:1329–1335.

Arahata, K., Ishuira, S., Ishiguro, T., Tsukahara, Y., Suhara, C., Eguchi, C., Ishihara, T., Nonaka, I., Ozawa, E., and Sugita, H. (1988). Immunostaining of skeletal and cardiac muscle surface membrane with antibody against Duchenne muscular dystrophy peptide. *Nature (London)* 333:861–866.

Artman, M. (1992). Sarcolemmal Na–Ca exchange activity and exchanger immunoreactivity in developing rabbits. *Am. J. Physiol.* 263:H1506–H1513.

Bacallao, R., and Garfinkel, A. (1994). Three dimensional volume reconstruction in confocal microscopy: Practical considerations. *In* "Three-Dimensional Confocal Microscopy" (Stevens, J.K., Mills, L.R., and Trogadis, J.E., eds.), pp. 169–180. Academic Press, New York.

Bennett, V. (1990). Spectrin-based membrane skeleton: A multipotential adaptor between plasma membrane and cytoplasm. *Physiol. Rev.* 70:1029–1065.

Bennett, V. (1992). Ankyrins: Adaptor between diverse plasma membrane proteins and the cytoplasm. *J. Biol. Chem.* 267:8703–8706.

Bennett, V., Baines, A.J., and Davis, J.Q. (1985). Ankyrin and synapsin: Spectrin-binding proteins associated with brain membranes. *J. Cell Biochem.* 29:157–169.

Block, B.A., Imagawa, T., Campbell, K., and Franzini-Armstrong, C. (1988). Structural evi-

dence for direct interaction between the molecular components of the transverse tubule/
sarcoplasmic reticulum junction in skeletal muscle. *J. Cell. Biol.* 107:2587–2600.

Bourguignon, L.Y.W., Jin, H., Lida, N., Brandt, N.R., and Zhang, S.H. (1993). The involvement
of ankyrin in the regulation of inositol 1,4,5-trophosphate receptor-mediated internal Ca^{2+}
release from Ca^{2+} storage vesicles in mouse T-lymphoma cells. *J. Biol. Chem.* 268:7290–
2797.

Brenman, J.E., Chao, D.S., Xia, H., Aldape, K., and Bredt, D.S. (1995). Nitric oxide synthase
complexed with dystrophin and absent from skeletal muscle sarcolemma in Duchenne
muscular dystrophy. *Cell* 82:743–752.

Byers, T., Kunkel, J., and Watkins, S.C. (1991). The subcellular distribution of dystrophin in
mouse skeletal, cardiac and smooth muscle. *J. Cell. Biol.* 115:411–421.

Campbell, K.P. (1995). Three muscular dystrophies—Loss of cytoskeleton–extracellular matrix linkage. *Cell* 80:675–679.

Carl, S.L., Felix, A., Caswell, R., Brandt, N.R., Ball, W.J., Vaghy, P.L., Meissner, G., and
Ferguson, D.G. (1995). Immunolocalization of sarcolemmal dehydropyridine receptor and
sarcoplasmic reticular triadin and ryanodine receptor in rabbit ventricle and atrium. *J.
Cell. Biol.* 129:673–682.

Caroni, P., and Carafoli, E. (1981). The Ca pumping ATPase of heart sarcolemma. *J. Biol.
Chem.* 256:3263–3270.

Chen, F., Mottino, G., Klitzner, T., Philipson, K., and Frank, J.S. (1995). Distribution of
the Na^+/Ca^{2+} exchange protein in developing rabbit myocytes. *Am. J. Physiol.* 268(Cell
37):C1126–C1132.

Chen, F., Mottino, G., Shin, V.Y., and Frank, J.S. (in press). Subcellular distribution of ankyrin
in developing rabbit heart relationship to the Na/Ca exchange.

Devarajan, P., Scararmuzzino, D.A., and Morrow, J.S. (1994). Ankyrin binds to two distinct
cytoplasmic domains of Na K ATPase α subunit. *Proc. Natl. Acad. Sci. USA* 91:2965–2969.

Eriksson, A., and Thornell, L.E. (1979). Intermediate filaments in heart Purkinje fibers. A
correlative morphological and biochemical identification with evidence of a cytoskeletal
function. *J. Cell. Biol.* 90:231–247.

Ervasti, J.M., and Campbell, K.P. (1993). Dystrophin and the membrane skeleton. *Curr. Opin.
Cell. Biol.* 5:82–87.

Flucher, B.E., and Daniels, M.P. (1989). Distribution of Na channels and ankyrin is neuromuscular junctions is complementary to that of actylcholine receptor in the 43 kd protein.
Neuron 3:163–175.

Flucher, B.E., Morton, M.E., Froehner, S.C., and Daniels, M.P. (1990). Localization of the
α1 and α2 subunits of the dihydropyridine receptor and ankyrin in skeletal muscle triads.
Neuron 5:339–351.

Frank, J.S. (1990). Ultrastructure of the unfixed myocardial sarcolemma. *In* "Calcium and
the Heart" (G.A. Langer, ed.) pp. 1–25. Raven Press, New York.

Frank, J.S., Beydler, S., and Mottino, G. (1987). Membrane structure in ultra-rapid frozen
unpretreated, freeze-fractured myocardium. *Circ. Res.* 61:141–147.

Frank, J.S., Mottino, G., Reid, D., Molday, R.S., and Philipson, K.D. (1992). Distribution of
the Na^+–Ca^{2+} exchange protein in mammalian cardiac myocytes: An immunofluorescence
and immunogold-labeling study. *J. Cell. Biol.* 117:337–345.

Frank, J.S., Mottino, G., Chen, F., Peri, V., Holland, P., and Juann, B.S. (1994). Subcellular
distribution of dystrophin in isolated adult and neonatal cardiac myocytes. *Am. J. Physiol.*
267(Cell):C1707–C1716.

Franzini-Armstrong, C. (1980). Structure of sarcoplasmic reticulum. *Fed. Proc.* 39:2403–2409.

Furst, D.O., Nave, R., Osborn, M., and Weber, K. (1989). Repetitive titin epitopes with a 42
nm spacing coinside in relative position with known A band striations also identified
by major myosin-associated proteins. An immunoelectron study of myofibrils. *J. Cell.
Sci.* 94:119.

Gelfand, V.I., and Bershadasky, A.D. (1991). Microtubule dynamics: Mechanism regulation and function. *Am. Rev. Cell. Biol.* 7:93–116.

Georgatos, S.D., and Blobel, G. (1988). Laminin B constitutes an intermediate filament attachment site at the nuclear envelope. *J. Cell. Biol.* 105:117–125.

Hemmings, L., Kuhlman, F.A., and Critchley, D.R. (1992). Analysis of the actinin-binding domain of α-actinin by mutagenesis and demonstration that dystrophin contains a functionally homologous domain. *J. Cell. Biol.* 116:1369–1380.

Hoffman, E.P., Brown, R.H., and Kunkel, L.M. (1987). Dystrophin: The protein product of the Duchenne muscular dystrophy locus. *Cell* 51:919–928.

Hoffman, E.P., and Kunkel, M. (1989). Dystrophin abnormalities in Duchenne/Becker muscular dystrophies. *Neuron* 2:1019–1029.

Jewett, P.H., Sommer, J.R., and Johnson, E.A. (1971). Cardiac muscle: Its ultrastructure in the finch and hummingbird with special reference to the sarcoplasmic reticulum. *J. Cell. Biol.* 49:50–69.

Jones, L.R. and Cala, S.E. (1981). Biochemical evidence for functional heterogeneity of cardiac sarcoplasmic reticulum vesicles. *J. Biol. Chem.* 256:11809–11819.

Jorgensen, A.O., Shen, C.Y., and Campbell, K.P. (1985). Ultrastructural localization of calsequestrin in adult rat atrial and ventricular muscle cells. *J. Cell Biol.* 101:257–268.

Jorgensen, A.O., Arnold, W., Shen, A.C., Yuan, S., Gaver, M., and Campbell, K.P. (1990). Identification of novel proteins unique to either transverse tubules (TS28) or the sarcolemma (SL50) in rabbit skeletal muscle. *J. Cell. Biol.* 110:1173–1185.

Jorgensen, A.O., Shen, A.C.Y., Arnold, W., McPherson, P.S., and Campbell, K.P. (1993). The Ca release channel/ryanodine receptor is localized in junctional and corbular sarcoplasmic reticulum in cardiac muscle. *J. Cell. Biol.* 120:969–980.

Joseph, S.K., and Samanta, S. (1993). Detergent solubility of the inositol triphosphate receptor in rat brain membranes. Evidence for association of the receptor with ankyrin. *J. Biol. Chem.* 268:6477–6486.

Juhaszova, M., Ambesi, A., Lindemayer, G., Block, R.J., and Blaustein, M. (1994). Na Ca exchanger in arteries: Identification by immunoblotting and immunofluorescence microscopy. *Am. J. Physiol.* 266C:C234–C242.

Kelly, D.E., and Kuda, A.M. (1979). Subunits of the triadic junction in fast skeletal muscle as revealed by freeze-fracture. *J. Ultrastructure Res.* 68:220–233.

Kieval, R.S., Block, R.J., Lindenmayer, G.E., Ambesi, A., and Lederer, W.J. (1992). Immunofluorescence localization of the Na–Ca exchanger in heart cells. *Am. J. Physiol.* 263(Cell 32):C545–C550.

Kijima, Y., Saito, A., Jitton, T.L., Magnuson, M., and Fleischer, S. (1993). Different intracellular localization of inositol 1,4,5-tusphosphate and ryanodine receptors in cardiomyocytes. *J. Biol. Chem.* 268:3499–3506.

Klietsch, R., Ewasti, J.M., Arnold, W., Campbell, K.P., and Jorgensen, A.O. (1993). Dystrophin–glycoprotein complex and laminin colocalize to the sarcolemma and transverse tubules of cardiac muscle. *Circ. Res.* 72:349–360.

Koenig, M., Monaco, A.P., and Kunkel, L.M. (1988). The complete sequence of dystrophin predicts a rod-shaped cytoskeletal protein. *Cell* 53:219–229.

Lai, F.A., Erickson, H.P., Rosseau, E., Lui, Q.Y., and Meissner, G. (1988). Purification and reconstitution of the calcium release channel from skeletal muscles. *Nature (London)* 331:315–319.

Langer, G.A. (1980). The role of calcium in the control of myocardial contractility: An update. *J. Mol. Cell Cardiol.* 12:231–239.

Langer, G.A., and Peskoff, A. (1996). Calcium concentration and movement in the diadic cleft space of the cardiac ventricular cell. *Biophys. J.* 70:1169–1182.

Lazarides, E. (1980). Intermediate filaments as mechanical integrators of cellular space. *Nature* 283:249–256.

FIGURE 2B Three-dimensional reconstruction of CRC/RR localization within a rat myocyte. Red "hot spots" seen here along the regularly arrayed cross bands represent areas of highest fluorescent intensity and presumably the highest number of CRC/RR sites. A series of confocal "slices" were taken at 0.5 μm intervals through an isolated rat myocyte exposed to antibodies against the CRC/RR as seen in 2A. These slices were used to construct this 3-D reconstruction. Volume rendering and gradient-based ray casting techniques were used to visualize the high-intensity regions (see Bacallao and Garfinkel, 1994, and Fig. 9, for more details).

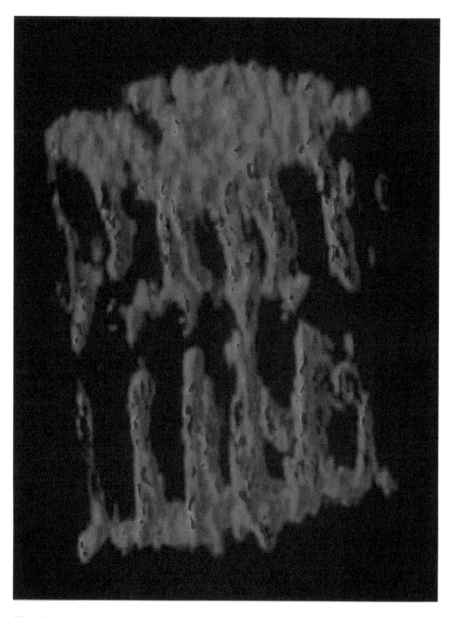

FIGURE 8 Three-dimensional reconstruction of T tubules and peripheral sarcolemma from a portion of a myocyte (guinea pig) immunolabeled for the Na^+–Ca^{2+} exchanger. Labeling was with R_3F_1 antibodies against the exchanger plus goat anti-mouse FITC secondary antibodies. Optical sections taken in the confocal microscope were subjected to visualization techniques based on volume-rendering, ray-casting methods and enhanced with gradient-based shading techniques. The highest intensity (and presumably the highest number of emitting fluorescent molecules at a point) were assigned the red color. The peripheral sarcolemma can be seen at the top (especially) and bottom of the figure. The T tubules project from the peripheral sarcolemma into the middle of the figure. The number and size of the red "hot spots" representing high density of exchanger sites are clearly most abundant and larger in size on the T-tubular membranes.

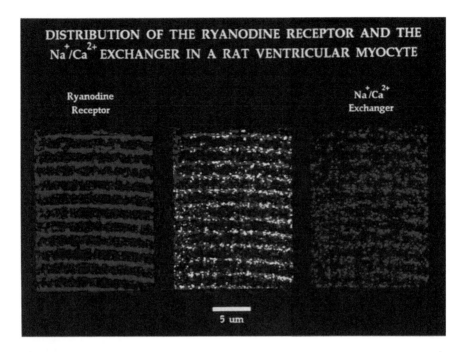

FIGURE 9 A pseudocolored stereo-pair image of a single rat cardiomyocyte that was dual-labeled for the Na+–Ca2+ exchanger with FITC (green) and the ryanodine receptor with Texas Red (red). The images have been superimposed and single voxels are 93 nm a side. White indicates that both proteins occupied the same voxel, black indicates that neither protein was present in that voxel. Both proteins can be seen located in spots running along the T tubule, with some of the Na+–Ca2+ exchanger being located in the longitudinal tubules (for details of deconvolution algorithm and regularization theory to remove out-of-focus light, see Moore *et al.*, 1993).

Lazarides, E. (1982). Intermediate filaments: A chemically heterogenous, developmentally regulated class of proteins. *Am. Rev. Cell. Biol.* 51:219.

Lederer, W.J., Niggli, E., and Hadley, R.W. (1990). Sodium–calcium exchange in excitable cells: Fuzzy space. *Science* 248:283.

Leung, A., Imagawa, T., Block, B., Franzini-Armstrong, C., and Campbell, K.P. (1988). Biochemical and ultrastructural characterization of the 1-4 dihydropyridine receptor from rabbit skeletal muscle. *J. Biol. Chem.* 263:994–1001.

Li, Z., Burke, E.P., Frank, J.S., Bennett, V., and Philipson, K.D. (1993). The cardiac Na/Ca exchanger binds to the cytoskeletal protein ankyrin. *J. Biol. Chem.* 268:11489–11491.

Lingrel, J.B. (1992). Na,K-ATPase: Isoform structure, function and expression. *J. Bioenerg. Biomembr.* 24:263–270.

Luna, E.J., and Hitt, A.L. (1992). Cytoskeleton–membrane interactions. *Science* 258:955–964.

Luther, P.W., Yip, R.K., Block, R.J., Ambesi, A., Lindemeyer, G., and Blaustein, M.P. (1992). Presynaptic localization of Na Ca exchanger in neuromuscular preparations. *J. Neurosci.* 12:4898–4904.

McDonough, A.A., Zhang, Y., Shin, V., and Frank, J.S. (1996). Subcellular distribution of sodium pump isoform subunits in mammalian cardiac myocytes. *Am. J. Physiol.* 270(Cell):C1221–C1227.

Meng, H., Leddy, J.J., Frank, J.S., Holland, P., and Juana, B.S. (1996). The association of cardiac dystrophin with myofibrils/Z—Disc regions in cardiac muscle suggest a novel role in the contractile apparatus. *J. Biol. Chem.* 271:12364–12371.

Miller, T., Roos, K.P., and Frank, J.S. (in press). Dystrophin localization in cardiac myocytes: Species variability.

Moore, E.D., Fogarty, K.E., and Fay, F.S. (1991). Role of Na/Ca exchanger in β-adrenergic relaxation of single smooth muscle cells. *Ann. N.Y. Acad. Sci.* 639:543–549.

Moore, E.D., Etter, E.F., Philipson, K.D., Carrington, W.A., Fogarty, K.E., Lifshitz, L.M., and Fay, F.S. (1993). Coupling of the $Na^+–Ca^{2+}$ exchanger, Na^+/K^+ pump and sarcoplasmic reticulum in smooth muscle. *Nature* 365:657–660.

Nelson, J.W., and Lazarides, E. (1984). Goblin (anhyrin) in striated muscle: Identification of the potential membrane receptor for erythroid spectrum in muscle cells. *Proc. Natl. Acad. Sci. USA* 81:3292–3296.

Noel, F., Wibo, M., and Godfraind, T. (1991). Distribution of $\alpha1$ and $\alpha2$ $Na^+K^+ATPase$ isoforms between junctional (T tubular) and non-junctional sarcolemmal domains in rat ventricle. *Biochem. Pharmacol.* 41:313–315.

Ogawa, Y. (1994). Role of ryanodine receptors. *Crit. Rev. Biochem. Mol. Biol.* 29:229–274.

Orlowski, J., and Lingrel, J.B. (1988). Tissue specific and developmental regulation of rat Na K ATPase catalytic α isoform and β subunit mRNAs. *J. Biol. Chem.* 263:10436–10442.

Pardo, J.V., Siliciano, J.D., and Craig, C.W. (1983). Vinculin is a component of an extensive network of myofibril–sarcolemma attachment regions in cardiac muscle fibers. *J. Cell. Biol.* 97:1081–1088.

Price, M. (1991). Striated muscle endosarcomeric and exosarcomeric lattices. *Adv. Strut. Biol.* 1:175–207.

Radermacher, M., Rao, V., Grassucci, R., Frank, J., Zimmerman, A.P., Fleischer, S., and Wagenknecht, J. (1994). Cryo-electron microscopy and three-D. Reconstruction of the Ca release channel/ryanodine receptor from skeletal muscle. *J. Cell. Biol.* 127:411–423.

Rappaport, L., and Samuel, J.L. (1988). Microtubules in cardiac myocytes. *Int. Rev. Cytol.* 113:101–143.

Rios, E., and Brum, G. (1987). Involvement of dihydropyridine receptor in excitation–contraction coupling in skeletal muscle. *Nature (London)* 325:717–720.

Smith, P., Saccomaini, G., Joe, E., Angelides, K., and Benos, D. (1991). Amiloride-sensitive Na channel is linked to the cytoskeleton in renal epithelial cells. *Proc. Natl. Acad. Sci. USA* 88:6971–6975.

Sommer, J.R., and Waugh, R.A. (1976). The ultrastructure of the mammalian cardiac muscle cell, with special emphasis on the tubular membrane system. A review. *Am. J. Path.* 82:192–232.

Srinivasan, Y., Elsner, L., Davis, J., Bennett, V., and Angelides, K. (1988). Ankyrin and spectrin associate with voltage dependent sodium channels in brain. *Nature* 333:177–180.

Steward, P.S., and MacLennan, D.H. (1974). Surface particles of sarcoplasmic reticulum membranes. Structural features of adenosine triphosphatase. *J. Biol. Chem.* 249:985–993.

Sun, X., Protasi, F., Takshashi, M., Takeshima, H., Ferguson, D., and Franzini-Armstrong, C. (1995). Molecular architecture of membranes involved in excitation–contraction coupling of cardiac muscle. *J. Cell Biol.* 129:659–671.

Suzuki, A., Yoshida, M., Hayschi, K., Mizuno, Y., Hagiwara, Y., and Ozawa, E. (1994). Molecular organization at the glycoprotein complex binding site of dystrophin. Three dystrophin-associated proteins bind directly to the carboxy-terminal portion of dystrophin. *Eur. J. Biochem.* 220:283–292.

Sweadner, K.J., Herrera, V.L.M., Amato, S., Moellmann, A., Gibbsons, D.K., and Repke, K.R. (1994). Immunologic identification of $Na^+ K^+$-ATPase isoforms in myocardium. *Circ. Res.* 74:669–678.

Tada, M., Yammamoto, T., and Tonomura, Y. (1978). Molecular mechanism of active Ca transport by sarcoplasmic reticulum. *Physiol. Rev.* 58:1–79.

Terracio, L., Simpson, D.G., and Hilenski, L.H. (1990). Distribution of vinculin in the Z-disc of striated muscle: Analysis by laser scanning confocal microscopy. *J. Cell. Physiol.* 145:78–87.

Terracio, L., Rubin, K., and Gullberg, D. (1991). Expression of collagen binding proteins during cardiac development and hypertrophy. *Circ. Res.* 68:734–744.

Wang, K., McClure, J., and Tu, A. (1979). Titan: Major myofibrillar components of striated muscle. *Proc. Natl. Acad. Sci. USA* 76:3698.

2

MYOCARDIAL CELLULAR DEVELOPMENT AND MORPHOGENESIS

HONG ZHU

I. INTRODUCTION

Mammalian heart is the first organ formed during embryonic development. Thus it can support growth and differentiation of the remaining organ systems. In human embryos, for instance, a fully functional primitive cardiovascular system is established by the end of third week (Moore, 1988). At this time, paired tubular heart primordia are contracting and connected to the intraembryonic arterial and venous blood vessels, as well as the extraembryonic blood vessels associated with the fetal membranes. Thus, development of the heart, including embryonic origin of the heart and the morphogenesis of various parts of the heart, has been the subject of intensive studies with conventional methods of embryology and histology for decades. The morphologic changes during heart development such as formation of the primitive straight heart tube, rightward looping of the heart tube, and septation of the ventricle have been systematically studied and well documented at the cellular level. However, until recently, the genetic cues for the commitment of cardiac progenitor cells to the cardiac muscle lineage, differentiation of cardiac muscle cells, diversification of various types of cardiac muscle cells, and growth of myocardium were completely unknown. As a result, it was virtually impossible to understand the etiology of congenital heart diseases at the molecular level. In the last 10 years or so, the advances of molecular biology and genetics have almost completely revolutionized the understanding of each individual aspect of heart development. One major outcome of the inroad made by the new techniques of molecular biology and genetics into heart development is that medical professionals have started to understand this extremely complex process at the molecular level. Genes that either control the onset of morphogenesis of various types of cardiac myocytes or serve as molecular markers for these myocytes have been isolated. These genes were soon proven to be essential to further the knowledge in both normal and abnormal heart development and growth. Alterations in the expression of these genes are now being associated with a number of congenital heart diseases. Therefore, with the advancing technology of gene therapy, intervention of or even prevention of congenital heart diseases will not remain forever at the stage of theory. This chapter focuses on the genes encoding either regulatory factors or specific molecular markers that are involved in various aspects of heart development.

II. EARLY CARDIAC MYOGENESIS

A. COMPARISON OF CARDIAC AND SKELETAL MYOGENESIS

Avian embryos are relatively easier to manipulate experimentally and, thus, have served as a fruitful model to study cardiac myogenesis for more than 50 years. The embryonic origin of cardiac progenitor cells was initially mapped in avian embryos by Rawles (1943), who mainly applied grafting techniques in his studies. Rawles demonstrated that at stage four of chick embryo, a small group of mesodermal cells located on both sides of Hensen's node would give rise to myocardial cells. Hensen's node is composed of a group of cells located at the anterior end of the primitive streak where mesodermal cells migrate from the epiblast to the blastocoel during gastrulation. Subsequently, Rosenquist (1966) confirmed Rawles' observation by autoradiographic techniques. The embryonic origin for myocardial cells in mammals turned out to be at a similar location to that of chick myocardial cells. In human embryonic development, for instance, the cardiac progenitor cells were identified within the anterior lateral plate mesoderm at early gastrulation (Colvin, 1990). Although the embryonic origin of myocardial cells have been mapped for more than 50 years, the molecular mechanisms that induce the commitment of cardiac progenitor cells to cardiac muscle lineage are still unknown. One major technical hurdle is the lack of a continuous cardiac muscle cell line. In contrast, the commitment of skeletal muscle precursor to muscle lineage has been well studied largely due to the availability of a number of continuous skeletal muscle cell lines. Skeletal muscle and cardiac muscle are the major striated muscle types in mammals and are both derived from mesoderm. They share many morphologic, electrophysiologic, and contractile properties. The commitment of skeletal muscle precursors to muscle lineage is regulated by the MyoD family of genes, which includes MyoD, myogenin, Myf5, and MRF4 (Edmondson and Olson, 1989; Lasser *et al.*, 1989; Miner *et al.*, Braun *et al.*, 1990). These four genes encode transcription factors that contain the highly conserved basic helix-loop-helix (bHLH) domain and are specifically expressed in skeletal muscle cells. These skeletal myogenic factors initiate the muscle program by activating the transcription of muscle-specific genes such as the contractile protein encoding genes and the muscle creatine kinase gene. Forced expression of one of these factors in certain types of fibroblasts induces a full spectrum of muscle phenotypes. Although functional redundancy exists for these factors in maintaining the muscle phenotypes, there may be differences in inducing commitment of skeletal muscle precursor cells to muscle lineage. Particularly, there appears a regulatory circuit among these four factors based on the temporal pattern of expression. For instance, the mRNA encoding myogenin is expressed prior to the mRNA encoding the other three factors, suggesting the myogenin may be involved in activating the expression of those three factors. This notion is supported by the observation that the homozygous mutation at the myogenin encoding gene results in severe deficiency in skeletal muscle development, whereas the homozygous mutation at the MyoD or Myf5 gene did not cause any phenotypic abnormality (Braun *et al.*, 1992; Rudnicki *et al.*, 1992; Hasty *et al.*, 1993). Furthermore, in the myogenin-/- mutant mice, the MyoD gene is not expressed. The MyoD family is probably not involved in cardiac myogenesis since the expression of the MyoD family members has never been detected in the cells of cardiac muscle lineage. Homozygous mutations at either the myogenin gene or MyoD gene do not affect heart development. Although a number of transcription factors have been identified that regulate expression of cardiac muscle genes, none of these factors seem to play the roles equivalent to those of the MyoD family members in skeletal muscle.

B. DEFINING CELLULAR SOURCE OF CARDIAC MUSCLE LINEAGE INDUCER

During gastrulation, the anterior endoderm is in close contact with the anterior lateral plate of mesoderm from which the heart is derived. Hence, it was speculated that the anterior endoderm might influence the commitment of cardiac muscle lineage. This speculation was first supported experimentally by Jacobson (1960, 1961), who demonstrated that the newt mesodermal explants could only be differentiated into beating heart muscle *in vitro* when coincubated with anterior endoderm. These results suggest that, in newt, the anterior endoderm produces diffusable signals that can initiate the cardiac myogenesis in anterior mesoderm. Similarly, Sugi and Lough (1994) have shown that the chick anterior lateral mesoderm at stage six can differentiate into cardiac muscle when cocultured with the anterior endoderm. This response of chick anterior lateral mesoderm seems specific for anterior endoderm because it does not differentiate into cardiac muscle when it is cocultured with ectoderm. However, based on the previous results, it is unclear whether the signals from anterior endoderm actually induce the commitment of the anterior mesodermal cells to cardiac muscle lineage or merely induce differentiation of the cells that are already committed. Schultheiss *et al.* (1995) have elegantly proven that in chick embryo, the signals from anterior endoderm actually induce the commitment of anterior lateral mesodermal cells to cardiac muscle lineage. These authors cocultured the chick anterior lateral endoderm with posterior primitive streak cells, which normally do not serve as cardiac progenitor cells, and induced cardiac muscle cells from these primitive streak cells. These results also indicate that the anterior endodermal signals do not require any specific receptors for anterior lateral mesoderm. The other piece of important evidence that indicates that the anterior endodermal signals induce commitment of cardiac muscle lineage came from the obervation that in chick embryo, the action of anterior lateral endoderm on the anterior lateral mesoderm seems transient (Gannon and Bader, 1995). Removal of anterior endoderm from the coculture prior to the differentiation of anterior lateral mesodermal cells into cardiac muscle did not affect the subsequent differentiation. Therefore, in newt and chick embryos, the anterior endoderm contains cells that produce diffusable signals that induce commitment of anterior lateral mesodermal cells to cardiac muscle lineage. The identities of these signals and the mechanisms by which they induce the commitment of anterior lateral mesoderm cells to cardiac muscle lineage are still unknown.

C. MOLECULAR MARKERS OF COMMITMENT AND EARLY DIFFERENTIATION OF CARDIAC MUSCLE LINEAGE

To define the anterior endodermal signals that induce commitment and early differentiation of cardiac muscle lineage, first identify molecular markers that are specifically expressed in cells committed to cardiac muscle lineage or cardiac myocytes at an early stage of differentiation. These markers could be either structurally or functionally important for the commitment and for early differentiation of cardiac muscle lineage. Furthermore, it is reasonable to assume that these molecular markers are the target genes of the endodermal signals and, therefore, they can be the critical clues to the molecular mechanisms by which the endodermal signals control cardiac commitment. With such molecular markers available, one would be able to define the regulatory factors that control the expression of those markers by biochemical and molecular techniques. Therefore, the anterior endodermal signals and their associated cellular signaling pathways could be defined as the regulators for the expression of those molecular markers.

In the following sections, four types of relatively well-studied molecular markers—the homeobox proteins, bHLH proteins, myocyte enhancer binding factor-2 (MEF2), and GATA transcription factors—are discussed.

1. Homeobox Proteins

Fruit fly, *Drosophila,* has been an extremely useful genetic model for screening a large number of mutations that lead to specific phenotypic changes. Today, numerous critical genes that control various aspects of development have been found highly conserved between *Drosophila* and mammals. This is why it is not surprising that many mammalian development controlling genes have been isolated based on their sequence similarities with their *Drosophila* counterparts. The other advantage of using the *Drosophila* model is that the fruit fly does not demonstrate redundancy for many regulatory genes as mammals do. Thus it serves as an excellent system to study the effects of loss-of-function mutations. Homeobox protein encoding genes were initially identified at the *Drosophila* homeotic loci and subsequently proven to be involved in determining the identity of the body structures in early development. Most known homeobox proteins are transcription factors and contain the highly conserved 60-amino acid homeo domain, which serves as sequence-specific DNA binding motif. Mammalian homeobox genes were soon identified and found to play very similar roles in mammalian embryonic development. Recently, it has been reported that one homeobox protein is involved in *Drosophila* "heart" development. Although *Drosophila* does not have the four-chamber heart of mammals, it has a dorsal vessel that is structurally similar to the primordial straight heart tube of mammalian embryos. The *Drosophila* dorsal vessel functions as the mammalian heart by contracting rhythmically and pumping hemolymph through an open circulatory system. Azpiazu and Frasch (1993) and Bodmer (1993) have discovered by genetic screening that a single gene plays a critical role in formation of the *Drosophila* dorsal vessel. The *Drosophila* strain that carries mutations at this gene completely lacks the dorsal vessel. Thus this gene was named tinman (TIN) after the movie character in *Wizard of Oz,* Tinman, who had no heart. These authors have also shown that in *Drosophila* embryonic development, the TIN gene is expressed in early mesoderm, which gives rise to the dorsal vessel structure, and its expression continues in the dorsal vessel. These results suggest that the TIN gene is probably not only involved in the commitment of cardiac muscle lineage but also in maintaining cardiac muscle phenotypes. Although sequence analysis of the TIN gene indicates that it encodes a homeobox protein and possibly functions as a transcription factor, the target genes of TIN still remain to be identified.

However, the encouraging results on the *Drosophila* TIN gene prompted the isolation of the mammalian counterpart of the *Drosophila* TIN gene. Komuro and Izumo (1993) and Lints *et al.* (1993) have isolated a TIN-related gene from mouse cDNA libraries, which is named as Nkx2.5 or Csx. The Nkx2.5 gene also encodes a homeobox protein that could act as a transcription factor. The temporal and spatial expression pattern of the Nkx2.5 gene during mouse embryonic development is similar to that of the *Drosophila* TIN gene. Nkx2.5 mRNA is expressed in cardiac progenitor cells in the anterior lateral mesoderm and its expression continues in fetal and adult hearts. The expression pattern of Nkx2.5 suggests a particularly important role of Nkx2.5 in both commitment of cardiac muscle lineage as well as in the maintenance of cardiac muscle phenotypes. To directly determine the roles of Nkx2.5 in commitment of cardiac muscle lineage, Lyons *et al.* (1995) generated a transgenic mouse line that carries null mutations at both alleles of the Nkx2.5 gene. The Nkx2.5-/- mutant mice died at the embryonic stage because they failed to complete looping of the primitive heart tube, although the primitive straight heart tube still formed. These results suggest that the Nkx2.5 gene must function at a stage after the commitment of cardiac muscle lineage is made and probably controls the growth of committed cardiac muscle cells in the straight primitive heart tube. However, recent evidence indicates that there may be a family of TIN-related genes in mice (Evans *et al.*, 1995). Thus redundancy for the putative Nkx2.5 function in inducing cardiac muscle lineage commitment may exist in this species. It would be interesting to assess the effects of homozygous mutations at other TIN-related genes on mouse heart development. The target genes of Nkx2.5 are still unknown. Since Nkx2.5 is the earliest molecular marker for cardiogenic lineage in mammals, identifying the mechanism by which the Nkx2.5 expression is activated will be critical for defining the signals that trigger the commitment and differentiation of cardiac muscle lineage.

2. Basic Helix-Loop-Helix Proteins

Although the four bHLH transcription factors of the MyoD family (MyoD, myogenin, Myf5, and MRF4) are not expressed in either cardiac progenitor cells or differentiated cardiac muscle cells, indirect evidence has suggested that other bHLH proteins are involved in muscle gene expression in cardiac myocytes. The MyoD family members activate the expression of muscle-specific genes in skeletal muscle cells by binding to a CAXXTG sequence, which is known as the E-box. The E-box seems required for the expression of several muscle genes in cardiac myocytes, suggesting bHLH proteins are involved in regulation of these genes (Sartorelli *et al.*, 1992; Evans *et al.*, 1993; Molkentin *et al.*, 1993; Moss *et al.*, 1994; Navankasattusas *et al.*, 1994). Litvin *et al.* (1993) have identified a nuclear protein in embryonic chick heart that reacts with an antiserum directed against the second helix domain of MyoD. Gel mobility shift assay indicates that this cardiac nuclear protein can specifically bind to the E-box element. However, direct evidence that this cardiac nuclear protein is involved in cardiac gene expression is missing partly because the cDNA encoding this putative bHLH protein has not been isolated. Therefore, until the cDNA is isolated, it would be impossible to assess functions of this putative cardiac bHLH protein in commitment or early differentiation of cardiac muscle lineage.

Recently, more direct evidence that bHLH proteins are involved in commitment and differentiation of cardiac muscle lineage has come from the isolation and characterization of two cDNA clones that encode two related cardiac bHLH proteins (Cserjesi *et al.*, 1995; Srivastava *et al.*, 1995). These two bHLH factors, named dHAND and eHAND, respectively, show a low degree of sequence similarities with the MyoD family members and, thus, possibly represent a subclass of bHLH proteins. Unlike the members of MyoD family, the dHAND and eHAND genes are expressed specifically in cardiac progenitor cells, the looping heart tube, and certain cardiac neurocrest-derived cells of both chick and mouse embryos. Therefore, these two genes could be involved in commitment or early differentiation of cardiac muscle lineage during chick and mouse embryonic development. A part of this speculation was supported by the *in vitro* studies, which showed that inhibition of the expression of both dHAND and eHAND in chicken embryos resulted in an arrest of cardiac morphogenesis at the looping stage (Srivastava *et al.*, 1995). These two genes probably play a similar role to that of the Nkx2.5 gene, which is controlling the growth of cardiac muscle cells in the primitive straight heart tube. At present, the target genes of these two bHLH proteins are unknown. To understand the functions of these two bHLH proteins, it is crucial to define their target genes in cardiac myocytes. However, identification of the factors that regulate the expression of these bHLH proteins will be useful to ultimately define the molecular cues for the commitment of cardiac muscle lineage.

3. Myocyte Enhancer Binding Factor-2

Mammalian myocyte enhancer binding factor-2 (MEF2) belongs to a MADS box family of transcription factors that includes the originally identified three transcription factors: **M**CM1 (yeast), **A**gamous **D**eficiens (plant), and **S**erum response factor (man) (Yu *et al.*, 1992). The MADS box family members all contain a homologous domain at the amino terminus (MADS box), which is involved in DNA binding. Unlike the two types of factors described previously, MEF2 activates a number of muscle-specific genes by binding to a conserved A–T-rich DNA sequence located in the promoter regions. So far, there are a total of four MEF2 genes identified in the mouse that are expressed in precursor cells of the cardiac, skeletal, and smooth muscle lineage, suggesting that these genes are involved in myogenesis of these muscle types (Edmondson *et al.*, 1994; Chambers *et al.*, 1994; Wong *et al.*, 1994). In particular, one of the four MEF2 genes, MEF2C, is first expressed among the four at day 7.5 postcoitum (p.c.) in the anterior lateral mesodermal cells that give rise to cardiac muscle lineage. These results strongly suggest that MEF2C may be essential for the commitment of cardiac muscle lineage. However, this hypothesis needs to be proven experimentally. In contrast, genetic studies on MEF2 in *Drosophila* have first shed light on MEF2 functions in commitment and differentiation of cardiac muscle lineage. *Drosophila* has served as a particularly useful model in understanding MEF2 functions *in vivo* since there is only one MEF2 gene in *Drosophila*, named D-MEF2 (Lilly *et al.*, 1994; Nguyen *et al.*, 1994). Similar to the mouse MEF2 genes, the D-MEF2 gene is expressed in the *Drosophila* precursor cells of cardiac muscle lineage, implying an important role played by D-MEF2 in commitment and differentiation of cardiac muscle lineage. To directly determine the role of D-MEF2 in cardiac commitment and differentiation, a *Drosophila* strain that contains null mutations at the D-MEF2 gene was created by a number of laboratories (Lilly *et al.*, 1995; Bour *et al.*, 1995; Ranganayakulu *et al.*, 1995). Mutations at the D-MEF2 gene blocked the differentiation of all muscle types in the embryos. Although the dorsal vessel still forms in the mutant embryos and the TIN gene is still expressed, none of the contractile protein encoding genes are expressed in dorsal vessel cells. These results indicate that D-MEF2 probably functions at a stage immediately after the commitment of cardiac muscle lineage has been made and probably before the initial differentiation of cardiac muscle cells. Since the mouse MEF2 factors and *Drosophila* D-MEF2 behave similarly with respect to expression pattern, DNA binding specificity, and transcriptional activator properties, the mouse MEF2 factors likely play the same roles in cardiac development. Although the mouse MEF2 factors probably function at a postcommitment stage as D-MEF2 does, they activate cardiac muscle genes. Thus, it is critical to define the upstream regulatory factors that turn on the expression of the MEF2 genes in the cardiac precursor cells.

4. GATA Transcription Factors

The GATA family of transcription factors activates transcription of a number of genes in various tissues by binding to a consensus GATA site, (A/T)GATA(G/A) (Leiden, 1993). The expression pattern of one member of the GATA family, GATA4, suggests that it may be involved in commitment and early differentiation of cardiac muscle lineage. During mouse embryonic development, GATA4 is detected in the anterior lateral mesoderm and subsequently in the endocardial and myocardial layers of the heart tube and developing heart (Heikinheimo *et al.*, 1994). The hypothesis that GATA4 is involved in commitment and early differentiation of cardiac muscle lineage was further supported by the observations that GATA4 regulates muscle gene expression in cardiac myocytes (Molkentin *et al.*, 1994; Thuerauf *et al.*, 1994; Grepin *et al.*, 1994). More direct evidence in support of this hypothesis came from *in vitro* antisense RNA studies in mouse p19 cells. The mouse pluripotent P19 cell line is derived from embryonic carcinoma and can be induced to differentiate into cardiac myocytes *in vitro*. Ip *et al.* (1994) and Grepin *et al.* (1995) have shown that inhibition of GATA4 expression in P19 cells by antisense RNA blocks the differentiation of these pluripotent cells into cardiac myocytes. Interestingly, Soudais *et al.* (1995) have shown that null mutations at both alleles of the mouse GATA4 gene disrupt differentiation of visceral endoderm *in vitro*. Thus, it is an intriguing speculation that GATA4 plays a critical role in the differentiation of the endoderm and thereby ultimately controls the generation of the cardiac muscle lineage inducer molecules. It would be interesting to assess the effects of null mutations at both alleles of the GATA4 gene on early cardiac development, particularly, the commitment of cardiac muscle lineage. Furthermore, the other two members of the GATA family, GATA5 and GATA6, may play the same roles that GATA4 does in the commitment and early differentiation of cardiac muscle lineage since they share a number of properties with GATA4. For instance, they bind to the same DNA sequence as GATA4, and their expression patterns during development are similar to that of GATA4. At present, however, in comparison with GATA4, much less is known about the functions of GATA5 and GATA6.

D. MOUSE EMBRYONIC STEM CELLS USED AS *IN VITRO* MODEL TO STUDY CARDIAC MYOGENESIS

The mouse embryonic stem (ES) cells are pluripotent cells derived from the inner cell mass of the preimplantation blastocyst. These cells in culture can form multicellular embroid bodies (EBs) through embryo-like aggregates and differentiate into several cell types, including spontaneously beating cardiac myocytes (Doetschman *et al.*, 1985). The identity of cardiac myocytes are confirmed by detecting the expression of cardiac muscle genes and electrophysiologic properties characteristic of cardiac myocytes (Miller-Hance *et al.*, 1993; Maltsev *et al.*, 1994; Metzger *et al.*, 1996). The change in contractile sensitivity of the cardiac myocytes isolated from EBs parallels that observed during mouse cardiac myocyte development *in vivo* (Metzger *et al.*, 1994). These results indicate that the signals inducing commitment and early differentiation of cardiac muscle lineage are generated in these EBs. The ES cell model for studying commitment and early differentiation of cardiac muscle lineage has a number of advantages in comparison with the *in vivo* embryo model. The first advantage is that it is relatively easy to obtain a large number of EBs for various analyses. Second, changes in morphology, gene expression, and electrophysiological properties can be monitored closely. Third, it is relatively easy to change culturing conditions for ES cells. Thus, the effects of hormones, growth factors, and chemicals on commitment and early differentiation can be studied. Fourth, the ES cells are amenable to genetic manipulations. Mutations can be introduced into specific genes through homologous recombination and fusion genes can be introduced into the genome of ES cells. The effects of mutating specific genes or expression fusion genes on commitment and early differentiation of cardiac muscle lineage can be assessed. If *in vivo* studies are required, the ES cells carrying mutant genes or fusion genes can be utilized to establish transgenic mouse lines. Although a number of new growth factors have been identified in EBs that stimulate growth of cardiac myocytes (Pennica *et al.,* 1995), the EB-derived factors that induce commitment and differentiation of cardiac muscle lineage in EBs remain to be identified.

In summary, mammalian cardiac and skeletal muscle cells are the major two types of striated muscle that are both derived from mesoderm. Cardiac muscle cells originate from a small number of cells located in the anterior lateral mesodermal plate and are committed to cardiac muscle lineage probably at the beginning of gastrulation. The commitment of cardiac progenitor cells is likely induced by diffusable factors produced by the anterior lateral endoderm, although the identities of the inducing factors remain to be determined. In contrast, it is well established that the four bHLH transcription factors of the MyoD family establish the skeletal muscle lineage in embryonic development. The expression of these bHLH transcription factors is not detected in cardiac progenitor cells or differentiated cardiac muscle cells. However, present evidence has indicated the involvement of four types of transcription factors—the homeobox proteins, bHLH proteins, MEF2 factors, and GATA factors—in cardiac growth and differentiation prior to looping of the primitive heart tube. These molecular markers will serve as critical tools in defining the inducing factors of cardiac progenitor cells. Since all of these markers are expressed in cardiac myocytes at an early stage, identification and characterization of the factors that regulate the activity or expression of these markers will bring researchers one step closer to identification of the inducing factors of cardiac progenitor cells.

III. DIVERSIFICATION AND SPECIFICATION OF MYOCYTES

The initially differentiated cardiac myocytes undergo a series of proliferation and migration steps to form the primitive straight heart tube. The straight heart tube first shows spontaneous depolarization in the caudal region that later gives rise to the sinoatrial (SA) node (DeHaan, 1963). The straight heart tube soon develops the ability of rhythmic contraction, suggesting that the genes encoding contractile proteins, ion channels, ion exchangers, and ion pumps are expressed in the myocytes. The straight heart tube then starts rightward looping and after a sequence of complicated looping and septation, the straight heart tube finally turns into a four-chamber heart (Colvin, 1990). During this time, not only the cardiac myocytes increase their sarcomeric content, but also commence diversification into three major types of cardiac myocytes—ventricular myocytes, atrial myocytes, and conduction system (CS) myocytes.

Although the molecular mechanisms that control the initiation of looping of the straight heart tube and the directionality of looping are not completely understood, factors involved in control of looping only start to emerge. As described previously, null mutations at the mouse Nkx2.5, chick dHAND and eHAND genes lead to incomplete looping. It is possible that these factors are involved in growth of the straight heart tube in an asymmetrical manner that leads to cardiac looping. The directionality of looping in chick embryos seems controlled primarily by two genes—the sonic hedgehog (Shh) gene and the activin receptor IIa (Act-RIIa) encoding gene (Levin et al., 1995). The Shh gene is expressed on the left side of the heart-forming region and induces the expression of the chick nodal-related morphogen (cNR1). Activin induces the expression of the Act-IIa gene and suppresses the expression of the Shh gene. Ectopic expression of Shh on the right side or Act-IIa on the left side of the embryo randomizes the direction of cardiac looping. In mouse embryos, at least two genes are involved in controlling the directionality of looping. In mice that carry mutations at both alleles of the iv gene on chromosome 12, the orientation of the heart and viscera is randomized (Brueckner et al., 1989). In mice that carry mutations at both alleles of the inv gene on chromosome 4, there is nearly 100% reversal of left–right asymmetry and cardiac looping (Yokoyama et al., 1993). However, further understanding of how these two genes control the directionality of heart looping is hampered by the fact that neither of these genes has been isolated. It is clear that both genes may function at least partly through controlling the spatial expression of the morphogen nodal gene (Lowe et al., 1996). Normally, the nodal gene is expressed on the left side of lateral mesoderm. In inv-/- mouse embryos, however, the nodal gene is expressed on the right side of the lateral mesoderm. In in-/- mouse embryos, the nodal expression is randomized.

As the straight heart tube starts looping, the atrial and ventricular chambers become morphologically distinguishable. The atrial and ventricular myocytes have different contractile, electrophysiological, and pharmacological properties that are attributed to specific subsets of genes expressed in these myocytes. The molecular mechanisms that regulate the diversification of atrial and ventricular myocytes are not known. The atrial and ventricular myocytes, as well as the SA nodal cells, are derived from distinct segments of the straight heart tube along the anteroposterior (AP) axis, suggesting that the fate for various myocytes may have been determined before looping. This determinative AP polarity has been demonstrated in chick, frog, and zebra fish embryos. Treatment of these embryos with retinoic acid results in lack of the anterior portion of the heart tube (Osmond *et al.*, 1991; Drysdale *et al.*, 1994; Stainier and Fishman, 1992). Anterior truncation of the entire embryo has been observed after treatment with retinoic acid. This suggests that the establishment of the AP polarity of the straight heart tube is under the same regulatory mechanisms that establish polarity of the entire body (Durston *et al.*, 1989; Altaba and Jessel, 1991). Retinoic acid regulates the expression of the homeobox genes that specify the patterning of development of the embryo along the AP axis (Cho and De Robertis, 1990; Sive and Cheng, 1991; Sundin and Eichele, 1992). Thus, the specification of the straight heart tube, as well as the entire embryo along the AP axis, are under control of the same homeobox proteins. It is possible that the homeobox proteins directly or indirectly activate transcription of chamber-specific or conduction system-specific genes that then confer contractile, electrophysiological, and pharmacological properties unique to various cell types. To define the specific homeobox genes that are involved in specification of the heart tube, it is key to identify chamber-specific or conduction system-specific molecular markers. Presumably, these markers are the target genes for the homeobox proteins. Understanding the mechanisms that regulate the expression of these markers will provide important clues to the regulation of myocyte specificity. Previous studies by a number of laboratories have indicated that the majority of atrial and ventricular specific genes are initially expressed in both atria and ventricles. Chamber-specific expression occurs at later stages (e.g., after completion of septation or even after birth). These observations suggest that negative regulation of these genes in response to anatomic and physiologic changes could also be critical in myocyte specification. However, there are also those markers that display chamber-specific expression at a relatively early stage (i.e., before birth or even septation). In the following sections, the relatively well-characterized markers for ventricular, atrial, and conduction system myocytes are discussed.

A. MYOSIN LIGHT CHAIN-2 AS A VENTRICULAR-SPECIFIC MARKER

The ventricular-specific myosin light chain-2 (MLC-2v) gene is expressed in ventricular muscle and slow-twitch skeletal muscle of adult rats and mice, but not in atrial muscle. The expression of MLC-2v is first detected in the ventricular region of the linear heart tube at day 8.0 p.c. in mouse embryo (O'Brien *et al.*, 1993). Much lower level of expression is detected in the proximal outflow tract of the heart tube and none in the atrial region of the heart tube. Prior to the completion of septation (day 11.0 p.c.), the expression of the MLC-2v gene becomes restricted to the ventricular region. Furthermore, the MLC-2v gene is only expressed in some of the embryoid body (EB)-derived myocytes (Miller-Hance *et al.*, 1993). These results clearly indicate that ventricular chamber specification could occur well before septation. In contrast, the remaining known ventricular-specific markers are initially expressed uniformly in the entire heart tube. Restriction of expression to ventricular muscle occurs at much later stages in comparison with MLC-2v (i.e., after septation or even after birth). For instance, the MLC-1v gene is expressed in both atrial and ventricular myocytes during entire prenatal development and becomes restricted to the ventricular muscle only at parturition. Similarly, the βMHC gene is initially expressed in both atrial and ventricular myocytes and becomes restricted to the ventricular muscle at day 9.5 p.c. It is speculated that the restricted expression of genes, such as MLC-1v and βMHC, in ventricular myocytes may represent the regulation of physiological stimuli for chamber specification, which occurs at a much later stage. In contrast, the ventricular-specific expression of the MLC-2v gene may represent basic genetic control of ventricular specification. Thus, understanding of the regulation of MLC-2v expression will help reveal the genetic cues for ventricular specification.

The control region in the MLC-2v gene and the transcription factors that are required for ventricular-specific expression have been extensively studied. DNA fragments of various lengths derived from the 5' flanking region of the rat MLC-2v gene were ligated to a firefly luciferase gene and tested for ventricular-specific promoter activity. A 250-base pair (bp) 5' flanking region of the MLC-2v gene can direct luciferase expression in cultured neonatal rat ventricular myocytes, but not in skeletal myoblasts or myotubes and nonmuscle cells, suggesting that the information for ventricular-specific expression may reside within the 250-bp DNA fragment (Henderson et al., 1989). The transgenic mouse line that carries the 250-bp MLC-2v promoter/luciferase fusion gene was established to test ventricular-specific promoter activity of the 250-bp region in vivo (Lee et al., 1992). In these transgenic mice, both luciferase mRNA and activity were detected exclusively in ventricular muscle, which supports the results of in vitro studies. Further studies have revealed a number of cis-acting elements within the 250-bp MLC-2v promoter that are essential for ventricular-specific expression (Zhu et al., 1991; Navankasattusas et al., 1994). A number of transcription factors, including MEF2, HF-1b, USF, and YB-1, activate the 250-bp promoter in ventricular myocyte content by binding to these cis elements (Navankasattusas et al., 1992, 1994; Zhu et al., 1993; Zou and Chien, 1995). Since none of these factors are cardiac specific, it is possible that the unique combination of transcription factors is responsible for ventricular-specific expression of the 250-bp MLC-2v promoter. In addition, a negative cis-acting element (HF3) could also play a critical role in ventricular-specific promoter activity of the 250-bp fragment. Mutations at the HF3 site within the 250-bp MLC-2v promoter result in a significant increase in luciferase expression in both atrial and skeletal muscle (Lee et al., 1994). However, the factor that binds to the HF3 site is not yet identified.

B. ATRIAL-SPECIFIC MARKERS

As most ventricular-specific markers, the known atrial-specific markers are initiallly expressed in both the ventricular and atrial portion of the linear heart tube and become restricted to atria at relatively late stages (i.e., after septation or even after birth). For instance, the atrial-specific myosin light chain-1 (MCL-1a) gene and the atrial natriuretic factor (ANF) gene are expressed in both atrial and ventricles during the entire prenatal development and become restricted to atria at parturition (Zeller et al., 1987). Thus, similar to the expression of most ventricular markers, the atrial-specific expression of these genes could depend on physiological and/ or hemodynamic changes within the various cardiac chambers for switching on the cardiac muscle gene program. However, the retinoic acid-induced truncation of the anterior portion of the straight heart tube suggests that the A-P polarity of the straight heart tube can be established as early as at the stage of commitment of cardiac muscle lineage. In addition, both atrial and ventricular chambers become morphologically identifiable soon after the straight heart tube starts looping. It is then reasonable to assume that atrial specification occurs at a very early stage, definitely before septation. Atrial markers whose expression becomes restricted to atria before septation will be particularly useful in identifying the genetic cues for atrial specification. Kubalak et al. (1994) has isolated and characterized the mouse atrial isoform of MLC-2 (MLC-2a), which could serve as an excellent atrial-specific marker. In adult mice, the MLC-2a gene is expressed in atria and at a much lower level in aorta, but not in ventricular or skeletal muscle. By using in situ hybridization technique, the MLC-2a message is detected in the atrial chamber at day 9.0 p.c. of mouse embryo at a much higher level than in the ventricular chamber. At day 12.0 p.c., the expression of the MLC-2a becomes restricted to atria before completion of septation (day 13.0 p.c.). The onset of MLC-2a expression in EB-derived cardiac myocytes precedes by several days the expression of the MLC-2v gene, confirming that chamber specification starts at a relatively early stage. Moreover, in the ES cell model, the MLC-2a gene can also serve as an excellent marker for early cardiac myogenesis. However, very little is known about the cis- and trans-acting factors that regulate atrial-specific expression of the MLC-2a gene.

C. CONDUCTION SYSTEM-SPECIFIC MARKERS

The specification of conduction system (CS) cells occurs at least as early as the specification of atrial and ventricular myocytes, since spontaneous depolarization is observed in the caudal region of the straight heart tube before contraction is visible (DeHaan, 1963). However, much less is known about the specification of CS cells during development, in part due to lack of specific molecular markers and limited material. The cardiac conduction system is made of a number of subtypes of cells including the SA nodal cells (pacemaker cells), the atrioventricular nodal cells, the His bundle cells, and the Purkinje cells. The origins of these various subtypes of conduction system cells are not completely understood. Since both muscle- and neuron-specific genes are coexpressed in these cells, two origins—myogenic and neural—have initially been suggested for these cells. Subsequent studies have indicated that the subtypes of CS cells may have distinct origins. By culturing explanted embryonic ventricles in *in vivo* culture, Tucker *et al.* (1988) have shown that the pacemaker cells arise from ventricular myocytes. Qin *et al.* (1995) have further shown that the pacemaker cells arise from the ventricular myocytes independently of extrinsic nerves and blood vessels. By using replication-defective retrovirus encoding β-galactosidase as the trace marker, Gourdie *et al.* (1995) have shown that Purkinje fibers arise in close association with coronary arterial blood vessels. The cells that give rise to Purkinje fibers do not contribute to the central conduction system (e.g., atrioventricular ring and bundles).

The HNK-1 protein of unknown functions has been studied as a molecular marker for CS cells primarily because of its temporal and spatial expression pattern during mouse development. At day 9.5 of rat embryo, the HNK-1 is detected along the endocardial surface of the fusing tubular heart. At day 12.5 in rat embryo, the HNK-1 is detected in the right and left bundle branches, primorida of the sinus node, and the transient left sinus node (Nagawa *et al.*, 1993). By immunoelectron microscopic studies, Sakai *et al.* (1994) have shown that at day 14.5 in rat embryo, HNK-1 is present on the surface and extracellular matrices of CS cells but not in other myocytes. Although the function of HNK-1 protein in specification of CS cells is unknown, it is speculated that it is possibly involved in cell–cell adhesion processes both temporally and spatially in the developing conduction system. Further studies of HNK-1 functions and its regulation in specification of CS cells have been hampered by the fact that the HNK-1 gene has not been isolated. The other early CS marker is the gap junction protein, connexin 40 (CX40), which is the only connexin expressed in adult mouse atrial and the proximal part of the ventricular conduction system. Delorme *et al.* (1995) has shown that at day 11.0 p.c., the CX40 mRNA is detected in both atrial and ventricular primordia. From day 14.0 p.c. and onward, CX40 expression in ventricles becomes preferentially distributed in ventricular conduction system although its expression in atria remains unchanged. The mechanisms that restrict CX40 expression to the conduction system in ventricles are not understood.

In summary, although the molecular mechanisms that control specification of various cardiac myocytes are not known, it has been shown that specification of these myocytes occurs at a very early stage, probably at the stage of formation of the straight heart tube. The straight tubular heart is polarized along the A-P axis: Various types of myocytes are derived from different segments of the heart tube. The concept of tubular heart polarization is also supported by the observations that (1) the anterior portion of the heart tube has different sensitivity to retinoic acid stimulation, and (2) the caudal part of the heart tube displays spontaneous depolarization even before signs of contraction. To identify the genetic cues for cardiac specification, molecular markers specific for various types of myocytes would be crucial. So far, the MLC-2v and MLC-2a genes appear to be good molecular markers specific for ventricular and atrial myocytes, respectively. The expression of both genes becomes restricted to either ventricles or atria before completion of septation. The regulation of the MLC-2v expression has been extensively studied in both *in vitro* and *in vivo* models. A number of *cis*- and *trans*-acting factors that are essential for ventricular-specific expression of the MLC-2v gene have been identified. Relatively little is known about the mechanisms that control atrial-specific expression of the MLC-2a gene. The specification of the CS myocytes is the least understood partly due to lack of specific molecular markers and very limited amount of tissue. The extracellular matrix protein, HNK-1, and the gap junction protein, CX40, have been utilized as the molecular markers for certain portions of the cardiac conduction system. However, the regulation of the genes encoding HNK-1 and CX40 in CS cells is completely unknown. Thus, future work may primarily involve identifying more specific molecular markers and understanding regulation of these markers in various cardiac myocytes.

IV. PROLIFERATION OF CARDIAC MYOCYTES

A. COMPARISON OF PROLIFERATION REGULATION BETWEEN CARDIAC AND SKELETAL MUSCLE CELLS

Mammalian skeletal and cardiac muscle are both derived from mesoderm and share many morphologic, physiologic, and biochemical properties. However, one of the long observed differences in embryonic development is that the proliferation and differentiation processes are exclusive in skeletal muscle but not in cardiac muscle (Claycomb, 1992). During embryonic development, the committed skeletal muscle myoblasts proliferate until a certain cell mass has been reached and then start fusing with each other to form multinucleated cells. The multinucleated myoblasts then withdraw permanently from the cell cycle and differentiate into contractile muscle cells. The subsequent growth of the differentiated skeletal muscle is through cell hypertrophy. In contrast, a small number of committed cardiac muscle progenitor cells of the linear heart tube differentiate into contractile cardiac muscle cells without withdrawing from the cell cycle. Instead, the differentiated cardiac muscle cells remain mitotically active throughout the prenatal period and the proliferation of cardiac myocytes is a key component of myocardial growth, morphogenesis, and differentiation. The fetal cardiac myocytes seem to have the ability to disassemble and reassemble the sarcomere during proliferation. Shortly after birth, cardiac myocytes irreversibly exit from the cell cycle and become terminally differentiated as skeletal muscle does at earlier stages (Claycomb, 1983, 1991; Rumyatsev, 1991). As for differentiated skeletal muscle, subsequent myocardial growth in adaptation to increasing workloads becomes dependent on hypertrophic growth. It has been shown that the skeletal myogenic factor, MyoD, induces permanent withdrawal of skeletal myoblasts from the cell cycle, independently of its myogenic functions. The genetic cues for exit of cardiac myocytes from the cell cycle are unknown.

B. REGULATION OF CARDIAC MYOCYTE PROLIFERATION BY GROWTH FACTORS AND HORMONES

A number of growth factors and their receptors are expressed in embryonic myocardium and the expression level generally decreases as the mitotic activity of cardiac myocytes decreases, suggesting that these growth factors are involved in supporting proliferation of cardiac myocytes. More direct evidence that these growth factors are involved in cardiac myocyte proliferation during embryonic development came from studies using both *in vitro* and *in vivo* models.

1. Insulin-like Growth Factor Receptor System

Insulin-like growth factor I (IGF1) and II (IGF2) stimulate mitosis in a number of cell types in culture by binding to the IGF1 receptor (IGF-1R), which is a tyrosine kinase receptor. IGF-2R, however, binds to IGF2 specifically and functions as an IGF-2 scavenge receptor, which binds to IGF2 and induces lysosomal degradation after endocytosis. Thus, the local IGF2 concentration is regulated by both the IGF2 synthesis rate and the concentration of IGF-2R. The genes encoding IGF1, IGF2, IGF-1R, and IGF-2R are all expressed in mammalian embryonic myocardium. The expression level of all four genes in myocardium is dramatically decreased during neonatal development (Cheng *et al.*, 1995; Liu *et al.*, 1996). Kajstura *et al.* (1994) have shown that activation of the IGF1/IGF-1R system can stimulate both DNA synthesis and proliferation in a portion of cultured neonatal rat ventricular myocytes. Similarly, Liu *et al.* (1996) have shown that DNA synthesis in cultured fetal (day 15.0 p.c.) rat ventricular myocytes can be stimulated by exogenous IGF2. The stimulatory effect of IGF2 also seems mediated by IGF-1R. Both IGF1 and IGF2 seem to function in an autocrine or paracrine fashion since inhibition of the expression of endogenous IGF1 or IGF2 decreases the basal level of DNA synthesis in those *in vitro* models. The transgenic mice that carry null mutations at both alleles of one of the IGF1, IGF2, and IGF-1R genes have been generated (DeChiara *et al.*, 1990; Baker *et al.*, 1993; Liu *et al.*, 1996). Although the mutant mice are smaller than the wild-type mice, they appear normal otherwise. Since IGF1 and IGF2 play their mitogenic roles by binding to the same receptor, there may be a functional redundancy. However, null mutations at the IGF-2R gene results in perinatal lethality (Lau *et al.*, 1994). Histologic examinations indicate that these mutant mice had ventricular hyperplasia as well as thinning of the intraventricular septum and an abnormal tricuspid valve. The concentration of circulating IGF2 is much higher in mutant mice than wild-type mice. These *in vivo* studies support that concept that IGF2 positively regulates proliferation of ventricular myocytes. Interestingly, it has been shown that the defects caused by deletion of a segment of chromosome 17 containing the IGF-2R gene could be rescued by a second mutation at IGF2 gene (Filson *et al.*, 1993).

2. Fibroblast Growth Factor/Fibroblast Growth Factor Receptor

The members of the fibroblast growth factor (FGF) family can be divided into two major classes: basic FGF (bFGF) and acidic FGF (aFGF), which all function as mitogens in a number of cell types by binding to FGF receptors (FGFRs). FGFRs are also tyrosine kinase receptors and catalyze a cascade of protein phosphorylation and ultimately activate the expression of proto-oncogenes. The genes encoding FGFs and FGFRs are expressed in embryonic myocardium, suggesting mitogenic roles for FGF. Sugi *et al.* (1993) have shown that inhibition of FGF-2 expression by a specific antisense oligonucleotide in cultured anterior lateral plate mesoderm of chicken embryos (stage six) inhibits proliferation of precardiac mesoderm cells. These results indicate that FGF-2 could play a critical role in the autocrine regulation of proliferation and, perhaps differentiation function, of embryonic cardiac myocytes. By using dominant-negative mutant FGFR1 or FGFR1 antisense RNA, Mima *et al.* (1992) have shown that during the first week of chick embryos the receptor-coupled FGF signaling pathway is essential for proliferation of myocytes in the tubular heart. After the second week in chick embryos, the proliferation of cardiac myocytes become FGF independent. Pasumarthi *et al.* (1996) have recently shown that the bFGF-2 gene is expressed as the high molecular weight and low molecular weight form in neonatal and adult rat hearts due to usage of different initiation codons. However, both forms can stimulate DNA synthesis and proliferation of neonatal rat cardiac myocytes in culture. Thus, FGF may play different roles in cardiac myocytes after terminal differentiation.

3. Retinoic Acid and Retinoic Acid Receptor System

Retinoic acids (RAs) are oxidation products of retinol (vitamin A) and function as steroid hormones that are involved in controlling pattern formation of the entire embryo and differentiation of many types of tissues during embryogenesis. RAs exert their effects on target cells by binding to the nuclear-located receptors that activate gene expression as transcription factors. The effects of RA/retinoic acid receptor (RAR) on myocardial development was initially observed by Wilson and Warkany in 1949. In their experiments, Wilson and Warkany have demonstrated that vitamin A deficiency resulted in various defects in a variety of tissues, including hypoplastic ventricular chamber and defective ventricular septum. These early observations strongly suggest that one of the functions of the RA/ RAR signaling pathways in myocardial development is to positively regulate proliferation of ventricular myocytes. The divergent effects of RA are likely mediated by various RA receptors, which can be classified into two subtypes—RAR (α, β, γ) and RXR (α, β, γ). Members of RAR form heterodimers with members of RXR, which function as heterodimeric transcription factors and recognize distinct *cis* regulatory elements. The RAR and RXR members display distinct temporal and spatial expression patterns. For instance, the RXRα gene is expressed in a high abundance in embryonic heart, skeletal muscle, liver, intestine, and kidney, whereas the RXRβ gene is expressed at a low level in all the tissues. The RXRγ gene is expressed at a high level in mesoderm and its derived tissues. Null mutations at both alleles of the RXRβ or RXRα genes did not cause embryonic lethality and only minor phenotypic changes in adult mice (Li *et al.*, 1993; Lohnes *et al.*, 1993; Lufkin *et al.*, 1993). In contrast, null mutations at both alleles of the RXRα gene did cause embryonic lethality (Sucov *et al.*, 1994). The RXRα-/- mouse embryos displayed defects identical to a subset of defects observed in vitamin A deficiency mouse embryos: hypoplastic development of the ventricular chambers and muscular ventricular septal malformation. These results indicate that RXRα is a component of the key receptor that mediates the mitogenic and morphogenic effects of retinoic acids in ventricular development. The target genes of RXRα in embryonic ventricular myocytes are not yet defined. Interestingly, similar ventricular defects observed in the RXRα-/- mouse embryos were also observed in those mouse embryos that carry null mutations at N-myc, TEF1, Wilms tumor (WT1), and neurofibromatosis (NF1) genes, which encode transcription factors (Rossant, 1996). Proliferation of embryonic ventricular myocytes seems to be a multifactorially regulated process that is sensitive to alterations in any of these factors.

4. Cytokines

Cytokines are involved in many processes in a variety of cells and the list of cytokines is still increasing. A novel cytokine, cardiotrophin-1 (CT-1) has been isolated from an *in vitro* embryonic cell system of cardiogenesis that can activate embryonic markers in neonatal rat cardiac myocytes (Pennica *et al.*, 1995; Wollert *et al.*, 1996). CT-1 shares a high degree of sequence similarity with the previously isolated cytokine, IL-6, and activates a hypertrophic signaling pathway in neonatal rat cardiac myocytes through the membrane receptor component gp130. Further studies by immuno *in situ* staining have demonstrated that the CT-1 gene is predominantly expressed in the embryonic mouse heart tube during day 8.5 p.c. and day 10.5 p.c. (Sheng *et al.*, 1996). Within the heart tube, CT-1 is specifically expressed in myocardial cells, and not in endocardial cushion or outflow tract tissues. These results, in combination with the fact that CT-1 is isolated from EBs where cardiac myocytes are proliferative, suggest that CT-1 could be involved in regulating proliferation of embryonic cardiac myocytes in a paracrine or autocrine fashion. This hypothesis is supported by the observation that the exogenous CT-1 can stimulate proliferation of embryonic cardiac myocytes in culture (Sheng *et al.*, 1996).

C. REGULATION OF GENES ENCODING CYCLINS, CYCLIN-DEPENDENT KINASES, AND TUMOR SUPPRESSORS IN CELL CYCLE REGULATION

Mammalian cell cycle regulation has been extensively studied, and numerous genes that are involved in cell cycle regulation have been identified and characterized. However, it is beyond the scope of this chapter to review the functions of every gene in cell cycle regulation in detail. Thus, I discuss the general picture of cell cycle regulation in mammalian cells and focus on what is known about cell cycle regulation in cardiac myocytes during development. In general, various extracellular mitogenic signals are transmitted into the nucleus through cytoplasmic signaling pathways, which are composed of a series of protein kinases. The nuclear mitosis control machinery basically consists of five types of factors: the cyclins, cyclin-dependent kinases (CDK), CDK inhibitors (CDKi), tumor suppressors [e.g., retinoblastoma protein (Rb)], and transcription factors (e.g., E2F). The Rb protein blocks G1/S transition by binding and thereby sequestering transcription factors such as E2F, which are essential for entry into the S phase. Cyclins activate CDKs by forming complexes with CDKs that then regulate the activities of downstream factors by phosphorylation. For instance, the complexes of D and E types of cyclins and CDKs (CDK2, CDK4, and CDK6) inactivate the Rb protein by phosphorylation so that the cells can enter the S phase. The activities of CDKs are inhibited by CDK inhibitors, which are sometimes also referred to as tumor suppressor proteins. Figure 1 illustrates the progression of the mammalian cell cycle, the positive regulation by the cyclin–CDK complexes, and the negative regulation by cyclin-dependent kinase inhibitors at different stages. There are a number of

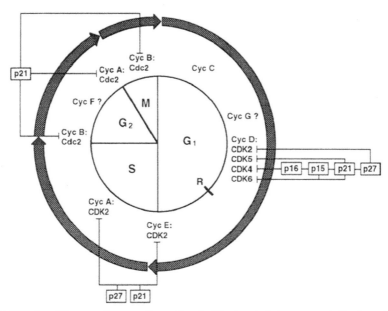

FIGURE 1   Diagram of mammalian cell cycle. The inner circle represents the four phases (G_1, S, G_2, and M) of one mitotic cycle and the outer circle shows the sequential progression of the four mitotic phases. The cyclins (Cyc) and cyclin-dependent protein kinases (CDK, cdc2 is also referred as CDK1) that positively regulate cell cycle progression at different stages are shown in the space between the inner and outer circle. The CDK inhibitors (boxed), which are named after their molecular weights, are shown outside the outer circle, and their specific targets are indicated. The complexes of Cyc E/CDK2, Cyc B/Cdc2, and Cyc A/Cdc2 are required for G_1/S, S/G_2, and G_2/M transitions, respectively. The complexes of Cyc D and its associated CDKs, Cyc A/CDK2, and Cyc B/Cdc2 are required for progression through G_1, S, and M phases, respectively. (Reprinted from *Cardiovasc. Res.* vol. 30, McGill, C.J., and Brooks, G., Cell cycle control mechanisms and their role in cardiac growth, pp. 567–569, Copyright 1995 with kind permission of Elsevier Science–NL, Sara Burgerhartstraat 25, 1055 KV Amsterdam, The Netherlands.)

excellent articles describing these factors (Grana and Reddy, 1995; McGill and Brooks, 1995).

In comparison with cardiac muscle cells, the mechanisms that regulate withdrawal of skeletal myoblasts from the cell cycle are better understood. It was first shown by Crescenzi *et al.* (1990) and Sorrentino *et al.* (1990) that MyoD can induce exit of skeletal myoblasts from the cell cycle independently of its myogenic activity. Since then, Halevy *et al.* (1995) and Guo *et al.* (1995) have shown MyoD induces withdrawal of skeletal myoblasts from the cell cycle by activating transcription of the gene encoding a CDK inhibitor (CDKi), p21. It has been shown in other cell types that p21 inhibits the activity of most CDKs that are essential for progression through the mitotic cell cycle (Gyuris *et al.*, 1993; Harper *et al.*, 1993; Xiong *et al.*, 1993). Thus, a part of p21 function is to inhibit the activities of CDK2, CDK4, and CDK6 so that the Rb protein remains active to block G1/S transition. Most of the identified factors including the Rb protein are expressed in cardiac myocytes, thus, it is likely that they play the same roles in cardiac myocytes.

The permanent withdrawal of cardiac myocytes from the cell cycle is a major step of terminal differentiation. However, little is known about the factors that induce cardiac myocytes to withdraw from the cell cycle during terminal differentiation and prevent them from reentering the cell cycle afterward. The majority of terminally differentiated cardiac myocytes are growth arrested at the G0 or G1 phase (Capasso *et al.*, 1992), suggesting that factors such as the Rb protein, which blocks G1/S transition, become constitutively active in these myocytes. As a result, the active Rb or related factors preclude E2F transcription factors from activating the expression of genes that are required for DNA synthesis. This speculation is supported by the observation that a number of genes required for DNA synthesis are down-regulated or turned off after myocardial terminal differentiation. For instance, the genes encoding DNA polymerase α-subunit, thymidine kinase, and proliferating cell nuclear antigen (PCNA) are turned off after terminal differentiation (Claycomb 1978; Marino *et al.*, 1991). Thus, the transcription factors such as E2F that activate these genes are probably inactivated. Tam *et al.* (1993) have reported that in neonatal rat cardiac myocytes, Rb protein is in both hypophosphorylated (active) and hyperphosphorylated (inactive) form. In adult rat cardiac myocytes, however, only hypophosphorylated Rb protein is detected. These results strongly suggest that Rb protein is deregulated and becomes constitutively active after myocardial terminal differentiation. Recently, McGill and Brooks (1995) reported that the activities of CDK1 and CDK2, which inactivate Rb protein by phosphorylation, are 5- and 9.5-fold higher in neonatal rat cardiac myocytes than in adult cardiac myocytes. The preceding results indicate that Rb protein and possibly its related proteins could play a role in preventing terminally differentiated cardiac myocytes from reentering the S phase. However, it is unclear how the expression of CDK genes are regulated during terminal differentiation.

D. VIRAL NUCLEAR ONCOGENES AS PROBES TO MECHANISMS THAT CONTROL MYOCYTE PROLIFERATION

As many oncoproteins, three viral oncoproteins—the SV40 large T antigen, adenovirus E1A, and human papilloma virus E7—can activate quiescent cells to reenter the cell cycle (Nevin, 1992; Moran, 1994; Weinberg, 1995). Although the molecular mechanisms by which these nuclear viral oncoproteins activate reentry into the cell cycle are not completely understood, it is suggested that these oncoproteins function by disrupting the functions of Rb and the related tumor suppressors, which are collectively named *pocket proteins*. These pocket proteins contain the *pocket domains*, which are involved in binding transcription factor E2F and thereby either inactivate E2F or convert E2F into a transcription suppressor. These three oncoproteins bind to pocket proteins at their pocket domains and thus, free E2F or prevent the pocket proteins from binding to E2F, which can then activate genes required for DNA synthesis. Therefore, these nuclear oncoproteins serve as useful probes for the cellular factors that prevent cells from reentering the cell cycle and provide the initial molecular framework for ultimately understanding the molecular mechanisms that regulate cell cycle. For instance, cellular factors such as Rb or its related proteins with which these nuclear oncoproteins interact can be identified by various biochemical and genetic methods.

Insight into the molecular mechanisms that induce cardiac myocytes to withdraw from the cell cycle and prevent them from reentering the cell cycle come from studies utilizing the SV40 large T antigen and the adenovirus E1A. Using the ANF promoter and α-MHC promoter to drive the expression of T antigen in transgenic mice, Field (1988) and Katz *et al.* (1992) have shown that T antigen can induce atrial and ventricular hyperplasia. Cardiac myocytes isolated from hyperplastic atria and ventricles have the properties of spontaneous contraction and proliferation. These results indicated that T antigen retarded withdrawal of cardiac myocytes from the cell cycle during terminal differentiation. Neither commitment nor differentiation of cardiac muscle lineage seems affected by the presence of large T antigen. Sen *et al.* (1988) have shown that ectopic expression of T antigen in cultured terminally differentiated neonatal ventricular myocytes can induce reentry of these myocytes into the cell cycle. Since majority of terminally differentiated cardiac myocytes are arrested at G0 or G1 phase (Capasso *et al.*, 1992), the previous results indicate that T antigen can possibly overcome the G1 blockade in these cardiac myocytes and thus stimulate G1/S transition. Similarly, Kirshebaum and Schneider (1995) have shown that overexpression of the retrovirus oncoprotein E1A in cultured neonatal rat ventricular myocytes increases 5-bromo-2'-deoxyuridine (BrdU) incorporation by these myocytes, suggesting that E1A can activate G1/S transition. In addition, Liu and Kitsis (1996) have reported that overexpression of E1A in cultured fetal rat cardiac myocytes (day 20.0 p.c.) increases BrdU positive myocytes from being nearly undetectable to 94%. These results strongly suggest that pocket proteins are involved in both inducing withdrawal of cardiac myocytes from the cell cycle and preventing them from reentering the cell cycle. Tam *et al.* (1993) have reported that two pocket proteins, Rb and the related p107 protein, are produced in proliferative cardiac myocytes and skeletal myoblasts, whereas only Rb is detected in terminally differentiated cardiac myocytes and skeletal myotubes. Hence, Rb protein could play a particularly important role during and after terminal differentiation. Mutations at both alleles of the Rb gene cause fetal lethality. Interestingly, in the Rb-/- mice, the gene encoding p107 remains expressed after terminal differentiation in skeletal muscle possibly to compensate the loss of Rb function. However, the terminally differentiated Rb-/- skeletal muscle cells in culture can reenter the cell cycle in response to growth factor stimulation (Schneider *et al.,* 1994) and reverse terminal differentiation. These processes were blocked by ectopic expression of wild-type Rb. These results indicate that although Rb and p107 are related structurally, their functions in cell cycle regulation and terminal differentiation are distinct. Furthermore, malignancies associated with mutations at p107 or another Rb-related protein, p130, have not been identified so far, which further supports the notion that Rb plays a dominant role in preventing terminally differentiated cells from reentering the cell cycle. Although Tam *et al.* (1993) have shown that Rb could play a key role in preventing terminally differentiated cardiac myocytes from reentering the cell cycle, it is possible that additional pocket proteins are involved.

E. PHYSIOLOGICAL SIGNALS FOR WITHDRAWAL OF CARDIAC MYOCYTES FROM CELL CYCLE

To cope with the changing demands for cardiac output after birth, the heart rate of mammals changes remarkably. For instance, the heart rate of rats increases from 300 to 500–600 beats/min during the first 3 weeks after birth, whereas the heart rate decreases after birth in larger mammals. The increase in the heart rate of rats appears to be the result of an increase in innervation of the adrenergic sympathetic nervous system (Adolf, 1967; Wekstein, 1965). As the activity of the parasympathetic nervous system subsequently increases, however, the heart rate of rats decreases to the steady-state heart rate of 300 beats/min. Accordingly, during the first week after birth, the concentration of norepinephrine increases in the myocardium, which leads to the increase in heart rate through adrenergic receptors and adenylyl cyclase pathway (see Chapter 3). During late fetal and early neonatal development, the cAMP concentration in cardiac myocytes increases as the cGMP concentration decreases (Claycomb, 1976). It has been shown that an increase in intracellular concentration of cAMP results in inhibition of DNA synthesis and subsequent terminal differentiation of cardiac myocytes (Claycomb, 1976, 1983). Thus, one possible mechanism by which cAMP inhibits DNA synthesis is to down-regulate the expression of genes required for DNA synthesis. Therefore, it was hypothesized by Claycomb that adrenergic innervation induces withdrawal of cardiac myocytes from the cell cycle using norepinephrine and cAMP as the chemical mediators. It would be interesting to determine whether norepinephrine or cAMP actually affects the expression of the nuclear cell cycle control factors, such as cyclins, CDKs, and CDK inhibitors in rat cardiac myocytes.

The physiological signals for withdrawal of cardiac myocytes from the cell cycle in other species are much less clear. To develop more force, the sarcomere content of cardiac myocytes in most mammals increases remarkably. The increased highly organized sarcomere content could physically block cytokenesis or even endomitosis. This hypothesis is indirectly supported by the observation that after terminal differentiation, cardiac myocytes of many mammals become multinucleated or contain nuclei of polyploid.

In summary, unlike skeletal muscle cells, differentiation and proliferation are not mutually exclusive for embryonic cardiac myocytes. Proliferation of embryonic cardiac myocytes is regulated by a number of growth factors and cytokines. Cardiac myocytes become permanently postmitotic shortly after birth. Since the same basic nuclear cell cycle control machinery is found in all the cell types studied so far, it is likely that the same machinery is involved in controlling cell cycle in cardiac myocytes. After myocardial terminal differentiation, most cardiac myocytes are arrested at the G0 and G1 phase, which suggests that the entry into the S phase is permanently blocked in these myocytes. Reentry of terminally differentiated cardiac myocytes into the S phase can be activated by viral nuclear oncoproteins, which interact with the tumor suppressors, the pocket proteins. Hence, the pocket proteins could play a critical role in blocking the entry into the S phase in these myocytes. This notion is further supported by the observation that the activity of CDKs that inactivates the pocket protein, Rb, is decreased in terminally differentiated cardiac myocytes. Thus, the key step during myocardial terminal differentiation seems to establish the blockade for the entry into the S phase.

V. HYPERTROPHIC RESPONSE TO INCREASED DEMAND

A. MORPHOLOGIC CHANGES IN HYPERTROPHIED CARDIAC MYOCYTES

Mammalian cardiac myocytes become postmitotic shortly after birth and, therefore, the subsequent growth of myocardium becomes dependent on hypertrophic growth of cardiac myocytes when adaptation to increasing workload is required. Cardiac myocytes can be more than 10-fold larger than the control myocytes through hypertrophy *in vitro*. The enlargement of cardiac myocytes is attributed to an increase in sarcomere concentration, which increases force development and cardiac output. However, hypertrophy can change from compensatory state to pathological state in certain cases such as hypertension. In pathologically hypertrophied hearts, cardiac myocytes display a disarray of myofibrils, myocyte apoptosis, and cardiac output decreases. Myocardial hypertrophy can eventually progress to heart failure. Much effort has been made to unravel the signaling pathways that mediate cardiac myocyte hypertrophy using genetic, cellular, and molecular methods. Today, a number of factors involved in mediating myocyte hypertrophy and their associated cellular signaling pathways have been identified and characterized in the cell culture model.

B. ALTERATIONS OF GENE EXPRESSION DURING MYOCYTE HYPERTROPHY

The core of cardiac myocyte hypertrophy at the molecular level is alteration in gene expression. The enlargement of the cardiac myocyte is largely attributed to the increase in the production of contractile proteins, suggesting that there is an increase in gene expression. There are two components of the increase in gene expression during cardiac myocyte hypertrophy: (1) reactivation of the embryonic gene program and (2) elevation in the expression of constitutively expressed contractile protein encoding genes. The reactivated embryonic genes in hypertrophied ventricular myocytes include those that encode ANF, βMHC, and skeletal α-actin (SkA). However, genes such as those encoding αMHC, MLC-2v, and troponin C, which are constitutively expressed in ventricular myocytes, are up-regulated during hypertrophy. The increased expression of genes from both classes contributes to the enlargement of cardiac myocytes. To understand the signaling pathways that mediate the up-regulation of gene expression during myocyte hypertrophy, genes of both classes have been extensively studied as the model systems. Up-regulation of all the genes studied so far seems mediated primarily at transcriptional level.

C. *IN VIVO* AND *IN VITRO* MODELS OF CARDIAC MYOCYTE HYPERTROPHY

Cardiac myocyte hypertrophy can be induced experimentally in both *in vivo* and *in vitro* model systems. For instance, in laboratory rats or mice, left ventricular hypertrophy can be induced by transverse aortic constriction, whereas right ventricular hypertrophy can be induced by transverse pulmonary artery constriction (Rockman *et al.*, 1994a; Adachi *et al.*, 1995). This type of *in vivo* model mimics well the ventricular hypertrophy induced by hypertension or atherosclerosis. However, both cultured neonatal and adult ventricular myocytes of a number of species can be induced to develop hypertrophy by a number of stimuli. Due to the relatively well-controlled conditions, the *in vitro* model has provided numerous critical clues as to the extracellular factors and their associated cellular signaling pathways, which mediate pressure-induced cardiac myocyte hypertrophy. As is discussed in the following sections, most of the stimuli defined in the *in vitro* hypertrophy model are involved in mediating pressure-induced ventricular hypertrophy *in vivo*. The hypertrophied ventricular myocytes from both *in vivo* and *in vitro* models share many properties in morphology, biochemistry, electrophysiology, and gene expression. For instance, they are all much larger in size than the corresponding control myocytes, and they reexpress embryonic genes such as the ANF, βMHC, and SkA genes. Also during the initial 30–60 minutes of stimulation, a number of proto-oncogenes collectively known as immediate early genes such as c-jun, c-fos, Egr-1, Nur77, and c-myc, are transiently activated in the ventricular myocytes. These genes encode transcription factors and, thus, are possibly involved in hypertrophy-associated transcriptional regulation. Therefore, it is reasonable to assume that in both *in vivo* and *in vitro* models, the same signaling pathways are involved in mediating cardiac myocyte hypertrophy. The *in vitro* model of myocyte hypertrophy has proven to be particularly powerful in defining cellular signaling pathways of individual hypertrophic stimuli and the proximal *cis/trans* regulatory factors that are required for alterations in gene expression. It is easy to envision that once these factors are identified by the *in vitro* model, their functions *in vivo* can be tested by the *in vivo* hypertrophy model in combination with the transgenic and gene "knockout" techniques.

D. SIGNALING PATHWAYS THAT MEDIATE MYOCYTE HYPERTROPHY

The signals generated by pressure or volume overload have to be transduced into nuclei to induce alterations in gene expression. Thus, there must be cellular signaling pathways that are involved in this signal transduction. Identification of key factors of such cellular signaling pathways is essential for understanding the regulation of myocardial hypertrophic growth and designing possible therapeutic reagents to prevent pathological hypertrophy. Activation of those signaling pathways in response to pressure or volume overload may involve an elevation in either the activity or expression of certain signaling factors, which help to identify those signaling pathways. The initial clues come from the *in vivo* observation that pressure or volume overload-induced ventricular hypertrophy is associated with an elevation in the local levels of vasoconstrictors, growth factors, cytokines, and adrenergic agonists. This suggests the involvement of these factors and their associated cellular signaling pathways in mediating hemodynamic change-induced ventricular myocyte hypertrophy. In addition, these factors may work through paracrine/autocrine mechanisms since aortic constriction only induces left ventricular hypertrophy, and pulmonary constriction primarily induces right ventricular hypertrophy. More direct evidence comes from studies utilizing the *in vitro* cell culture model, which demonstrates that these factors can induce the same morphologic, biochemical, electro-physiological, and gene expression changes in cultured cardiac myocytes as observed in the myocytes of hypertrophied hearts. In the following sections, the results on four types of well-studied factors, vasoconstrictors (e.g., angiotensin II, endothelin I), α-adrenergic agonists, growth factors, and cytokines, are described. Although these individual factors are studied separately in the *in vitro* model, it is possible that they act synergistically *in vivo*.

1. Vasoconstrictors

a. Angiotensin II

Angiotensin II is a peptide vasoconstrictor and is synthesized as an inactive precursor (angiotensinogen) predominantly in hepatocytes. Angio-tensinogen is then converted to angiotensin I by renin, and then to active angiotensin II by angiotensin-converting enzyme (ACE) in the lungs. Therefore, the level of angiotensin II is directly determined by ACE activity. The role of angiotensin II in mediating cardiac myocyte hypertrophy was first indicated by the observations that both ACE inhibitors and angiotensin II antagonists can cause regression of cardiac myocyte hypertrophy *in vivo* (Linz *et al.*, 1989; Baker *et al.*, 1990). Furthermore, the ACE mRNA level is increased fourfold in hypertrophied ventricles (Lindpaintner and Ganten, 1991; Dostal and Baker, 1992). More direct evidence came from the studies with the *in vitro* model, which shows that exogenous angiotensin II can induce hypertrophic responses of cultured neonatal rat ventricular myocytes (Sadoshima and Izumo, 1993). In these myocytes, there was a characteristic transient activation of the immediate early genes (c-fos, c-jun, junB, Egr-1, and c-myc) at the initial stage of angiotensin II stimulation, which was followed by the reactivation of embryonic genes (ANF and SkA). Mechanical stretching of cultured ventricular myocytes can induce a number of morphologic and genetic properties found in myocytes from hypertrophied ventricles, and thus has been widely utilized as an *in vitro* model for pressure-induced myocyte hypertrophy (Sadoshima *et al.*, 1992). The connection between mechanical stretching and angiotensin II has been demonstrated by the observation that mechanical stretching of cultured neonatal rat ventricular myocytes can induce release of angiotensin II (Sadoshima *et al.*, 1993; Yamazaki *et al.*, 1994). These observations strongly support the hypothesis that angiotensin II plays a critical role in mediating pressure overload-induced ventricular hypertrophy *in vivo*. Angiotensin II works at least partially by a paracrine or autocrine mechanism since the genes encoding angiotensinogen, renin, and ACE are expressed in all four chambers of the heart (Lindpaintner and Ganten, 1991; Dostal and Baker, 1992). This theory was further supported by electron microscopic studies that show that angiotensin II is packed in structures similar to secretary granules in ventricular myocytes (Sadoshima and Izumo, 1993).

There are two types of angiotensin II receptors identified in cardiac myocytes: type I (AT1) and type II (AT2). The cDNA-encoding AT1 and AT2 have been isolated (Murphy *et al.*, 1991; Mukoyama *et al.*, 1993), and both encode seven-transmembrane receptors. AT1 mediates angiotensin II-induced ventricular hypertrophy by activating a G protein-coupled signaling pathway (Murphy *et al.*, 1991; Mukoyama *et al.*, 1993). Subsequently, both *in vivo* and *in vitro* studies indicate that AT1 is the predominant receptor that mediates the hypertrophic effects of angiotensin II in cardiac myocytes, whereas the functions of AT2 remain unclear. Rockman *et al.* (1994b) have shown that the AT1-specific antagonist but not the AT2-specific antagonist can attenuate left ventricular hypertrophy in response to transverse aortic constriction. Similarly, Miyata and Haneda (1994) and Kojima *et al.* (1994) have shown that the hypertrophic effects of angiotensin II on cultured ventricular myocytes can be blocked by the AT1-specific but not AT2-specific antagonists. To identify factors acting downstream of the AT1 receptor, Yamazaki *et al.* (1995a, 1995b) have demonstrated that mechanical stretching of cultured neonatal rat ventricular myocytes induces hypertrophic responses possibly by activating mitogen-activated protein kinase (MAPK). The stretching-induced activation of MAPK can be inhibited by the AT1-specific antagonists, suggesting that MAPK is a downstream factor of angiotensin II. MAPK is a serine/threonine kinase and was initially identified as a key protein kinase in response to mitogenic stimulation in many cell types. Further studies indicated that MAPK has rather a wide range of functions including regulation of mitosis and differentiation (Seger and Krebs, 1995). The results of Yamazaki *et al.* (1995a, 1995b) indicate that MAPK is also involved in mediating angiotensin II-induced hypertrophy in cardiac myocytes. This notion was further supported by the observation that exogenous angiotensin II leads to a rapid activation of MAPK and its downstream protein kinase, 90 kD S6 kinase (RSK) in cultured neonatal rat ventricular myocytes (Sadoshima *et al.*, 1995). The relationships between MAPK activity and cardiac myocyte hypertrophy were established by the introduction into cultured cardiac myocytes of the constitutively active MAPK or MAPK kinase (MEK1), which activates MAPK by phosphorylation (Gillespie-Brown *et al.*, 1995). Gillespie-Brown *et al.* (1995) have shown that the ectopic expression of the constitutively active MEK1 or MAPK in cultured adult rat ventricular myocytes results in the reexpression of the ANF, βMHC, and SkA genes.

Since the expression of the angiotensinogen gene is activated by exogenous angiotensin II, angiotensin II possibly initiates a positive feedback regulation of myocyte hypertrophy (Sadoshima and Izumo, 1993). In addition, angiotensin II may also mediate cardiac hypertrophic response by activating other autocrine/paracrine mechanisms. For instance, the expression of TGFβ1 gene in neonatal rat ventricular myocytes can be induced by angiotensin II (Sadoshima and Izumo, 1993; Lee *et al.*, 1995).

b. Endothelin-1

The other peptide vasoconstrictor, endothelin-1 (ET-1), also plays a critical role in mediating pressure overload-induced cardiac myocyte hypertrophy. ET-1 is synthesized as a 200-amino acid precursor, primarily in vascular endothelial cells, and processed by multiple proteolytic steps to become an active 21-amino acid peptide. Stimulation of cultured neonatal rat ventricular myocytes with exogenous ET-1 induces hypertrophic responses including activation of immediate early genes, reexpression of the embryonic markers, enlargement of cell size, and organization of sarcomere (Shubeita *et al.*, 1990). The ET-1 mRNA level in left ventricles was transiently increased after transverse aortic constriction (Ito *et al.*, 1994), and an ET-1 antagonist (BQ123) blocked ventricular hypertrophy in response to transverse aortic constriction. Interestingly, the ET-1 antagonist (BQ123) can partially block the angiotensin II-induced cardiac hypertrophic responses (Ito *et al.*, 1993a). Furthermore, stimulation with exogenous ET-1 resulted in a three- to four-fold increase in the activity of MAPK in cultured neonatal rat ventricular myocytes (Bogoyevitch *et al.*, 1994). Recently, Yamazaki *et al.* (1996) have shown that expression of the ET-1 gene is induced in cardiac myocytes by mechanical stretching. The ET-1 antagonist (BQ123) can effectively block MAPK activation by mechanical stretching. Thus, it is possible that ET-1 and angiotensin II work synergistically in inducing hypertrophic responses in cardiac myocytes.

2. Adrenergic Agonists

Cultured neonatal rat ventricular myocytes can be induced to develop hypertrophy by both α- (e.g., phenylephrine) and β-adrenergic (e.g., isoproterenol) agonists (Simpson, 1983; Starksen *et al.*, 1986; Bishopric *et al.*, 1987; Long *et al.*, 1989). Although the end phenotypes of hypertrophied neonatal rat ventricular myocytes are very similar, the patterns of immediate early gene activation vary between stimulation by α- and β-adrenergic agonists (Iwaki *et al.*, 1990). Thus, it is possible that differences exist between the signaling pathways that mediate the hypertrophic effects of α- and β-adrenergic agonists. The model of α-adrenergic agonist-induced neonatal rat ventricular myocyte hypertrophy has been extensively studied and thus well characterized. Knowlton *et al.* (1993) have shown that the α-adrenergic (e.g., phenylephrine) agonist-induced hypertrophy of cultured neonatal rat ventricular myocytes can only be blocked by antagonists specific for α1A but not α1B subtype of α1-adrenergic receptor. The α1A-adrenergic receptor is coupled to a G protein, Gq, which is required for α1A receptor to mediate α-adrenergic agonist-induced hypertrophy. LaMorte *et al.* (1994) have shown that a neutralizing antibody against Gq can block phenylephrine-induced hypertrophy, whereas a constitutively active Gq protein can induce the same hypertrophic response as phenylephrine does. The cytoplasmic proto-oncoprotein, Ras, seems to be a downstream factor of Gq in mediating phenylephrine-induced ventricular myocyte hypertrophy. A dominant-negative Ras protein can block phenylephrine-induced reexpression of the ANF gene in neonatal rat ventricular myocytes (Thorburn *et al.*, 1993). However, a constitutively active Ras protein can induce both reexpression of the ANF gene and morphologic changes observed in phenylephrine-stimulated neonatal rat ventricular myocytes (Thorburn *et al.*, 1993). Activation of the Ras-dependent signaling pathway involves sequential activation of Ras, Raf-1, MEK1, and MAP kinases. Activation of the Raf-1 gene in cultured neonatal rat ventricular myocytes induces activation of MAP kinases and luciferase reporter expression from the ANF and MLC-2v promoters (Thorburn *et al.*, 1994). The dominant-negative Raf-1 can block the activation of the ANF and MLC-2v promoters by phenylephrine (Thorburn *et al.*, 1994). Recently, Glennon *et al.* (1996) have reported that phenylephrine-induced hypertrophic responses of rat cardiac myocytes can be significantly down-regulated by an antisense oligonucleotide that inhibits MAPK expression. As previously described, the constitutively active MEK1 and MAPK1 can activate expression from the ANF, βMHC, and SkA promoters in cultured adult rat ventricular myocytes (Gillespie-Brown *et al.*, 1995). Transgenic mice that carry the MLC-2v promoter-Ras transgene display ventricular hypertrophy (Hunter *et al.*, 1995). Therefore, it is possible that angiotensin II and an α-adrenergic agonist share certain cellular signaling pathways, such as the Ras/MAPK signaling pathway. Thus, phenylephrine possibly induces hypertrophy in neonatal rat ventricular myocytes through the following factors: A1 subtype of α-adrenergic receptor $\Rightarrow$ Gq protein $\Rightarrow$ Ras $\Rightarrow$ Raf $\Rightarrow$ MEK1 $\Rightarrow$ MAPK $\Rightarrow$ alterations in gene expression. This sequence may not be complete, and more factors are likely involved.

3. Growth Factors

The levels of mRNA encoding a number of growth factors are increased in the heart at an initial stage of pressure overload-induced hypertrophy, suggesting that these growth factors mediating hypertrophy through a paracrine/autocrine mechanism. The functions of three types of growth factors—transforming growth factor β1 (TGFβ1), insulin-like growth factors I and II (IGF1 and IGF2), and basic fibroblast growth factor (bFGF)—in mediating hypertrophic responses have been relatively well studied.

Since TGFβ1 is produced in the heart by both cardiac myocytes and nonmyocytes, it is speculated that TGFβ1 is involved in myocardial growth. The role of TGFβ1 in mediating cardiac myocyte hypertrophy was demonstrated by the observation that exogenous TGFβ1 can induce the reexpression of embryonic muscle genes in cultured neonatal rat ventricular myocytes (Parker *et al.*, 1990). Furthermore, transverse aortic constriction resulted in a three- to fourfold increase in the TGFβ1 mRNA level in ventricular myocytes, but not in the nonmyocytes (Takahashi *et al.*, 1994).

In addition, TGFβ1 expression in cardiac myocytes can be activated by subcutaneous norepinephrine infusion, and TGFβ1 expression in cardiac fibroblasts can be induced by angiotensin II (Lee *et al.*, 1995). It is thus likely that TGFβ1 mediates cardiac myocyte hypertrophy by autocrine and paracrine mechanisms in cooperation with angiotensin II and adrenergic systems. There are at least three subtypes of TGFβ1 receptors, type I, II, and III, that are serine and threonine kinases. However, it is not clear which subtype of TGFβ1 receptor mediates hypertrophic responses in cardiac myocytes.

IGF1 and IGF2 are potent mitogens that promote mitosis of a number of cell types by binding to the IGF-1R on plasma membrane. IGF-1R contains tyrosine kinase activity in the cytoplasmic domain, which is activated by binding of IGF1 or IGF2 to the extracellular domain. Once activated by IGF1 or IGF2 binding, the tyrosine kinase activity of IGF-1R activates the Ras/MAPK signaling pathway, which mediates cardiac myocyte hypertrophy. The role of IGF1 in mediating cardiac hypertrophic responses was first indicated by the observation that stimulation with exogenous IGF1 induces enlargement of neonatal rat ventricular myocytes and up-regulation of genes encoding βMHC, MLC-2, and troponin I (Ito *et al.*, 1993b). In addition, the IGF1-stimulated adult cardiac myocytes showed a dramatic increase in nascent myofibrils compared with the control (Donath *et al.*, 1994). The expression of the IGF1 gene in left ventricles, at both the mRNA and protein levels, is increased in response to pressure overload in the left ventricles (Hanson *et al.*, 1993; Donohue *et al.*, 1994). In addition, the increase in IGF1 mRNA level is also detected in volume-overloaded rat hearts (Isgaard *et al.*, 1994). Administration of IGF1 resulted in hypertrophy of cardiac myocytes *in vivo* (Duerr *et al.*, 1995). These results support the notion that IGF1 mediates hypertrophy in response to both pressure and volume overload. IGF-II seems to induce similar hypertrophic responses in myocytes to those of IGF1, probably by binding to IGF-1R. For instance, IGF2 can induce the expression of genes encoding ANF, βMHC, SkA, and troponin I, increase the rate of protein synthesis, and increase the organized myofibrils and the size of cultured neonatal rat ventricular myocytes (Adachi *et al.*, 1994; Liu *et al.*, 1996).

The bFGF gene encodes another potent mitogen and is expressed in the heart at all developmental stages (Liu and Kitsis, 1996), suggesting that bFGF could play an important role in controlling myocardial growth and development. As IGF-1R, the bFGF receptor contains a tyrosine kinase activity in the cytoplasmic domain that is activated by bFGF binding. The activated bFGF receptor activates the Ras/MAPK signaling pathway. Kaye *et al.* (1996) have reported that increased mechanical activity induced hypertrophic responses in cultured adult rat ventricular myocytes. In those hypertrophied myocytes, both sarcolemma permeability and release of cytosolic bFGF were increased, indicating an important role of bFGF in mediating ventricular myocyte hypertrophy by an autocrine or paracrine mechanism.

4. Cytokines

Although cytokines were initially identified as peptides that primarily mediate either anti- or pro-inflammatory responses, it has become evident that these peptides are actually involved in a variety of biologic processes. Recent evidence has indicated that certain types of cytokines are also involved in mediating cardiac myocyte hypertrophy. Not only are these cytokines expressed in the heart, but they are also capable of inducing hypertrophy of cardiac myocytes *in vitro*. For instance, interleukin 1β (IL-1β) was first recognized as an inflammatory cytokine for myocardium because it becomes prevalent in myocardial inflammation such as occurs during cardiac transplant rejection or congestive heart failure. However, stimulation of neonatal rat ventricular myocytes with IL-1β *in vitro* induces hypertrophic responses, including reexpression of the embryonic muscle genes (e.g., ANF, βMHC) and an increase in protein synthesis rate (Thaik *et al.*, 1995; Palmer *et al.*, 1995). It has been shown in other cell types that IL-1β1 exerts its effects by activating the ubiquitous transcription factor, NF-κB, and the MAPK signaling pathway (Akira *et al.*, 1990; Bird *et al.*, 1991). Thus, it is likely that IL-1β induces cardiac myocyte hypertrophy through activation of NF-κB and the MAPK signaling pathway. Similarly, another cytokine, IL-6, can also induce hypertrophic responses in cultured cardiac myocytes. Recent, a cDNA encoding a new member of the interleukine family, cardiotrophin-1 (CT-1), has been isolated from EBs by expression cloning based on its ability to induce hypertrophy in cultured neonatal rat ventricular myocytes (Pennica *et al.*, 1995). CT-1 shares sequence similarities with IL-6 and acts as a more potent hypertrophy inducer than IL-6, phenylephrine, and endothelin-1. Similar to IL-6, CT-1 induces hypertrophic responses possibly through the gp130/LITRβ heterodimer receptor, which activates the Raf-1/MAPK pathway (Wollert *et al.*, 1996).

E. TRANSCRIPTIONAL REGULATION DURING CARDIAC MYOCYTE HYPERTROPHY

Although the extracellular factors that mediate pressure or volume overload-induced cardiac myocyte hypertrophy can be quite divergent, they all seem to converge at MAPK and finally result in alteration in gene expression, primarily at the transcriptional level. Thus, identification of both *cis* and *trans* regulatory factors that are required for transcriptional activation during hypertrophy is essential to completely understand the molecular mechanisms by which the mechanical signals are transduced into alterations in gene expression. So far, most results of mapping *cis* and *trans* regulatory factors come from studies in cell culture systems.

1. Reexpression of Embryonic Cardiac Muscle Genes

As described in the preceding section, the genes encoding ANF, SkA, and βMHC are normally expressed in ventricular myocytes during prenatal development and turned off shortly after birth. In hypertrophied ventricular myocytes, the expression of these genes is reactivated. Thus, these genes have been widely used as an critical molecular markers for ventricular myocyte hypertrophy. To connect the regulation of these genes with extracellular signals, a number of laboratories have focused on defining the *cis* regulatory sequences within the promoter regions and their associated *trans*-acting factors that are required for reexpression. Knowlton *et al.* (1991) have shown that a 638-bp DNA fragment from the 5' flanking region of the ANF gene can confer a 10-fold induction of luciferase reporter gene in cultured neonatal rat ventricular myocytes in response to α-adrenergic stimulation. Interestingly, this DNA fragment is distinct from the *cis* elements, which are located upstream of the 638-bp fragment, required for tissue-specific expression. These results suggest that transcription factors involved in α-adrenergic-inducible expression of the ANF gene could be distinct from those required for cardiac muscle-specific expression. Ardati and Nemer (1993) have identified a smaller region within the 638-bp ANF promoter that is required for α-adrenergic-inducible expression. Furthermore, this site interacts specifically with a number of cardiac nuclear zinc-dependent proteins whose expressions are induced in neonatal rat ventricular myocytes by α-adrenergic stimulation. However, the functions of these *cis* and *trans* regulatory factors *in vivo* remain to be determined.

Similarly, Kariya *et al.* (1993, 1994) have reported that a 215-bp promoter region of the rat βMHC gene is capable of mediating α-adrenergic-inducible expression of a reporter gene in cultured neonatal rat ventricular myocytes. A binding site for transcription enhancer factor-1 (TEF-1) within the 215-bp promoter is essential for the inducible expression from the 215-bp βMHC promoter. Thus, TEF-1 could be a key factor in activating reexpression of the βMHC gene during hypertrophy.

Karns *et al.* (1995) have identified a 67-bp DNA fragment in the promoter of mouse SkA gene that is required for α-adrenergic-inducible expression of a reporter gene in cultured neonatal rat ventricular myocytes. Three known *cis* regulatory elements, M-CAT, CArG, and Sp1 binding sites, are identified within this region and are required for α-adrenergic-inducible expression. Interestingly, TEF-1 binds the M-CAT site in the SkA promoter region, suggesting that TEF-1 could be also involved in activating SkA reexpression in response to α-adrenergic stimulation. Paradis *et al.* (1996) have shown that the transcription factor, AP-1, which is a heterodimer of c-jun and c-fos, activates the SkA promoter possibly by activating the expression of CArG binding protein in neonatal rat ventricular myocytes. Thus, one function of the immediate early genes could be activating the expression of the transcriptional factors that directly activate the expression of embryonic cardiac muscle genes. Interestingly, the activity of TEF-1 and a CArG binding protein is induced by fivefold in adult rat hearts in response to pressure overload, confirming that these factors could mediate reexpression of embryonic muscle genes *in vivo* (Molkentin and Markham, 1994).

2. Activation of Contractile Protein Encoding Genes

In addition to reexpression of embryonic cardiac muscle genes in hypertrophic responses, the contractile protein encoding genes that are expressed in adult cardiac myocytes are also activated. The ventricular-specific myosin light chain-2 gene (MLC-2v) has been extensively studied as such a model gene. MLC-2v expression begins in ventricular myocytes before completion of septation and its expression remains active throughout the entire life span. The MLC-2v mRNA level in ventricular myocytes is increased in both *in vitro* and *in vivo* hypertrophy models. A 250-bp MLC-2v promoter, which confers cardiac specific expression (Henderson *et al.,* 1989), also mediates both α-adrenergic and ET-1-inducible expression in cultured neonatal rat ventricular myocytes (Zhu *et al.,* 1991). Two known *cis* regulatory elements, Y-Box and MEF2 site, were identified within the 250-bp MLC-2v promoter and required for both cardiac-specific and -inducible expression. However, it remains to be determined whether factors required for cardiac-specific and -inducible expression are the same.

In summary, shortly after birth, mammalian cardiac myocytes become postmitotic and the myocardial growth in response to increased workload is hypertrophic. Hypertrophied cardiac myocytes are characterized by increases in cell size and sarcomeric content. At the molecular level, the core of cardiac myocyte hypertrophy is alteration in gene expression—principally, increases in expression of constitutively active contractile protein encoding genes and reactivation of the embryonic gene program. Cardiac hypertrophy in response to increased workload is a complex process and involves multiple regulatory factors. It is clear that a number of paracrine or autocrine mechanisms are activated in response to the mechanical stimulus, including vasoconstrictors, adrenergic agonists, growth factors, and cytokines. Evidence from both *in vitro* and *in vivo* studies have indicated that these autocrine or paracrine systems are involved in mediating hypertrophic responses in cardiac myocytes. Despite the diversity in the properties of these ligands and their receptors, their cellular signaling pathways all seem to converge at MAPK, suggesting MAPK plays a critical role in hypertrophic responses (Fig. 2). The role of MAPK in hypertrophic responses was further supported by the observation that introduction of a constitutively active MAPK into neonatal myocytes induces reexpression

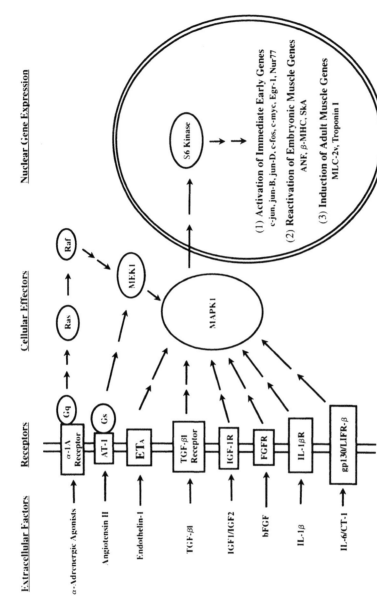

FIGURE 2 The extracellular factors and their associated cellular signaling pathways mediating alterations in gene expression during myocardial hypertrophic responses. The mitogen-activated protein kinase 1 (MAPK1) seems the converging point of the listed extracellular factors. The receptors for angiotensin II, endothelin-1, TGF-β1, IGF1/IGF2, bFGF, IL-1β, and IL-6/CT-1 may also activate MAPK1 through the upstream factors, such as Ras, Raf, and MEK1. However, this still remains to be proven experimentally.

of embryonic muscle genes. The connections between MAPK activation and alterations in gene expression still remain to be determined.

Alterations in gene expression in response to hemodynamic changes are primarily at the transcriptional level. A number of *cis* and *trans* regulatory factors have been identified that are required for inducible expression of embryonic muscle genes or constitutively active muscle genes. At the present, there does not seem to be a common master transcription factor that regulates the inducible expression of all the target genes in hypertrophic responses. However, results on hypertrophy-associated transcriptional regulation are relatively primitive, and more transcription factors involved in this regulation will be likely identified. The *cis* and *trans* regulatory factors identified, or to be identified, will be useful tools for completing the signaling pathways from hemodynamic changes to alterations in gene expression.

Cardiac myocyte hypertrophy is beneficial initially since it increases the force development required to increase cardiac output. However, when hypertrophy changes from compensatory to pathologic, heart failure occurs. During the switch from compensatory hypertrophy to pathological hypertrophy, cardiac myocytes undergo many molecular and cellular changes, including myofibril disarray, apoptosis and necrosis, and fibrosis. The molecular cues for this transition are unknown.

ACRONYMS AND ABBREVIATIONS

Act-RIIa activin receptor IIa

aFGF acidic fibroblast growth factor

ANF atrial natriuretic factor

AP-1 activator protein-1 (a ubiquitous transcription factor)

AT1 angiotensin receptor type I

AT2 angiotensin receptor type II

bFGF basic fibroblast growth factor

bHLH basic helix-loop-helix

BrdU 5-bromo-2'-deoxyuridine

CArG a *cis* regulatory element for gene transcription

CDK cyclin-dependent kinase

CDKi cyclin-dependent kinase inhibitor

CS conduction system

Csx cardiac-specific homeobox gene; also known as Nkx2.5

CT-1 cardiotrophin-1 (a cytokine)

CX40 connexin 40

dHAND   a bHLH protein expressed in deciduum, heart, autonomic nervous system, neural crest derivatives

D-MEF2   *Drosophila* myocyte enhancer binding factor-2

E1A   an adenovirus early gene

E2F   a ubiquitous transcription factor essential for G1/S transition

eHAND   a bHLH protein expressed in extraembryonic tissues, heart, autonomic nervous system, neural crest derivatives

ET-1   endothelin-1

HF-1b   heart factor 1b

HF3   heart factor 3

HNK-1   an early molecular marker for cardiac conduction system

IGF1   insulin-like growth factor I

IGF2   insulin-like growth factor II

IGF-1R   insulin-like growth factor I receptor

IGF-2R   insulin-like growth factor II receptor

IL-1β1   Interleukin-β1 (a cytokine)

IL-6   interleukin-6 (a cytokine)

inv   inversion of embryonic turning

iv   inversus viscerum

MADS box   a conserved domain originally identified in three transcription factors: MCM1 (yeast), Agamous Deficiens (plant), and Serum response factor (man)

MAPK   mitogen-activated protein kinase

MEF2   myocyte enhancer binding factor 2

MEK1   mitogen-activated protein kinase kinase

αMHC α-myosin heavy chain

βMHC β-myosin heavy chain

MLC-1a   atrial-specific myosin light chain-1

MLC-1v   ventricular-specific myosin light chain-1

MLC-2a   atrial-specific myosin light chain-2

MLC-2v   ventricular-specific myosin light chain-2

MRF4   myogenic regulatory factor 4 (also known as herculin)

Myf5   myogenic factor gene 5

MyoD   myoblast determination gene

Nkx2.5   a homeobox gene of the NK class (Kim and Nirenberg, 1989); also known as Csx

P19 cells   a mouse embryonic carcinoma cell line

PCNA proliferating cell nuclear antigen
RAR retinoic acid receptor
Rb retinoblastoma protein
RXR retinoic acid X receptor
SkA skeletal α-actin
Sp1 stimulator protein 1 (a ubiquitous transcription factor)
TEF-1 transcriptional enhancer factor-1
TGFβ1 transforming growth factor β1
TIN tinman gene
USF upstream stimulating factor
YB-1 a ubiquitous transcription factor

REFERENCES

Adachi, S., Ito, H., Tanaka, M., Fujisaki, H., Marumo, F., and Hiroe, M. (1994). Insulin-like growth factor-II induces hypertrophy with increased expression of muscle specific genes in cultured rat cardiomyocytes. *J. Mol. Cell. Cardiol.* 26:789–795.

Adachi, S., Ito, H., Ohta, Y., Tanaka, M., Ishiyama, S., Nagata, M., Toyozaki, T., Hirata, Y., Marumo, F., and Hiroe, M. (1995). Distribution of mRNA for natriuretic peptides in RV hypertrophy after pulmonary arterial banding. *Am. J. Physiol.* 268:H162–H169.

Adolf, E.F. (1967). Range of heart rates and their regulations at various ages (rat). *Am. J. Physiol.* 212:595–602.

Akira, S., Isshiki, H., Sugita, T., Tanabe, O., Kinoshita, S., Nishio, Y., Nakajima, T., Hirano, T., and Kishimoto, T. (1990). A nuclear factor for IL-6 expression (NF-IL6) is a member of a C/EBP family. *EMBO J.* 9:1897–1906.

Altaba, A.R., and Jessell, T. (1991). Retinoic acid modifies mesodermal patterning in early Eenopus embryos. *Genes Dev.* 5:174–187.

Ardati, A., and Nemer, M. (1993). A nuclear pathway for α1-adrenergic receptor signaling in cardiac cells. *EMBO J.* 12:5131–5139.

Azpiazu, N., and Frasch, M. (1993). Tinman and bagpipe: Two homeo box genes that determine cell fates in the dorsal mesoderm of *Drosophila. Genes Dev.* 7:1325–1340.

Baker, K.M., Cherin, M.I., Wixon, S.K., and Aceto, J.F. (1990). Renin angiotensin system involvement in pressure-overload cardiac hypertrophy in rats. *Am. J. Physiol.* 259:H324–H332.

Baker, J., Liu, J.-P., Robertson, E.J., and Efstratiadis, A. (1993). Role of insulin-like growth factors in embryonic and postnatal growth. *Cell* 75:73–82.

Bird, T.A., Sleath, P.R., deRoos, P.C., Dower, S.K., and Virca, G.D. (1991). Interleukin-1 represents a new modality for the activation of extracellular signal-regulated kinases/microtubule-associated protein-2 kinases. *J. Biol. Chem.* 266:22661–22670.

Bishopric, N.H., Simpson, P.C., and Ordahl, C.P. (1987). Induction of the skeletal α-actin gene in α1-adrenoceptor-mediated hypertrophy of rat cardiac myocytes. *J. Clin. Invest.* 80:1194–1199.

Bodmer, R. (1993). The gene tinman is required for specification of the heart and visceral muscle in *Drosophila. Development* 118:719–729.

Bogoyevitch, M.A., Glennon, P.E., Andersson, M.B., Clerk, A., Lazou, A., Marshall, C.J., Parker, P.J., and Sugden, P.H. (1994). Endothelin-1 and fibroblast growth factors stimulate the mitogen-activated protein kinase signaling cascade in cardiac myocytes. The potential role of the cascade in the integration of two signaling pathways leading to myocyte hypertrophy. *J. Biol. Chem.* 269:1110–1119.

Bour, B., O'Brien, M.A., Lockwood, W.L., Goldstein, E.S., Bodmer, R., Taghert, P.H., Abmayr, S.M., and Nguten, H.T. (1995). *Drosophila* MEF2, a transcription factor that is essential for myogenesis. *Genes Dev.* 9:730–741.

Braun, T., Bober, E., Winter, B., Rosenthal N., and Arnold, H.H. (1990). Myf-6, a new member of the human gene family of myogenic determination factors: Evidence for a gene cluster on chromosome 12. *EMBO J.* 9:821–831.

Braun, T., Rudnicki, M.A., Arnold, H.H., and Jaenisch, R. (1992). Targeted inactivation of the muscle regulatory gene Myf-5 results in abnormal rib development and perinatal death. *Cell* 71:369–382.

Brueckner, M., D'Eustachio, P., and Horwich, A. (1989). Linkage mapping of a mouse gene, iv, that controls left–right asymmetry of the heart and viscera. *Proc. Natl. Acad. Sci. USA* 86:5035–5038.

Capasso, J.M., Bruno, S., Cheng, W., Li, P., Rodgers, R., Darzynkiewicz, Z., and Anversa, P. (1992). Ventricular loading is coupled with DNA synthesis in adult cardiac myocytes after acute and chronic myocardial infarction in rats. *Circ. Rev.* 71:1379–1389.

Chambers, A.E., Loga, M., Kotacha, S., Towers, N., Sparrow, D., and Mohun, T.J. (1994). The RSRF/MEF2 protein SL1 regulates cardiac muscle-specific transcription of a myosin light-chain gene in Xenopus embryos. *Genes Dev.* 8:1324–1334.

Cheng, W., Reiss, K., Kajstura, J., Kowal, K., Quaini, F., and Anversa, P. (1995). Downregulation of the IGF-1 system parallels the attenuation in the proliferation capacity of rat ventricular myocytes during postnatal development. *Lab. Invest.* 72:646–655.

Cho, K.W.Y., and De Robertis, E.M. (1990). Differential activation of Xenopus homeo box genes by mesoderm-inducing growth factors and retinoic acid. *Genes Dev.* 4:1910–1916.

Claycomb, W.C. (1976). Biochemical aspects of cardiac muscle differentiation: Possible control of deoxyribonucleic acid synthesis and cell differentiation by adrenergic innervation and cyclic adenosine $3':5'$-monophosphate. *J. Biol. Chem.* 251:6082–6089.

Claycomb, W.C. (1978). Biochemical aspects of cardiac muscle differentiation. *Biochem. J.* 171:289–298.

Claycomb, W.C. (1983). Cardiac muscle cell proliferation and cell differentiation *in vivo* and *in vitro*. *Adv. Exp. Med. Biol.* 161:249–265.

Claycomb, W.C. (1991). Proliferative potential of the mammalian ventricular cardiac muscle cell. *In* "The Developmental and Regenerative Potential of Cardiac Muscle" (Oberpriller, J.O., Oberpriller, J.C., and Mauro, A., eds.), pp. 351–363. Harwood, New York.

Claycomb, W.C. (1992). Control of cardiac muscle cell division. *Trends Cardiovasc. Med.* 2:231–236.

Colvin, E.V. (1990). Cardiac embryology. *In* "The Science and Practice of Pediatric Cardiology" (A. Garson, J.T. Bricker, and D.G. McNamara, eds.), pp. 71–108. Lea and Febiger, Philadelphia.

Crescenzi, M., Fleming, T.P., Lassar, A.B., Weintraub, H., and Aaronson, S.A. (1990). MyoD induces growth arrest independent of differentiation in normal and transformed cells. *Proc. Natl. Acad. Sci. USA* 87:8842–8846.

Cserjesi, P., Brown, D., Lyons, G.E., and Olson, E.N. (1995). Expression of the novel basic helix-loop-helix gene eHAND in neural crest derivatives and extraembryonic membranes during mouse development. *Dev. Biol.* 170:664–678.

DeChiara, T.M., Efstratiadis, A., and Robertson, E.J. (1990). A growth-deficiency phenotype in heterozygous mice carrying an insulin-like growth factor II gene disrupted by targeting. *Nature* 345:78–80.

DeHaan, R.L. (1963). Regional organization of pre-pacemaker cells in the cardiac primordia of the early chick embryo. *J. Embryol. Exp. Morphol.* 11:65–76.

Delorme, B., Dahl, E., Jerry-Guichard, T., Marics, I., Briand, J.P., Willecke, K., Gros, D., and Theveniau-Ruissy, M. (1995). Developmental regulation of connexin 40 gene expression in mouse heart correlates with the differentiation of the conduction system. *Dev. Dyn.* 204:358–371.

Doetschman, T.C., Eistetter, H., Katz, M., Schmidt, W., and Kemler, R. (1985). The *in vitro* development of blastocyst-derived embryonic stem cell line: Formation of visceral yolk sac, blood islands and myocardium. *J. Embryol. Exp. Morphol.* 87:27–45.

Donath, M.Y., Zapf, J., Eppenberger-Eberhardt, M., Frosch, E.R., and Eppenberger, H.M. (1994). Insulin-like growth factor-1 stimulates myofibril development and decreases smooth muscle α-actin of adult cardiomyocytes. *Proc. Natl. Acad. Sci. USA* 91:1686–1690.

Donohue, T.J., Dworkin, L.D., Lango, M.N., Fliegner, K., Lango, R.P., Benstein, J.A., Slater, W.R., and Catanese, V.M. (1994). Induction of myocardial insulin-like growth factor-I gene expression in left ventricular hypertrophy. *Circulation* 89:799–809.

Dostal, D.E., and Baker, K.M. (1992). Angiotensin II stimulation of left ventricular hypertrophy in adult rat heart. Mediation by the AT1 receptor. *Am. J. Hypertens.* 5:276–280.

Drysdale, T.A., Tonissen, K.F., Patterson, K.D., Crawdord, M.J., and Krieg, P. (1994). Cardiac troponin I is a heart-specific marker in the Xenopus embryo: Expression during abnormal heart morphogenesis. *Development* 165:432–441.

Duerr, R.L., Huang, S., Miraliakbar, H.R., Clark, R., Chien, K.R., and Ross, J., Jr. (1995). Insulin-like growth factor-1 enhances ventricular hypertrophy and function during the onset of experimental cardiac failure. *J. Clin. Invest.* 95:619–627.

Durston, A.J., Timmermans, J.P.M., Hage, W.J., Hendricks, H.F.J., de Vries, N.J., and Nieuwkoop, P.D. (1989). Retinoic acid causes an anteroposterior transformation in the developing central nervous system. *Nature* 340:140–144.

Edmondson, D.G., and Olson, E.N. (1989). A gene with homology to the myc similarity region of MyoD1 is expressed during myogenesis and is sufficient to activate the muscle differentiation program. *Genes Dev.* 3:628–640.

Edmondson, D.G., Lyons, G.E., Martin, J.F., and Olson, E.N. (1994). MEF2 gene expression marks the cardiac and skeletal muscle lineages during mouse embryogenesis. *Development* 120:1251–1263.

Evans, S.M., Walsh, B.A., Newton, C.B., Thorburn, J.S., Gardner, P.D., and van Bilsen, M. (1993). Potential role of helix-loop-helix proteins in cardiac gene expression. *Circ. Res.* 73:569–578.

Evans, S.M., Yan, W., Murillo, M.P., Ponce, J., and Papalopulu, N. (1995). Tinman, a *Drosophila* homeobox gene required for heart and visceral mesoderm specification, may be represented by a family of genes in vertebrates: Xnkx-2.3, a second vertebrate homologue of tinman. *Development* 121:3889–3899.

Field, L.J. (1988). Atrial natriuretic factor-SV40 T antigen transgenes produce tumors and cardiac arrhythmia in mice. *Science* 239:1029–1033.

Filson, A.J., Louvi, A., Efstratiadis, A., and Robertson, E.J. (1993). Rescue of the T-associated maternal effect in mice carrying null mutations in IGF-2 and IGF-2R, two reciprocally imprinted genes. *Development* 118:731–736.

Gannon, M., and Bader, D. (1995). Initiation of cardiac differentiation occurs in the absence of anterior endoderm. *Development* 121:2439–2450.

Gillespie-Brown, J., Fuller, S.J., Bogoyevitch, M.A., Cowley, S., and Sugden, P.H. (1995). The mitogen-activated protein kinase kinase MEK1 stimulates a pattern of gene expression typical of the hypertrophic phenotype in rat ventricular cardiomyocytes. *J. Biol. Chem.* 270:28092–28096.

Glennon, P.E., Kaddoura, S., Sale, E.M., Sale, G.J., Fuller, S.J., and Sugden, P.H. (1996). Depletion of mitogen-activated protein kinase using an antisense oligonucleotide approach downregulates the phenylephrine-induced hypertrophic response in rat cardiac myocytes. *Circ. Res.* 78:954–961.

Gourdie, R.G., Mima, T., Thompson, R.P., and Mikawa, T. (1995). Terminal diversification of the myocyte lineage generates Purkinje fibers of the cardiac conduction system. *Development* 121:1423–1431.

Grana, Z., and Reddy, E.P. (1995). Cell cycle control in mammalian cells: Role of cyclins, cyclin dependent kinases (CDKs), growth suppressor genes and cyclin-dependent kinase inhibitors (CKIs). *Oncogene* 11:211–219.

Grepin, C., Dagnino, L., Robitaille, L., Haberstron, L., Antakly, T., and Nemer, M. (1994). A hormone-encoding gene identifies a pathway for cardiac but not skeletal muscle gene transcription. *Mol. Cell. Biol.* 14:3115–3129.

Grepin, C., Robitaille, L., Antakly, T., and Nemer, M. (1995). Inhibition of transcription factor GATA-4 expression blocks *in vitro* cardiac muscle differentiation. *Mol. Cell. Biol.* 15:4095–4102.

Guo, K., Wang, J., Andres, V., Smith, R.C., and Walsh, K. (1995). MyoD-induced expression of p21 inhibits cyclin-dependent kinase activity upon myocyte terminal differentiation. *Mol. Cell. Biol.* 15:3823–3829.

Gyuris, J., Golemis, E., Chertkov, H., and Brent, R. (1993). Cdi1, a human G1 and S phase protein phosphatase that associates with CDK2. *Cell* 75:791–803.

Halevy, O., Vovitch, B.G., Spicer, D.B., Skapek, S.X., Rhee, J., Hannon, G.J., Beach, D., and Lassar, A.B. (1995). Correlation of terminal cell cycle arrest of skeletal muscle with induction of p21 by MyoD. *Science* 267:1018–1021.

Hanson, M.C., Fath, K.A., Alexander, R.W., and Delafontaine, P. (1993). Induction of cardiac insulin-like growth factor I gene expression in pressure overload hypertrophy. *Am. J. Med. Sci.* 306:69–74.

Harper, J.W., Adami, G.R., Wei, N., Keyomarsi, K., and Elledge, S.J. (1993). The p21 CDK-interacting protein Cip1 is a potent inhibitor of G1 cyclin-dependent kinase. *Cell* 75:805–816.

Hasty, P., Bradley, A., Morris, J.H., Edmondson, D.G., Venuti, J.M., Olson, E.N., and Klein, W.H. (1993). Muscle deficiency and neonatal death in mice with a targeted mutation in the myogenin gene. *Nature* 364:501–506.

Heikinheimo, M., Scandrett, J.M., and Wilson, D.B. (1994). Localization of transcription factor GATA-4 to region of the mouse embryo involved in cardiac development. *Dev. Biol.* 164:361–373.

Henderson, S.A., Spencer, M., Sen, A., Kumar, C., Siddiqui, M.A.Q., and Chien, K.R. (1989). Structure, organization, and expression of the rat cardiac myosin light chain-2 gene. *J. Biol. Chem.* 264:18142–18148.

Hunter, J.J., Tanaka, N., Rockman, H.A., Ross, J., Jr., and Chien, K.R. (1995). Ventricular expression of a MLC-2v-Ras fusion gene induces cardiac hypertrophy and selective diastolic dysfunction in transgenic mice. *J. Biol. Chem.* 270:23173–23178.

Ip, H.S., Wilson, D.B., Heikinheimo, M., Tang, Z., Ting, C.N., Simon, M.C., Leiden, J.M., and Parmacek, M.S. (1994). The GATA-4 transcription factor transactivates the cardiac muscle-specific troponin C promoter-enhancer in nonmuscle cells. *Mol. Cell. Biol.* 14:7517–7526.

Isgaard, J., Wahlander, H., Adams, M.A., and Friberg, P. (1994). Increased expression of growth hormone receptor mRNA and insulin-like growth factor-I mRNA in volume-overloaded hearts. *Hypertension* 23:884–888.

Ito, H., Hirata, Y., Adachi, S., Tanaka, M., Tsujino, M., Koike, A., Nogami, A., Murumo, F., and Hiroe, M. (1993a). Endothelin-1 is an autocrine/paracrine factor in the mechanism of angiotensin II-induced hypertrophy in cultured rat cardiomyocytes. *J. Clin. Invest.* 92:398–403.

Ito, H., Hiroe, M., Hirata, Y., Tsujino, M., Adachi, S., Shichiri, M., Koike, A., Nogami, A., and Marumo, F. (1993b). Insulin-like growth factor-I induces hypertrophy with enhanced expression of muscle specific genes in cultured rat cardiomyocytes. *Circulation* 87:1715–1721.

Ito, H., Hiroe, M., Hirata, Y., Fujisaki, H., Adachi, S., Akimoto, H., Ohta, Y., and Marumo, F. (1994). Endothelin ETA receptor antagonist blocks cardiac hypertrophy provoked by hemodynamic overload. *Circulation* 89:2198–2203.

Iwaki, K., Sukhatme, V.P., Shubeita, H.E., and Chien, K.R. (1990). α- and β-Adrenergic stimulation induces distinct patterns of immediate early gene expression in neonatal rat myocardial cells. *J. Biol. Chem.* 265:13809–13817.

Jacobson, A.G. (1960). Influences of ectoderm and endoderm on heart differentiation in the newt. *Dev. Biol.* 2:138–154.

Jacobson, A.G. (1961). Heart determination in the newt. *J. Exp. Zool.* 146:139–151.

Kajstura, J., Cheng, W., Reiss, K., and Anversa, P. (1994). The IGF-1-IGF1 receptor system modulates myocyte proliferation but not myocyte cellular hypertrophy *in vitro. Exp. Cell Res.* 215:273–283.

Kariya, K., Farrance, I.K., and Simpson, P.C. (1993). Transcriptional enhancer factor-1 in cardiac myocytes interacts with an α1-adrenergic- and β-protein kinase C-inducible element in the rat β-myosin heavy chain promoter. *J. Biol. Chem.* 268:26658–26662.

Kariya, K., Karns, L. R., and Simpson, P. C. (1994). An enhancer core element mediates stimulation of the rat β-myosin heavy chain promoter by an α1-adrenergic agonist and activated β-protein kinase C in hypertrophy of cardiac myocytes. *J. Biol. Chem.* 269:3775–3782.

Karns, L.R., Kariya, K., and Simpson, P.C. (1995). M-CAT, CArG, and Sp1 elements are required for β1-adrenergic induction of the skeletal α-actin promoter during cardiac myocyte hypertrophy. Transcriptional enhancer factor-1 and protein kinase C as conserved transducers of the fetal program in cardiac growth. *J. Biol. Chem.* 270:410–417.

Kaye, D., Pimental, D., Prasad, S., Maki, T., Berger, H.J., McNeil, P.L., Smith, T.W., and Kelly, R.A. (1996). Role of transiently altered sarcolemmal membrane permeability and basic fibroblast growth factor release in the hypertrophic response of adult rat ventricular myocytes to increased mechanical activity *in vitro. J. Clin. Invest.* 97:281–291.

Katz, E.B., Steinhelper, M.E., Delcarpio, J.B., Daud, A.I., and Claycomb, W.C. (1992). Cardiomyocyte proliferation in mice expressing α-cardiac myosin heavy chain-SV40 T-antigen transgenes. *Am. J. Physiol.* 262:H1867–H1876.

Kim, Y., and Nirenberg, M. (1989). *Drosophila* NK-homeobox genes. *Proc. Natl. Acad. Sci. USA* 86:7716–7720.

Kirshenbaum, L.A., and Schneider, M.D. (1995). Adenovirus E1A receptor represses cardiac gene transcription and reactivates DNA synthesis in ventricular myocytes, via alternative pocket protein- and p300-binding domains. *J. Biol. Chem.* 270:7791–7794.

Knowlton, K.U., Baracchini, E., Ross, R.S., Harris, A.N., Henderson, S.A., Evans, S.M., Glembotski, C.C., and Chien, K.R. (1991). Co-regulation of the atrial natriuretic factor and cardiac myosin light chain-2 genes during α-adrenergic stimulation of neonatal rat ventricular cells. *J. Biol. Chem.* 266:7759–7768.

Knowlton, K.U., Michel, M.C., Itani, M., Shubeita, H.E., Ishihara, K., Brown, J.H., and Chien, K.R. (1993). The α1A-adrenergic receptor subtype mediates biochemical, molecular, and morphologic features of cultured myocardial cell hypertrophy. *J. Biol. Chem.* 268:15374–15380.

Kojima, M., Shiojima, I., Yamazaki, T., Komuro, I., Zou, Z., Wang, Y., Mizuno, T., Ueki, K., Tobe, K., and Kadowaki, T. (1994). Angiotensin II receptor antagonist TCV-116 induces regression of hypertensive left ventricular hypertrophy *in vivo* and inhibits the intracellular signaling pathway of stretch-mediated cardiomyocyte hypertrophy *in vitro. Cir.* 89:2204–2211.

Komuro, I., and Izumo, S. (1993). Csx: A murine homeobox-containing gene specifically expressed in the developing heart. *Proc. Natl. Acad. Sci. USA* 90:8145–8149.

LaMorte, V.J., Thorburn, J., Absher, D., Spiegel, A., Brown, J.H., Chien, K.R., Feramisco, J.R., and Knowlton, K.U. (1994). Gq- and Ras-dependent pathways mediate hypertrophy of neonatal rat ventricular myocytes following α1-adrenergic stimulation. *J. Biol. Chem.* 269:13490–13496.

Lassar, A.B., Buskin, J.N., Lockshon, D., Davis, R.L., Apone, S., Hauschka, S.D., and Weintraub, H. (1989). MyoD is a sequence-specific DNA binding protein requiring a region of myc homology to bind to the muscle creatine kinase enhancer. *Cell* 58:823–831.

Lau, M.M.H., Stewart, C.E.H., Liu, Z., Bhatt, H., Rotwein, P., and Stewart, C.L. (1994). Loss of the imprinted IGF2/cation-independent mannose 6-phosphate receptor results in fetal overgrowth and perinatal lethality. *Genes Dev.* 8:2953–2963.

Lee, K.J., Ross, R.S., Rockman, H.A., Harris, A.N., O'Brien, T.X., van Bilsen, M., Shubeita, H.E., Kandolf, R., Brem, G., Price, J., Evans, S.M., Zhu, H., Franz, W.-M., and Chien, K.R. (1992). Myosin light-chain-2 luciferase transgenic mice reveal distinct regulatory program for cardiac and skeletal muscle-specific expression of a single contractile protein gene. *J. Biol. Chem.* 267:15875–15885.

Lee, K.J., Hickey, R., Zhu, H., and Chien, K.R. (1994). Positive regulatory elements (HF-1a and HF-1b) and a novel negative regulatory element (HF-3) mediate ventricular muscle-specific expression of myosin light-chain 2-luciferase fusion genes in transgenic mice. *Mol. Cell. Biol.* 14:1220–1229.

Lee, A.A., Dillman, W.H., McCulloch, A.D., and Villarreal, F.J. (1995). Angiotensin II stimulates the autocrine production of transforming growth factor-β1 in adult rat cardiac fibroblasts. *J. Mol. Cell. Cardiol.* 27:2347–2357.

Leiden, J.M. (1993). Transcriptional regulation of T cell receptor genes. *Annu. Rev. Immunol.* 11:539–570.

Levin, M., Johnson, R.L., Stern, C.D., Kuahn, M., and Tabin, C. (1995). A molecular pathway determining left–right asymmetry in chick embryogenesis. *Cell* 82:803–814.

Li, E., Sucov, H.M., Lee, K.-F., Evans, R.M., and Jaenisch, R. (1993). Normal development and growth of mice carrying a targeted disruption of the α1 retinoic acid receptor gene. *Proc. Natl. Acad. Sci. USA* 90:1590–1594.

Lilly, B., Zhao, B., Ranganayakulu, G., Paterson, B.M., Schulz, R.A., and Olson, E.N. (1995). Requirement of MADS domain transcription factor D-MEF2 for muscle formation in *Drosophila. Science* 267:688–693.

Lindpaintner, K., and Ganten, D. (1991). The cardiac renin-angiotensin system: An appraisal of present experimental and clinical evidence. *Circ. Res.* 68:905–921.

Lints, T.J., Parsons, L.M., Hartley, L., Lyons, I., and Harvey, R.P. (1993). Nkx-2.5: A novel murine homeobox gene expressed in early heart progenitor cells and their myogenic descendants. *Development* 119:419–431.

Litvin, J., Montgomery, M.O., Goldhamer, D.J., Emerson, C.P., Jr., and Bader, D.M. (1993). Identification of DNA-binding protein(s) in the developing heart. *Dev. Biol.* 156:409–417.

Liu, Q., Yan, H., Dawes, N.J., Mottino, G.A., Frank, J.S., and Zhu, H. (1996). Insulin-like growth factor II (IGF2) induces DNA synthesis in fetal ventricular myocytes *in vitro. Circ. Res.* (in press).

Liu, Y., and Kitsis, R.N. (1996). Induction of DNA synthesis and apoptosis in cardiac myocytes by E1A oncoprotein. *J. Cell Biol.* 133:325–334.

Lohnes, D., Kastner, P., Dierich, A., Mark, M., LeMeur, M., and Chambon, P. (1993). Functions of retinoic acid receptor γ in the mouse. *Cell* 73:643–658.

Long, C.S., Ordahl, C.P., and Simpson, P.C. (1989). α1-Adrenergic receptor stimulation of sarcomeric action isogene transcription in hypertrophy of cultured rat heart muscle cells. *J. Clin. Invest.* 83:1078–1082.

Lowe, L.A., Supp, D.M., Sampath, K., Yokoyama, T., Wright, C.V., Potter, S.S., Overbeek, P., and Kuehn, M.R. (1996). Conserved left–right asymmetry of nodal expression and alterations in murine situs inversus. *Nature* 381:158–161.

Lufkin, T., Lohnes, D., Mark, M., Dierich, A., Gorry, P., Gaub, M.-P., LeMeur, M., and Chambon, P. (1993). High postnatal lethality and testis degeneration in retinoic acid receptor α mutant mice. *Proc. Natl. Acad. Sci. USA* 90:7225–7229.

Lyons, I., Parsons, L.M., Hartley, L., Li, R., Andrews, J.E., Robb, L., and Harvey, R.P. (1995). Myogenic and morphogenic defects in the heart tubes of murine embryos lacking the homeo box gene Nkx2-5. *Genes Dev.* 9:1654–1666.

Marino, T.A., Halder, S., Williamson, E.C., Beaverson, K., Walter, R.A., Marino, D.R., Beatty, C., and Lipson, K.E. (1991). Proliferating cell nuclear antigen in developing and adult rat cardiac muscle cells. *Circ. Res.* 69:1353–1360.

McGill, C.J., and Brooks, G. (1995). Cell cycle control mechanisms and their role in cardiac growth. *Cardiovasc. Res.* 30:557–569.

Metzger, J.M., Lin, W.T., and Samuelson, L.C. (1994). Transition in cardiac contractile sensitivity to calcium during the *in vitro* differentiation of mouse embryonic stem cells. *J. Cell Biol.* 126:701–711.

Metzger, J.M., Lin, W.T., and Samuelson, L.C. (1996). Vital staining of cardiac myocytes during embryonic stem cell cardiogenesis *in vitro*. *Circ. Res.* 78:547–552.

Miller-Hance, W.C., LaCorbiere, M., Fuller, S.J., Evans, S.M., Lyons, G., Schmidt, C., Robbins, J., and Chien, K.R. (1993). *In vitro* chamber specification during embryonic stem cell cardiogenesis. *J. Biol. Chem.* 268:25244–25252.

Mima, T., Veno, H., Fischman, D.A., Williams, L.T., and Mikawa, T. (1992). Fibroblast growth factor receptor is required for *in vivo* cardiac myocyte proliferation at early embryonic stages of heart development. *Proc. Natl. Acad. Sci. USA* 92:467–471.

Molkentin, J.D., Brogan, R.S., Jobe, S.M., and Markham, B.E. (1993). Expression of the α-myosin heavy chain gene in the heart is regulated in part by an E-box-dependent mechanism. *J. Biol. Chem.* 268:2602–2609.

Molkentin, J.D., Kalvakolanu, D.V., and Markham, B.E. (1994). Transcription factor GATA-4 regulates cardiac muscle-specific expression of the α-myosin heavy-chain gene. *Mol. Cell. Biol.* 14:4947–4957.

Molkentin, J.D., and Markham, B.E. (1994). An M-CAT binding factor and an RSRF-related A-rich binding factor positively regulate expression of the α-cardiac myosin heavy-chain gene *in vivo*. *Mol. Cell. Biol.* 14:5056–5065.

Moore, K.L. (1988). The cardiovascular system. *In* "The developing human, 4th ed.," pp. 286–333. W.B. Saunders, Philadelphia.

Moran, E. (1994). Mammalian cell growth controls reflected through protein interactions with adenovirus E1A gene products. *Semin. Virol.* 5:327–340.

Moss, J.B., McQuinn, T.C., and Schwartz, R.J. (1994). The avian cardiac α-actin promoter is regulated through a pair of complex elements composed of E boxes and serum response elements that bind both positive- and negative-acting factors. *J. Biol. Chem.* 269:12731–12740.

Mukoyama, M., Nakajima, M., Horiuchi, M., Sasamura, H., Pratt, R.E., and Dzau, V.J. (1993). Expression cloning of type 2 angiotensin II receptor reveals a unique class of seven-transmembrane receptors. *J. Biol. Chem.* 268:24539–24542.

Murphy, T.J., Alexander, R., Griedling, K.K., Runge, M.S., Bernstein, K.E. (1991). Isolation of a cDNA encoding the vascular type-I angiotensin II receptor. *Nature* 351:233–236.

Nagawa, M., Thompson, R.P., Terracio, L., and Borg, T.K. (1993). Developmental anatomy of HNK-1 immunoreactivity in the embryonic rat heart: Co-distribution with early conduction tissue. *Anat. Embryol.* 187:445–460.

Navankasattusas, S., Zhu, H., Garcia, A.V., Evans, S.M., and Chien, K.R. (1992). A ubiquitous factor (HF-1a) and a distinct muscle factor (HF-1b/MEF-2) form an E-box-independent pathway for cardiac muscle gene expression. *Mol. Cell. Biol.* 12:1469–1479.

Navankasattusas, S., Sawadogo, M., van Bilsen, M., Dang, C.V., and Chien, K.R. (1994). The basic helix-loop-helix protein upstream stimulating factor regulates the cardiac ventricular myosin light chain-2 gene via independent *cis* regulatory elements. *Mol. Cell. Biol.* 14:7331–7339.

Nevin, J.R. (1992). E2F: A link between the Rb tumor suppressor protein and viral oncoproteins. *Science* 258:424–429.

Nguyen, H.T., Bodmer, R., Abmayr, S.M., McDermott, J.C., and Spoerel, N.A. (1994). D-mef2: A *Drosophila* mesoderm-specific MADS box-containing gene with a biphasic expression profile during embryogenesis. *Proc. Natl. Acad. Sci. USA* 91:7520–7524.

O'Brien, T.X., Lee, K.J., and Chien, K.R. (1993). Positional specification of ventricular myosin light chain-2 expression in the primitive murine heart tube. *Proc. Natl. Acad. Sci. USA* 90:5157–5161.

Osmond, M.K., Butler, A.J., Voon, F.T., and Bellairs, R. (1991). The effects of retinoic acid on heart formation in the early chick embryo. *Development* 113:1405–1417.

Palmer, J.N., Hartogensis, W.E., Patten, M., Fortuin, F.D., and Long, C.S. (1995). Interleukin-1β induces cardiac myocyte growth but inhibits cardiac fibroblast proliferation in culture. *J. Clin. Invest.* 95:2555–2564.

Pennica, D., King, K.L., Shaw, K.J., Luis, E., Rullamas, J., Luoh S., Darbonne, W.C., Knutzon, D.S., Yen, R., Chien, K.R., Baker, J.B., and Wood, W.I. (1995). Expression cloning of cardiotrophin 1, a cytokine that induces cardiac myocytes hypertrophy. *Proc. Natl. Acad. Sci. USA* 92:1142–1146.

Paradis, P., MacLellan, W.R., Belaguli, N.S., Schwartz, R.J., and Schneider, M.D. (1996). Serum response factor mediates AP-1-dependent induction of the skeletal α-actin promoter in ventricular myocytes. *J. Biol. Chem.* 271:10827–10833.

Parker, T.G., Packer, S.E., and Schneider, M.D. (1990). Peptide growth factors can provoke "fetal" contractile protein gene expression in rat cardiac myocytes. *J. Clin. Invest.* 85:507–514.

Pasumarthi, K.B., Kardami, E., and Cattini, P.A. (1996). High and low molecular weight fibroblast growth factor-2 increase proliferation of neonatal rat cardiac myocytes but have differential effects on binucleation and nuclear morphology. Evidence for both paracrine and intracrine action of fibroblast growth factor-2. *Circ. Res.* 78:126–136.

Qin, W., Woods, C.G., Schneider, J.A., and Woods, W.T., Jr. (1995). Organization and fine structure of a pacemaker derived from fetal rat myocardium. *Pediatr. Res.* 37:283–388.

Ranganayakulu, G., Zhao, B., Dokidis, A., Molkentin, J.D., Olson, E.N., and Schulz, R.A. (1995). A series of mutations in the D-MEF2 transcription factor reveal multiple functions in larval and adult myogenesis in *Drosophila*. *Dev. Biol.* 171:169–181.

Rawles, M.E. (1943). The heart-forming regions of the early chick blastoderm. *Physiol. Zool.* 16:22–42.

Rockman, H.A., Ono, S., Ross, R.S., Jones, L.R., Karimi, M., Bhargava, V., Ross, J., Jr., and Chien, K.R. (1994a). Molecular and physiological alterations in murine ventricular dysfunction. *Proc. Natl. Acad. Sci. USA* 91:2694–2698.

Rockman, H.A., Wachhorst, S.P., Mao, L., and Ross, J., Jr. (1994b). ANG II receptor blockade prevents ventricular hypertrophy and ANF gene expression with pressure overload in mice. *Am. J. Physiol.* 266:H2468–2475.

Rosenquist, G.C. (1966). A radioautographic study of labeled grafts in the chick blastoderm. Development of primitive-streak stages to stage 12. (Carnegie Inst. Wash. Publ. 625) *Embryol.* 38:111–121.

Rossant, J. (1996). Mouse mutants and cardiac development: new molecular insights into cardiogenesis. *Circ. Res.* 78:349–353.

Rudnicki, M.A., Braun, T., Hinuma, S., and Jaenisch, R. (1992). Inactivation of MyoD in mice leads to up-regulation of the myogenic HLH gene Myf-5 and results in apparently normal muscle development. *Cell* 71:383–390.

Rumyatsev, P.P. (1991). "Growth and Hyperplasia of Cardiac Muscle Cells." (Soviet Medical Review Supplement Series Cardiology, Vol. 3) Harwood, New York.

Sadoshima, J., Jahn, L., Takahashi, T., Kulik, T.J., and Izumo, S. (1992). Molecular characterization of the stretch-induced adaptation of cultured cardiac cells. An *in vitro* model of load-induced cardiac hypertrophy. *J. Biol. Chem.* 267:10551–10560.

Sadoshima, J., and Izumo, S. (1993). Molecular characterization of angiotensin II-induced hypertrophy of cardiac myocytes and hyperplasia of cardiac fibroblasts. Critical role of the AT1 receptor subtypes. *Circ. Res.* 73:413–423.

Sadoshima, J., Qiu, Z., Morgan, J.P., and Izumo, S. (1995). Angiotensin II and other hypertrophic stimuli mediated by G protein-coupled receptors activated protein kinase, and 90-kD S6 kinase in cardiac myocytes. The critical role of Ca^{2+}-dependent signaling. *Circ. Res.* 76:1–15.

Sakai, H., Ikeda, T., Ito, H., Nakamura, T. Shimokawa, I., and Matsuo, T. (1994). Immunoelectron microscopic localization of HNK-1 in the embryonic rat heart. *Anat. Embryol.* 190:13–20.

Sartorelli, V., Hong, N.A. Bishopric, N.H., and Kedes, L. (1992). Myocardial activation of the human cardiac α-actin promoter by helix-loop-helix proteins. *Proc. Natl. Acad. Sci. USA* 89:4047–4051.

Schneider, J.W., Gu, W., Zhu, L., Mahdavi, V., and Nadal-Ginard, B. (1994). Reversal of terminal differentiation mediated by p107 in Rb-/- muscle cells. *Science* 264:1467–1470.

Schultheiss, T.M., Xydas, S., and Lassar A.B. (1995). Induction of avian cardiac myogenesis by anterior endoderm. *Development* 121:4203–4214.

Seger, R., and Krebs, E.G. (1995). The MAPK signaling cascade. *FASEB J.* 9:726–735.

Sen, D., Dunnmon, P., Henderson, S.A., Gerard, R.D., and Chien, K.R. (1988). Terminally differentiated neonatal rat myocardial cells proliferate and maintain specific differentiated functions following expression of SV40 large T antigen. *J. Biol. Chem.* 263:19132–19136.

Sheng, Z., Pennica, D., Wood, W.I., and Chien, K.R. (1996). Cardiotrophin-1 displays early expression in the murine heart tube and promotes cardiac myocyte survival. *Development* 122:419–428.

Shubeita, H.E., McDonough, P.M., Harris, A.N., Knowlton, K.U., Glembotski, C.C., Brown, J.H., and Chien, K.R. (1990). Endothelin induction of inositol phospholipid hydrolysis, sarcomere assembly, and cardiac gene expression in ventricular myocytes. *J. Biol. Chem.* 265:20555–20562.

Simpson, P. (1983). Norepinephrine-stimulated hypertrophy of cultured rat myocardial cells is an α1 adrenergic response. *J. Clin. Invest.* 72:732–738.

Sive, H.L., and Cheng, P.-F. (1991). Retinoic acid perturbs the expression of Xhox.lab genes and alters mesodermal determination in Xenopus laevis. *Genes Dev.* 5:1312–1332.

Sorrentino, R., Pepperkok, R., David, R.L., Anorge, W., and Philipson, L. (1990). Cell proliferation inhibited by MyoD1 independently of myogenic differentiation. *Nature* 345:813–815.

Soudais, C., Bielinska, M., Heikinheimo, M., MacArthur, C.A., Narita, N., Saffitz, J.E., Simon, M.C., Leiden, J.M., and Wilson, D.B. (1995). Targeted mutagenesis of the transcription factor GATA-4 gene in mouse embryonic stem cells disrupts visceral endoderm differentiation *in vitro. Development* 121:3877–3888.

Srivastava, D., Cserjesi, P., and Olson, E.N. (1995). A subclass of bHLH proteins required for cardiac morphogenesis. *Science* 270:1995–1999.

Stainier, D.Y.R., and Fishman, M.C. (1992). Patterning of the zebrafish heart tube: Acquisition of anteroposterior polarity. *Dev. Biol.* 153:91–101.

Starksen, N.F., Simpson, P.C., Bishopric, N., Coughlin, S.R., Lee, W.M.F., and Williams, L.T. (1986). Cardiac myocyte hypertrophy is associated with c-myc proto-oncogene expression. *Proc. Natl. Acad. Sci. USA* 83:8348–8350.

Sucov, H.M., Dyson, E., Gumeringer, C.L., Price, J., Chien, K.R., and Evans, R.M. (1994). RXRα mutant mice establish a genetic basis for vitamin A signaling in heart morphogenesis. *Genes Dev.* 8:1007–1018.

Sudin, O., and Eichele, G. (1992). An early marker of axial pattern in the chick embryo and its respecification by retinoic acid. *Development* 114:841–852.

Sugi, Y., Sasse, J., and Lough, J. (1993). Inhibition of precardiac mesoderm cell proliferation by antisense oligonucleotide complementary to fibroblast growth factor-2 (FGF-2). *Dev. Biol.* 157:28–37.

Sugi, Y., and Lough, J. (1994). Anterior endoderm is a specific effector of terminal cardiac myocyte differentiation of cells from the embryonic heart forming region. *Dev. Dyn.* 200:155–162.

Takahashi, N., Calderone, A., Izzo, N.J., Jr., Maki, T.M., Marsh, J.D., and Colucci, W.S. (1994). Hypertrophic stimuli induce transforming growth factor-β1 expression in rat ventricular myocytes. *J. Clin. Invest.* 94:1470–1476.

Tam, S.K.C., Gu, W., Mahdavi, V., and Nadal-Ginard, B. (1993). Tumor suppressor genes and cell cycle regulators expression in cardiac development and hypertrophy. *Circ.* 88:1–24.

Thaik, C.M., Calderone, A., Takahashi, N., and Colucci, W.S. (1995). Interleukin-1β modulates the growth and phenotype of neonatal rat cardiac myocytes. *J. Clin. Invest.* 96:1093–1099.

Thorburn, A., Thorburn, J., Chen, S.-Y., Powers, S., Shubeita, H.E., Feromisco, J.R., and

Chien, K.R. (1993). Hras-dependent pathways can activate morphological and genetic markers of cardiac muscle cell hypertrophy. *J. Biol. Chem.* 268:2244–2249.

Thorburn, J., McMahon, M., and Thorburn, A. (1994). Raf-1 kinase activity is necessary and sufficient for gene expression changes but not sufficient for cellular morphology changes associated with cardiac myocyte hypertrophy. *J. Biol. Chem.* 269:30580–30586.

Thuerauf, D.J., Hanford, D.S., and Glembotski, C.C. (1994). Regulation of rat brain natriuretic peptide transcription. A potential role for GATA-related transcription factors in myocardial cell gene expression. *J. Biol. Chem.* 269:17772–17775.

Tucker, D.C., Snider, C., and Wood, W.T., Jr. (1988). Pacemaker development in embryonic rat heart cultured in oculo. *Pediat. Res.* 23:637–642.

Weinberg, R.A. (1995). The retinoblastoma protein and cell cycle control. *Cell* 81:323–330.

Wekstein, D.R. (1965). Heart rate of the preweanling rat and its autonomic control. *Am. J. Physiol.* 208:1259–1262.

Wilson, J.G., and Warkany, J. (1949). Aortic-arch and cardiac anomalies in the offspring of vitamin A deficient rats. *Am. J. Anat.* 85:113–155.

Wollert, K.C., Taga, T., Saito, M., Narazaki, M., Kishimoto, T., Glembotski, C.C., Vernallis, A.B., Heath, J.K., Pennica, D., Wood, W.L., and Chien, K.R. (1996). Cardiotrophin-1 activates a distinct form of cardiac muscle cell hypertrophy. *J. Biol. Chem.* 271:9535–9545.

Wong, M.-W., Pisegna, M., Lu, M.-F., Leibham, D., and Perry, M. (1994). Activation of Xenopus MyoD transcription by members of the MEF2 protein family. *Dev. Biol.* 166:683–695.

Xiong, Y., Hannon, G.J., Zhang, H., Casso, D., Kobayashi, R., and Beach, D. (1993). p21 Is a universal inhibitor of cyclin kinase. *Nature* 366:701–704.

Yamazaki, T., Shiojima, I., Komuro, I., Nagai, R., and Yazaki, Y. (1994). Involvement of the renin-angiotensin system in the development of left ventricular hypertrophy and dysfunction. *J. Hypertens.* 12:S153–S157.

Yamazaki, T., Komuro, I., Kudoh, S., Zou, Y., Shiojima, I., Mizuno, T., Takano, H., Hiroi, Y., Ueki, K., Tobe, K., Kadowaki, T., Nagai, R., and Yazaki, Y. (1995a). Angiotensin II partly mediates mechanical stress-induced cardiac hypertrophy. *Circ. Res.* 77:258–265.

Yamazaki, T., Komuro, I., Kudoh, S., Zou, Y., Shiojima, I., Mizuno, T., Yakano, H., Hiroi, Y., Ueki, K., Tobe, K., Kadowaki, T., Nagai, R., and Yazaki, Y. (1995b). Mechanical stress activates protein kinase cascade of phosphorylation in neonatal rat cardiac myocytes. *J. Clin. Invest.* 96:438–446.

Yamazaki, T., Komuro, I., Kudoh, S., Zou, Y., Shiojima, I., Hiroi, Y., Mizuno, T., Maemura, K., Kurihara, H., and Aikawara, R. (1996). Endothelin-1 is involved in mechanical stress-induced cardiocyte hypertrophy. *J. Biol. Chem.* 271:3221–3228.

Yokoyama, T., Copeland, N.G., Jenkins, N.A., Montgomery, C.A., Elder, F.F.B., and Overbeek, P.A. (1993). Reversal of left–right asymmetry: A situ inversus mutation. *Science* 260:679–682.

Zeller, R., Bloch, K.D., Williams, B.S., Arceci, R.J., and Seidman, C.E. (1987). Localized expression of the atrial natriuretic factor gene during cardiac embryogenesis. *Genes Dev.* 1:693–698.

Zhu, H., Garcia, A.V., Ross, R.S., Evans, S. M., and Chien, K.R. (1991). A conserved 28-base-pair element (HF-1) in the rat cardiac myosin light chain-2 gene confers cardiac-specific and α-adrenergic-inducible expression in cultured neonatal rat myocardial cells. *Mol. Cell. Biol.* 11:2272–2281.

Zhu, H., Nguyen, V.T.B., Brown, A.B., Pourhosseini, A., Garcia, A.V., van Bilsen, M., and Chien, K.R. (1993). A novel, tissue-restricted zinc finger protein (HF-1b) binds to the cardiac regulatory element (HF-1b/MEF-2) in the rat myosin light chain-2-gene. *Mol. Cell. Biol.* 13:4432–4444.

Zou, Y., and Chien, K.R. (1995). EFI_A/YB-1 is a component of cardiac HF-1A binding activity and positively regulates transcription of the myosin light-chain 2v gene. *Mol. Cell. Biol.* 15:2972–2982.

3

ION CHANNELS IN
CARDIAC MUSCLE

JAMES N. WEISS

I. INTRODUCTION

Ion channels are the major transducers for communicating physiological signals within and between cells. At the cellular level, the formation of the heartbeat, as well as cardiac force development, are both directly regulated by the cardiac action potential, which depends on the coordinated actions of a large number of distinct sarcolemmal ion channels. At the subcellular level, the changes in membrane potential and flow of ions through sarcolemmal ion channels regulate ion channels in intracellular organelles, which determine contractile force by controlling Ca release from internal stores, and modulate energy production by the metabolic machinery in the cell. At the supercellular level, the staged activation sequence of the heart, progressing from the sinoatrial (SA) node to ventricle, is critically mediated by intercellular ion channels (connexons) that connect adjacent myocytes. As sensors of the extracellular environment, ion channels also play important but less understood roles in cardiac cell volume regulation, endocrine function, and gene regulation in development and hypertrophy. This chapter provides an overview of cardiac ion channel function in these various roles.

Section II of this chapter focuses on various ion channels identified in the heart (Table 1), including their physiological functions, biophysical properties, pharmacology, structure–function relationships, and molecular identities, if known. Section III deals with how ion channels function as an integrated system to produce the cardiac action potential, including differences between the cardiac action potential in various regions of the heart, regulation by autonomic tone, and aspects of impulse propagation.

TABLE 1   Cardiac Ionic Currents/Channels

Voltage gated	Na	I_{Na}
	Ca	$I_{Ca(L)}$, $I_{Ca(T)}$
	K	I_{to1}, I_{to2}, I_{Ks}, I_{Kr}, I_{Kur}, I_{Kp}
	Nonselective cation	I_f
Ligand gated	K	I_{K1}, $I_{K(Ach)}$, $I_{K(ATP)}$, $I_{K(Ca)}$, $I_{K(Na)}$, $I_{K(AA)}$
	Ca	RyR2
	Nonselective cation	$I_{NSC(Ca)}$
	Cl	$I_{Cl(cAMP)}$, $I_{Cl(Ca)}$
Mechanosensitive	K	I_{Ks}, $I_{K(ATP)}$
	Cl	$I_{Cl(SA)}$
	Nonselective cation	$I_{NSC(SA)}$
Background/leak	Na	$I_{Na(bg)}$
	Ca	$I_{Ca(bg)}$
	Nonselective cation	$I_{NSC(bg)}$
	Cl	$I_{Cl(bg)}$

II. CARDIAC ION CHANNELS

From a biophysical standpoint, ion channels are characterized by three key properties: permeation, selectivity, and gating. Ion channels are described by the type of ion they allow to permeate (e.g., K, Na, Ca, nonselective cation) and the gating mechanism that regulates the flow of the permeant ions. The gating mechanisms fall into four general categories: voltage gated, ligand gated, mechanosensitive, and "nongated" background, or leak, channels. The ionic currents/channels in these various categories that have been identified in cardiac tissue are listed in Table 1. The voltage-gated ion channels have been the most extensively investigated, and many of the cardiac isoforms of these channels have been cloned. The marriage of patch clamp and molecular biology techniques has been a particularly powerful combination for studying their structure–function relationships, and has led to a detailed conceptual understanding of how they operate at the molecular level. More recently, many cardiac ligand-gated ion channels have also been cloned, and studies unraveling the molecular basis for their properties have closely followed. The mechanosensitive and background ion channels in the heart are the least well characterized, and their molecular identities are still unknown.

A schematic representation of an ion channel is shown in Fig. 1. The ion-conducting pathway, or pore, has *inner* and *outer vestibules,* often lined with charges of opposite polarity to the permeant ion(s) to concentrate them. The narrowest region of the pore is the *selectivity filter,* which determines the type of ion that can permeate. An *activation gate,* which senses the transmembrane voltage, binding of a ligand to the channel, or mechanical tension in the sarcolemma, controls access of the permeant ions to the

CLOSED   OPEN

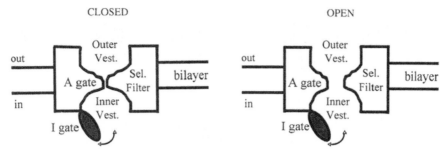

FIGURE 1   Schematic diagram of the general structure of an ion channel. Inner and outer vestibules (vest.) provide access to the selectivity (sel.) filter. The activation gate (A gate) opens and closes the pore. An inactivation gate (I gate) may also be present, often in a cytoplasmic region of the channel.

pore. Many channels also have an *inactivation gate,* which shuts off ion permeation and prevents the ion channel from reopening.

A. SARCOLEMMAL VOLTAGE-GATED ION CHANNELS

1. General Molecular Structure–Function Relationships

The classic voltage-gated Na, Ca, and K channels were cloned in the early 1980s (Noda *et al.,* 1984; Tanabe *et al.,* 1987; Papazian *et al.,* 1987), and their cardiac counterparts were identified shortly thereafter. The α-subunits of voltage-gated Na or Ca channels consist of four homologous domains, each with six transmembrane-spanning segments, which are connected by cytoplasmic linkers (Fig. 2). The four domains form a tetrameric array, with the ion-conducting pathway (pore) located in the center. In contrast, the α-subunit of a voltage-gated K channel is a smaller protein with six transmembrane-spanning segments, similar to a single domain of a Na or Ca channel. A functional K channel is formed by the tetrameric array of four α-subunits, with the pore in the center (see Fig. 2).

a. Permeation

When the first voltage-gated ion channels were cloned, it became apparent that the region between the fifth and sixth transmembrane segments (S_5 and S_6) was highly conserved, suggesting that it may serve a common function. With site-directed mutagenesis, this region (the P region) was identified as the selectivity filter of the pore. Mutations in this area markedly affect the permeation characteristics and selectivity of the ion channel. For example, the Na channel can be made to preferentially conduct Ca over Na by changing a single amino acid at position 1422 from lysine to glutamate, the amino acid present at the analogous site in the cardiac L-type Ca

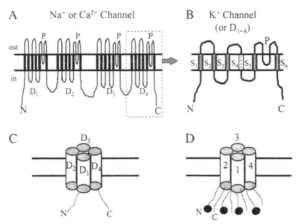

FIGURE 2   Predicted topology of the classic voltage-gated Na or Ca and K channels. (A) The α-subunits of voltage-gated Na or Ca channels are composed of four domains (D₁–D₄) linked together by cytoplasmic linkers. (B) Each domain has six transmembrane-spanning segments (cylinders, S₁–S₆), and a short segment (P) between S₅ and S₆ that dips into the membrane to form the pore. A K channel α-subunit is a monomer equivalent to one domain of a Na or Ca channel, with four K channel α-subunits combining to form a functional channel. (C and D) The four domains of a Na or Ca channel α-subunit (C) or K channel α-subunit (D) are arranged with tetrameric symmetry. The P regions (not shown) dip into the center to line the ion-conducting pathway.

channel (Heinemann *et al.*, 1992). Also, the affinities of various toxins (e.g., tetrodotoxin for Na channels, charybdotoxin for K channels) and blocking molecules (e.g., tetraethylammonium for K channels), which act by directly plugging up the pore of the channel, are very sensitive to mutations in this region (Fig. 3). The P region is believed to project into the pore region to form its narrowest region, the selectivity filter. However, other transmembrane segments, such as S₆, also affect ion permeation and are believed to form part of the ion-conducting pathway (Lopez *et al.*, 1994). Extensive electrophysiological evidence strongly supports the notion that the narrow region of the pore accommodates multiple (2–4) ions simultaneously, which pass through this region in single file. Yet studies probing the deep pore structure of voltage-gated ion channels indicate that rather large molecules applied from the cytoplasmic or extracellular compartments can readily gain access to residues in this putatively narrow space. Since this seems to exclude the notion of the selectivity filter as a long pore, how multiple permeant ions can coexist within such a narrow region remains an important paradox (Miller, 1996).

b. Activation Gating

Voltage-gated ion channels open in response to a change in membrane potential. In their study, Hodgkin and Huxley (1952) postulated that a

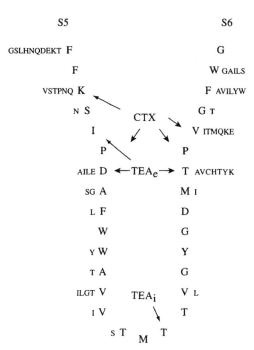

FIGURE 3 The variation in the amino acid sequence of the P region among voltage-gated K channels, with the *Shaker* sequence indicated in larger type. Residues that alter charybdotoxin (CTX) and internal and external tetraethylammonium (TEA) binding are indicated by arrows. (Reprinted from Bogusz *et al.*, 1992, by permission of Oxford University Press.)

charged gating particle moved across the membrane in response to the change in membrane potential, thereby inducing a conformational change that opened the ion channel. The hypothesis was validated in 1972 by Armstrong and Bezanilla (1973), who measured a transient current associated with the movement of these gating charges across the membrane (Fig. 4). They found that upon membrane depolarization, most of the gating charge moved over a more negative voltage range than the flow of ionic current, implying that the Na channel moved through a series of closed states associated with movement of gating charges before making the final transition to the open state. When the voltage-dependent ion channels were cloned, the S_4 transmembrane segments were recognized as likely candidates for the voltage sensor. The S_4 segments are highly conserved between voltage-gated Na, Ca, and K channels and, in all cases, contain a series of positively charged amino acids at every third position in the putative transmembrane α-helix. Studies have elegantly confirmed that amino acids in the S_4 segments of voltage-dependent Na and K channels physically move across the membrane to the aqueous extracellular environment during membrane depolarization. Horn and colleagues (Yang *et al.*, 1996) mutated

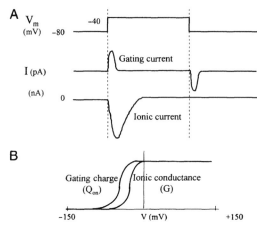

FIGURE 4   (A) The simulated course of gating current and ionic current during activation of Na channels when the voltage is clamped from −80 to −40 mV. Gating current is caused by the movement of the voltage sensors (S_4 segments) relative to the membrane voltage field. (B) The gating charge (Q_{on}) saturates at a more hyperpolarized membrane potential than the ionic conductance (G).

the outermost positively charged arginines in the S_4 segment of the fourth domain of Na channels to cysteine, which is capable of reacting with sulfhy-dryl-reducing agents. They found that when Na channels were in the closed state, the cysteines in these positions reacted with impermeant sulfhydryl-reducing agents applied intracellularly but not extracellularly. However, when the channels were open, they interacted only with extracellular sulfhy-dryl-reducing agents, documenting their movement across the membrane voltage field. Isacoff and colleagues (Mannuzzu *et al.*, 1996) further refined this technique in *Shaker* K channels by labeling the cysteines with a fluoro-phore. By measuring changes in fluorescence, they showed that, when the K channels in these cysteine-mutant *Shaker* K channels were opened during depolarization, a stretch of at least seven amino acids moved from a buried position in the membrane to an extracellular environment.

A picture of the molecular mechanism of activation derived from these and other studies is schematically illustrated in Fig. 5. Each tetrameric Na, Ca, or K channel has four positively charged S_4 segments. When the membrane is hyperpolarized, the S_4 positive charges are attracted to the inside of the cell membrane and are in the "down" position. With depolar-ization, the positive charges are electrostatically attracted to the "up" posi-tion. The movement of each S_4 segment from "down" to "up" causes a transient outward current (the gating current) due to the movement of positive charges. Five distinct closed states (C_0 to C_4) are defined by the number of S_4 segments in the up position. When all the S_4s are up, the channel can undergo a final transition from the last closed state (C_4) to

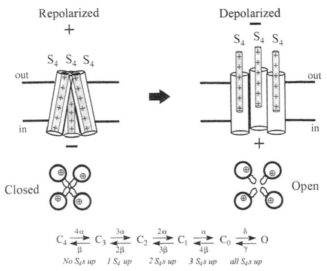

FIGURE 5 Hypothetical model of activation gating. In the most closed state (C_4), all of the S_4 segments are in the "down" position. During depolarization, gating charge movement corresponds to S_4 segments moving to the "up" position (toward the extracellular face). Five closed states (C_4–C_0) correspond to the number of S_4s in the "up" position (only three domains are shown for illustrative purposes). When all S_4s are "up," the channel undergoes the final voltage-independent transition to the open state (from C_0 to O). α is the rate constant for movement of an S_4 segment from "down" to "up," and β from "up" to "down."

the open state (O). This mechanism explains why gating charge–voltage (Q–V) relationship is shifted negatively with respect to the ionic conductance–voltage (G–V) relationship (see Fig. 4) (i.e., most of the gating charge moves before ionic current flows, consistent with the experimental findings). The steepness of the Q–V relationship indicates that a total of 12–14 charges move across the membrane voltage field during activation, consistent with 3–4 charges per each S_4 segment. Detailed electrophysiological analysis of gating charge movements in *Shaker* also indicates that there is some cooperativity in the movement of the S_4 segments through the various closed states (Bezanilla *et al.*, 1991).

Findings from cyclic-nucleotide-gated nonselective cation channels, which have a very similar pore structure to voltage-gated K channels, suggest that the P region may form both the selectivity filter and the activation gate (see Fig. 1). Sun *et al.* (1996) used the cysteine mutagenesis technique to probe the structure of the P region and found that when the channel was in the closed state, certain amino acids that were mutated to cysteine in the P region could still react with membrane-impermeant sulfhydryl-reducing agents applied to either the intracellular or extracellular side of the membrane. Since the reagents chosen were too large to pass through the selectivity filter, this finding excludes the possibility that there is a region of the pore between the closed activation gate and the selectivity filter that is inaccessible to both intracellular and extracellular aqueous solutions. They therefore argued that the selectivity filter must also be the activation gate (i.e., the P loops are constricted in the closed conformation of the channel) and must widen to allow selective ion flow only when the channel adopts the open conformation (see Fig. 5).

c. Inactivation Gating

A number of the cardiac voltage-gated ion channels inactivate after they open, including Na channels, Ca channels, transient outward K channels, and the rapid component of the delayed rectifier K channel. Early electrophysiological results indicated that the inactivation process is not intrinsically very voltage dependent. Its apparent voltage dependence arises from the voltage dependence of the activation process (for review, see Patlak, 1991); that is, the channel can readily inactivate only from the activated state, so even though inactivation is nearly voltage independent, it follows the same pattern of voltage dependence as activation. Furthermore, treatment of the intracellular side of the membrane with proteolytic enzymes was found to remove inactivation of both Na and Ca currents (Armstrong et al., 1973; McDonald et al., 1994), suggesting that the inactivation gates were formed by cytoplasmic regions of the channels. Armstrong and Bezanilla (1977) proposed a "ball-and-chain" mechanism for the Na channel, consisting of a "ball" floating in the cytoplasm tethered to the channel by a "chain." When the channel opened, the ball would bounce around until it collided with and plugged up the pore. With the cloning of the voltage-gated channels, the regions of the proteins that composed the inactivation gates have been identified. In *Shaker* K channels, the ball-and-chain mechanism was elegantly validated by Aldrich and colleagues (Hoshi et al., 1990), who found that the first 46 amino acids of the N-terminus constitute the ball and the remaining amino acids in the N-terminus constituted the chain (Fig. 6). They showed that in mutant *Shaker* channels with the N-terminus deleted, fast inactivation was completely eliminated, but could be restored by applying the synthesized peptide corresponding to the ball portion of the N-terminus to the cytoplasmic surface of mutant *Shaker* channels. Other studies demonstrated that lengthening or shortening the chain segment slowed or accelerated, respectively, the kinetics of fast inactivation and that only a single ball and chain attached to any one of the subunits was sufficient to confer fast inactivation (MacKinnon et al., 1993). In *Shaker,* the docking site to which the "ball" binds to plug up the pore of the channel was localized to the cytoplasmic linker between S_4 and S_5 (see Fig. 2) (Isacoff et al., 1991). This same ball-and-chain-type mechanism is responsible for fast inactivation of the transient outward current (I_{to}) in cardiac muscle. In a number of mammalian K channels, however, the ball is located on

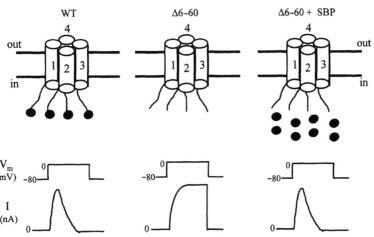

FIGURE 6 "Ball-and-chain" mechanism of fast inactivation. Wild-type (WT) *Shaker* K channels inactivate rapidly during a depolarizing voltage-clamp pulse (*left lower panel*). Deletion of a portion of the N-terminus (amino acids 6–60) eliminates the fast inactivation (*middle panel*). Adding a *Shaker* ball peptide (SBP) corresponding to the deleted amino acids to the solution that bathes the intracellular membrane surface restores fast inactivation, indicating that these peptides form the ball that inactivates the channel (*right panel*).

a β-subunit of channel, rather than on the N-terminus of the α-subunit (Adelman, 1995).

A second type of inactivation in *Shaker* K channels is known as C-type, or slow, inactivation and occurs over a much slower time course (many seconds) than fast inactivation (milliseconds). The C-terminus and the C-terminal end of the P region are important in regulating C-type inactivation. Recent evidence suggests that C-type inactivation occurs when the channel undergoes a conformational change to a nonconducting state facilitated by the ion-conducting pathway being empty (i.e., when no K ions are actually in the pore) (Baukrowitz and Yellen, 1996). The rapid inactivation process in the K channel *HERG,* which is the rapid component of the delayed rectifier K channel in the heart, is believed to be a rapid variant of C-type inactivation (Smith *et al.,* 1996).

In Na and Ca channels, the voltage-sensitive inactivation gates are believed to be more analogous to a "hinged door" than a ball and chain. In Na channels, the region important for inactivation is the cytoplasmic linker between the third and fourth domains (see Fig. 2) (Stuhmer *et al.,* 1989). Mutations in this region interfere with Na current inactivation; in the heart, one variant of the long QT syndrome, LQT3, is caused by mutations in this region leading to noninactivating Na current during the plateau of the action potential, which prolongs its duration (Bennett *et al.,* 1995). In skeletal muscle Na channels, mutations in this region underlie the defective Na channel inactivation causing myotonia in periodic paralyses (Barchi, 1996).

In L-type Ca channels, the voltage-sensitive inactivation gate has been localized to a different region, the S_6 segment of the first domain (Zhang et al., 1994) (see Fig. 2). L-type Ca channels also exhibit a second type of inactivation, Ca-dependent inactivation, which is of greater physiological importance. This gating mechanism has been localized to a region of the C-terminus containing a classic EF-hand Ca-binding site (de Leon et al., 1995). Whether this EF-hand site is responsible for Ca-induced inactivation, however, is still controversial (Olcese et al., 1996).

In summary, current evidence indicates that classic voltage-gated Na, Ca, and K channels have a similar overall topological structure consisting of a tetrameric array of domains (for Na or Ca channels) or subunits (for K channels), each with six transmembrane-spanning segments (see Fig. 2). The ion-conducting pathway is located in the center of the tetrameric array, into which the P loops (between the S_5 and S_6 transmembrane segments) project to form the selectivity filter (see Fig. 2). Each S_4 transmembrane segment contains a large number of positive charges, and is composed of voltage sensors, which physically move with respect to the transmembrane voltage field during depolarization. Their movement, detected as gating current (see Fig. 4), induces a conformational change in the protein that opens the pore to the flow of ions (see Fig. 5). Inactivation is attributed to a ball-and-chain or hinged-door mechanism; cytoplasmic regions of the channel protein bind to a docking site that becomes exposed when the channel is activated, thereby blocking ion flow through the pore (see Fig. 6).

2. Na Channels

The first Na channel to be cloned was from the electric eel electroplax (Noda et al., 1984), and the α-subunit of the human cardiac Na channel was subsequently reported (Gellens et al., 1992). The α-subunit is a $\sim$230-kD protein consisting of 2016 amino acids. When heterologously expressed, the α-subunits form functional voltage-gated Na channels that are sensitive to block by tetrotoxin and other Na channel blockers, and have qualitatively similar but not identical gating properties to native Na channels. In addition to the α-subunit, $\beta1$- and $\beta2$-subunits of the Na channel have also been identified (Table 2) (Adelman, 1995). Neither of the β-subunits form functional channels by themselves, but when coexpressed with the α-subunit, increase the number of functional channels and modify the gating properties to more closely resemble those of native Na channels.

The main physiological role of the Na channel in a myocyte is to generate a sufficiently large enough inward current to depolarize the adjacent myocytes to the threshold of their Na channels, ensuring rapid propagation of the cardiac impulse. Once this task is completed, rapid inactivation of the Na current is important to minimize unnecessary Na influx, whose removal involves an energy cost by the Na–K pump (see Chapter 4). In single-

TABLE 2 Candidate Clones for Cardiac Ion Channels

Native current	Candidate clone(s)	Subunits
I_{Na}	SCN5A	$\beta 1, \beta 2$
$I_{Ca(L)}$	$\alpha 1$ DHP receptor	$\alpha 2, \beta, \gamma, \delta$
$I_{Ca(T)}$	?	?
I_{to1}	?	?
I_{to2}	Kv4.2, ?Kv1.4	β
I_{Ks}	minK (sK)/K_vLQT1	?
I_{kr}	HERG	?
I_{Kur}	Kv1.5, ?Kv1.2	β
I_{Kp}	?	?
I_{K1}	Kir2.1	?
$I_{K(ACh)}$	Kir3.1/Kir3.4 heteromer	?
$I_{K(ATP)}$	Kir6.1	?SUR
$I_{K(Ca)}, I_{K(Na)}, I_{K(AA)}$	?	?
$I_{Cl(cAMP)}$	CFTR	?
$I_{Cl(Ca)}$	?	?
SR CRC	RyR2	FKBP12
SR IP_3-gated CRC	IP_3R	?
Mechanosensitive	?	?
Background/leak	?	?

channel recordings, Na channels have a unitary conductance of 10–15 pS, and typically only open briefly early in the depolarizing pulse before entering the inactivated state. However, occasionally, Na channels may enter modes characterized by persistent late openings (Zilberter *et al.*, 1994; Saint *et al.*, 1992). A noninactivating mode of Na channel behavior is in part responsible for persistent Na currents during the action potential plateau, which plays a secondary role in regulating action potential duration. The genetic defect in the LQT3 variant of the long QT syndrome, which involves a 3-amino acid deletion in the cytosolic linker between domains 3 and 4 (the inactivation gating region), enhances the probability of the Na channels entering a noninactivating mode (Bennett *et al.*, 1995). This increases inward current during the action potential plateau, prolonging its duration. The enhanced inward current also predisposes the heart to early afterdepolarizations, which have been implicated in the pathogenesis of the form of ventricular tachycardia known as *torsades de pointes*. This life-threatening arrhythmia is the cause of sudden cardiac death in individuals afflicted with this genetic disease.

The noninactivating mode of the Na channel may also be important during myocardial ischemia. Hypoxia enhances this mode of Na channel behavior (Jue *et al.*, in press). Lysophospholipids, which accumulate in ischemic myocardium, also cause Na channels to open at hyperpolarized membrane potentials (Undrovinas *et al.*, 1992). This inappropriate activation of Na channels may contribute to arrhythmias by promoting membrane depolarization in ischemic cardiac tissue.

The cardiac Na channel is regulated directly by G proteins (Schubert *et al.*, 1989) and by both protein kinase A (PKA) and protein kinase C (PKC) (for review, see Cukierman, 1996). The α-subunit contains multiple consensus sites for phosphorylation by PKA and PKC. Activation of PKA (also known as cAMP-dependent protein kinase) is the major effector used by the β-adrenergic system through the β-receptor/G_s protein/adenylate cyclase/cAMP/PKA signaling pathway. Phosphorylation by PKA increases Na current amplitude, but also shifts the steady-state inactivation curve to more negative potentials. When the Na current is elicited from a normal resting membrane potential (approximately -80 mV), the net effect of PKA is a decrease in the amplitude of the peak Na current due to the greater percentage of channels in the inactivated state at the resting potential. The physiological importance of this effect is uncertain. It is possible, however, that the facilitation of inactivation by PKA may suppress Na window currents during the action potential plateau. (Window currents are defined by the voltage range over which the steady-state activation and inactivation curves overlap.) This prevents excessive action potential prolongation and also minimizes potentially arrhythmogenic inward currents during the plateau phase.

PKC activation is linked to M_1 and M_3 muscarinic, α-adrenergic, angiotensin, and some purinergic receptors. The effects of PKC on voltage-gated Na channels are variable and depend on the method used to stimulate PKC activity (Cukierman, 1996). The physiological significance of Na channel phosphorylation by PKC is unclear.

The Na channel is the target of Type 1 antiarrhythmic drugs, the most common clinically used antiarrhythmic agents. Type 1 agents are divided into three subtypes, 1A (e.g., quinidine, procainamide, disopyramide), 1B (e.g., lidocaine, mexiletine, dilantin), and 1C (e.g., flecainide, encainide, propafenone), depending on whether they prolong, shorten, or do not change the action potential duration, respectively. Use and voltage dependence are important properties relating to the antiarrhythmic efficacy of these agents. Use dependence refers to the property that the more the channel is used, the more effectively the drug blocks the channel (Fig. 7). This is a desirable property for a Type 1 antiarrhythmic drug, since it potentiates block of the Na current at fast heart rates during tachycardia and minimizes block at slower heart rates during sinus rhythm. Voltage dependence refers to the property that the efficacy of the drug varies with membrane potential; in the case of Type 1 antiarrhythmic agents, the more depolarized the tissue, the more effective the drug at blocking Na channels. This is also a desirable property of Type 1 agents, since depolarized tissue (e.g., during acute myocardial ischemia) is often the source of arrhythmias.

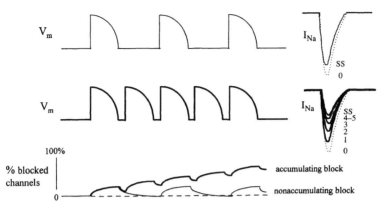

FIGURE 7   Illustration of use-dependent block by a drug that preferentially blocks the open or inactivated state over the resting state of the Na channel. At the slow heart rate (*upper trace*), the drug unbinds nearly completely during diastole (*thin line, lower trace*). At the fast heart rate (*middle trace*), the drug does not have time to unbind completely during the short diastolic interval, and the degree of block accumulates (*thick line, lower trace*). The magnitude of I_{Na} for successive beats relative to I_{Na} in the absence of the drug (trace labeled 0) is indicated at the left. SS, steady state in the presence of the drug.

Two hypotheses have been proposed to account for use and voltage dependence, the modulated receptor hypothesis (Hondeghem, 1987) and the guarded receptor hypothesis (Starmer *et al.*, 1984). The former proposes that the affinity of the binding site for the drug differs with the conformational state of the channel (closed, open, or inactivated). The latter proposes that the affinity of the binding site is unchanged, but the physical access of the drug to its binding site is altered by the conformational state of the channel. For example, if a hydrophilic drug can only gain access to its binding site through the aqueous (hydrophilic) pathway of the pore channel, then a closed gate might physically limit its access. In general, hydrophilic antiarrhythmic agents show more use and voltage dependence than the hydrophobic drugs (e.g., the polar lidocaine molecule shows much greater use and voltage dependence than its lipid-soluble derivative mexiletine).

In summary, the cardiac Na channel is a classic voltage-gated ion channel (see Fig. 2). The fundamental characteristics of ion permeation, ion selectivity, gating and pharmacological properties are contained in the α-subunit, and β-subunits modify gating properties. Physiologically, the Na current is essential for ensuring rapid conduction of the cardiac impulse and also influences action potential duration. A genetic defect in Na channel inactivation is one cause of the long QT syndrome. The Na channel is regulated by protein kinases. It is also the major target of Type 1 antiarrhythmic drugs used clinically.

3. Ca Channels

a. L-Type Ca Channels

The L-type Ca channel, also known as the dihydropyridine receptor, was first cloned from skeletal muscle by Numa and coworkers in 1987 (Tanabe *et al.*, 1987), and the cardiac isoform was later identified by the same group (Mikami *et al.*, 1989). The $\alpha 1$-subunit consists of a 155–200-kD protein (see Fig. 2). Like the α-subunit of the Na channel, the $\alpha 1$-subunit of the L-type Ca channel expressed by itself yields fully functional voltage-gated Ca channels sensitive to Ca channel antagonists and agonists. Additional Ca channel subunits have also been cloned, including $\alpha 2$-, β-, γ-, and δ-subunits (see Table 2). Like Na channel subunits, they do not form functional ion channels when expressed alone, but modify gating properties when coexpressed with the $\alpha 1$-subunit (Adelman, 1995). The β-subunit, for example, markedly increases the availability of Ca channels by inhibiting the inactive mode (mode 0, Fig. 8) (Neely *et al.*, 1993).

The L-type Ca channel is the major regulator of Ca release, and hence contractile force, on a beat-to-beat basis. L-type channels in the T-tubular sarcolemma are colocalized with the cardiac ryanodine receptors, or Ca release channels, in the junctional sarcoplasmic reticulum (jSR). Because of this special compartmentalized arrangement, it has been hypothesized that Ca influx through L-type Ca channels locally controls release of intracellular Ca from the sarcoplasmic reticulum (SR) stores (Stern, 1992; Cheng *et al.*, 1993). This local control mechanism (discussed in detail in Chapter 5) allows contractile force to be graded on a beat-to-beat basis by the magnitude of the L-type Ca current. Although other Ca influx pathways, such as Na–Ca exchange, also can modulate the beat-to-beat response (LeBlanc and Hume, 1990), quantitatively they play a less important role. However, these other processes strongly regulate cardiac inotropy over a time course of a number of beats by affecting the degree of Ca loading of the SR.

In single-channel recordings, L-type Ca channels have a unitary conductance of 8 pS in 100 mM [Ca]$_o$ and exhibit prominent modal gating, as illustrated in Fig. 8 (for review, see McDonald *et al.*, 1994). Normally, channels alternate between Mode 0, in which no or very brief (mean open time, ~0.15 ms) openings separated by long, closed periods are observed, and Mode 1, characterized by frequent brief openings (mean open time, ~1 ms). Mode 2, characterized by long openings (mean open time, ~25 ms), is only rarely observed under normal conditions. However, Mode 2 is facilitated by exposure to the Ca channel agonist BayK 8644 and also to isoproterenol, by low-frequency pulsing and by pulsing to high positive potentials.

As might be anticipated from its key role in modulating contractile force, the L-type Ca current is highly regulated by hormone-receptor signaling

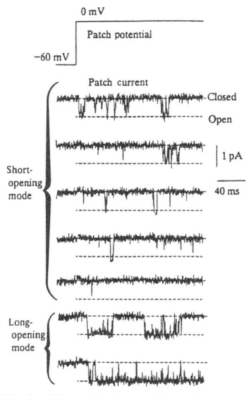

FIGURE 8 Modal gating of L-type Ca channels. Mode 1 (short-opening mode) is characterized by brief openings, and Mode 2 (long-opening mode) is characterized by much longer openings. In Mode 0 (not shown), no, or extremely brief, openings occur. (Reprinted with permission from *Nature* 348:192–193 [Bean, 1990]. Copyright 1990 Macmillan Magazines Limited.)

pathways. Its most prominent regulation is by PKA. The α1-subunit contains a number of consensus sites for phosphorylation by PKA, and phosphorylation of these site(s) is strongly implicated in modulating channel gating. In response to PKA stimulation, the majority of the increase in the magnitude of the L-type Ca current is attributable to shift in modal gating favoring Mode 1 (brief openings) over Mode 0 (no openings) (McDonald *et al.*, 1994). Transitions to Mode 2 (long-lasting openings) are also occasionally observed. Agents that up-regulate the L-type Ca current, by increasing intracellular cAMP and enhancing PKA in the heart, include β-receptor agonists, histamine, serotonin, glucagon, and calcitonin-related gene peptide (in atria only) (McDonald *et al.*, 1994). Agents that down-regulate the L-type Ca current by receptor-mediated reduction in intracellular cAMP include acetylcholine, adenosine, ATP, neuropeptide Y, and nitric oxide (NO). In addition to the PKA pathway, L-type Ca channels have also been reported to be directly stimulated by a membrane-delimited G protein pathway. Brown and colleagues (Yatani *et al.*, 1987) have presented evidence that $G_{\alpha s}$ binds to and directly increases the open probability of L-type Ca channels, although their findings have been contested (Hartzell *et al.*, 1991).

L-type Ca channels are also regulated by PKC. As with Na channels, the effects of PKC on the L-type Ca current are variable (McDonald *et al.*, 1994). Cytosolic Ca also regulates the amplitude of the L-type Ca current, having a stimulatory effect at low concentrations and an inhibitory effect at higher concentrations (McDonald *et al.*, 1994). The stimulatory effect may be mediated by Ca-calmodulin-dependent protein kinase. The mechanism of the inhibitory effect is unknown. Cytosolic Ca also enhances inactivation of Ca channels, probably by a direct allosteric effect induced by Ca binding to the C-terminus of the α1-subunit (de Leon *et al.*, 1995). This Ca-induced inactivation of the L-type Ca current is a major factor regulating the cardiac action potential duration. MgATP also has a stimulatory effect on the L-type Ca current, which is not phosphorylation mediated (O'Rourke *et al.*, 1992). Cytoskeletal elements may also be important in maintaining L-type Ca channels in a functional state (Johnson and Byerly, 1993). In excised membrane patches, Ca channels typically run down irreversibly, even in the presence of MgATP and the catalytic subunit of PKA. Disruption of cytoskeletal elements may play a role in this process.

b. T-Type Ca Channels

The T-type Ca current (T for "transient") was originally described in cardiac muscle (Nilius *et al.*, 1985; Bean, 1985) and is distinguished from the L-type current (L for "long-lasting") by its more rapid inactivation kinetics, as well as by a more negative membrane potential for its activation threshold (-50 mV vs -30 mV), its peak current (-30 mV vs 0 mV), and steady-state inactivation ($V_{1/2}$; -70 vs -20 mV) (Fig. 9). In addition, the T-type Ca current is not sensitive to block by dihydropyridines, has a different profile of susceptibility to block by divalent cations, and is not significantly up-regulated by β-adrenergic stimulation. The molecular identity of the cardiac T-type Ca channel is still unknown, although a low-threshold Ca channel with somewhat similar properties has been cloned from brain (Soong *et al.*, 1993). The density of the T-type Ca current in cardiac muscle is variable from species to species. The physiological role of the T-type Ca current in cardiac muscle is not entirely clear, although evidence suggests that it contributes to pacemaking currents in SA cells (Hagiwara *et al.*, 1988) and may act as an ancillary trigger for SR Ca release in ventricular myocytes (Sipido and Carmeliet, 1996).

In summary, two types of Ca channels are described in the heart, the L-type and the T-type. The L-type Ca channel is a classic voltage-gated

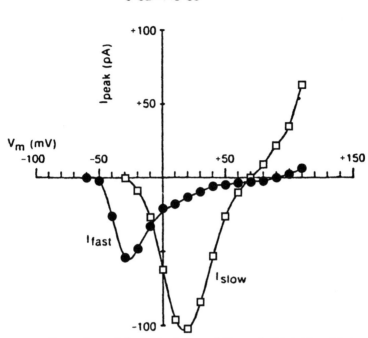

5 Ca//145 Cs

FIGURE 9 Comparison of the I–V relations of T-type (solid dots) and L-type (open squares) Ca currents in canine atrium. (Reproduced from *The Journal of General Physiology* by copyright permission of The Rockefeller University Press.)

ion channel (see Fig. 2). The fundamental characteristics of ion permeation, ion selectivity, gating, and pharmacological properties are contained in the $\alpha 1$-subunit. $\alpha 2$-, β-, γ-, and δ-subunits have also been cloned, and some of these are important for modifying gating properties. The L-type Ca current plays a key role in regulating cardiac contractility by regulating SR Ca release, and also contributes to cardiac pacemaking and regulation of action potential duration. It is highly regulated by autonomic signaling pathways, particularly PKA. The cardiac T-type channel has not yet been definitively cloned. Its physiological role is less certain, but it may play a role in pacemaking in SA nodal cells.

4. K Channels

K channels are the most diverse family of cardiac ion channels, fitting with the multiple roles they play in setting the resting membrane potential, modulating heart rate, and controlling action potential duration. Unlike Na or Ca channel α-subunits, a single K channel protein encodes only one-quarter of a K channel. This facilitates diversity by allowing for the formation of heterotetrameric K channels whose four subunits are derived from different genes. Thus, depending on the subunit arrangement, a few genes can encode a considerable variety of functionally different K channels. This property yields considerable plasticity to the voltage-gated K channel family, but makes it daunting to identify which native cardiac channels correspond to which K channel homo- and heterotetramers. The situation is complicated further by the existence of K channel β-subunits, which, when coexpressed with the α-subunits, modify their properties, although they do not form functional ion channels when expressed alone (for review, see Adleman, 1995). The same channel may be expressed in both atrium and ventricle, but have markedly different properties depending on which β-subunits are also expressed. Deal *et al.* (1996) have summarized a set of strict criteria to decide unequivocally which cloned channels correspond to which native cardiac channels. These include (1) isoform-specific antibodies made against the cloned channel subunits must confirm the presence of the channel protein in the expected cardiac cells, (2) isoform-specific antibodies that alter channel function of the cloned channel must be shown to alter the function of the native current under study, (3) affinity purification from native tissue must confirm the α- and β-subunit protein composition of the native channel, and (4) deletion of the cloned channel gene in a transgenic animal should eliminate the native current under study. Keeping in mind that all four criteria have not generally been met, Table 2 summarizes the current state of knowledge about the identity of native cardiac voltage-dependent K channels.

The largest family of cardiac voltage-gated K (Kv) channels are the mammalian homologues of *Drosophila* K channels, which include *Shaker* (Kv1), *Shab* (Kv2), *Shaw* (Kv3), and *Shal* (Kv4). The less closely related *Drosophila* K channel *eag* (for "*ether à go-go*") also has a mammalian cardiac counterpart called *HERG* (for "*human ether à go-go related gene*"), which has been identified as the rapid component of the delayed rectifier K current (I_{Kr}) (Sanguinetti *et al.*, 1995). The unrelated minK (or sK) channel, originally cloned from rat kidney, may be a subunit of the slow component (I_{Ks}) (Deal *et al.*, 1996).

a. Transient Outward K Channels

The transient outward current (I_{to}) is most prominently expressed in atrium and ventricular epicardium. It is responsible for the notch in the early portion of action potential plateau (phase 1) and contributes to the overall shorter action potential duration in these regions. I_{to} has two components—a Ca_i-independent (I_{to1}) and a Ca_i-dependent (I_{to2}) component. I_{to1} is a classic voltage-gated K channel that activates in a voltage-dependent manner and then inactivates rapidly by the ball-and-chain mechanism similar to *Shaker*. The most well-known blocker of I_{to1} is 4-aminopyridine (4-AP), but it is also blocked by flecainide, quinidine, and other drugs. The major candidate clones for I_{to1} are Kv1.4 and Kv4.2, with the latter clone showing the greatest similarity to the native current (Deal *et al.*, 1996).

I_{to2} is activated by the rise in cytosolic Ca during the Ca transient and can be eliminated by buffering cardiac cells with a high concentration of a Ca chelator, such as EGTA. However, the relative contributions of Ca-activated K channels, Ca-activated Cl channels, Ca-activated nonselective cation channels, and the Na–Ca exchange current to I_{to2} have been difficult to sort out, and this current remains poorly characterized at the present time (Hume and Harvey, 1991). Since the candidate ion channels for I_{to2} are ligand gated, they are discussed in more detail in Section B.

b. Delayed Rectifier K Channels

Delayed rectifier K channels play a very key role in regulating action potential duration. The delayed rectifier K current (I_K) has been found to have three components: slow (I_{Ks}), rapid (I_{Kr}), and ultrarapid (I_{Kur}). I also include I_{Kp} in this group because of its overall similarity to I_{Kur}. I_{Ks} and I_{Kr} are present to varying extents in all the cardiac tissues. I_{Kur} has been described only in atrium, and I_{Kp} only in ventricle.

i. Slowly Activating Component (I_{Ks}) I_{Ks} activates slowly with depolarization, over a time course of many seconds, and does not inactivate during maintained depolarization. The single-channel conductance is too small to be directly measured, but the channels have sufficiently high density to record macroscopic currents in excised membrane patches (Walsh *et al.*, 1991). I_{Ks} activates slowly during the action potential plateau and makes a major contribution to the slow phase of repolarization (phase 2). I_{Ks} is sensitive to block by La and clofilium, but not by sotalol, E4031, or other methanosulfonanilides (Sanguinetti and Jurkiewicz, 1990).

I_{Ks} is highly regulated by both the PKA and PKC pathways (Walsh and Kass, 1988). During β-receptor stimulation, the activation of PKA markedly increases I_{Ks} amplitude, offsetting the tendency of the simultaneously potentiated L-type Ca current to prolong action potential duration. PKC stimulation activates I_{Ks} in guinea pig but inhibits it in rat and mouse (Deal *et al.*, 1996). PKC effects are also temperature dependent (Walsh and Kass, 1988).

Recent evidence indicates that I_{Ks} is a heteromeric channel consisting of the minK protein and K_vLQT1 (Barhanin *et al.*, 1996; Sanguinetti *et al.*, 1996). The minK protein is very different from the classic voltage-dependent channels, consisting of only 130 amino acids with a single predicted transmembrane-spanning segment. The transmembrane region has no significant homology to either the S4 segment or P region of voltage-gated ion channels (Deal *et al.*, 1996). However, mutations to the transmembrane region of minK affect channel permeation and selectivity, and the protein placed in lipid bilayers by itself forms channels, although not K selective. Like I_{Ks}, minK is also regulated by PKC, and a mutation of the asparigine-102 found in the guinea pig minK protein to the serine found in the rat and mouse converts the PKC effect from stimulation to inhibition, analogous to the effects of PKC on native I_{Ks} in these species (Deal *et al.,* 1996). Thus this atypical channel protein regulates a number of key properties of I_{Ks}. K_vLQT1, however, has the predicted topology of classic voltage-gated K channels, with six membrane-spanning segments and a P region. However, it forms functional homomeric channels poorly. K_vLQT1 was originally discovered because it is the genetic abnormality causing the LQT1 variant of the long QT syndrome. It is presumed that mutations in K_vLQT1 render I_{Ks} defective in these patients, causing prolongation of the action potential duration that predisposes affected individuals to early afterdepolarizations and often lethal ventricular arrhythmias (*torsades des pointes*).

ii. Rapidly Activating Component (I_{Kr}) In contrast to I_{Ks}, I_{Kr} activates much more rapidly and also inactivates. Inactivation is quasi-instantaneous at positive voltages, giving a bell-shaped I–V curve that peaks near 0 mV (Fig. 10C). The negative slope region of the I–V curve progressively increases membrane K conductance as the plateau voltage becomes more negative during phase 2 of the action potential, leading to a positive feedback in repolarization rate facilitating the onset the rapid repolarization phase (phase 3). I_{Kr} is insensitive to block by La and clofilium, but highly sensitive to sotalol, E4031, dofetilide and other methanosulfonanilides (Sanguinetti and Jurkiewicz, 1990). I_{Kr} has an intermediate sensitivity to 4-AP ($K_{0.5} \sim 2$ mM).

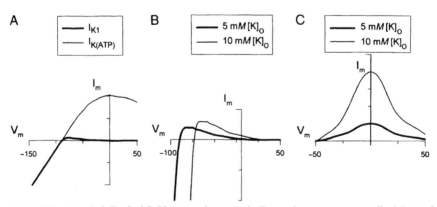

FIGURE 10 (A) Typical I–V curves for a weak ($I_{K(ATP)}$) versus a strong (I_{K1}) inward rectifier K current. (B) The I–V curve of I_{K1} at two different $[K]_o$. The reversal potential follows the shift in E_K ($\sim +18$ mV), and the maximum outward current increases with the square root of $[K]_o$. (C) The I–V curve of I_{Kr} at two different $[K]_o$. The current amplitude increases with the square of $[K]_o$.

Recently, *HERG* has been identified as the channel corresponding to native cardiac Kr channels (Sanguinetti *et al.*, 1995). *HERG* is unrelated to the other Kv channels, but has a similar overall topological structure with six transmembrane-spanning regions, a positively charged S4 segment, and a similar P region. Mutations in *HERG* are the cause of the LQT2 variant of the long QT syndrome (Sanguinetti *et al.*, 1995). A striking property of I_{Kr} is the strong dependence of its conductance on the square of the extracellular K concentration ($[K]_o$). Thus I_{Kr} plays a major role causing the shortening of the action potential duration when $[K]_o$ is elevated and, conversely, its prolongation when $[K]_o$ is reduced. This interaction is especially important for Type 3 antiarrhythmic drugs, which prevent arrhythmias by blocking K currents and prolonging action potential duration. The action potential prolongation by Type 3 antiarrhythmic agents that block I_{Kr}, such as quinidine, sotalol, and dofetilide, can be markedly potentiated in the setting of low $[K]_o$, causing undesirable proarrhythmic effects. The ventricular arrhythmias in the setting of excessive action potential prolongation by I_{Kr} blockers are identical to the *torsades des pointes* form of ventricular tachycardia seen in the LQT2 form of congenital long QT syndrome. Nonantiarrhythmic drugs, such as the antibiotic erythromycin, also occasionally cause *torsades des pointes*, and the mechanism is probably related to block of I_{Kr} (Daleau *et al.*, 1995).

iii. Ultrarapidly Activating Component (I_{Kur}) I_{Kur} is an ultrarapidly activating, noninactivating delayed rectifier K channel that has been described in rat and human atria (Boyle and Nerbonne, 1991; Wang *et al.*, 1993). I_{Kur} activates considerably more rapidly than I_{Kr}, with a time constant of activation of 2–20 ms (Wang *et al.*, 1993) and is much more sensitive to block by 4-AP ($K_{0.5}$, ~50 μM). It is insensitive to TEA and α-dendrotoxin. I_{Kur} makes an important contribution to atrial repolarization; when I_{Kur} was selectively blocked in human atrial myocytes, the action potential duration was prolonged by 66% (Wang *et al.*, 1993). Current evidence favors Kv1.5 as the candidate clone for I_{Kur}, based on its similar activation kinetics and pharmacological profile (see Table 2). Kv1.2 has similar kinetics, but differs from I_{Kur} in its sensitivity to α-dendrotoxin and TEA, and lack of sensitivity to 4-AP (Deal *et al.*, 1996).

iv. Plateau K Current (I_{Kp}) I_{Kp} is a voltage-activated K current with rapid activation kinetics (time constant < 10 ms) and no inactivation, similar to the atrial I_{Kur}, but present in the ventricle (Backx and Marban, 1993). However, I_{Kp} differs from I_{Kur} in a higher sensitivity to block by Ba (1 mM) and its relatively linear I–V relation (i.e., lack of outward rectification). I_{Kp} contributes to ventricular repolarization. Its molecular identity is unknown, although the rapidly activating candidates for I_{Kur}, Kv1.2 and Kv1.5, are expressed in the ventricle and atrium (Deal *et al.*, 1996).

In summary, voltage-gated K channels are the most diverse voltage-gated ion channels in the heart (see Table 1). The fundamental characteristics of ion permeation, ion selectivity, gating, and pharmacological properties are contained in the α-subunits (see Fig. 2). Different α-subunits may combine to form heterotetramers with properties that are different from the corresponding homotetrameric channels. β-subunits have also been cloned and can substantially modify gating properties when coexpressed with α-subunits. This diversity has hampered attempts to identify unequivocally which native cardiac K current corresponds to which cloned K channel in many cases (see Table 2). Voltage-gated K currents fall into two general classes depending on whether they demonstrate time-dependent inactivation— transient outward K currents (which inactivate) and delayed rectifier K currents (which do not activate). Physiologically, these K currents play a key role in regulating repolarization of the cardiac action potential. The marked regional variation in action potential configuration throughout the heart is largely attributed to differential expression of these various K channels.

5. Pacemaker Channels (I_f)

I_f is a hyperpolarization-activated current that is permeable to both Na and K ions, and is present in SA nodal, AV nodal, and Purkinje cells (DiFrancesco, 1995b) (Fig. 11). Unlike most nonselective cation channels, f channels have a very small single-channel conductance of ~ 1 pS. I_f is activated at membrane potentials negative to -50 mV and is sensitive to block by Cs and by selective inhibitors such as alinidine and zatebradine. The role of I_f versus deactivating K currents as the dominant pacemaker

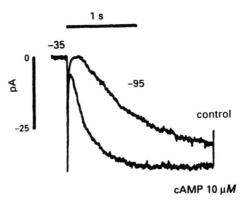

FIGURE 11   Recording of I_f in an excised inside-out macropatch. I_f activates progressively when membrane potential is hyperpolarized from -35 to -95 mV (control). Application of cAMP to the bath solution markedly increases the rate of activation and amplitude of I_f. (Reprinted from *Cardiovasc. Res.* **29**, D.DiFrancesco, The onset and autonomic regulation of cardiac pacemaker activity: Relevance of the f current, pp. 449–456, Copyright 1995 with kind permission of Elsevier Science–NL, Sara Burgerhartstraat 25, 1055 KV Amsterdam, The Netherlands.)

current in the SA node and Purkinje fibers is still somewhat controversial (DiFrancesco, 1995a; Vassalle, 1995) (see p. 125). However, regulation of I_f by neurotransmitters is very consistent with an important role in cardiac pacemaking. Evidence suggests that I_f is directly activated by G protein subunits during β-receptor stimulation with isoproterenol and decreased by G protein subunits during muscarinic receptor stimulation by acetylcholine (Yatani and Brown, 1990). In addition to this membrane-delimited G protein pathway, I_f is also regulated by intracellular cAMP levels. cAMP enhances I_f by shifting its activation curve to more positive potentials. This appears to involve a direct interaction of cAMP with the channels rather than PKA-mediated phosphorylation (DiFrancesco and Tortora, 1991). At the present time, the molecular identity of f channel is unknown.

B. SARCOLEMMAL LIGAND-GATED ION CHANNELS

1. Inward Rectifier K Channels

a. General Properties

Cardiac inward rectifier K channels consist of two groups: strong and weak inward rectifiers (Doupnik et al., 1995). The strong inward rectifiers include background inward rectifier (K1) channels and the muscarinic acetylcholine-activated K (K_{ACh}) channels; the weak inward rectifiers include ATP-sensitive K (K_{ATP}) channels, Na-activated K (K_{Na}) channels, Ca-activated K (K_{Ca}) channels, and arachidonate-activated (K_{AA}) channels. Although K1 channels are not classic ligand-gated ion channels, they belong to the same gene family as the other inward rectifier ligand-gated K channels that have been cloned, and for this reason I have included them in this group. In addition to inward rectification, several general features of inward rectifier K channels include a sensitivity to voltage-dependent block by extracellular Cs and Ba, a dependence of channel conductance on the square root of $[K]_o$, and a requirement for cytosolic MgATP to prevent rundown of channel activity.

Inward (or anomalous) rectification refers to the property of these channels that they pass current in the inward direction more readily than in the outward direction (see Fig. 10). Since under physiological conditions, the membrane potential (E_m) is positive to the K equilibrium potential (E_K), the channels pass predominantly outward current. Unlike voltage-gated ion channels, in which the gating mechanism senses the absolute value of E_m, the rectification process in inward rectifier K channels follows ($E_m - E_K$). The more positive this difference (i.e., the more depolarized the membrane potential is relative to E_K), the lower the conductance (see Fig. 10A). Under physiological conditions, inward rectification is predominantly due to voltage-dependent open channel block by intracellular Mg and Ca (Vandenberg, 1987; Mazzanti and DiFrancesco, 1989), and by intracellular polyamines such as spermine, spermidine, and putrescine (Lopatin et al., 1994;

Fickler *et al.*, 1994). Polyamines are polycations (spermine +4, spermidine +3, putrescine +2), which are products of ornithine metabolism. For the strong inward rectifiers, the $K_{0.5}$'s for block are in the μM range for Mg and Ca, and in the nM range for polyamines. The weak inward rectifiers require mM Mg to produce partial rectification and are generally insensitive to polyamines. Physiologically, the intracellular concentrations of free Mg (0.5–1 mM), free Ca (0.1–1 μM), and polyamines (0.1–1 mM) in cardiac cells are well above the $K_{0.5}$'s of the strong inward rectifiers. Intracellular divalent cations and polyamines are believed to block outward current by binding to a site(s) within the pore to which they can gain access only from the intracellular side, as external divalent cations and polyamines have no effect. The binding site(s) are partially through the transmembrane voltage field, accounting for the voltage dependence of block. By entering the pore from the outside, extracellular K destabilizes divalents or polyamines bound in the pore. Thus, when E_m is negative to E_K so that entry of extracellular K into the pore is favored, any divalents or polyamines residing in the channel are destabilized and ejected, allowing inward K current to flow readily. When the driving force for K flux through the channel is reversed, however, the likelihood of extracellular K ejecting the blocking agent decreases, resulting in little outward current flow (see Fig. 10A).

The molecular basis for the differences between the strong and weak inward rectifiers has been elucidated by mutational analysis of cloned channels from the two groups. The first inward rectifier K channel to be cloned (ROMK1 or Kir1.1) was a weak rectifier isolated from renal tissue (Ho *et al.*, 1993). Although it shows weak cardiac expression (Shuck *et al.*, 1994), its function in the heart, if any, is unknown. Following this discovery, the primary structure of inward rectifiers corresponding to cardiac K1 channels (IRK1 or Kir2.1) (Kubo *et al.*, 1993a), K_{ACh} channels (GIRK or Kir3.1) (Kubo *et al.*, 1993b; Dascal *et al.*, 1993), and K_{ATP} channels (μKATP or Kir6.1) (Inagaki *et al.*, 1995a, 1995b) were reported. All of these channels are of similar size (~400 amino acids) and have a similar predicted overall topological structure, consisting of cytoplasmic N- and C-termini and two transmembrane-spanning regions (M1 and M2) linked extracellularly by a highly conserved P region whose sequence is homologous to the P region of voltage-gated K channels (Fig. 12). Like the voltage-gated K channels, Kir proteins assemble in a tetrameric array, with the P regions dipping into the central ion-conducting pathway to form the selectivity filter (Yang *et al.*, 1995a). Evidence suggests that regions of the M2 segment and the C-terminus also line the ion-conducting pathway. Two negatively charged amino acids in the strong rectifier Kir2.1, Asp-172 in the M2 segment, and Glu-224 in the C-terminus are responsible for high-affinity voltage-dependent block by Mg and polyamines (Yang *et al.*, 1995b). Mutating these amino acids to uncharged residues, which are present at the corresponding sites in weak rectifiers such as Kir1.1, converted Kir2.1 into a

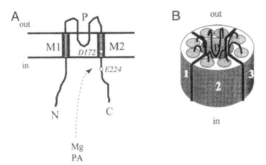

FIGURE 12   Predicted topology of members of the inward rectifier K channel family. (A) Two transmembrane segments (M1 and M2) are linked by a P region, which forms the selectivity filter of the ion translocation pathway. In the strong inward rectifiers (Kir2 and Kir3), negatively charged amino acids in the M2 and C-terminus (*D172* and *E224*) are critical determinants of high-affinity block of outward current by intracellular Mg and polyamines (PA). (B) The channel has a tetrameric structure, with the P loops lining the ion-conducting pathway, similar to voltage-gated ion channels.

weak rectifier with low sensitivity to block by Mg and polyamines. These mutations also altered single-channel permeation and gating properties of Kir2.1, consistent with these sites being located in the ion-conducting pathway. Recently, we have described a third, apparently intrinsic gating mechanism of inward rectification in Kir2.1, which was observed in the absence of intracellular polyamines or Mg (Shieh *et al.*, 1996). The intrinsic gating mechanism had much slower kinetics and less voltage dependence than divalent cation- or polyamine-induced rectification, making it unlikely to be physiologically important on its own. However, this gating mechanism was eliminated by mutating the same two charged amino acids that confer divalent and polyamine sensitivity to Kir2.1 to neutral residues (Lee *et al.*, 1997). One possibility currently under evaluation is that this intrinsic gating process is the fundamental mechanism of inward rectification in Kir2.1, and that polyamines and/or divalents act by binding to and enhancing the effectiveness of the intrinsic gate.

Like voltage-gated K channels, some of the inward rectifier K channels form heterotetramers composed of more than one protein (e.g., Kir3). So far, only one β-subunit for Kir channels has been identified, the sulfonylurea receptor (SUR), which does not form a channel by itself but coassociates with Kir6 (Inagaki *et al.*, 1995a) and Kir1 channels (Ammala *et al.*, 1996).

b. I_{K1}

I_{K1} is the background strong inward rectifier K current present in the heart and shares similar properties to strong inward rectifiers in many other tissues. As discussed in the previous section, the candidate clone for the native cardiac K1 channels is the Kir2.1 (or IRK1) channel. In the heart,

K1 channels stabilize the resting membrane potential near E_K, due to their high membrane conductance when E_m is near E_K. As E_m is driven positive to E_K during depolarization, K1 channel conductance progressively decreases due to inward rectification, increasing the effectiveness of the Na current at depolarizing the cell, and minimizing the amount of net cellular K loss during the plateau when the driving force for K efflux ($E_m - E_K$) is very large. Preventing unnecessary net K loss during the long cardiac action potential plateau is important from an energetic standpoint because of the need to pump the lost K back into the cell, using NaKATPase, in the steady state. The negative slope region (anomalous rectification) in the I–V curve of I_{K1} (see Fig. 10B), which results from the voltage dependence of inward rectification, is also extremely important for cardiac repolarization. As the membrane potential gradually repolarizes during phase 2 of the action potential, the negative slope region produces a positive feedback in outward K current, accelerating the rate of repolarization in its final phase (phase 3).

There is no evidence that the native cardiac I_{K1} is regulated to an important extent by PKA or PKC pathways. Kir2.1, however, is regulated by both PKA and PKC (Fakler *et al.*, 1994). I_{K1} is regulated by intracellular MgATP levels and usually runs down quickly in excised membrane patches when MgATP is removed. I_{K1} also decreases markedly during metabolic inhibition, at the same time as K_{ATP} channels are activated (Trube and Hescheler, 1984). The mechanism for this sensitivity to metabolic factors is unclear at the present time.

c. Muscarinic K Channels

Cardiac K_{ACh} channels are present in high density in the atria, SA node, and AV node, and are a major target of the vagal autonomic system in the heart. Release of ACh from vagal nerve fibers causes slowing of sinus rate, prolongation of A-V conduction time, and shortening of atrial refractoriness, in part mediated by activation of K_{ACh} channels. By binding to cardiac M_2 receptors, ACh activates the GTPase activity of trimeric G_i proteins, causing dissociation of $G_{\beta\gamma}$ from $G_{\alpha i}$ subunits. Although there was an initial controversy over whether $G_{\beta\gamma}$ or $G_{\alpha i}$ subunits stimulated the K_{ACh} channel, it is now clear that $G_{\beta\gamma}$ subunits directly bind to K_{ACh} channels to cause an increase in their open probability (Reuveny *et al.*, 1994). One study shows that specific $G_{\alpha i}$ subunits (e.g., $G_{\alpha i1}$) potently inhibit the activation of K_{ACh} channels by $G_{\beta\gamma}$ subunits (Schreibmayer *et al.*, 1996). Upon activation of K_{ACh} channels, the resulting increase in membrane K conductance causes a hyperpolarization of resting membrane potential, which shifts the site of the dominant pacemaker to slower beating cells in the SA node (Hoffman and Cranefield, 1960), shortens the atrial action potential duration, and slows conduction speed through the AV node. K_{ACh} channels have a single-channel conductance of $\sim$25 pS and are strong inward rectifiers blocked with high affinity by intracellular divalent cations and polyamines. Their open probability is relatively voltage independent.

Current evidence indicates that the native cardiac K_{ACh} is a heterotetramer consisting of two Kir3.1 (GIRK1) and two Kir3.4 (CIR or rcKATP) molecules (Krapivinsky et al., 1995). $G_{\beta\gamma}$ subunits appear to bind to both protein molecules, and, in Kir3.1, a binding region has been localized to the C-terminus (Reuveny et al., 1994). It is presumed that the binding of $G_{\beta\gamma}$ to the Kir3 subunits induces a conformational change favoring the open state; however, evidence for an indirect stimulation involving phospholipase A_2 activation has also been proposed (Kurachi et al., 1989).

d. ATP-Sensitive K Channels

Although first discovered in the heart (Noma, 1983), ATP-sensitive K channels are present in many different tissues where they play key roles in the regulation of insulin release, vasomotor and smooth muscle tone, neuroexcitability, skeletal muscle excitability, and cation transport (Ashcroft and Ashcroft, 1990). These channels are closed by cytosolic ATP and act as a metabolic sensor for the cell by opening when cellular ATP levels fall. However, their activity is also modulated by a variety of hormones, especially in vascular and smooth muscle. In the heart, K_{ATP} channels are normally closed and play little role in regulating action potential duration or other physiological functions, except in atria where they can be activated by muscarinic stimulation under special conditions (Wang and Lipsius, 1995). During myocardial ischemia, hypoxia, or metabolic inhibition, K_{ATP} channels open in response to the decrease in the cytosolic ATP/ADP ratio and the release of adenosine and other factors, and cause progressive shortening of the action potential duration, eventually leading to inexcitability. Because of the high density of cardiac K_{ATP} channels and their property of weak rectification (see Fig. 10), they pass considerable outward current at plateau potentials; an increase of their open probability to <1% of maximum is sufficient to shorten the ventricular action potential duration by 50% (Nichols et al., 1991; Weiss et al., 1992). Shortening of the APD during ischemia has a cardioprotective effect by reducing contractile force and hence metabolic energy consumption in the ischemic region. Activation of K_{ATP} channels plays an important role in the phenomenon of ischemic preconditioning, which can be mimicked by preexposure to K_{ATP} channel agonists in some species (Yao et al., 1993).

In the heart, K_{ATP} channels have a single-channel conductance of ~40 pS with physiological K concentrations, and show weak inward rectification in response to mM intracellular Mg and no sensitivity to polyamines. K_{ATP} channels are suppressed by ATP in the presence or absence of Mg and by nonmetabolizable analogues of ATP such as AMP-PNPP, suggesting that the mechanism of suppression is not dependent on a phosphorylation step

(Ashcroft and Ashcroft, 1990). The $K_{0.5}$ for suppression of K_{ATP} channels by intracellular MgATP is 15–50 μM, well below the 5–10 mM cytosolic [ATP] normally present. However, μM MgADP significantly shifts the $K_{0.5}$ to higher values (~100 μM), and the cytosolic ATP/ADP ratio is physiologically more important than ATP alone as the regulator of the channel open probability. The open probability of K_{ATP} channels is also increased by adenosine receptor stimulation via direct stimulation by $G_{\alpha i}$ (Kirsch *et al.*, 1990), and in atria by acetylcholine, via activation of phospholipase C in the presence of adequate intracellular Ca stores (Wang and Lipsius, 1995). In the heart, there is no apparent regulation of K_{ATP} channels by PKA. However, there is evidence that PKC may desensitize K_{ATP} channels to ATP, providing a possible link between PKC activation and $I_{K(ATP)}$ activation during ischemic preconditioning (Hu *et al.*, 1996; Liu *et al.*, 1996). Metabolic inhibition can also directly desensitize K_{ATP} channels to suppression by ATP, probably due to proteolytic digestion of the cytosolic regions by Ca-dependant proteases, since treatment of membrane patches with trypsin has a similar effect (Deutsch and Weiss, 1993). The sensitivity of K_{ATP} channels to suppression by ATP appears to be modulated by the density of negative surface charges near the ATP-binding sites, since surface charge screening agents have marked effects on ATP sensitivity (Deutsch *et al.*, 1994).

Like other inward rectifier K channels, K_{ATP} channels run down rapidly in the absence of cytosolic MgATP. The mechanism of rundown does not appear to involve dephosphorylation of the channel, since phosphatase inhibitors do not protect against rundown and protein kinases do not restore K_{ATP} channel activity. Recent evidence indicates that MgATP prevents rundown by preventing actin depolymerization (Furukawa *et al.*, 1996); PIP$_2$ metabolism appears to play a key role, possibly by maintaining cytoskeletal elements (Hilgemann and Ball, 1996). Indirect evidence that cytoskeleton may be important in K_{ATP} channel function also derives from the observation that atrial K_{ATP} channels are activated by stretch (i.e., they are mechanosensitive) (van Wagoner, 1993). Partially rundown K_{ATP} channels can also be reactivated by nucleotide diphosphates (i.e., UDP, ADP) in the presence of Mg (Tung and Kurachi, 1991).

K_{ATP} channels are blocked by Ba and Cs, like other inward rectifiers. Specific antagonists include sulfonylureas (i.e., glibenclamide, glipizide, tolbutamide) and 5-hydroxyl-decanoate; specific agonists are K channel openers (i.e., cromakalim, lemakalim, pinacidil, aprikalim, nicorandil) and diazoxide.

A candidate clone for the cardiac K_{ATP} channel (Kir6.1 or μKATP) was reported by Seino and coworkers (Inagaki *et al.*, 1995b). The expressed channel was suppressed by ATP and activated by K channel openers, but not sensitive to sulfonylureas. Recently, the pancreatic β cell K_{ATP} channel was reconstituted by coexpressing the sulfonylurea receptor (SUR) with a putative β cell inward rectifier K channel (Kir6.2 or BIR), which did not form an active channel when expressed alone (Inagaki et al., 1995a). These results suggest that the cardiac K_{ATP} is likely to be a heteromultimer of SUR and another Kir subunit such as Kir6.1. SUR is a member of a super-family of 12 transmembrane-spanning proteins known as ABC-binding cassette proteins, which also includes CFTR (cystic fibrosis transmembrane regulator) and multidrug resistance P-gp (Higgins, 1995). Since SUR does not form functional channels when expressed alone, it has been classified as a β-subunit of the Kir channels. Recent evidence indicates that SUR can form heteromultimers conferring sulfonylurea sensitivity to Kir1, as well as to Kir6 channels (Ammala et al., 1996). SUR modifies the gating properties Kir6.2; a mutation in the nucleotide binding region of the SUR protein found in children with persistent hyperinsulinemic hypoglycemia of infancy syndrome was shown to remove the ability of MgADP to desensitize reconstituted SUR/Kir6.2 channels to suppression by ATP (Nichols et al., 1996). As a result, pancreatic β cell K_{ATP} channels fail to activate appropri-ately when serum glucose concentration falls and the cytosolic ATP/ADP ratio decreases. The persistent suppression of K_{ATP} channels promotes β cell depolarization and consequent insulin release at inappropriately low serum glucose concentrations in these children.

e. Na-Activated K Channels

A Na-activated K channel has been described in cardiac muscle (Kame-yama et al., 1984; Wang et al., 1991), which is activated by increases in intracellular Na, with half-maximal activation at $[Na]_i = 66$ mM and a Hill coefficient of 2.8. The single-channel conductance at physiological K concentrations is ~180 pS, with weak Mg_i- and Na_i-dependent inward rectification. The channel is blocked by external Cs and Ba, and by the selective agent R56865. The physiological and pathophysiological role of K_{Na} channels is uncertain due to the high level of $[Na]_i$ required to activate them. However, evidence has been presented that during Na-pump inhibi-tion, subsarcolemmal [Na] can rise sufficiently to activate these channels, despite only modest increases in bulk cytosolic [Na] (Carmeliet, 1992). The molecular identity of this channel is unknown.

f. Arachidonic Acid-Activated K Channels

Outwardly rectifying K channels activated by arachidonic acid and phos-photidylcholine have been described in rat atrial and ventricular cells, with single-channel conductances of 160 and 60 pS, respectively (Kim and Clap-ham, 1989; Kim and Duff, 1990). It has been speculated that they may be activated during myocardial ischemia, although significant accumulation of nonesterified fatty acids does not occur until late (>30 minutes) into isch-emia (Chien et al., 1984). Neither of these channels has yet been cloned.

g. Ca-Activated K Channels

Ca-activated K channels, also known as maxi-K channels because of their large single-channel conductance, are present in a variety of excitable cells including neural tissue, smooth muscle, endothelium, and pancreatic β cells. They play a key role as modulators of excitation–secretion coupling. In the heart, a large conductance Ca-activated K channel has been described in Purkinje cells, with a single-channel conductance of 120 pS (Callewaert *et al.*, 1986), and may be a component of I_{to2}. Although the maxi-K channels have been cloned from other tissues, the cardiac form has not yet been identified.

In summary, inward rectifier K channels, most of which are ligand-gated, consist of a diverse group of cardiac K channels. A number of these channels have been cloned and have a similar predicted topological structure, with two membrane-spanning segments connected extracellularly by a P region similar to the P region of voltage-gated ion channels (see Fig. 12). Whether a channel demonstrates strong or weak inward rectification (see Fig. 10) is determined by specific negatively charged amino acid residues in the second transmembrane (M2) region and in the C-terminus. These residues regulate the sensitivity of the channels to voltage-dependent block by intracellular Mg and polyamines, the primary agents responsible for inward rectification. Strong inward rectifiers, K1 and K_{ACh} channels, regulate resting membrane potential and action potential duration in heart. K_{ACh} channels mediate many of the effects of vagal stimulation. Their open probability is directly regulated by G proteins coupled to muscarinic and purinergic receptors. Weak inward rectifiers, including K_{ATP}, K_{Na}, K_{AA}, and K_{Ca} channels, primarily act to shorten the cardiac action potential and/or reduce excitability. They are primarily activated under pathophysiological conditions and may be important during ischemia, reperfusion, and intracellular Ca or Na overload.

2. Cl Channels

a. cAMP-Regulated Cl Channels

β-adrenergic receptor stimulation has been shown to activate a Cl current ($I_{Cl(cAMP)}$) in the heart via the cAMP–PKA pathway (Hume and Harvey, 1991). Since E_{Cl} is normally about -50 mV in heart, activation of this current has two important effects in the β-adrenergic response: (1) depolarization of the resting membrane potential, which enhances excitability; and (2) shortening of action potential duration, which preserves diastole (and coronary blood flow) at faster heart rates. The regulation of $I_{Cl(cAMP)}$ involves phosphorylation by PKA, rather than a direct G protein interaction. A PKC-activated Cl current, with similar but not identical properties, has also been described in the heart (Walsh and Long, 1994).

The cardiac Cl_{cAMP} channel is an alternatively spliced isoform of CFTR (Horowitz *et al.*, 1993), which was originally cloned from epithelium and contains the genetic defect in patients with cystic fibrosis (Hanrahan *et al.*, 1995). CFTR is a member of a super-family of proteins with 12 transmembrane-spanning segments that share the property of regulating other ion channels; other members include SUR and multidrug resistance P-gp (Higgins, 1995). Each CFTR channel has five domains: two transmembrane domains, each with six putative transmembrane-spanning segments, form the anion channel; two nucleotide-binding domains interact with and likely hydrolyze ATP, which is required for channel activity; and a fifth regulatory domain contains the consensus PKA phosphorylation sites, which are important for activating the channel by PKA (Rich *et al.*, 1993). The single-channel conductance of cardiac CFTR channels is 13 pS in symmetrical [Cl]. In addition to Cl, ATP also has been shown to permeate CFTR channels in some studies. When coexpressed with an amiloride-sensitive Na channel, CFTR modified its properties (Gabriel *et al.*, 1993), analogous to the effects of SUR on Kir channels. However, there is no evidence that CFTR modifies the properties of any other cardiac ion channels. $I_{Cl(cAMP)}$ is blocked by disulfonic stilbene compounds such as DIDS, SITS, and 9-anthracine carboxylic acid (9-AC). Some reports have also indicated a sensitivity to sulfonylureas (Tominaga *et al.*, 1995), reinforcing its genetic super-family kinship to SUR.

b. Ca-Activated Cl Channels

A Ca-activated Cl current has been described in rabbit and canine ventricular myocytes (Zygmunt and Gibbons, 1991; Collier *et al.*, 1996), which has been proposed to be a major component of the Ca-sensitive I_{to} (I_{to2}). With I_{to1}, I_{to2} contributes to phase 1 early rapid repolarization of the action potential and may also contribute to the arrhythmogenic transient inward current (I_{ti}) under conditions of Ca overload (Hume and Harvey, 1991). The whole cell current is outwardly rectifying due to the normally asymmetrical [Cl] gradient and is sensitive to block by disulfonide stilbenes (e.g., DIDS and SITS) and nifumic acid. Recent single-channel studies have shown a linear I–V relationship in symmetrical [Cl] and a single-channel conductance at 1–1.3 pS (Collier *et al.*, 1996). Half-maximal activation occurred at a [Ca]$_i$ of ~150 μM, with a Hill coefficient of 3. Similar Ca-activated Cl channels have also been characterized in oocytes, endocrine, and smooth muscle cells. No candidate clones have yet been identified.

In summary, two types of ligand-gated Cl channels have been described in the heart, gated by cAMP or by intracellular Ca, respectively. The cAMP-gated Cl channel is an alternatively spliced variant of CFTR. It plays an important role in regulating action potential duration and excitability during β-adrenergic stimulation. The Ca-activated Cl current is a component of

I_{to2}, which regulates cardiac action potential duration, and may contribute to the arrhythmogenic transient inward current I_{ti} in the setting of intracellular Ca overload.

3. Ca-Activated Nonselective Cation Channels

Ca-activated nonselective cation channels (NSC_{Ca}) were first discovered in cardiac cells (Colquhoun *et al.*, 1981) and subsequently characterized in detail (Ehara *et al.*, 1988). The I–V relationship is linear with a single-channel conductance of ~15 pS and with nearly equal permeability to Na, K, and Cs. Half-maximal activation occurs at 1.2 μM Ca_i, with a Hill coefficient of 3. Along with the Na–Ca exchange current (I_{NaCaX}) and possibly the Ca-activated Cl current, $I_{NSC(Ca)}$ has been implicated as a component of the transient inward current (I_{ti}), which causes delayed and early afterdepolarizations and triggered activity in the setting of intracellular Ca overload. In simulation studies, the relative contributions of $I_{NSC(Ca)}$ and I_{NaCaX} to I_{ti} are variable depending on the setting (Luo and Rudy, 1994). $I_{NSC(Ca)}$ may also contribute to the background "leak" cation conductance in cardiac cells. The molecular identity of NSC_{Ca} channels is unknown.

C. SARCOLEMMAL MECHANOSENSITIVE ION CHANNELS

A number of mechanosensitive ion channels have been characterized in cardiac tissue, predominantly in atrial tissue (for reviews, see Sachs, 1989; Morris, 1990; Sackin, 1995). These include stretch-activated nonselective cation and K and Cl channels. Stretch-inactivated K channels have also been described (Morris and Sigurdson, 1989) and were identified in atrial cells in one preliminary report (van Wagoner, 1991). Mechanosensitive channels may be important in a variety of cellular functions, including osmoregulation, endocrine function (e.g., ANP release from the atria), and gene regulation in growth and hypertrophy. Stretch-activated nonselective cation and Cl channels may also promote stretch-induced arrhythmias by causing membrane depolarization. It is likely that mechanosensitivity is conferred by the cytoskeletal elements, which transduce changes in membrane tension via direct attachments to the channel protein. Microtubular agents were found to have no effect on stretch-activated channels in chick muscle. Actin was also not the primary mechanotransducer, since agents causing actin depolymerization increased mechanosensitivity, suggesting that the actin microfilaments acted as a secondary support structure resisting changes in membrane tension. Intermediate filaments such as spectrin are more likely candidates. Although no specific inhibitors of intermediate filaments are available, in chick muscle, the appearance of spectrin and stretch-activated channels paralleled each other during development, suggesting a possible causal relationship (Sachs, 1989). Fatty acids also have activated some mechanosensitive ion channels, possibly through altering membrane tension by changing membrane curvature (Sackin, 1995).

Among the cardiac mechanosensitive currents, a nonselective cation current permeable to divalent as well as monovalent cations has been described (Stern, 1992). $I_{K(ATP)}$ in atrial myocytes (van Wagoner, 1993) and I_{Ks} in ventricular myocytes (Wang *et al.*, 1996) have been activated by cell swelling or membrane stretch. A cell swelling-induced Cl current has also been described in SA node and atrium (Sorota, 1992; Hagiwara *et al.*, 1992).

D. SARCOLEMMAL BACKGROUND LEAK CHANNELS

Among cardiac ion channels, leak and background channels are the least well characterized. However, background conductances are significant, since the resting membrane potential is at least 10–15 mV positive to the K equilibrium potential even in atrial and ventricular myocardium. A background tetrodotoxin-sensitive Na conductance is present in the heart (Saint *et al.*, 1992), and other background Na currents may also be present. Ca-activated nonselective cation channels may be partially active at physiological diastolic Ca levels (Ehara *et al.*, 1988). Background Cl channels have been described in rabbit atrium (Duan and Nattel, 1994). A Ca leak channel is present in ventricle, which may influence resting Ca levels (Coulombe *et al.*, 1989), especially during metabolic poisoning where they have been implicated as a cause of cellular Ca overload (S.Y. Wang *et al.*, 1995). In the SA node, a sustained inward current modulated by isoproterenol and inhibited by high $[Ca]_0$, and also, paradoxically, by Ca channel blockers, has been described that contributes to the maintenance of the low maximum diastolic membrane potential in these cells (Guo *et al.*, 1995).

E. ION CHANNELS IN INTRACELLULAR ORGANELLES

1. Sarcoplasmic Reticulum

Intracellular ion channels play key roles in signaling pathways involving intracellular organelles. Of paramount importance in the heart are the ion channels in the SR involved in Ca release. The cardiac ryanodine receptor (RyR2), the major Ca release channel of the SR, is activated primarily by Ca influx through L-type Ca channels. RyR2 receptors in the junctional SR membrane are physically located adjacent to Ca channels in the T tubules. This arrangement provides a local control mechanism for Ca release, as discussed in Chapter 5. RyR2 receptors are also functionally coupled to calsequestrin, a major Ca-binding protein in the SR lumen, presumably to facilitate Ca release from the SR lumen (Guo et al., 1996). It has been proposed that triadin, another SR protein, mediates this coupling.

The cardiac RyR2 is one of three mammalian RyR isoforms, the other two corresponding to skeletal muscle (RyR1) and brain (RyR3) (Meissner, 1995). The cardiac RyR2 has been cloned and is a ~5000 amino acid protein with a molecular weight of ~565 kD (Nakai et al., 1990). RyRs assemble as homotetramers with a quadrafoil structure to form functional Ca release channels (Fig. 13). The pore is in the C-terminal region, and the large extramembrane N-terminus forms the foot structure in the SR–T tubule junction. The channel is permeable to both divalent and monovalent cations and has a very large single-channel conductance (~150 and 750 pS for Ca and K, respectively). The native Ca release channel also contains a sub-

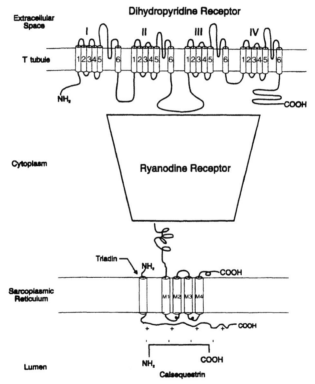

FIGURE 13   Schematic of the ryanodine receptor (RyR) in the SR–T tubule junction in the heart muscle. The foot structure of the RyR is in close proximity to the L-type Ca channel in the T tubular membrane. Associated sarcoplasmic reticulum (SR) proteins include triadin and calsequestrin. FKBP12 (not shown) also binds to the RyR. (Reproduced from *The Organellar Ion Channels and Transporters, 49th Annual Symposium of The Society of General Physiologists*, pp. 19–28, 1996, by copyright permission of The Rockefeller University Press.)

unit, FKBP12, bound to the RyR2 tetramer. In bilayers, FKBP12 modifies gating properties of RyR2 to resemble more closely the native Ca release channel, by reducing the frequency of subconductance states (Ondrias *et al.*, 1996).

The ability of cytosolic Ca to activate the channel is modulated by cytosolic ATP, Mg, and pH. Steady-state activation by Ca is biphasic, reaching a peak at μM and declining at mM Ca. However, in experiments using photorelease of caged Ca to track the kinetic response to rapid increases in Ca, RyR2s demonstrated adaptation (Gyorke and Fill, 1993). That is, in response to a rapid change in cytosolic Ca, the open probability increased transiently before declining to a steady-state level, at which point a further increase in Ca was capable of inducing a new transient increase in channel activity. Adaptation may be an important mechanism for preventing positive feedback of Ca release from SR (see Chapter 5).

Ca release channels are activated by caffeine and ryanodine (see Chapter 5). At low concentrations, ryanodine stabilizes a long-lasting open subconductance state of the RyR2, making the SR leaky and unable to retain stored Ca. At higher concentrations, ryanodine blocks the channels. Ruthenium red also blocks SR Ca release channels.

Another Ca release channel present in the SR and endoplasmic reticulum is the inositol-triphosphate receptor (IP$_3$R). A family of IP$_3$Rs has been cloned whose members are ~2700 amino acid proteins with molecular weights around 300 kD (Berridge, 1993). Like RyRs, they assemble as homotetramers with a quadrafoil structure. In many different cell types, IP$_3$Rs are the major mediators of intracellular Ca release that trigger excitation–secretion coupling. IP$_3$ production is coupled to activation of phospholipase C by various receptors in the cell membrane, including α-adrenergic and muscarinic receptors. In the heart, however, α-adrenergic stimulation has small inotropic effects, and IP$_3$Rs appear to play only a minor modulatory role in cardiac excitation–contraction (E–C) coupling.

Other ion channels in the cardiac SR include a large conductance K channel and a Cl channel. These channels play important roles in balancing charge movement associated with Ca efflux (Meissner, 1995).

2. Mitochondria and Nucleus

In addition to many enzymes and transporters required for ATP synthesis by the electron transport chain, mitochondria also contain a number of ion channels. The voltage-dependent anion channel (VDAC) located in the outer membrane is a high-conductance channel (single-channel conductance ~4 nS in 1 M KCl), which weakly favors anions over cations (2:1 Cl vs K permeability), and is permeable to molecules up to a molecular weight of ~1000 (Forte *et al.*, 1996). The VDAC channel is believed to function as the primary pathway for movement of adenine nucleotides and other metabolites. The VDAC channel in yeast has been cloned. In the inner mitochondrial membrane, at least five types of ion channels have been discovered. These include alkaline-activated anion (AAA) and cation (ACA) channels, the anionic centum pico-siemens channel (mCS), the mitochondrial megachannel/multiconductance channel (MMC/MCC), and an ATP-sensitive K channel (K_{ATP}). The physiological roles of these channels are still largely speculative (for a review of their properties, see Ballarin *et al.*, 1996).

A variety of ion channels has also been described in the nuclear envelope and may be important for transducing signals from the cytosol to the nucleus regulating gene function. The nuclear envelope consists of an outer membrane facing the cytosol and an inner membrane that interacts with the nuclear lamina. The space between the two membranes is continuous with the endoplasmic reticulum. The ion channels that have been described in the nuclear membranes include a variety of K channels, nonselective cation channels, Cl channels, and IP_3R Ca release channels (for review, see Stehno-Bittel *et al.*, 1996).

In summary, ion channels in intracellular organelles play important roles in intracellular signaling. In the heart, the best-understood organellar ion channel is the RyR2 Ca release channel located in the SR, which plays a key role in E–C coupling. RyR2 channels are localized in close proximity to L-type Ca channels in the SR–T tubule junction (see Fig. 13). This specialized arrangement is believed to provide a local control mechanism for regulating SR Ca release (see Chapter 5). IP_3R Ca release channels are also located in cardiac SR, but appear to play only a minor role in modulating E–C coupling. K and Cl channels in SR are important for balancing charge movement during Ca release. Numerous ion channels are also present in mitochondrial and nuclear membranes in the heart, although currently their roles are not completely understood.

F. INTERCELLULAR ION CHANNELS (CONNEXINS)

Low-resistance pathways (gap junctions) connecting adjacent myocytes are essential for propagation of the action potential from cell to cell. Most cardiac gap junctions are located in the intercalated discs, which mechanically couple adjacent myocytes. The gap junction itself consists of closely apposed sections of the membranes from adjacent myocytes, separated by a narrow 20 Å extracellular gap filled with hexagonally arrayed protein subunits. The regional density of gap junctions varies considerably, from only a few in SA nodal cells to ~100 per cell in ventricular myocytes. Compared to transsarcolemmal resistivity, gap junctional resistivity is low, but it is higher than cytoplasmic resistivity, and so constitutes the major component of intercellular resistivity. Cardiac cells have a cylindrical shape, so that the number of gap junctions per unit length in the direction transverse to fiber orientation is significantly higher than in the longitudinal direction. This geometry results in a higher intercellular resistivity in the transverse rather than the longitudinal direction. Conduction velocity is therefore slower in the transverse than the longitudinal direction, a property known as *anisotropy*. In ventricular myocardium, in which gap junctions are distributed fairly evenly over the surface of the cell, the anisotropy ratio is about 3 to 1. In the atrial crista terminalis, in which 80% of the gap junctions are at the ends of the cells, the anisotropy ratio is 10 to 1 (Saffitz *et al.*, 1994).

Gap junction proteins belong to a multigene family known as connexins (for review, see Beyer *et al.*, 1995; Jongsma and Rook, 1995). About 15 connexins have been described, based on their molecular weight (in kD). In the heart, connexin43 is the most prevalent, with smaller amounts of connexin40 and connexin45, and perhaps several others. There are regional differences in the relative level of expression of different connexins. Connexin43 expression was ubiquitous except for the SA node, His bundle, and bundle branches in one study. Connexin40 is expressed at higher levels in the His-Purkinje system than in the ventricle. Each connexin has four putative transmembrane-spanning segments and a large C-terminus. Six connexins assemble to form a hemichannel, known as a *connexon*. Two connexons from apposed membranes form the aqueous pore between the cytoplasm of the adjacent cells (Fig. 14). Single-channel studies show multiple subconductance states with single-channel conductances ranging from 20–150 pS. One model hypothesizes that gating between open and closed conformations is related to the degree of twist of the transmembrane segments, acting like the iris of a camera. The permeability of the pore to different-size molecules varies between connexins; connexin43 is permeable to both ions and fluorescent dyes like Lucifer yellow, whereas connexin45 only allows passage of ions. The open probability of connexins is influenced by transjunctional voltage to varying extents; connexin40 shows pronounced voltage-dependent inactivation, whereas connexin43 does not. However, significant voltage differences between adjacent cardiac myocytes exist only briefly during the upstroke, so that the physiological significance of transjunctional voltage effects on gap junction conductance is uncertain.

Physiologically important mediators of gap junction conductance include intracellular [H] and [Ca] (Jongsma and Rook, 1995). Both acidosis and elevated $[Ca]_i$ reduce gap junction conductance and are important causes of increased intercellular resistance during acute myocardial ischemia (Cascio *et al.*, 1990). Ultimately, this serves a protective function, by electrically isolating arrhythmogenic Ca overloaded myocytes from their normal neighbors. Lipophilic agents such as halothane, hepatanol, and octanol reduce gap junctional resistance.

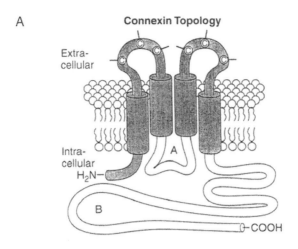

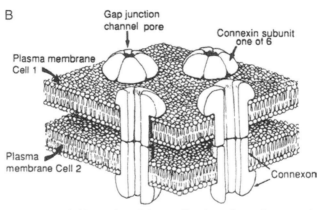

FIGURE 14   Schematic illustration of the predicted topology of a connexin. (A) Highly conserved cysteine residues (C) in the extracellular loops are essential for (B) the hexameric assembly of connexins into connexons. Each connexon is a hemichannel and docks with connexon in the apposing membrane to form an intercellular ion channel. (Adapted from Beyer, 1993, and Jongsma and Rook, 1995, with permission.)

In summary, conduction of the cardiac impulse is critically dependent on low-resistance pathways between adjacent cardiac myocytes. These low-resistance pathways are provided by connexins, a family of proteins that form hexameric hemichannels called connexons (see Fig. 14). Connexons in one myocyte link with connexons in an adjacent myocyte at gap junctions, where the sarcolemmal membranes are closely apposed, providing a continuous aqueous (low-resistance) pathway between the cells. Several different connexins are expressed in the heart. Although functional differences exist between them, the consequences with respect to cardiac function are not well understood.

III. CARDIAC ACTION POTENTIAL:
FORMATION AND PROPAGATION

A. GENESIS OF CARDIAC ACTION POTENTIAL

The normal cardiac action potential is produced by the coordinated interaction of the many ion channels discussed in the previous section, as well as currents generated from electrogenic transport proteins such as Na–Ca exchange and the Na–K pump. Consistent with the heterogeneous expression of ion channels in different regions of the heart, the morphology of the action potential varies markedly. Even within the same region, significant variations are present. From the dominant pacemaker cells in the SA node to perinodal transitional cells to atrium is a continuous gradation of action potential characteristics (Fig. 15). The AV node is subdivided into AN, N, and NH regions, and at least six different cell types distinguished by different action potential characteristics have been identified (Shrier *et al.*, 1995) (Fig. 16). His-Purkinje action potentials differ from ventricular action potentials, and within ventricular myocardium, three distinct types of action potentials exist in the epicardium, midmyocardium, and endocardium

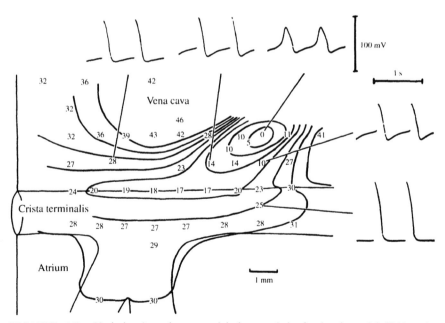

FIGURE 15 Variation in action potential characteristics in the sinoatrial (SA) node. The sinus impulse originates at 0 and conducts outward along the isochrone lines indicated by the numbers. Action potentials are shown at selected distances as the impulse conducts through perinodal tissue into the crista terminalis and atrium, as labeled. (Reprinted from Sano and Yamagishi, 1965, with permission.)

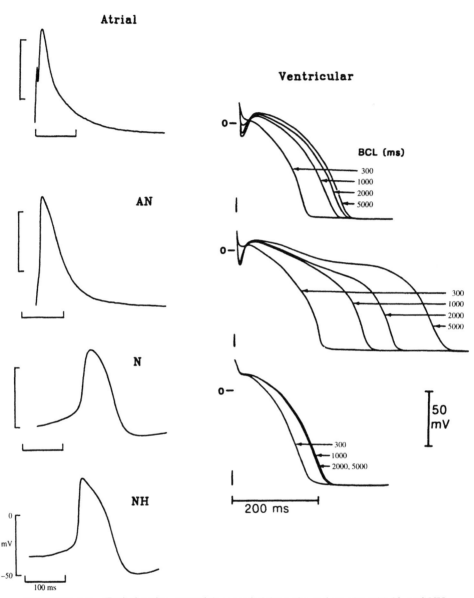

FIGURE 16 Typical action potentials recorded from the atrium, the AN, N, and NH regions of the AV node, and from the ventricular epicardium (*top*), midmyocardium (*middle*), and endocardium (*bottom*). The ventricular action potentials at several different pacing cycle lengths (300, 1000, 2000, and 5000 ms) are superimposed, illustrating the marked prolongation of the action potential in the midmyocardium at slow heart rates. (Adapted from Shrier *et al.*, 1995, and Antzelovitch *et al.*, 1995, with permission.)

(see Fig. 16). The time course of various currents during a typical endocardial ventricular action potential is shown in Fig. 17, using the Luo–Rudy
action potential model (Zeng et al., 1995). The action potential is classically
divided into five phases (see Fig. 17) as follows.

1. Phase 0 (Action Potential Upstroke)

In cardiac cells that are well polarized, including atrial, His-Purkinje,
and ventricular myocytes, the upstroke of the action potential is carried
predominantly by the Na current. As the membrane potential depolarizes,
outward I_{K1} becomes progressively smaller, allowing membrane potential
to overshoot 0 mV and approach E_{Na}. The quasi-instantaneous turning off
of I_{K1} is important for maximizing the rate of depolarization (dV/dt), which
is a major determinent of conduction velocity (0.2–0.6 m/s in atrial and
ventricular muscle, and up to 4 m/s in the His-Purkinje system). At the
normal resting potential of these cells near -80 mV, Na channels are partly
inactivated. Application of hyperpolarizing current removes inactivation
in many of these channels, so that when the hyperpolarizing current is
terminated, the larger number of available Na channels can lead to spontaneous activation of the Na current, known as anodal break excitation.

In less polarized SA nodal and AV nodal cells whose maximum diastolic
potential is -50 to -60 mV, the action potential upstroke is carried predominantly by the L-type Ca current. The rate of membrane depolarization is
much slower, contributing to the very slow conduction velocity in these
tissues (0.01–0.1 m/s). In perinodal regions of the SA and AV nodes, the
upstroke is carried by a mixture of Ca and Na current.

2. Phase 1 (Early Rapid Repolarization)

After the action potential reaches its peak amplitude (or maximum
overshoot), a rapid early repolarization phase occurs that is attributed
primarily to the activation of the Ca-independent and Ca-dependent components of the transient outward current I_{to}. The magnitude of I_{to} varies
markedly at different locations in the heart and also is modulated dynami-
cally by heart rate and other factors. I_{to} density is highest in atrial myocytes,
accounting for their short spikey atrial action potential, and in His-Purkinje
and epicardial ventricular myocytes, producing a characteristic spike-and-
dome appearance of the plateau (see Fig. 16). I_{to} density is intermediate
in midmyocardial ventricular myocytes (M cells) and almost absent in
endocardial myocytes. The presence of I_{to} in ventricular epicardium is a
major reason for the shorter epicardial than endocardial action potential
duration. Under conditions in which I_{to} is potentiated (e.g., hypothermia,
hypercalcemia), the accentuated spike-and-dome becomes manifest in the
surface electrocardiogram as the J wave (Antzelevitch et al., 1995).

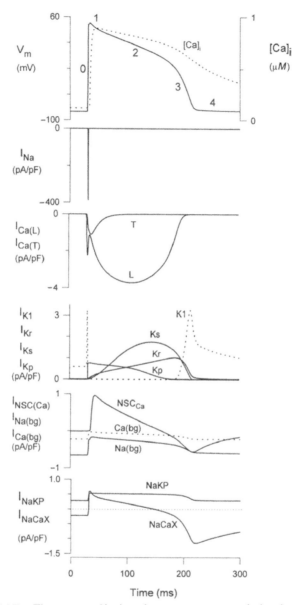

FIGURE 17   Time course of ionic and transporter currents during the ventricular cardiac action potential, simulated using the Luo–Rudy model (Zeng *et al.*, 1995). I_{NaCaX}, Na–Ca exchange current; I_{NaKP}, Na–K pump current; bg, background. Phases 0–4 of the action potential are indicated in the top trace.

3. Phase 2 (Slow Repolarization)

The slow repolarization phase occupies the majority of the action potential plateau, except in atrium where this phase is very short (see Fig. 16). The major dynamic determinants of the rate of slow repolarization during phase 2 are the inactivation of the L-type Ca current and the activation of delayed rectifier K currents. The dynamic influence of Ca-induced inactivation of the L-type Ca current is manifestly evident during mechanical alternans, in which the magnitude of the Ca transient varies between alternate beats. For example, a sudden stepwise increase in heart rate often produces a transient alternans pattern in which the beats with large Ca transients are associated with shorter action potentials due to the enhanced Ca-induced inactivation of $I_{Ca(L)}$.

The relative contributions of I_{Ks} and I_{Kr} to repolarization vary considerably between different regions of the heart and among different species. In the ventricular wall, decreased expression of I_{Ks} in the midmyocardium relative to the epi- and endocardium is the major reason for the longer action potential duration of the M cells (Liu and Antzelevitch, 1995). Delayed repolarization of M cells is now believed to be the explanation for the U wave on the surface electrocardiogram, rather than delayed repolarization of the His-Purkinje system, as previously believed (Antzelevitch *et al.,* 1995). Also contributing to plateau currents during phase 2 are I_{Kp}, I_{Kur} (in atria), Na window currents, I_{Cl}, $I_{NSC(Ca)}$, and the Na–Ca exchange current (see Fig. 17). Along with $I_{Ca(L)}$ and I_{Ks}, I_{Cl} is important in the β-adrenergic regulation of action potential duration. Na window currents, $I_{NSC(Ca)}$, and I_{NaCaX} can have important effects on phase 2 (and phase 3) under pathophysiological conditions. $I_{K(ATP)}$, $I_{K(Na)}$, and $I_{K(AA)}$ also can markedly shorten the duration of phase 2 when activated under pathophysiological conditions.

4. Phase 3 (Late Rapid Repolarization)

Toward the end of the plateau phase, the rate of repolarization accelerates markedly, due primarily to the rapid increase in I_{Kr} and I_{K1}. The I–V curves of both of these currents have negative slope regions (see Fig. 10); that is, as the membrane potential repolarizes, outward current through these channels increases to create a positive feedback situation that causes progressively faster repolarization. The negative slope region for I_{Kr} extends from approximately +30 to 0 mV and, for I_{K1}, from approximately 0 to −50 mV; thus, I_{Kr} controls the initial onset of the rapid repolarization phase, whereas I_{K1} controls the latter portion. As noted earlier, the conductances of I_{Kr} and I_{K1} are proportional to the square and square root, respectively, of $[K]_0$, which accounts in large part for the sensitivity of action potential duration to changes in $[K]_0$. In the case of I_{K1}, elevated $[K]_0$ also shifts the negative slope region of the I–V curve in the positive direction, facilitating repolarization. Drugs that block I_{Kr}, such as sotalol and dofetilide, are particularly effective at prolonging action potential duration. Recently, it has been shown that these drugs also block I_{K1} in a use-dependent manner (Kiehn *et al.*, 1995), which may contribute to their potency in prolonging action potential duration.

5. Phase 4 (Diastolic Depolarization)

Spontaneous diastolic depolarization is a normal property of SA nodal, AV nodal, and Purkinje cells, but not atrial or ventricular myocytes except under pathophysiological conditions. Because SA and AV nodal cells do not express K1 channels to an appreciable extent and contain a nonselective background conductance (Guo *et al.*, 1995), their maximum diastolic potential is low and less sensitive to depolarization by elevated $[K]_0$ than atrial and ventricular myocardium. The maximum diastolic potential (MDP) typically ranges from -50 to -60 mV, although more negative MDPs are seen in perinodal transitional cells. K_{ACh} channels are present in SA and AV nodal cells, and cause membrane hyperpolarization in response to vagal stimulation. The SA nodal cells with the most rapid diastolic depolarization appear to be more sensitive to the effects of ACh than slower latent pacemaker cells, so that upon vagal stimulation, the dominant pacemaker focus shifts location to the slower cells (Hoffman and Cranefield, 1960). In contrast, the maximum diastolic potential in pacemaking Purkinje cells is considerably more negative (~ -85 mV) due to the significant density of I_{K1} in these cells. K_{ACh} channels are largely absent in Purkinje cells.

The mechanism of diastolic depolarization is complex, and different ionic conductances predominate depending on the voltage range (Fig. 18). Membrane depolarization requires a net inward current, achieved by both an increase in inward currents and a decrease in outward (K) currents superimposed on the time-independent background currents. Following rapid repolarization to the maximum diastolic potential (near -60 mV in SA and AV nodal cells, and -85 mV in Purkinje cells), three time-dependent

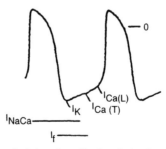

FIGURE 18 Currents underlying diastolic depolarization in SA nodal cells. I_f, I_{NaCa}, and I_K contribute to the first phase; the T-type Ca current contribute to the second, and the L-type Ca current contribute to the third. (Reprinted from Anunumwa and Jalife, 1995, with permission.)

currents are changing. An inward Na–Ca exchange current progressively decays as $[Ca]_i$ declines, promoting further hyperpolarization. Opposing this effect, the delayed rectifier K current activated during the plateau gradually deactivates (time constant ~ 300 ms), and the pacemaker current I_f activates (time constant 1–2 s), both favoring net inward current. The relative importance of I_K deactivation versus I_f activation in this phase of diastolic depolarization is controversial (DiFrancesco, 1995a; Vassalle, 1995). The disagreement about the role of I_f centers on whether its theshold of activation is low enough to contribute significantly over the range of voltages relevant to diastolic depolarization in the SA node (-60 to -35 mV), and whether the time constant of activation is sufficiently rapid (Vassalle, 1995). It has been argued that the activation range of I_f is more appropriate for driving diastolic depolarization in Purkinje cells because of their more negative maximum diastolic potential, or in transitional cells surrounding the SA node to assist in impulse propagation to the atria (see p. 129).

In the midregion of diastolic depolarization around -50 mV in SA and AV nodal cells, the T-type Ca current is activated, providing an additional source of depolarizing current (Hagiwara *et al.*, 1988) and compensating for progressive inactivation of I_f with membrane depolarization. In the late phase of diastolic depolarization, the L-type Ca current is progressively activated, increasing regeneratively beyond -30 mV to cause the action potential upstroke.

B. AUTONOMIC REGULATION OF CARDIAC ACTION POTENTIAL

The autonomic nervous system is the most important regulator of cardiac function on a beat-to-beat basis. Opposing effects are mediated by β-adrenergic receptors and M2 muscarinic receptors (Fig. 19). Their mechanisms of action on the myocardial function are described briefly as follows.

1. β-Adrenergic Stimulation

β-adrenergic stimulation speeds heart rate, increases contractility, and enhances relaxation. The signaling pathway involves dissociation of $G_{\alpha s}$ from $G_{\beta \gamma}$ subunits promoted by β-receptor occupancy. $G_{\alpha s}$ directly increases $I_{Ca(L)}$ and I_f, and also stimulates adenylyl cyclase to increase production of cAMP. cAMP directly stimulates I_f and binds to the regulatory subunit of PKA, releasing the catalytic PKA subunit, which then is available to phosphorylate and modify the function of a variety of proteins.

a. Increased Heart Rate

The increase in heart rate by β-adrenergic stimulation is mediated primarily by an increased rate of diastolic depolarization of dominant pace-

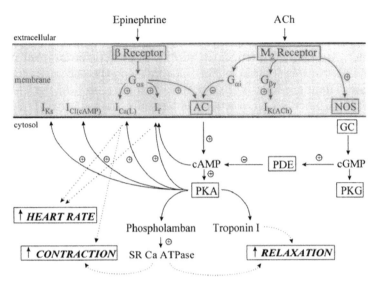

FIGURE 19   Schematic of cardiac autonomic regulation in cardiac muscle. AC, adenylyl cyclase; NOS, nitric oxide synthase; PDE, phosphodiesterase.

makers in the SA node. The activation of I_f is shifted to less negative membrane potentials by direct actions of $G_{\alpha s}$ and cAMP (Brown, 1990; DiFrancesco and Tortora, 1991), enhancing the rate of diastolic depolarization. The cAMP responsiveness of I_f is a major argument in favor of its making an important contribution to diastolic depolarization in the SA node, since the deactivation kinetics of K_{Ks} are only slightly affected by cAMP-dependent phosphorylation, at least in ventricular myocytes (Walsh and Kass, 1991). β-receptor stimulation also leads to a potentiation of the L-type Ca current, facilitating its contribution to the late stage of diastolic depolarization. The enhancement of $I_{Ca(L)}$ also increases the rate of rise of the upstroke (dV/dt), increasing conduction velocity through the SA and AV nodes.

As heart rate increases and $I_{Ca(L)}$ is potentiated in response to β-receptor stimulation, it is important that the action potential duration is shortened to preserve diastole for coronary blood flow. The increase in $I_{Ca(L)}$ tends to prolong action potential duration, and this is offset by PKA-mediated increase in two outward currents—I_{Ks} and I_{Cl}. Because E_{Cl} is normally ~ -50 mV, the increase in Cl conductance generates an outward hyperpolarizing current during the plateau that accelerates repolarization and an inward, depolarizing current during diastole that increases excitability.

b. Inotropic and Relaxant Effects

The positive inotropic and relaxant effects of β-receptor stimulation are mediated by three factors. The enhancement of the L-type Ca current by

$G_{\alpha s}$ and PKA results in increased Ca influx and triggered Ca release from the SR. PKA also phosphorylates the SR protein phospholamban. When unphosphorylated, phospholamban bound to SR Ca ATPase exerts an inhibitory effect. Phosphorylation of phospholamban removes its inhibitory effect on the SR Ca ATPase, increasing its Ca affinity and Ca pumping rate (see Chapter 4). The stimulation of SR Ca ATPase accelerates Ca removal, speeding relaxation rate. Also, the SR Ca stores are filled to a greater extent, potentiating Ca release on the subsequent beat. A third action of β-receptor stimulation is PKA-mediated phosphorylation of troponin I, which contributes to the relaxant effect by reducing myofilament Ca sensitivity (see Chapter 6).

2. Vagal Stimulation

By causing release of ACh from vagal afferents, vagal stimulation slows the heart rate by both hyperpolarizing and reducing the rate of diastolic depolarization of SA nodal cells, increases AV nodal conduction time, shortens the atrial action potential, and antagonizes the effects of β-receptor stimulation. These acute effects are mediated primarily by the binding of ACh to M_2 muscarinic receptors, which are coupled to pertussis toxin-sensitive G_i proteins. ACh binding to M_2 receptors causes dissociation of $G_{\alpha i}$ from $G_{\beta \gamma}$ subunits. $G_{\beta \gamma}$ subunits then directly activate K_{ACh} channels, which hyperpolarize SA nodal, AV nodal, and atrial cells to slow heart rate, increase AV nodal conduction time, and shorten the atrial action potential duration, respectively. These effects are further amplified by the indirect effects of M_2-receptor stimulation, which decreases intracellular cAMP levels augmented by β-receptor stimulation. Although $G_{\alpha i}$ subunits in high concentrations have an inhibitory effect on adenylyl cyclase, the major mechanism by which M_2-receptor stimulation decreases cAMP levels in the heart has been shown to involve an obligatory role of NO (Han et al., 1995). M_2-receptor activation leads to stimulation of nitric oxide synthase (NOS), which enhances cGMP production via NO-dependent stimulation of guanylyl cyclase. cGMP then activates a cGMP-sensitive cAMP phosphodiesterase that decreases cAMP. The decrease in cAMP inhibits I_f and $I_{Ca(L)}$, which slows diastolic depolarization in the SA node and conduction velocity through the AV node, and contributes to the shortening of atrial action potential. Because $I_{K(ACh)}$ is absent in the ventricle, vagal stimulation has little effect on the ventricular action potential under basal conditions, but opposes the effects of β-receptor stimulation by antagonizing the elevation in cAMP.

C. PROPAGATION OF CARDIAC ACTION POTENTIAL

1. Source–Sink Relationships

The source–sink concept is central to an understanding of electrical propagation of the cardiac impulse. For the action potential upstroke of one myocyte to be conducted to adjacent cells, it must generate enough depolarizing current to bring the adjacent cells to their threshold. Because the heart is a syncytium, the inward current from the depolarized cell is distributed (and diluted) among its nondepolarized neighbors. Stated more generally, the depolarized cell is a current source, and the neighboring resting cells are a current sink. If the sink is too large for the source, a source–sink mismatch is present and the cardiac impulse cannot propagate. A single myocyte cannot generate nearly enough current to bring its nearest neighbors to their threshold if all the neighbors are in the resting state. However, if some of the neighbors are also depolarized, they contribute to the current source, rather than act as part of the sink. The ratio of the source current to the current sink is determined by the curvature of the activation wavefront (Fig. 20). The cardiac impulse can only propagate if the activation wavefront exceeds a certain critical threshold of curvature (κ_c), at which the ratio of source current to the current sink is sufficient to

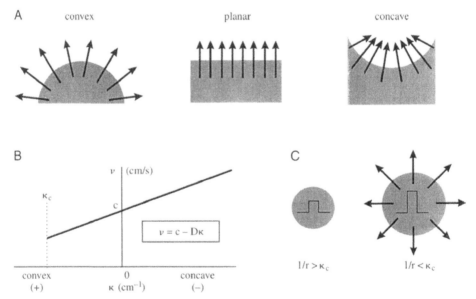

FIGURE 20   Source–sink relationships and conduction velocity in cardiac tissue. (A) Compared to a planar wavefront (*middle*), the depolarizing current generated by a convex wavefront (*left*) is diluted into a greater amount of undepolarized tissue (as illustrated by the spacing of the arrowheads), slowing the propagation speed. Conversely, the depolarizing current generated by a concave wavefront (*right*) is concentrated into the undepolarized tissue, increasing propagation speed. (B) The relationship between wavefront curvature and conduction velocity (v), known as the eikonal–curvature relationship. When the convexity of the wavefront reaches a critical value (κ_c), propagation fails because the sink (undepolarized tissue) is too large for the source (depolarized tissue). c, conduction velocity of a planar wavefront; D, diffusion coefficient. (C) A subthreshold stimulus does not depolarize enough tissue to exceed the critical curvature κ_c (*left*), whereas a suprathreshold stimulus does (*right*).

bring the adjacent resting cells to their threshold. This ratio determines the speed with which the adjacent cells reach their activation threshold, so that conduction velocity is related directly to the curvature of the activation wavefront (Fig. 20B). Convex wavefronts propagate slower than concave wavefronts. The dependence of conduction velocity on wavefront curvature is described by the eikonal–curvature equation (Keener, 1991) and is a generic property of many other excitable media, including reaction–diffusion systems, such as the Belousov–Zhabatinsky chemical reaction (Winfree, 1972) and intracellular Ca waves due to agonist-induced Ca release (Lechleiter et al., 1991). The following examples illustrate the importance of the source–sink concept.

a. Propagation of Sinus Impulse

The slow $I_{Ca(L)}$-dependent upstroke of the SA nodal action potential is poorly suited for depolarizing adjacent resting cells, and the SA node would be completely ineffective at propagating to the atrium due to a source–sink mismatch were it not surrounded by perinodal transitional cells with intermediate properties between SA nodal and atrial cells (Joyner and Van capelle, 1986) (see Fig. 15). The transitional cells have a partially depolarized resting membrane potential and exhibit spontaneous diastolic depolarization at a slower rate than the dominant SA nodal pacemakers. Their rate of diastolic depolarization is accelerated by the more rapid diastolic depolarization rate of the dominant SA nodal pacemaker cells, so that by the time the activation wavefront originating in the SA node reaches atrial tissue in the crista terminalis, it is broad enough to exceed the critical curvature necessary to propagate successfully through the atrium. The controversy over the role of I_f as the major pacemaker current in SA nodal cells centers on the observation that I_f activates at membrane potentials somewhat more negative than optimal for driving diastolic depolarization in these cells (Vassalle, 1995). An alternative possibility is that the major role of I_f is to drive diastolic depolarization in transitional cells, which have a more negative maximum diastolic potential, and so assist the successful propagation of the SA nodal impulse.

The propagation of the impulse formed in the SA node to the atria is quite slow, due both to the slow upstroke of the SA nodal action potential as well as the special architecture of the perinodal tissues. The conduction time from the SA node through the perinodal tissue to the atrium (<125 ms in humans) is comparable to the conduction time through the AV node (80–150 ms). Similar considerations apply to the AV node, where the cells in the lower AV junction have intermediate transitional properties between those in the AV node and His-Purkinje cells. In the ventricle, the branching structure of the Purkinje system is also important for successful propagation at the Purkinje-myocardial junction.

b. Pacing Threshold

A propagated cardiac impulse can be elicited by applying current exoge-
nously to the heart, which is the principle used by all external electronic
cardiac pacemakers. The pacing threshold is the minimum amount of exogenous current that must be applied to elicit a propagated response. The
threshold is not determined by the amount of current required to depolarize
a single myocyte to its activation threshold. Rather, it is determined by the
amount of current required to depolarize all the myocytes within a radius
sufficiently large to exceed the critical curvature necessary for conduction
(see Fig. 20C). Only then can the impulse propagate successfully throughout
the tissue.

c. Why Is the Cardiac Action Potential So Long?

It is usually assumed that the major reason for the long duration of the
cardiac action potential is to regulate contractility. However, a fundamen-
tally more important reason is that the cardiac impulse needs to propagate
through a three-dimensional syncytium. Imagine, for example, that the
cardiac ventricular action potential had a 4-ms duration, similar to a nerve
action potential. For a normal conduction velocity of 0.4 m/s, the excitation
wavelength (i.e., the distance between the activation wavefront and the
repolarization waveback, which equals the product of conduction velocity
times action potential duration) would be 1.6 mm, comparable to the space
constant of ventricular tissue (Levine *et al.*, 1987). Hence, the depolarizing
current from the activation wavefront would now be drawn into two current
sinks: in front by the resting cells, which are not yet activated, and in back
by the recently repolarized cells. This creates a source–sink mismatch, and
in computer simulations, a 5-ms action potential cannot propagate in a two-
or three-dimensional syncytium (B. Kogan, personal communication, 1996).
With a long action potential duration for which the excitation wavelength
exceeds the tissue space constant considerably, the contribution of the
repolarization waveback as a current sink is minimized, allowing all the
depolarizing current to flow in the direction of wavefront activation. The
effects of the repolarization waveback on conduction are even more critical
in a three-dimensional syncytium, which may explain teleologically why
the atrial action potential (propagating through an approximately two-
dimensional medium) can be much shorter than the ventricular action
potential. For a nerve, a short action potential is not a problem, since action
potential needs only to propagate down a one-dimensional cable as a planar
wavefront (i.e., no curvature).

d. Arrhythmias

The source–sink relationship and dependence of conduction velocity on
wavefront curvature has important implications for both focal and reentrant
cardiac arrhythmias. Although it is beyond the scope of this chapter to

discuss cardiac arrhythmias in detail, several general points are worth consideration. As should be apparent from the previous discussion of how the special architecture of the SA node facilitates propagation to the atria, similar considerations make it difficult for a focal excitation to propagate throughout the atria or ventricle unless a significant amount of tissue (exceeding the critical curvature requirement) is involved. The syncytial nature of cardiac tissue therefore protects it from the emergence of focal arrhythmias due to abnormal automaticity or triggered activity. This may partially account for why reentrant arrhythmias (with large wavefronts) are more common clinically than focal arrhythmias.

The possibility of reentry, however, is facilitated by the relationship between wavefront curvature and conduction velocity in the heart, because it provides a mechanism for reentry in the absence of any anatomic electrophysiological heterogeneity in the tissue (e.g., an infarct scar). This mechanism is the spiral wave (or scroll wave in three dimensions), which is a generic property of excitable media (Winfree, 1972; Lechleiter et al., 1991; Siegert and Weijer, 1992; Lipp and Niggli, 1993), and has been demonstrated in cardiac tissue (Davidenko et al., 1992). A spiral wave is generated when a break in the activation wavefront occurs, as might be produced by the collision of two perpendicular planar wavefronts during a premature beat (Fig. 21). The curvature at the broken end of the wavefront is extremely high and exceeds the critical curvature required for successful propagation so that the wavefront bends around the tip, forming the curved arm of a spiral wave. The tip of the spiral wave then precesses around a radius equivalent to the critical curvature, defining the core of the spiral wave. The curvature of the arm of the spiral wave self-adjusts so that the circumferential component of its velocity is proportional to the distance from the center of the core. At the point at which the spiral arm can no longer reduce its curvature further, the spiral arm begins to wrap around itself, producing the characteristic spiral pattern. Depending on the electrophysiological properties, mass, and geometry of the cardiac tissue, spiral and scroll waves can be stationary, or can meander and break up to form multiple daughter spiral waves that interact in complex patterns. Spiral/scroll wave phenomenology may underlie many types of functional reentry (i.e., reentry not dependent on an anatomic obstacle), including atrial flutter and some cases of stable monomorphic atrial or ventricular tachycardia (i.e., stationary spiral/scroll waves) (Davidenko et al., 1992); polymorphic ventricular tachycardia including torsades des pointes (i.e., meandering spiral/scroll waves) (Pertsov et al., 1993); and atrial and ventricular fibrillation (i.e., multiple meandering spiral/scroll waves) (Karma, 1993; Garfinkel et al., 1996). Although other mechanisms for functional reentry have been proposed, such as the leading circle mechanism (Allessie et al., 1977) and anisotropic reentry (Allessie et al., 1989), none have as firm a theoretical underpinning as the spiral/scroll wave mechanisms.

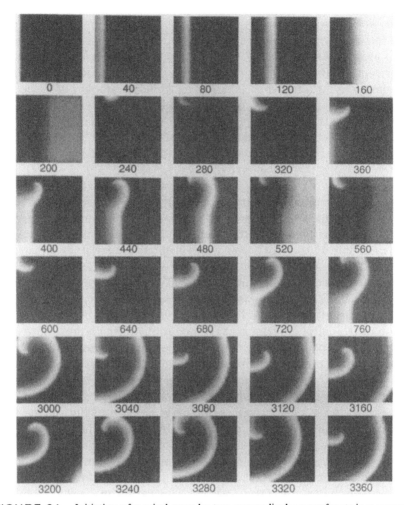

FIGURE 21   Initiation of a spiral wave by two perpendicular wavefronts in a computer model of cardiac propagation. The panels show successive snapshots of the position of the activation wavefront (white is depolarized, black is resting) at the time indicated in milliseconds. A stimulus generating a wavefront that travels from left to right (delivered at 0 ms) is followed by a second stimulus generating a wavefront from top to bottom (delivered at 240 ms). Since the right half has not yet recovered excitability, the wavefront initially propagates down the left side. At the broken end, however, the curvature of the wavefront is extremely high, slowing conduction velocity. The wavefront therefore bends to form the curved arm of the spiral wave. The tip precesses around a core whose radius is equivalent to the critical curvature κ_c (see Fig. 20). The curvature of the spiral arm decreases (and, therefore, conduction velocity increases) progressively with distance from the tip. (Reprinted from Mercader *et al.*, 1995, with permission.)

IV. FUTURE CHALLENGES

In the nearly 50 years since the glass microelectrode was invented (Ling and Gerard, 1949), making intracellular recordings from cardiac cells possible, tremendous progress has been made in the understanding of the electrophysiology of cardiac muscle. The major physiological and pathophysiological ionic currents have been identified, the cell signaling pathways that regulate them have been explored in detail, and their pharmacology has been investigated to develop many clinically useful treatments of cardiovascular diseases. The primary molecular structure of many ion channels has now been described, and we are in the process of gaining a detailed conceptual understanding of their structure–function relationships at the molecular level and of understanding how genetic defects and adaptive alterations in ion channels lead to cardiovascular disease.

This reductionist approach has been very successful for understanding the molecular basis of the cardiac action potential. However, there is still much to learn about how the heart functions as a three-dimensional organ. The marked heterogeneity of ion channel expression in different regions, the complex interwoven geometric arrangements of fibers and sheets of myocytes, and the higher order self-organizing principles that govern propagation in an excitable medium are all key issues to understand. For example, the elegant studies unraveling the basis of congenital long QT syndrome have identified specific genetic abnormalities in voltage-gated Na and K channels (Q. Wang *et al.,* 1995; Sanguinetti *et al.,* 1995). However, at the organ level, how these defects cause *torsades des pointes* and sudden cardiac death is still largely a matter of speculation. The disappointing results of the CAST trial (The Cardiac Arrhythmia Suppression Trial (CAST) Investigators, 1989) have illustrated the dangers of extrapolating a seemingly very reasonable hypothesis (frequent premature ventricular beats are associated with increased mortality after a myocardial infarction; therefore, preventing premature ventricular beats should reduce postinfarct mortality) to the treatment of real-life patients. Understanding the higher order principles governing cardiac function is a major undertaking, requiring a joint focus that synthesizes molecular and supramolecular approaches.

REFERENCES

Adelman J.P. (1995). Proteins that interact with the pore-forming subunits of voltage-gated ion channels. *Curr. Opin. Neurobiol.* 5:286–295.

Allessie, M., Bonke, F., and Shopman, F. (1977). Circus movement in rabbit atrial muscle as a mechanism of tachycardia. *Circ. Res.* 41:9–18.

Allessie, M.A., Schalij, M.J., Kirchhof, C.J.H.J., Boersma, L., Huybers, M., and Hollen, J. (1989). Experimental electrophysiology and arrhythmogenicity. Anisotropy and ventricular tachycardia. *Eur. Heart J.* 10:2–8.

Ammala, C., Moorhouse, A., Gribble, F., Ashfield, R., Proks, P., Smith, P.A., Sakura, H., Coles, B., Ashcroft, S.J.H., and Ashcroft, F.M. (1996). Promiscuous coupling between the sulphonylurea receptor and inwardly rectifying potassium channels. *Nature* 379:545–548.

Antzelevitch, C., Sicouri, S., Lukas, A., Nesterenko, V.V., Liu, D., and Di Diego, J.M. (1995). Regional differences in the electrophysiology of ventricular cells: Physiological and clinical implications. *In* "Cardiac Electrophysiology: From Bench to Bedside" (Zipes, D.P., and Jalife, J., eds.), pp. 228–245. W.B. Saunders, Philadelphia.

Anumonwa, J.M.B., and Jalife, J. (1995). Cellular and subcellular mechanisms of pacemaker activity initiation and synchronization in the heart. *In* "Cardiac Electrophysiology: From Cell to Bedside" (Zipes, D.P., and Jalife, J., eds.), pp. 151–164. W.B. Saunders, Philadelphia.

Armstrong, C.M., and Bezanilla, F. (1973). Currents related to movement of the gating particles of sodium channels. *Nature* 242:459–461.

Armstrong, C.M., and Bezanilla, F. (1977). Inactivation of the sodium channel. *J. Gen. Physiol.* 70:567–590.

Armstrong, C.M., Bezanilla, F., and Rojas, E. (1973). Destruction of the sodium conductance inactivation in squid axons perfused with pronase. *J. Gen. Physiol.* 62:375–391.

Ashcroft, S.J.H., and Ashcroft, F.M. (1990). Properties and functions of ATP-sensitive K-channels. *Cell. Signal.* 2:197–214.

Backx, P.H., and Marban, E. (1993). Background potassium current active during the plateau of the action potential in guinea pig ventricular myocytes. *Circ. Res.* 72:890–900.

Ballarin, C., Bertoli, A., Wojcik, G., and Sorgato, M.C. (1996). Michondrial inner membrane channels in yeast and mammals. *In* "Organellar Ion Channels and Transporters" (Clapham, D.E., and Erhlich, B.E., eds.), pp. 155–171. The Rockefeller University Press, New York.

Barchi, R.L. (1996). Molecular pathology of the skeletal Na channel. *Annu. Rev. Physiol.* 57:355–385.

Barhanin, J., Lesage, F., Guillemare, E., Fink, M., Lazdunski, M., and Romey, G. (1996). K_vLQT1 and IsK (minK) proteins associate to form the I_{Ks} cardiac potassium current. *Nature* 384:78–80.

Baukrowitz, T., and Yellen, G. (1996). Use-dependent blockers and exit rate of the last ion from the multi-ion pore of a K channel. *Science* 271:653–656.

Bean, B.P. (1985). Two kinds of calcium channels in canine atrial cells. *J. Gen. Physiol.* 86:1–30.

Bean, B.P. (1990). Gating for the physiologist. *Nature* 348:192–193.

Bennett, P.B., Yazawa, Y., Makita, N., and George, A.L. (1995). Molecular mechanism for an inherited cardiac arrhythmia. *Nature* 376:683–685.

Berridge, M.J. (1993). Inositol trisphosphate and calcium signalling. *Nature* 361:315–325.

Beyer, E.C. (1993). *In* "Receptor and Membrane Proteins: Molecular Biology of Membrane Transport" (Mueckler, M., and Friedlander, M., eds.), pp. 1–37. Academic Press, San Diego.

Beyer, E.C., Veenstra, R.D., Kanter, H.L., and Saffitz, J.E. (1995). Molecular structure and patterns of expression of cardiac gap junction proteins. *In* "Cardiac Electrophysiology: From Cell to Bedside" (Zipes, D.P., and Jalife, J., eds.), pp. 31–38. W.B. Saunders, Philadelphia.

Bezanilla, F., Perozo, E., Papazian, D.M,., and Stefani, E. (1991). Molecular basis of gating charge immobilization in *Shaker* potassium channels. *Science* 254:679–683.

Bogusz, S., Boxer, A., and Busath, D.D. (1992). An SS1-SS2 β-barrel structure for the voltage-gated potassium channel. *Protein Eng.* 5:285–293.

Boyle, W.A., and Nerbonne, J.M. (1991). A novel type of depolarization-activated K current in isolated adult rat atrial myocytes. *Am. J. Physiol.* 260:H1236–H1247.

Brown, A.M. (1990). Regulation of heartbeat by G protein-coupled ion channels. *Am. J. Physiol.* 259:H1621–H1628.

Callewaert, G., Vereecke, J., and Carmeliet, E. (1986). Existence of a calcium-dependent potassium channel in the membrane of cow cardiac Purkinje cells. *Pflugers Arch.* 406:424–426.

Carmeliet, E. (1992). A fuzzy subsarcolemmal space for intracellular Na in cardiac cells? *Cardiovasc. Res.* 26:433–442.

Cascio, W.E., Yan, G.X., and Kleber, A.G. (1990). Passive electrical properties, mechanical activity, and extracellular potassium in arterially perfused and ischemic rabbit ventricular muscle. *Circ. Res.* 66:1461–1473.

Cheng, H., Lederer, W.J., and Cannell, M.B. (1993). Calcium sparks—Elementary events underlying excitation–contraction coupling in heart muscle. *Science* 262:740–744.

Chien, K.R., Han, A., Sen, A., Buja, L.M., and Willerson, J.T. (1984). Accumulation of unesterified arachidonic acid in ischemic canine myocardium. *Circ. Res.* 54:313–322.

Collier, M.L., Levesque, P.C., Kenyon, J.L., and Hume, J.R. (1996). Unitary Cl channel activated by cytoplasmic Ca in canine ventricular myocytes. *Circ. Res.* 78:936–944.

Colquhoun, D., Neher, E., Reuter, H., and Stevens, C.F. (1981). Inward current channels activated by intracellular calcium in cultured cardiac cells. *Nature* 294:752–754.

Coulombe, A., Lefevre, I.A., Baro, I., and Coraboeuf, E. (1989). Barium- and calcium-permeable channels open at negative membrane potentials in rat ventricular myocytes. *J. Membr. Biol.* 111:57–67.

Cukierman, S. (1996). Regulation of voltage-dependent sodium channels. *J. Membr. Biol.* 151:203–214.

Daleau, P., Lessard, E., Groleau, M., and Turgeon, J. (1995). Erythromycin blocks the rapid component of the delayed rectifier potassium current and lengthens repolarization of guinea pig ventricular myocytes. *Circulation* 91:3010–3016.

Dascal, N., Lim, N.F., Schreibmayer, W., Wang, W.Z., Davidson, N., and Lester, H.A. (1993). Expression of an atrial G-protein-activated potassium channel in *Xenopus* oocytes. *Proc. Natl. Acad. Sci. USA* 90:6596–6600.

Davidenko, J.M., Pertsov, A.V., Salomonsz, J.R., Baxter, W., and Jalife, J. (1992). Stationary and drifting spiral waves of excitation in isolated cardiac muscle. *Nature* 355:349–351.

de Leon, M., Wang, Y., Jones, L., Perezreyes, E., Wei, X.Y., Soong, T.W., Snutch, T.P., and Yue, D.T. (1995). Essential Ca-binding motif for Ca-sensitive inactivation of L-type Ca channels. *Science* 270:1502–1506.

Deal, K.K., England, S.K., and Tamkun, M.M. (1996). Molecular physiology of cardiac potassium channels. *Physiol. Rev.* 76:49–67.

Deutsch, N., and Weiss, J.N. (1993). ATP-sensitive K channel modification by metabolic inhibition in isolated guinea-pig ventricular myocytes. *J. Physiol.* 465:163–179.

Deutsch, N., Matsuoka, S., and Weiss, J. (1994). Surface charge and properties of cardiac ATP-sensitive K channels. *J. Gen. Physiol.* 104:773–800.

DiFrancesco, D. (1995a). The pacemaker current (I_f) plays an important role in regulating SA node pacemaker activity. *Cardiovasc. Res.* 30:307–308.

DiFrancesco, D. (1995b). The onset and autonomic regulation of cardiac pacemaker activity: Relevance of the f current. *Cardiovasc. Res.* 29:449–456.

DiFrancesco, D., and Tortora, P. (1991). Direct activation of cardiac pacemaker channels by intracellular cyclic AMP. *Nature* 351:145–147.

Doupnik, C.A., Davidson, N., and Lester, H.A. (1995). The inward rectifier potassium channel family. *Curr. Opin. Neurobiol.* 5:268–277.

Duan, D., and Nattel, S. (1994). Properties of single outwardly-rectifying Cl channel in heart. *Circ. Res.* 75:789–795.

Ehara, T., Noma, A., and Ono, K. (1988). Calcium-activated non-selective cation channel in ventricular cells isolated from adult guinea-pig hearts. *J. Physiol.* 403:117–133.

Fakler, B., Brandle, U., Zenner, Z., and Ruppersberg, J.P. (1994). Kir2.1 inward rectifier K channels are regulated independently by protein kinases and ATP hydrolysis. *Neuron* 13:1413–1420.

Fickler, E., Taglialatela, M., Wible, B., Henley, C., and Brown, A. (1994). Spermine and spermidine as gating molecules for inward rectifier K channels. *Science* 266:1068–1071.

Forte, M., Blachly-Dyson, E., and Colombini, M. (1996). Structure and function of the yeast outer mitochondrial membrane channel, VDAC. In "Organellar Ion Channels and Transporters" (Clapham, D.E., and Erhlich, B.E., eds.), pp. 145–154. The Rockefeller University Press, New York.

Furukawa, T., Yamane, T., Terai, T., Katayama, Y., and Hiraoka, M. (1996). Functional linkage of the cardiac ATP-sensitive K channel to the actin cytoskeleton. *Pflugers Arch.- Eur. J. Physiol.* 431:504–512.

Gabriel, S.E., Clarke, L.L., Boucher, R.C., and Stutts, M.J. (1993). CFTR and outward rectifying chloride channels are distinct proteins with a regulatory relationship. *Nature* 363:263–266.

Garfinkel, A., Chen, P.-S., Walter, D.O., Karagueuzian, H.S., Kogan, B., Evans, S.J., Karpoukhin, M., Hwang, C., Uchida, T., Gotoh, M., Nwasokwa, O., Sager, P., and Weiss, J.N. (1997). Quasiperiodicity and chaos in fibrillation. *J. Clin. Invest.* 99:305–314.

Gellens, M.E., George, A.L., Jr., Chen, L., Chahine, M., Horn, R., Barchi, R.L., and Kallen, R.G. (1992). Primary structure and functional expression of the human cardiac tetrodotoxin-insensitive voltage-dependent sodium channel. *Proc. Natl. Acad. Sci. USA* 89:554–558.

Guo, J.Q., Ono, K.I., and Noma, A.N. (1995). A sustained inward current activated at the diastolic potential range in rabbit sino-atrial node cells. *J. Physiol.* 483:1–13.

Guo, W., Jorgensen, A.O., and Campbell, K.P. (1996). Triadin, a linker for calsequestrin and the ryanodine receptor. In "Organellar Ion Channels and Transporters" (Clapham, D.E., and Ehrlich, B.E., eds.), pp. 19–28. The Rockefeller University Press, New York.

Gyorke, S., and Fill, M. (1993). Ryanodine receptor adaptation—Control mechanism of Ca-induced Ca release in heart. *Science* 260:807–809.

Hagiwara, N., Irisawa, H., and Kameyama, M. (1988). Contribution of two types of calcium currents to the pacemaker potentials of rabbit sino-atrial node cells. *J. Physiol.* 395:233–253.

Hagiwara, N., Masuda, H., Shoda, M., and Irisawa, H. (1992). Stretch-activated anion currents of rabbit cardiac myocytes. *J. Physiol.* 456:285–302.

Han, X., Shimoni, Y., and Giles, W.R. (1995). A cellular mechanism for nitric oxide-mediated cholinergic control of mammalian heart rate. *J. Gen. Physiol.* 106:45–65.

Hanrahan, J.W., Tabcharani, J.A., Becq, F., Matthews, C.J., Augustinas, O., Jensen, T.J., Change, X.-B., and Riordan, J.R. (1995). Function and dysfunction of the CFTR chloride channel. In "Ion Channels and Genetic Diseases" (Dawson, D.C., and Frizzell, R.A., eds.), pp. 125–137. The Rockefeller University Press, New York.

Hartzell, H.C., Mery, P.F., Fischmeister, R., and Szabo, G. (1991). Sympathetic regulation of cardiac calcium current is due exclusively to cAMP-dependent phosphorylation. *Nature* 351:573–576.

Heinemann, S.H., Teriau, H., Stuhmer, W., Imoto, K., and Numa, S. (1992). Calcium channel characteristics conferred on the sodium channel by single mutations. *Nature* 356:441–443.

Higgins, C.F. (1995). The ABC of channel regulation. *Cell* 82:693–696.

Hildemann, D.W., and Ball, R. (1996). Regulation of cardiac Na–Ca exchange and K_{ATP} potassium channels by PIP_2. *Science* 273:956–959.

Ho, K., Nichols, C.G., Lederer, W.J., Lytton, J., Vassilev, P.M., Kanazirska, M.V., and Hebert, S.C. (1993). Cloning and expression of an inwardly rectifying ATP-regulated potassium channel. *Nature* 362:31–38.

Hodgkin, A.L., and Huxley, A.F. (1952). A quantitative description of membrane current and its application to conduction and excitation in nerve. *J. Physiol.* 117:500–544.

Hoffman, B.F., and Cranefield, P.F. (1960). "Electrophysiology of the Heart." McGraw-Hill, New York, pp. 104–106.

Hondeghem, L.M. (1987). Antiarrhythmic agents: Modulated receptor applications. *Circulation* 75:514–520.

Horowitz, B., Tsung, S.S., Hart, P., Levesque, P.C., and Hume, J.R. (1993). Alternative splicing of CFTR Cl channels in heart. *Am. J. Physiol.* 264:H2214–H2220.

Hoshi, T., Zagotta, W.N., and Aldrich, R.W. (1990). Biophysical and molecular mechanisms of *Shaker* potassium channel inactivation. *Science* 250:533–538.

Hu, K.L., Duan, D.Y., Li, G.R., and Nattel, S. (1996). Protein kinase C activates ATP-sensitive K current in human and rabbit ventricular myocytes. *Cir. Res.* 78:492–498.

Hume, J.R., and Harvey, R.D. (1991). Chloride conductance pathways in heart. *Am. J. Physiol.* 261:C399–C412.

Inagaki, N., Gonoi, T., Clement, J.P., Namba, N., Inazawa, J., Gonzalez, G., Aguilarbryan, L., Seino, S., and Bryan, J. (1995a). Reconstitution of I_{KATP}: An inward rectifier subunit plus the sulfonylurea receptor. *Science* 270:1166–1170.

Inagaki, N., Tsuura, Y., Namba, N., Masuda, K., Gonoi, T., Horie, M., Seino, Y., Mizuta, M., and Seino, S. (1995b). Cloning and functional characterization of a novel ATP-sensitive potassium channel ubiquitously expressed in rat tissues, including pancreatic islets, pituitary, skeletal muscle, and heart. *J. Biol. Chem.* 270:5691–5694.

Isacoff, E.Y., Jan, N.Y., and Jan, L.Y. (1991). Putative receptor for the cytoplasmic inactivation gate in the *Shaker* K channel. *Nature* 353:86–90.

Johnson, B.D., and Byerly, L. (1993). A cytoskeletal mechanism for Ca channel metabolic dependence and inactivation by intracellular Ca. *Neuron* 10:797–804.

Jongsma, H.J., and Rook, M.B. (1995). Morphology and electrophysiology of cardiac gap junction channels. *In* "Cardiac Electrophysiology: From Cell to Bedside" (Zipes, D.P., and Jalife, J., eds.), pp. 115–126. W.B. Saunders, Philadelphia.

Joyner, R.W., and Van capelle, F.J.L. (1986). Propagation through electrically coupled cells— How a small SA node drives a large atrium. *Biophys. J.* 50:1157–1164.

Jue, Y., Saint, D.A., and Gage, P.W. (1996). Hypoxia increases persistent sodium current in rat ventricular myocytes. *J. Physiol.* 497:337–347.

Kameyama, M., Kakei, M., Sato, R., Shibasaki, T., Matsuda, H., and Irisawa, H. (1984). Intracellular Na activates a K channel in mammalian cardiac cells. *Nature* 309:354–356.

Karma, A. (1993). Spiral breakup in model equations of action potential propagation in cardiac tissue. *Phys. Rev. Lett.* 71:1103–1106.

Keener, J. (1991). An eikonal–curvature equation for action potential propagation in myocardium. *J. Math. Biol.* 29:629–651.

Kiehn, J., Wible, B., Ficker, E., Taglialatela, M., and Brown, A.M. (1995). Cloned human inward rectifier K channel as a target for Class III methanesulfonanilides. *Circ. Res.* 77:1151–1155.

Kim, D., and Clapham, D.E. (1989). Potassium channels in cardiac cells activated by arachidonic acid and phospholipids. *Science* 244:1174–1176.

Kim, D., and Duff, R.A. (1990). Regulation of K channels in cardiac myocytes by free fatty acids. *Circ. Res.* 67:1040–1046.

Kirsch, G.E., Codina, J., Birnbaumer, L., and Brown, A.M. (1990). Coupling of ATP-sensitive K channels to A1 receptors by G proteins in rat ventricular myocytes. *Am. J. Physiol.* 259:H820–H826.

Krapivinsky, G., Gordon, E.A., Wickman, K., Velimirovic, B., Krapivinsky, L., and Clapham, D.E. (1995). The G-protein-gated atrial K channel I_{KACh} is a heteromultimer of two inwardly rectifying K channel proteins. *Nature* 374:135–141.

Kubo, Y., Baldwin, T., Jan, Y., and Jan, L. (1993a). Primary structure and functional expression of a mouse inward rectifier potassium channel. *Nature* 362:127–133.

Kubo, Y., Reuveny, E., Slesinger, P., Jan, Y., and Jan, L. (1993b). Primary structure and functional expression of a rat G-protein-coupled muscarinic potassium channel. *Nature* 364:802–806.

Kurachi, Y., Ito, H., Sugimoto, T., Shimizu, T., Miki, I., and Ui, M. (1989). Arachidonic acid metabolites as intracellular modulators of the G protein-gated cardiac K channel. *Nature* 337:555–557.

LeBlanc, N., and Hume, J.R. (1990). Sodium current-induced release of calcium from cardiac sarcoplasmic reticulum. *Science* 248:372–376.

Lechleiter, J., Girard, S., Peralta, E., and Clapham, D. (1991). Spiral calcium wave propagation and annihilation in *Xenopus laevis* oocytes. *Science* 252:123–126.

Lee, J.-K., John, S.A., Lu, Y., Shieh, R., and Weiss, J.N. (1997). The intrinsic gating mechanism is fundamental to inward rectification of the IRK1 channel. *Biophys. J.* 72:A253.

Levine, J.H., Moore, E.N., Weisman, H.F., Kadish, A.H., Becker, L.C., and Spear, J.F. (1987). Depression of action potential characteristics and a decreased space constant are present in postischemic, reperfused myocardium. *J. Clin. Invest.* 79:107–116.

Ling, A., and Gerard, R.W. (1949). The normal action potential of frog sartorius fibers. *J. Cell. Comp. Physiol.* 34:382–396.

Lipp, P., and Niggli, E. (1993). Microscopic spiral waves reveal positive feedback in subcellular calcium signaling. *Biophys. J.* 65:2272–2276.

Liu, D.W., and Antzelevitch, C. (1995). Characteristics of the delayed rectifier current (I-Kr and I-Ks) in canine ventricular epicardial, midmyocardial, and endocardial myocytes—A weaker I-Ks contributes to the longer action potential of the M cell. *Circ. Res.* 76:351–365.

Liu, Y.G., Gao, W.D., O'Rourke, B., and Marban, E. (1996). Synergistic modulation of ATP-sensitive K currents by protein kinase C and adenosine—Implications for ischemic preconditioning. *Circ. Res.* 78:443–454.

Lopatin, A., Makhina, E., and Nicols, C. (1994). Potassium channel block by cytoplasmic polyamines as the mechanism of intrinsic rectification. *Nature* 372:366–371.

Lopez, G., Jan, Y., and Jan, L. (1994). Evidence that the S6 segment of the *Shaker* voltage-gated K channel comprises part of the pore. *Nature* 367:179–182.

Luo, C.H., and Rudy, Y. (1994). A dynamic model of the cardiac ventricular action potential. 2. After depolarizations, triggered activity, and potentiation. *Circ. Res.* 74:1097–1113.

MacKinnon, R., Aldrich, R.W., and Lee, A.W. (1993). Functional stoichiometry of *Shaker* potassium channel inactivation. *Science* 262:757–759.

Mannuzzu, L.M., Moronne, M.M., and Isacoff, E.Y. (1996). Direct physical measure of conformational rearrangement underlying potassium channel gating. *Science* 271:213–216.

Mazzanti, M., and DiFrancesco, D. (1989). Intracellular Ca modulates K-inward rectification in cardiac myocytes. *Pflugers Arch.* 413:322–324.

McDonald, T.F., Pelzer, S., Trautwein, W., and Pelzer, D.J. (1994). Regulation and modulation of calcium channels in cardiac, skeletal, and smooth muscle cells. *Physiol. Rev.* 74:365–507.

Meissner, G. (1995). Sarcoplasmic reticulum ion channels. *In* "Cardiac Electrophysiology: From Cell to Bedside" (Zipes, D.P., and Jalife, J., eds.), pp. 49–56. W.B. Saunders, Philadelphia.

Mercarder, M.A., Michaels, D.C., and Jalife, J. (1995). Reentrant activity in the form of spiral waves in mathematical models of the sinoatrial node. *In* "Cardiac Electrophysiology: From Cell to Bedside" (Zipes, D.P., and Jalife, J., eds.), pp. 389–403. W.B. Saunders, Philadelphia.

Mikami, A., Imoto, K., Tanabe, T., Niidome, T., Mori, Y., Takeshima, H., Narumiya, S., and Numa, S. (1989). Primary structure and functional expression of the cardiac dihydropyridine-sensitive calcium channel. *Nature* 340:230–233.

Miller, C. (1996). The long pore gets molecular. *J. Gen. Physiol.* 107:445–447.

Morris, C.E. (1990). Mechanosensitive ion channels. *J. Membr. Biol.* 113:93–107.

Morris, C.E., and Sigurdson, W.J. (1989). Stretch-inactivated ion channels coexist with stretch-activated ion channels. *Science* 243:807–809.

Nakai, J., Imagawa, T., Hakamata, Y., Shigekawa, M., Takeshima, H., and Numa, S. (1990). Primary structure and functional expression from cDNA of cardiac muscle ryanodine receptor/calcium release channel. *FEBS Lett.* 271:169–177.

Neely, A., Wei, X.Y., Olcese, R., Birnbaumer, L., and Stefani, E. (1993). Potentiation by the beta-subunit of the ratio of the ionic current to the charge movement in the cardiac calcium channel. *Science* 262:575–578.

Nichols, C.G., Ripoll, C., and Lederer, W.J. (1991). ATP-sensitive potassium channel modulation of the guinea pig ventricular action potential and contraction. *Circ. Res.* 68:280–287.

Nichols, C.G., Shyng, S.-L., Nestorowicz, A., Glaser, B., Clement, J.P., Gonzalez, G., Aguilar-Bryan, L., Permutt, M.A., and Bryan, J. (1996). Adenosine diphosphate as an intracellular regulator of insulin secretion. *Science* 272:1785–1787.

Nilius, B., Hess, P., Lansman, J.B., and Tsien, R.W. (1985). A novel type of cardiac calcium channel in ventricular cells. *Nature* 316:443–446.

Noda, M., Shimizu, S., Tanabe, T., Takai, T., Kayano, T., Ikeda, T., Takahashi, H., Nakayama, H., Kanaoka, N., Minamino, N., Kangawa, K., Matsuo, H., Raferty, A., Hirose, T., Inayama, S., Hayashida, H., Miyata, T., and Numa, S. (1984). Primary structure of *Electrophorus electricus* sodium channel deduced from cDNA sequence. *Nature* 312:121–127.

Noma, A. (1983). ATP-regulated K channels in cardiac muscle. *Nature* 305:147–148.

O'Rourke, B., Backx, P.H., and Marban, E. (1992). Phosphorylation-independent modulation of L-type calcium channels by magnesium-nucleotide complexes. *Science* 257:245–248.

Olcese, R., Zhou, J., Qin, L., Birnbaumer, H., and Stefani, E. (1996). Transferring Ca-dependent inactivation property from cardiac to neuronal Ca channels. *Biophys. J.* 70:A186(abstract).

Ondrias, K., Brillantes, A.B., Scott, A., Ehrlich, B.E., and Marks, A.M. (1996). Single channel properties and calcium conductance of the cloned ryanodine receptor/calcium-release channel. *In* "Organellar Ion Channels and Transporters" (Clapham, D.E., and Ehrlich, B.E., eds.), pp. 29–45. The Rockefeller University Press, New York.

Papazian, D.M., Timpe, L.C., Jan, Y.N., and Jan, L.N. (1987). Cloning and genomic and complementary DNA from *Shaker*, a putative potassium channel gene from *Drosophila*. *Science* 237:749–753.

Patlak, J. (1991). Molecular kinetics of voltage-dependent Na channels. *Physiol. Rev.* 71:1047–1079.

Pertsov, A.M., Davidenko, J.M., Salomonsz, R., Baxter, W.T., and Jalife, J. (1993). Spiral waves of excitation underlie reentrant activity in isolated cardiac muscle. *Circ. Res.* 72:631–650.

Reuveny, E., Slesinger, P.A., Inglese, J., Morales, J.M., Iniguezlluhi, J.A., Lefkowitz, R.J., Bourne, H.R., Jan, Y.N., and Jan, L.Y. (1994). Activation of the cloned muscarinic potassium channel by G protein beta gamma subunits. *Nature* 370:143–146.

Rich, D.P., Berger, H.A., Cheng, S.H., Travis, S.M., Saxena, M., Smith, A.E., and Welsh, M.J. (1993). Regulation of the cystic fibrosis transmembrane conductance regulator Cl channel by negative charge in the R-domain. *J. Biol. Chem.* 268:20259–20267.

Sachs, F. (1989). Ion channels as mechanical transducers. *In* "Cell Shape: Determinants, Regulation and Regulatory Role" (Bonner, F., and Stein, W., eds.), pp. 63–92. Academic Press, New York.

Sackin, H. (1995). Mechanosensitive channels. *Annu. Rev. Physiol.* 57:333–353.

Saffitz, J.E., Kanter, H.L., Green, K.G., Tolley, T.K., and Beyer, E.C. (1994). Tissue-specific determinants of anisotropic conduction velocity in canine atrial and ventricular myocardium. *Circ. Res.* 74:1065–1070.

Saint, D.A., Ju, Y., and Gage, P.W. (1992). A persistent sodium current in rat ventricular myocytes. *J. Physiol.* 453:219–231.

Sanguinetti, M.C., and Jurkiewicz, N.K. (1990). Two components of cardiac delayed rectifier K current. *J. Gen. Physiol.* 96:195–215.

Sanguinetti, M.C., Jiang, C., Curran, M.E., and Keating, M.T. (1995). A mechanistic link between an inherited and an acquired cardiac arrhythmia: *HERG* encodes the I_{Kr} potassium channel. *Cell* 81:299–307.

Sanguinetti, M.C., Curran, M.E., Zou, A., Shen, J., Spector, P.S., Atkinson, D.L., and Keating, M.T. (1996). Coassembly of K$_v$LQT1 and minK (IsK) proteins to form cardiac I_{Ks} potassium channel. *Nature* 384:80–83.

Sano, T., and Yamagishi, S. (1965). Spread of excitation from the sinus node. *Circ. Res.* 16:423–430.

Schreibmayer, W., Dessauer, W., Vorobiov, D., Gilman, A.G., Lester, H.A., Davidson, N., and Dascal, N. (1996). Inhibition of an inwardly-rectifying K channel by G-protein α-subunits. *Nature* 380:624–627.

Schubert, B., Vandongen, A.M.J., Kirsch, G.E., and Brown, A.M. (1989). β-adrenergic inhibition of cardiac sodium channels by dual G-protein pathways. *Science* 245:516–519.

Shieh, R., John, S.A., Lee, J.K., and Weiss, J.N. (1996). Inward rectification of IRK1 expressed in *Xenopus* oocytes: Effects of intracellular pH reveal an intrinsic gating mechanism. *J. Physiol.* 494:363–376.

Shrier, A., Adjemian, R.A., and Munk, A.A. (1995). Ionic mechanisms of atrioventricular nodal cell excitability. *In* "Cardiac Electrophysiology: From Cell to Bedside" (Zipes, D.P., and Jalife, J., eds.), pp. 164–173. W.B. Saunders, Philadelphia.

Shuck, M.E., Bock, J.H., Benjamin, C.W., Tsai, T.D., Lee, K.S., Slightom, J.L., and Bienkowski, M.J. (1994). Cloning and characterization of multiple forms of the human kidney ROMK potassium channel. *J. Biol. Chem.* 269:24261–24270.

Siegert, F., and Weijer, C. (1992). Three-dimensional scroll waves organize Dictyostelium slugs. *Proc. Natl. Acad. Sci. USA* 89:6437.

Sipido, K.R., and Carmeliet, E. (1996). Can Ca entry through T-type Ca channels trigger Ca release from the sarcoplasmic reticulum in guinea-pig ventricular myocytes? *Biophys. J.* 70:A245 (abstract).

Smith, P.L., Baukrowitz, T., and Yellen, G. (1996). The inward rectification mechanism of the *HERG* cardiac potassium channel. *Nature* 379:833–836.

Soong, T.W., Stea, A., Hodson, C.D., Dubel, S., Vincent, S.R., and Snutch, T.P. (1993). Structure and functional expression of a member of the low voltage-activated Ca channel family. *Science* 260:1133–1136.

Sorota, S. (1992). Swelling-induced chloride-sensitive current in canine atrial cells revealed by whole-cell patch-clamp method. *Circ. Res.* 70:679–687.

Starmer, C.F., Grant, A.O., and Strauss, H.C. (1984). Mechanisms of use-dependent block of sodium channels in excitable membranes by local anesthetics. *Biophys. J.* 46:15–27.

Stehno-Bittel, L., Perez-Terzic, C., Luckhoff, A., and Clapham, D.E. (1996). Nuclear ion channels and regulation of the nuclear pore. *In* "Organellar Ion Channels and Transporters" (Clapham, D.E., and Ehrlich, B.E., eds.), pp. 195–207. The Rockefeller University Press, New York.

Stern, M.D. (1992). Theory of excitation–contraction coupling in cardiac muscle. *Biophys. J.* 63:497–517.

Stuhmer, W., Conti, F., Suzuki, H., Wang, X.D., Noda, M., Yahagi, N., Kubo, H., and Numa, S. (1989). Structural parts involved in activation and inactivation of the sodium channel. *Nature* 339:597–603.

Sun, Z., Akabas, M.H., Goulding, E.H., Karlin, A., and Siegelbaum, S.A. (1996). Exposure of residues in the cyclic nucleotide-gated channel pore: P region structure and function in gating. *Neuron* 16:141–149.

Tanabe, T., Takeshinma, H., Mikami, A., Flockerzi, V., Takahashi, H., Kangawa, K., Kojima, M., Matsuo, H., Hirose, T., and Numa, S. (1987). Primary structure of the receptor for calcium channel blockers from skeletal muscle. *Nature* 328:313–318.

The Cardiac Arrhythmia Suppression Trial (CAST) Investigators. (1989). Preliminary report: Effect of encainide and flecainide on mortality in a randomized trial of arrhythmia suppression after myocardial infarction. *N. Engl. J. Med.* 321:406–410.

Tominaga, M., Horie, M., Sasayama, S., and Okada, Y. (1995). Glibenclamide, an ATP-sensitive K channel blocker, inhibits cardiac cAMP-activated Cl conductance. *Circ. Res.* 77:417–423.

Trube, G., and Hescheler, J. (1984). Inward-rectifying channels in isolated patches of the heart cell membrane: ATP-dependence and comparison with cell-attached patches. *Pflugers Arch.* 401:178–184.

Tung, R.T., and Kurachi, Y. (1991). On the mechanism of nucleotide diphosphate activation of the ATP-sensitive K channel in ventricular cell of guinea pig. *J. Physiol.* 437:239–256.

Undrovinas, A. I., Fleidervish, I.A., and Makielski, J.C. (1992). Inward sodium current at resting potentials in single cardiac myocytes induced by the ischemic metabolite lysophosphatidylcholine. *Circ. Res.* 71:1231–1241.

van Wagoner, D.R. (1991). Mechanosensitive ion channel in atrial myocytes. *Biophys. J.* 59:546a(abstract).

van Wagoner, D.R. (1993). Mechanosensitive gating of atrial ATP-sensitive potassium channels. *Circ. Res.* 72:973–983.

Vandenberg, C. (1987). Inward rectification of a potassium channel in cardiac ventricular cells depends on internal magnesium ions. *Proc. Natl. Acad. Sci. USA* 84:2560–2564.

Vassalle, M. (1995). The pacemaker current (I_f) does not play an important role in regulating SA node pacemaker activity. *Cardiovasc. Res.* 30:309–310.

Walsh, K.B., and Kass, R.S. (1988). Regulation of a heart potassium channel by protein kinase A and C. *Science* 242:67–69.

Walsh, K.B., and Kass, R.S. (1991). Distinct voltage-dependent regulation of a heart-delayed I_K by protein kinases A and C. *Am. J. Physiol.* 261:C1081–C1090.

Walsh, K.B., and Long, K.J. (1994). Properties of a protein kinase C-activated chloride current in guinea pig ventricular myocytes. *Circ. Res.* 74:121–129.

Walsh, K.B., Arena, J.P., Kwok, W., Freeman, L., and Kass, R.S. (1991). Delayed-rectifier potassium channel activity in isolated membrane patches of guinea pig ventricular myocytes. *Am. J. Physiol.* 260:H1390–H1393.

Wang, S.Y., Clague, J.R., and Langer, G.A. (1995). Increase in calcium leak channel activity by metabolic inhibition or hydrogen peroxide in rat ventricular myocytes and its inhibition by polycation. *J. Mol. Cell Cardiol.* 27:211–222.

Wang, Q., Shen, J., Splawski, I., Atkinson, D., Li, Z., Robinson, J.L., Moss, A.J., Towbin, J.A., and Keating, M.T. (1995). *SCN5A* mutation associated with an inherited cardiac arrhythmia, long QT syndrome. *Cell* 80:805–811.

Wang, Y.G., and Lipsius, S.L. (1995). Acetylcholine activates a glibenclamide-sensitive K current in cat atrial myocytes. *Am. J. Physiol.* 268:H1321–H1334.

Wang, Z., Kimitsuki, T., and Noma, A. (1991). Conductance properties of the Na-activated K channel in guinea-pig ventricular cells. *J. Physiol.* 433:241–257.

Wang, Z., Fermini, B., and Nattel, S. (1993). Sustained depolarization-induced outward current in human atrial myocytes: Evidence for a novel delayed rectifier K current similar to Kv1.5 cloned channel currents. *Circ. Res.* 73:1061–1076.

Wang, Z.R., Mitsuiye, T., and Noma, A. (1996). Cell distension-induced increase of the delayed rectifier K current in guinea pig ventricular myocytes. *Circ. Res.* 78:466–474.

Weiss, J.N., Venkatesh, N., and Lamp, S.T. (1992). ATP-sensitive K channels and cellular K loss in hypoxic and ischaemic mammalian ventricle. *J. Physiol.* 447:649–673.

Winfree, A. (1972). Spiral waves of chemical activity. *Science* 175:634–636.

Yang, J., Jan, N., and Jan, L. (1995a). Determination of the subunit stoichiometry of an inwardly-rectifying potassium channel. *Neuron* 15:1441–1447.

Yang, J., Jan, Y.N.. and Jan, L.Y. (1995b). Control of rectification and permeation by residues in two distinct domains in an inward rectifier K channel. *Neuron* 14:1047–1054.

Yang, N., George, A.L., and Horn, R. (1996). Molecular basis of charge movement in voltage-gated sodium channels. *Neuron* 16:113–122.

Yao, Z.H., Cavero, I., and Gross, G.J. (1993). Activation of cardiac K_{ATP} channels—An endogenous protective mechanism during repetitive ischemia. *Am. J. Physiol.* 264:H495–H504.

Yatani, A., and Brown, A.M. (1990). Regulation of cardiac pacemaker current I_f in excised membranes from sinoatrial node cells. *Am. J. Physiol.* 258:H1947–H1951.

Yatani, A., Codina, J., Imoto, Y., Reeves, J.P., Birnbaumer, L., and Brown, A.M. (1987). A G protein directly regulates mammalian cardiac calcium channels. *Science* 238:1288–1292.

Zeng, J., Laurita, K.R., Rosenbaum, D.S., and Rudy, Y. (1995). Two components of the delayed rectifier K current in ventricular myocytes of the guinea pig type—Theoretical formulation and their role in repolarization. *Circ. Res.* 77:140–152.

Zhang, J.F., Ellinor, P.T., Aldrich, R.W., and Tsien, R.W. (1994). Molecular determinants of voltage-dependent inactivation in calcium channels. *Nature* 372:97–100.

Zilberter, Y.I., Starmer, C.F., Starobin, J., and Grant, A.O. (1994). Late Na channels in cardiac cells: The physiological role of background Na channels. *Biophys. J.* 67:153–160.

Zygmunt, A.C., and Gibbons, W.R. (1991). Calcium-activated chloride current in rabbit ventricular myocytes. *Circ. Res.* 68:424–437.

4

MYOCARDIAL ION TRANSPORTERS

KENNETH D. PHILIPSON

I. INTRODUCTION

The contractility of cardiac muscle is directly controlled by the level of Ca^{2+} in the myoplasm. The Ca^{2+} level, in turn, is a function of many factors: Ca^{2+} fluxes across the sarcolemma (SL), fluxes across the membranes of internal organelles, intracellular Ca^{2+} buffering, and rates of Ca^{2+} diffusion from regions of limited access. This chapter focuses on those ion tarnsporters that mediate Ca^{2+} fluxes and have the greatest role in myocardial excitation–contraction (E–C) coupling. Of interest are three SL transporters (the Na^+–Ca^{2+} exchanger, an ATP-dependent Ca^{2+} pump, and the ATP-dependent Na^+ pump) and the ATP-dependent Ca^{2+} pump of the sarcoplasmic reticulum (SR). The cardiac myocyte has two distinctly different ATP-dependent Ca^{2+} pumps: One resides in the SR and the other is found in the SL. One of the included transporters, the ATP-dependent Na^+ pump, does not directly transport Ca^{2+}; nevertheless, it has an indirect, but important, role in the regulation of Ca^{2+}.

This chapter focuses on recent molecular concepts on the structure, function, and kinetics of these transporters. It also relates the molecular information to modern ideas on the role of each transporter in E–C coupling.

This chapter is not heavily referenced, but readers are directed to appropriate review articles or to essential primary references as appropriate.

II. THE PLAYERS

A brief overview of each transporter in the order in which they are discussed follows:

1. The ATP-dependent Na^+ pump is present in the plasma membrane (PM) of almost all mammalian cells. With each reaction cycle, the pump extrudes three Na^+ ions from the cell in exchange for two K^+ ions at the expense of one ATP. The enzymatic manifestation of the Na^+ pump is the Na^+,K^+-ATPase.

2. The plasma membrane ATP-dependent Ca^{2+} pump (PMCA) extrudes one Ca^{2+} ion from the cell with the hydrolysis of one ATP. Also known as the plasma membrane Ca^{2+}ATPase, this pump is subject to a high degree of regulation.

3. The SR ATP-dependent Ca^{2+} pump (SERCA) is more efficient than the PM Ca^{2+} pump, and it pumps two Ca^{2+} ions into the lumen of the SR for each ATP consumed.

4. The Na^+–Ca^{2+} exchanger catalyzes the countertransport of three Na^+ ions on one side of the PM for one Ca^{2+} ion on the other side. In general, the major role of the Na^+–Ca^{2+} exchanger is to extrude Ca^{2+} from cells. The exchanger can be considered to be a secondary active transporter: The energy of the Na^+ gradient, rather than ATP, is used to expel Ca^{2+} from a cell.

As is described, each of these transporters is a member of a multigene family. In each case, cardiac muscle expresses specific isoforms.

III. P-TYPE ION PUMPS

Three of the transporters that are discussed, the ATP-dependent Na^+ pump and the two ATP-dependent Ca^{2+} pumps, are members of a class of ion translocators known as P-type ion pumps. Both the kinetics and molecular properties of these ion pumps have some similarities, and it is instructive to consider these common features. The P-type pumps are primary active transporters directly converting the energy of ATP into an ion gradient. The P-type pumps, as implied by their name, all form phosphorylated intermediates at an aspartyl residue. The phosphate is donated by an ATP molecule and forces the pump into a high-energy intermediate state. The energy stored in the high-energy form of the protein is then used to translocate ions. The presence of a high-energy phosphorylated state of the pump protein is verified by the reversibility of the reaction at this step. The phosphate from the protein can be readily transferred back to ADP to reform ATP. Since energy is required to resynthesize ATP, this demonstrates that the energy state of the pump has been increased by phosphorylation.

How does a P-type pump transport ions? A detailed molecular answer is not known for any P-type pump. Models in which ion binding sites have alternating access to the aqueous environment on the two sides of the membrane are likely correct. This is exemplified schematically in Fig. 1. In this model, a monovalent cation binds to the protein at the internal surface. The protein undergoes a conformational change to occlude the cation. In the occluded state, the cation has no access to either membrane surface. After a second conformational change, the cation is now exposed to the external medium and can dissociate from the ion pump. Thus, the ion binding sites on the pump are exposed to only one side of the membrane at a time. This simplified model does not show the coupling of ATP to the translocation process but, nevertheless, demonstrates some important points. First, in the occluded state, the cation is trapped within the protein. This is not high-affinity binding, but rather a state in which high-energy barriers or gates prevent movement of the ion in either direction. Clear experimental evidence for occluded states has been obtained for both the SR Ca^{2+} pump and for the Na^+ pump. Radioactive cations occluded in isolated pump proteins remain associated with the protein for extended times.

Second, the cation does not have to be physically transported across the entire length of the ion pump. As modeled in Fig. 1, the ion first diffuses through a wide access channel to reach a binding site near the center of the protein. After the conformational changes, only a small physical movement of the ion has occurred, but now the ion has access to a diffusion channel at the opposite membrane surface.

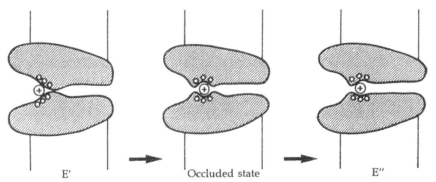

FIGURE 1   Alternating access model of ion pumps. E' and E" are states with the ion binding sites facing the internal and external solution, respectively. In the occluded state, ions are trapped within the protein and are unable to exchange with media on either side of the membrane. Open circles represent ligand groups of the protein. (Reprinted from Läuger, 1991, with permission of Sinauer Associates, Inc.)

As shown, the access channels are either wide or narrow. With a narrow access channel, there can be a drop in the transmembrane voltage along the length of the channel. In this case, the concentration of an ion within the access channel will be a function of the membrane potential, and the apparent affinity will vary with potential. Some of the behavior of the Na^+ pump can be modeled to result from electrostatic effects on ions diffusing through a narrow access channel on the extracellular side of the pump.

In this alternating access model, the pumps cycle between two states with the ion binding sites alternating between being inward facing and outward facing. These are commonly known as the E_1 and E_2 states, respectively. Formerly, the P-type pumps were referred to as E_1,E_2-type pumps.

Some of the preceding discussion applies to transporters, such as the Na^+–Ca^{2+} exchanger, as well as to pumps. The Na^+–Ca^{2+} exchanger does not use ATP, but can be modeled to involve E_1 and E_2 states with an alternating access of binding sites on the two sides of the transport protein. This is discussed in more detail later.

Pumps and other transporters have much lower transport rates than channels. Up to 10^7 ions may pass through a channel per second. Transporters typically translocate 10^2–10^4 ions/s. Different biophysical techniques are applied to the study of transporters and channels because of this quantitative difference in flux rates. For example, the opening and closing of individual channel proteins can be observed by patch clamp techniques. In contrast, the current passed by a single electrogenic transporter is below detection levels. An excellent discussion of the energetics and biophysical properties of P-type pumps can be found in the monograph by Läuger (1991).

IV. THE ATP-DEPENDENT Na^+ PUMP

A. CELL PHYSIOLOGY

The ATP-dependent Na^+ pump has a central role in the cellular physiology of almost all tissues including the myocardium. Cells maintain gradients of both Na^+ and K^+ across the plasma membrane created by the ATP-dependent Na^+ pump. Cellular Na^+ and K^+ levels are low and high, respectively, due to the Na^+ pump. In each reaction cycle, the Na^+ pump extrudes three Na^+ ions from a cell and two K^+ ions are taken up in exchange. These events require energy, as the Na^+ pump can transport both Na^+ and K^+ against their electrochemical gradients. The energy is provided by the hydrolysis of one ATP molecule per reaction cycle. The enzymatic equivalent of the ATP-dependent Na^+ pump can be measured in plasma membrane preparations as an ATPase activity that requires the presence of both Na^+ and K^+ (i.e., the Na^+,K^+-ATPase).

The Na^+ pump creates the ionic environment that supports and sustains electrical activity in the heart and other excitable cells. The action potential is based on the movement of ions through channels. No net movement through a channel will occur, however, in the absence of an electrochemical gradient for that ion. Thus, the central roles of Na^+ and K^+ currents in the genesis of the action potential are made possible through the Na^+ pump. The Na^+ and K^+ gradients must be maintained to preserve excitability. The small movements of Na^+ and K^+ through channels, which accompany each action potential, must be balanced by equal and opposite movements of Na^+ and K^+ catalyzed by the Na^+ pump to maintain steady-state cellular ionic levels.

The inwardly directed Na^+ gradient created by the active transport of the Na^+ pump has another essential role. The energy stored in the Na^+ gradient can be utilized by a cell to drive other processes. For example, the transport of other substances can be coupled to the downhill movement of Na^+. This can be accomplished through the action of cotransporters or exchangers. An example of a cotransporter that makes use of the Na^+ gradient is the Na^+/glucose cotransporter found in the intestinal brush border. This transporter efficiently facilitates the absorption of intestinal glucose from the gut. The prime example of a Na^+-gradient driven exchanger of great importance to cardiac function is the Na^+-Ca^{2+} exchanger. Essentially, the Na^+-Ca^{2+} exchanger uses the Na^+ gradient to extrude Ca^{2+} from myocardial cells. In this way, the activity of the Na^+-Ca^{2+} exchanger is directly coupled to the Na^+ pump as a means to regulate intracellular Ca^{2+} and hence contractility. The exchanger is discussed in further detail later.

As first shown by Schatzmann (1953), the Na^+ pump is the receptor for the drug digitalis. Digitalis is given to patients with failing hearts as a positive inotropic agent. Digitalis is a specific inhibitor of the Na^+ pump and acts by binding to a site on the extracellular surface of the pump. Partial inhibition of the Na^+ pump leads to a small rise in the intracellular Na^+ concentration. Because of the high level of Na^+-Ca^{2+} exchange activity in heart tissue, a rise in internal Na^+ leads to a higher internal Ca^{2+} level and hence a positive inotropic effect. The actual action of digitalis is somewhat more complicated than just described. The Ca^{2+} transport systems of myocardial cells interact with one another. When the efficiency of Ca^{2+} extrusion by the Na^+-Ca^{2+} exchange system is decreased due to digitalis, more Ca^{2+} is sequestered in the intracellular organelle, the SR, via the SR ATP-dependent Ca^{2+} pump. More Ca^{2+} is then available for release from the SR following cell excitation. Thus, the sequence of events upon myocardial application of digitalis is as follows: (1) partial inhibition of the Na^+ pump, (2) a rise in intracellular Na^+, (3) less Ca^{2+} extrusion by the SL Na^+-Ca^{2+} exchanger, (4) greater SR Ca^{2+} uptake, and (5) increased SR Ca^{2+} release and positive inotropy (see Chapter 5).

Related topics that are discussed in the following sections include progress on identification of the digitalis binding site on the Na^+ pump, Na^+ pump isoforms and their digitalis affinities, and the dependence on internal Na^+ of the Na^+–Ca^{2+} exchanger.

B. REACTION MECHANISM

Biochemical research on the Na^+ pump began with the major discovery by Skou (1957) that crab nerve plasma membranes contained an ATPase activity with the properties expected of an ATP-driven Na^+ pump. Kinetics could now be assessed in cell-free preparations obviating, to a large degree, reliance on more difficult experiments using Na^+ and K^+ isotopes. Since 1957, much detailed information has accumulated; the Na^+ pump has become a system of intense interest to physiologists, biochemists, and, more recently, molecular biologists.

Much information about the Na^+ pump is summarized in what is known as the Post-Albers scheme for the pumping cycle of the Na^+,K^+-ATPase (Fig. 2). Several salient features of the Na^+ pump are displayed. The ion

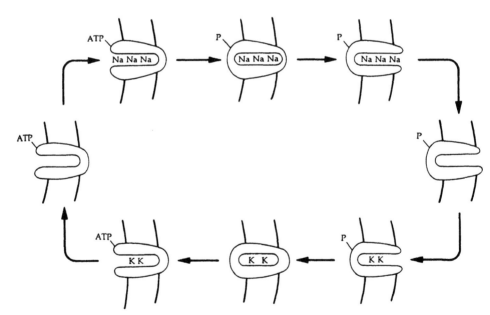

FIGURE 2   Post-Albers scheme for the reaction cycle of the Na,K-ATPase. Starting at the left, three Na^+ bind to the ATP-bound E' state of the protein. A phosphorylated intermediate forms that occludes the ions. After a conformational change, the E" state forms and the Na^+ ions are released allowing two K^+ to bind. Dephosphorylation and occlusion occur. The release of the K^+ is then facilitated by the binding of ATP. (Reprinted from Läuger, 1991, with permission of Sinauer Associates, Inc.)

binding sites have alternating access to either the intracellular or extracellular medium (see Section III). In the state, E_1ATP, for example, the sites face the cell interior. Three Na_i^+ ions can bind to this site, triggering the hydrolysis of ATP, phosphorylation of the enzyme, and occlusion of the Na^+ ions in the interior of the protein. Subsequently, the enzyme converts to the E_2 configuration with access of the ion binding sites to the external medium. The Na^+ ions diffuse away and are replaced by two K^+ ions. Binding of the K^+ ions stimulates dephosphorylation and K^+ occlusion. Binding of ATP stimulates conversion back to an E_1 configuration and release of K^+ internally. The enzyme is now ready to begin a new cycle of Na^+ extrusion and K^+ uptake. Many experiments support the general framework of this scheme. General aspects of this scheme also apply to other P-type ion pumps, such as the SL and SR Ca^{2+} pumps.

In each reaction cycle, the Na^+ pump extrudes three positive charges, carried by Na^+, in exchange for two positive charges, carried by K^+. Thus, there is net movement of one positive charge in the outward direction; that is, the pump is electrogenic, and will generate a current and be affected by membrane potential. Following the influx of Na^+, which initiates the cardiac action potential, Na^+ extrusion by the Na^+ pump will tend to repolarize myocardial cells by a few millivolts.

Elegant studies have characterized the electrogenicity of the Na^+ pump. Transport by the Na^+ pump is affected by transmembrane voltage. From the reaction scheme (see Fig. 2), it is unclear which specific steps will be influenced by voltage. This will be determined by the movement of intrinsic charges on the protein itself, by the location of ion binding sites with respect to the electric field, and by other factors. It appears that release of Na^+ at the extracellular surface of the pump is the predominant voltage-sensitive step. This model has been refined; it may be specifically the first Na^+ to be released that is the major charge-carrying step. Relevant references include Gadsby *et al.* (1993), Hilgemann (1994), and Wuddel and Apell (1995).

C. THE MOLECULE

The Na^+,K^+-ATPase is composed of two subunits. The α-subunit contains about 1000 amino acids with a molecular weight of about 110 kDa, while the β-subunit is much smaller with a molecular weight of about 35 kDa. All catalytic activity appears to reside on the α-subunit. The α-subunit contains the ATP binding site, the aspartate residue involved in the phosphorylated intermediate, and the binding site for ouabain, a digitalis-like compound. The β-subunit is heavily glycosylated, contains one transmembrane segment, and is of unknown function. It has not yet been unequivocally resolved whether the functional Na^+ pump in a biological membrane is an $\alpha\beta$ monomer or a $\alpha_2\beta_2$ dimer. A β-subunit is tightly bound to each α-subunit, and functional α-subunits cannot be isolated or expressed

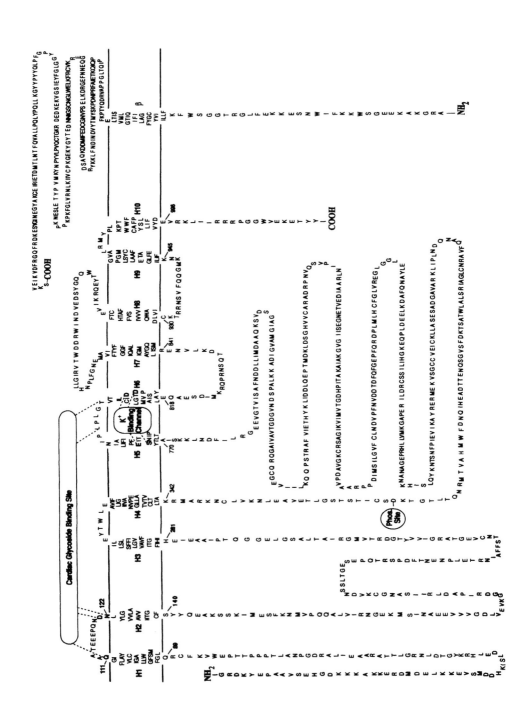

in the absence of β-subunits. It has been proposed that the main function of the β-subunit is to ensure proper processing and targeting of the α-subunit, although the β-subunit may also contribute to the overall structure of the α-subunit. Detailed reviews on possible roles of the β-subunit can be found in Fambrough et al. (1994) and Chow and Forte (1995).

The α-subunit of Na^+,K^+-ATPase was first cloned in the laboratories of Lingrel (Shull et al., 1985) and Numa (Kawakami et al., 1985). This accomplishment greatly expanded the molecular approaches available to investigate the Na^+ pump. A current secondary structure model of the Na^+ pump is shown in Fig. 3. The model shows the pump with 10 putative transmembrane segments with both the amino and carboxyl termini located intracellularly.

Many of the molecular aspects of the Na^+ pump are detailed in recent reviews. These include Lingrel et al. (1990), Lingrel and Kuntzweiler (1994), Canessa et al. (1994), and Horisberger (1994). Much of the material in the following sections is also reviewed in these articles.

D. TOPOLOGY

The topological arrangement of the α-subunit of the Na^+ pump across the plasma membrane has been an area of intense investigation. Interpretation of data on transporter topology is rarely unequivocal. Models have been proposed with 6–10 α-helical transmembrane segments. Only recently has a consensus begun to build that indeed the Na^+ pump has 10 transmembrane segments. Results from several laboratories suggest that the 10-transmembrane segment model may be correct. This conclusion appears to apply to other members of the P-type ATPase family as well. Simple topological models may yet prove to be incorrect: All transmembrane regions may not be α-helical. Also, some parts of the protein may move in or out of the membrane during different portions of the reaction cycle.

E. ISOFORMS

cDNA's coding for three different α-subunits has been cloned. These cDNAs are the products of distinct genes. The encoded proteins are referred to as the $\alpha 1$-, $\alpha 2$-, $\alpha 3$-subunits and have 1023, 1020, and 1013 amino acids,

FIGURE 3   Primary sequence and secondary structure model of the sheep Na,K-ATPase. Shown are both the α-subunit with 10 transmembrane spans and the β-subunit (right) with a single transmembrane segment. The extracellular surface is on the top. Some features of note include residues involved in glycoside binding and K^+ translocation. Also indicated is the aspartate (D) phosphorylated by ATP during the reaction cycle. (Courtesy of Drs. T. Kuntzweiler and J. Lingrel.)

respectively. The different α-isoforms are about 85% identical to each another at the amino acid level. The isoforms have distinct tissue distributions; for example, human kidney cortex appears to express only the $\alpha1$-isoform, whereas human heart expresses all three isoforms. A putative fourth Na^+,K^+-ATPase α-subunit is expressed in testis but has not yet been functionally analyzed (Shamraj and Lingrel, 1994). Two isoforms of the β-subunit ($\beta1$ and $\beta2$) of unknown functional difference are known to exist. Only the $\beta1$ protein has been detected in the heart. The kinetic behaviors of the different α-subunits appear to be generally similar. However, the $\alpha3$-isoform has been observed to have a threefold higher affinity for extracellular K^+ and a fourfold lower affinity for cytoplasmic Na^+ than observed for $\alpha1$ and $\alpha2$. These differences may partly depend on the presence of the $\beta1$- and $\beta2$-subunit (Munzer et al., 1994; Therien et al., 1996). Further research on possible differences in the kinetic characteristics of the α- and β-subunits is necessary.

The physiological rationale for the existence of multiple isoforms of the Na^+ pump is not completely clear. Expression of the α-subunits is tissue specific, and perhaps different cell types require pumps with different kinetic properties for optimal physiological function. The presence of different isoforms also allows for the possibility for differential regulation, either acutely or long term. For example, the activity of Na^+ pump isoforms could be regulated by phosphorylation in different manners. In fact, it is known that up- and down-regulation of α-subunit protein levels can be controlled by different humoral factors (see p. 154). Targeting of Na^+ pumps to the proper membrane may also be affected by isoform composition. Targeting issues would be most relevant to epithelial cells where Na^+ pumps are often present only in the basolateral membrane.

In cardiac tissue, the presence of specific α-subunits has much pharmacological significance. The three α-subunits have differential sensitivities to digitalis. The differences are most substantial in the rat; the rat $\alpha1$-isoform has a very low ouabain affinity (50 μM), whereas the rat $\alpha2$- and $\alpha3$-isoforms have much higher affinities (1–100 nM). Thus, therapeutic doses of digitalis would inhibit high-affinity forms of the Na^+ pump but would not affect low-affinity forms. The human heart expresses all three isoforms of the α-subunit of the Na^+ pump. However, the ouabain affinities of each of the α-subunit isoforms have not been determined. The values cannot be extrapolated from the rat data as substantial species differences exist. Nevertheless, relative ouabain affinities of pump isoforms are of clinical relevance. The relative expression of pump isoforms may change during heart failure, affecting the appropriate digitalis dose for these patients. Review articles on the relationship between the ATP-dependent Na^+ pump and digitalis treatment include Medford (1993) and McDonough et al. (1995).

F. STRUCTURE AND FUNCTION

1. ATP Binding Site

An aspartate residue on the large intracellular loop between transmembrane segments 4 and 5 is phosphorylated by ATP as an integral part of the catalytic cycle (see Fig. 3). In the sheep α1-subunit of the Na$^+$ pump, this is asp-369. The location of the phosphorylated residue suggests that the binding site for ATP is on the cytoplasmic loop. Indeed, many experiments using ATP analogues to label the Na$^+$ pump have confirmed this possibility. Somehow the binding of ATP and phosphorylation in the cytoplasmic loop affects the translocation of Na$^+$ and K$^+$ through the transmembrane segments. Long-range interactions must exist whereby conformational changes in the cytoplasmic region are transduced to the ion translocation pathway. All the P-type ATPases are homologous in the ATP binding domain, implying similar protein conformations and reaction mechanisms.

2. Cation Binding Sites

It is assumed for the ATP-dependent Na$^+$ pump (and for other cation transporters) that the cations must interact with acidic and hydrophilic groups to traverse the membrane. Two general approaches have been applied to assess this hypothesis. First, chemical modifying agents have been used to react with anionic residues and, more recently, site-directed mutagenesis has been utilized. So far, the data indicate that aspartate and glutamate residues located in at least four different transmembrane segments are candidates to form portions of the ion binding sites. Thus, it is likely that a specific three-dimensional arrangement of the transmembrane segments is required for ion binding. More progress has been made in identifying the Ca^{2+} binding site within the SR Ca^{2+}ATPase (see p. 161).

3. Ouabain Binding Site

The ATP-dependent Na$^+$ pump is the pharmacological receptor for cardiac glycosides (e.g., ouabain or digoxin, which bind from the extracellular surface). Initial cross-linking studies suggested that the binding site was on the α-subunit but were unable to define the binding site. Much recent progress has been made using site-directed mutagenesis. The technique is to transfect cells with DNA coding for a ouabain-sensitive Na$^+$,K$^+$-ATPase. If the transfected Na$^+$ pump is first mutated, the cells will survive in the presence of ouabain only if the mutation disrupts the ouabain sensitivity of the Na$^+$ pump. Most mutations that decrease ouabain sensitivity are located in the first extracellular loop and in the frist transmembrane segment (see Fig. 3). However, mutations to residues in other portions of the

α-subunit also affect ouabain affinity. Thus, it appears that multiple locations form the ouabain binding site on the Na^+ pump.

G. REGULATION OF Na^+ PUMP ACTIVITY

Both short- and long-term regulation of the Na^+ pump are described. Short-term regulation is the modulation of existing plasma membrane Na^+ pumps on a short time scale (minutes), often as the result of a second messenger. Long-term regulation occurs as an adaptation to a chronic change in some humoral or vascular parameter. Changes in protein synthesis or mRNA metabolism are usually involved.

1. Short-Term Regulation

Short-term regulation of the Na^+ pump by hormonal factors can be demonstrated in intact cells but is often absent in subcellular preparations. Conditions that promote phosphorylation tend to inhibit the Na^+ pump. In general, these effects have not been extensively studied in the heart, and physiological significance in this organ are unclear. Short-term regulation of the Na^+ pump has been reviewed by Bertorello and Katz (1995) and McDonough and Farley (1993).

2. Long-Term Regulation

The Na^+ pump isoforms are regulated differentially in cardiac muscle. This has been studied most extensively in the rat heart, which expresses the $\alpha1$- and $\alpha2$-isoforms in about a 10 to 1 ratio at the protein level. The $\alpha1$-subunit of the Na^+ pump is constitutively present in all tissues. Most long-term regulation is due to changes in expresion levels of the $\alpha2$- and $\alpha3$-isoforms. Examples of interventions that alter cardiac Na^+ pump levels are thyroid status, hypokalemia, and hypertension. Thyroid hormone up-regulates Na^+ pump levels in several tissues as part of a general metabolic response. The hypo- or hyperthyroid heart has many altered contractile and electrophysiological properties. It is difficult to determine which alterations are consequences of Na^+ pump regulation. Nevertheless, effects can be quite dramatic. For example, protein levels of the $\alpha2$-subunit are 15 times higher in the hyperthyroid rat heart than in the hypothyroid rat heart.

The major finding in hypokalemia is a down-regulation of the $\alpha2$-subunit in cardiac and, especially, skeletal muscle. The down-regulation of the Na^+ pump in skeletal muscle may be an adaptive response to help shift K^+ from an intracellular store into the plasma. The $\alpha2$-subunit is also down-regulated in various models of hypertension. Long-term regulation of the Na^+ pump has been reviewed in McDonough et al. (1992a, 1992b), McDonough and Farley (1993), and Clausen (1996).

In summary, regulation of intracellular Na^+ by the ATP-dependent Na^+ pump has a key role in cardiac contractility. The Na^+ extrusion and K^+ influx mediated by the Na^+ pump are essential in maintaining cellular excitability. In addition, the Na^+ pump indirectly regulates contractility by modulating Na^+–Ca^{2+} exchange activity via its effects on intracellular Na^+ levels.

V. THE PLASMA MEMBRANE ATP-DEPENDENT Ca^{2+} PUMP

Readers are referred to any of several excellent reviews (Carafoli, 1992; Wang *et al.,* 1992; Carafoli, 1994; Strehler, 1996; Penniston and Enyedi, 1996) for more detail on the plasma membrane Ca^{2+} pump than presented in this chapter.

A. CELL PHYSIOLOGY

The PM ATP-dependent Ca^{2+} pump is an ubiquitous Ca^{2+} extrusion mechanism. Ca^{2+} extrusion is essential for all cells to maintain Ca^{2+} homeostasis. The ability to expel Ca^{2+} is most important in those cells in which substantial fluxes of Ca^{2+} across the PM occur as part of repetitive signaling pathways. Examples would be neurons and myocardial cells.

The PM Ca^{2+} pump and the Na^+–Ca^{2+} exchanger "compete" as cellular Ca^{2+} efflux pathways. In many cell types, the ATP-dependent Ca^{2+} pump is the prime mover of Ca^{2+} out of the cell. In myocardial cells, however, the Na^+–Ca^{2+} exchanger is by far the dominant Ca^{2+} efflux mechanism. Experimentally, it is difficult to demonstrate that the PM (sarcolemmal) Ca^{2+} pump has any functional significance in cardiac muscle. Nevertheless, the Ca^{2+} pump has a much higher affinity for Ca^{2+} than the Na^+–Ca^{2+} exchanger. In addition, as described later, it is subject to a high degree of regulation. Thus, the SL ATP-dependent Ca^{2+} pump may be responsible for "fine tuning" Ca^{2+} levels in myocardial cells. Such a role would be difficult to detect and may be unappreciated.

B. BIOCHEMISTRY

The presence of an ATP-dependent Ca^{2+} pump in red cell membranes was first detected by Schatzmann in 1966. The PM Ca^{2+} pump is not as amenable to biochemical analysis as some of the other P-type pumps. First, the PM Ca^{2+} pump is present in only low abundance. There is no tissue in which this pump is highly concentrated. This contrasts with the Na^+ pump and the SR Ca^{2+} pump, which can be found in high concentration in kidney and skeletal muscle, respectively. Second, in most cell types, there is always the problem that measurements of the Ca^{2+} pump in PM preparations can be compromised by the presence of Ca^{2+} pumps in contaminating membranes from intracellular organelles. This second point explains why the red cell plasma membrane has been the system of choice for those in the field. The plasma membrane is the only membrane present in erythrocytes, thus contaminating membranes from internal organelles are not a problem.

Clear demonstration of an SL ATP-dependent Ca^{2+} pump in cardiac muscle was not readily possible. It was difficult to distinguish the SL and SR Ca^{2+} pumps. Two studies overcame this problem (Caroni and Carafoli, 1980; Trumble et al., 1980). These studies took advantage of the fact that SL membrane vesicles would contain both the ATP-dependent Ca^{2+} pump and the $Na^+–Ca^{2+}$ exchanger. SL vesicles were loaded with Ca^{2+} via an ATP-dependent pumping mechanism. The Ca^{2+} could then be released from the vesicles via the $Na^+–Ca^{2+}$ exchanger by the addition of Na^+ to the medium. Thus, the vesicles loaded with Ca^{2+} had to be sarcolemmal. This provided the first unequivocal evidence that an ATP-dependent Ca^{2+} pump was inherent to the SL membrane, and was not due to SR contamination.

The Ca^{2+} pump comprises only 0.1% of the total membrane protein of a red cell and is difficult to purify. Successful purification was achieved by Niggli et al. (1979) through the use of a calmodulin-affinity column and shown to have an apparent molecular mass of about 140 kDa. Like other P-type ion pumps, the transfer of the γ-phosphate from ATP to an aspartyl residue is an essential step in the reaction cycle. The PM Ca^{2+} pump extrudes one Ca^{2+} with the consumption of one ATP molecule. This stoichiometry provides sufficient energy for the pump to extrude Ca^{2+} against a 1000-fold Ca^{2+} gradient. The affinities for Ca^{2+} on the two sides of the Ca^{2+} pump are asymmetrical. On the cytosolic side, the affinity is less than 1 μM (subject to regulation), while on the extracellular side Ca^{2+} affinity is in the millimolar range. This asymmetry facilitates the transfer of Ca^{2+} against a Ca^{2+} gradient.

The PM Ca^{2+} pump operates as a cation exchanger. One H^+ is taken up into a cell in exchange for each Ca^{2+}. This stoichiometry of 1 Ca^{2+}/ 1 H^+ implies that the PM Ca^{2+} pump should be electrogenic. However, the electrogenicity of the Ca^{2+} pump has been apparent in some, but not all, investigations.

C. MOLECULAR BIOLOGY

Two groups reported the cloning of PM Ca^{2+} pumps in 1988 (Shull and Greeb, 1988; Verma et al., 1988). The Ca^{2+} pumps have about 1200 amino acids with a predicted molecular mass of 135 kDa. The PM Ca^{2+} pumps, as expected, have overall similarities to other P-type pumps. The sequence identity between PM and SR Ca^{2+} pumps is, for example, about 30%. Regions that are most highly conserved include the nucleotide binding site and the site of formation of the phosphorylated intermediate. A schematic representation of the PM Ca^{2+} pump is shown in Fig. 4.

The PM Ca^{2+} pump is modeled to have 10 transmembrane segments. The nucleotide binding site and the phosphorylation site are located in a large cytoplasmic loop between transmembrane segments 4 and 5. Most of the mass of the protein is intracellular with only short loops exposed on the extracellular surface. A striking feature of the PM Ca^{2+} pump is the long carboxyl-terminal extension. This cytoplasmic domain is approximately 150 amino acids and is responsible for many regulatory features as is discussed in the next section.

At least four separate genes code for PM Ca^{2+} pumps. These isoforms are known as PMCA1, -2, -3, and -4. In addition, each of these four isoforms undergoes alternative splicing. With this rich diversity, about 20 different variants of PM Ca^{2+} pumps have been described. A study (Stauffer *et al.*, 1995) has examined isoform distribution using isoform-specific antibodies. Two isoforms, PMCA1 and PMCA4, are expressed in all tissues, including the heart. PMCA2 and PMCA3 were only found in neuronal tissues. Alternative splicing occurs at two locations in PMCA transcripts. One location is in the cytoplasmic loop between transmembrane segments 2 and 3 and the other is in the calmodulin-binding domain near the carboxyl terminus of the protein. The different splice variants have different regulatory properties as is described as follows.

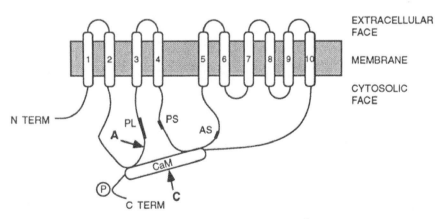

FIGURE 4   Topological model of the plasma membrane Ca^{2+} ATPase. As with all the P-type pumps, most of the bulk of the protein is on the cytosolic surface. The pump is shown in the autoinhibited state with the calmodulin-binding domain (CaM) bound to two intracellular sites. The catalytic phosphorylation site (PS) and a portion of the ATP-binding site (AS) are indicated. Other sites that are noted are a site of regulatory phosphorylation (circled P), a phospholipid interaction site (PL), and sites of alternative splicing A and C. (Courtesy of Dr. E. Strehler.)

D. REGULATION

A prominent feature of the PM Ca^{2+} pump is the multiplicity of mechanisms involved in regulation of activity. The rationale for the high degree of regulation is at least twofold. First, in the absence of regulation, the Ca^{2+} pump might deplete cells of Ca^{2+}. Second, regulation allows the pump to be activated, as necessary, during signaling processes. Mechanisms and molecules that activate the PM Ca^{2+} pump include calmodulin, acidic lipids, proteolysis, phosphorylation, and dimerization. Much molecular information on these regulatory pathways is now available. Most of the regulation involves the long cytoplasmic carboxyl terminus not present on other P-type pumps. Regulatory properties differ somewhat for the Ca^{2+} pump isoforms.

In general, activation of the pump both increases the V_{max} and decreases the K_m (Ca^{2+}). The latter effect is more dramatic. For example, calmodulin lowers the K_m (Ca^{2+}) from about 10 to 0.5 μM.

1. Calmodulin

The PM Ca^{2+} pump was one of the first enzymes to be found to be regulated by Ca^{2+}/calmodulin. Calmodulin binds to the pump in a Ca^{2+}-dependent manner with high affinity (K_D as low as 1 nM). Information on the calmodulin binding site on the pump was initially obtained with detailed analyses of pump fragments created by proteolysis. The full-length isolated pump is in an inactivated state but can be activated by the addition of calmodulin. Mild proteolysis activates the pump by reducing the size to 81 or 76 kDa. These activated forms of the pump no longer bind calmodulin. The experiments led to the determination that proteolysis removed an autoinhibitory portion of the carboxyl terminus of the pump, which included a calmodulin binding site.

A model has evolved in which the calmodulin binding site is an autoinhibitory portion of the protein. In the absence of calmodulin, this region interacts with another region of the protein to produce inhibition. The binding of calmodulin eliminates this interaction and relieves the inhibition. Consistent with this model, a 28-residue peptide that mimics the sequence of the calmodulin binding domain is able to inhibit the proteolyzed, activated Ca^{2+} pump by reintroducing an autoinhibitory domain.

Cross-linking experiments (Falchetto et al., 1991, 1992) have identified sites that interact with the calmodulin binding domain of the carboxyl terminus. Interestingly, one of the sites of interaction is located on the large intracellular loop between transmembrane segments 4 and 5 not far from the aspartic acid phosphorylation site (see Fig. 4). Thus, the calmodulin binding domain may autoinhibit by binding near the phosphorylation site and preventing access of substrate to the active site.

The preceding discussion should have made clear the mechanism of activation by limited proteolysis. Proteolysis cleaves off the calmodulin binding domain and thus removes autoinhibition. It is unclear whether this mechanism of activation ever has a functional role *in vivo*.

Interestingly, substantial alternative splicing occurs in the carboxyl-terminal regions of the PM Ca^{2+} pump. Some splice variants have sequences that alter the calmodulin binding site and have altered regulatory properties. For example, some splice variants are more basic than others and subsequently have higher affinities for calmodulin. Also, a splice variant of PMCA1 has some basic residues replaced with histidines and, in this case, calmodulin binding is pH dependent.

2. Other Activators

a. Phosphorylation

The Ca^{2+} pump of both red cell membranes and cardiac SL is stimulated by cAMP-dependent kinase (PKA). The site of PKA phosphorylation is on the carboxyl terminus. Phosphorylation may exert a stimulatory effect by decreasing the potency of the autoinhibitory domain. In this way, the mechanism of action would be analogous to that of calmodulin. The Ca^{2+} pump can also be phosphorylated by protein kinase C (PKC). Activation by PKC is modest, however, and does not occur in all cells.

b. Dimerization

The purified PM Ca^{2+}ATPase aggregates to form dimers at high concentration. The dimerized pump is fully activated. Dimerization involves interaction between carboxyl-terminal segments and thus activation may again be mediated through the autoinhibitory domain. The concentration of pumps is quite low in the plasma membrane, however, and there is no evidence that dimerization occurs *in vivo*.

c. Ca^{2+} Binding

A recent finding is that the carboxyl terminal tail of the Ca^{2+} pump is capable of binding up to three Ca^{2+} ions (Hofmann *et al.*, 1993). One of these Ca^{2+} binding sites has extremely high affinity and is probably constitutively occupied by Ca^{2+}. The other two sites have affinities of about 30 and 300 nM and may be involved in regulation. Ca^{2+} bound to these sites is not transported as this cytoplasmic Ca^{2+} binding site can be removed by proteolysis with maintenance of Ca^{2+} transport function.

d. Lipids

The PM Ca^{2+} pump can be activated by the presence of acidic phospholipids. Optimal levels of acidic phospholipids can stimulate the pump to an even greater degree than calmodulin. Two sites on the Ca^{2+} pump appear to interact with phospholipids. One is located on the carboxyl terminus whereas the other is located between transmembrane segments 2 and 3.

Levels of common acidic phospholipids such as phosphatidylserine are probably invariant in the cell membrane. However, levels of phospatidylinositol (PI) and phosphorylated products of PI may change rapidly during cell signaling processes and may be important in regulating pump activity.

In summary, the sarcolemmal ATP-dependent Ca^{2+} pump appears to have a modest role in myocardial Ca^{2+} homeostasis. The Ca^{2+} pump is subject to a high degree of regulation, which has been characterized in detail. A possible regulatory role for the Ca^{2+} pump in cardiac muscle has perhaps been overlooked.

VI. THE SARCOPLASMIC RETICULAR ATP-DEPENDENT Ca^{2+} PUMP

General mechanistic features of the other P-type cation pumps also apply to the SR Ca^{2+} pump. In this section, the role of the SR Ca^{2+} pump in E–C is discussed as well as aspects unique to this transporter. First, large quantities of this protein can be purified facilitating structural studies. Second, initial mutagenesis work has been successful in obtaining information on the nature of the Ca^{2+} binding site within the transmembrane helices. Third, the SR Ca^{2+} pump is regulated by a separate protein, phospholamban. Appropriate detailed reviews include Lytton and MacLennan (1992), Inesi *et al.* (1992), Lompré *et al.* (1994), and Tada and Kadoma (1995).

A. EXCITATION–CONTRACTION COUPLING

Contractile Ca^{2+} has two sources: cellular Ca^{2+} via L-type Ca^{2+} channels (see Chapter 3) entry and SR Ca^{2+} release. The amount of Ca^{2+} that enters the myoplasm from the SR must be subsequently resequestered to bring about relaxation. This component of Ca^{2+} is 80–90% of the total contractile Ca^{2+} in cardiac muscle. An ATP-dependent Ca^{2+} pump (Ca^{2+}ATPase) accomplishes this task. The Ca^{2+} pump is localized in the longitudinal SR and is much less evident in the terminal cisternae. The Ca^{2+} pump is highly enriched in the SR membrane and can compose up to 50% of SR protein.

Myocardial contractility is highly dependent on SR Ca^{2+} load. Higher loads lead to greater Ca^{2+} release and increased contractility. Several factors influence SR Ca^{2+} load, including the activities of all other Ca^{2+} channels and transporters. The SR Ca^{2+} pump of cardiac muscle, however, is also subject to substantial regulation. Regulation is mediated by the phosphorylation of a second protein, phospholamban, discussed in more detail in this section.

B. BIOCHEMISTRY

The SR Ca^{2+} pump is composed of a single polypeptide chain with an apparent molecular weight of 110 kDa. The enzyme catalyzes the transfer of two Ca^{2+} ions into the lumen of the SR at the expense of one ATP molecule. Protons are exchanged for the Ca^{2+} but this is not obligatory. The Ca^{2+}/H^{+} stoichiometry varies with pH and has a maximum value of 1. The enzyme was first purified by MacLennan (1970).

A low-resolution structural model of the SR Ca^{2+} pump has been obtained by applying electron microscopy and electron diffraction techniques to two-dimensional crystalline arrays and to three-dimensional microcrystals of the isolated Ca^{2+}ATPase. A model has emerged in which a pear-shaped cytoplasmic domain is attached to an intramembrane stalk of transmembrane helices (Fig. 5). The general features of the model are consistent with the amino acid sequence of the Ca^{2+} pump and with a variety of biochemical and biophysical data.

C. SR Ca^{2+} PUMP GENE FAMILY

MacLennan *et al.* (1985) were the first to clone the cDNA for an SR Ca^{2+} pump. Ca^{2+} pumps are encoded by three distinct genes and are known as SERCA1, SERCA2, and SERCA3. In addition, two splice variants of SERCA1 (SERCA1a and SERCA1b) and SERCA2 (SERCA2a and SER-CA2b) are known. The predominant Ca^{2+} pumps in adult myocardium and fast-twitch skeletal muscle are SERCA2a and SERCA1a, respectively. When expressed in cultured cells, SERCA1a and SERCA2a have very similar kinetic properties.

A model of the topological arrangement of the Ca^{2+} pump is shown in Fig. 5. As with the other P-type pumps, 10 transmembrane segments are shown (SERCA2b has additional amino acids and is modeled to include an eleventh transmembrane segment). The nucleotide binding and phosphorylation domains, between transmembrane segments 4 and 5, show substantial similarity to the homologous domains of other pumps.

D. Ca^{2+} BINDING SITES

It is difficult to determine the binding sites for transported Ca^{2+} within the transmembrane segments of the Ca^{2+} pump. First, the three-dimensional arrangement of the helices is unknown. Second, the binding sites are likely to involve amino acids on adjacent helices that can be widely separated in the linear sequence. Thus, one cannot search for known Ca^{2+} binding motions such as EF hands. An elegant study by Clarke *et al.* (1989) provided evidence that six polar residues, all located toward the center of four different transmembrane segments, are involved in the binding and translo-

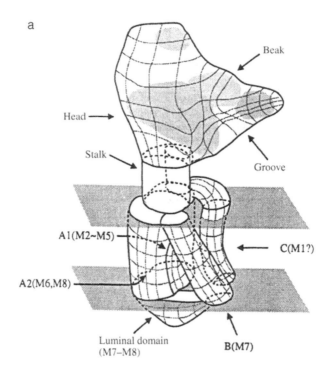

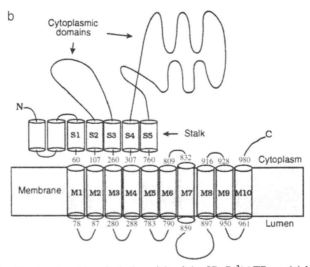

FIGURE 5   Structural and topological models of the SR Ca^{2+}ATPase. (a) Model of the Ca^{2+} pump based on three-dimensional cryo-electron microscopy. (b) Predicted secondary structure of the Ca^{2+} pump. (For more detail, see Toyoshima *et al.*, 1993.) [Reprinted with permission from Toyoshima *et al.* (1993). *Nature* 362:471. Copyright 1993 Macmillan Magazines Limited.]

cation of Ca^{2+}. The critical residues are located on transmembrane segments 4, 5, 6, and 8. Biochemical data indicate that pumps mutated at any one of these six polar residues are unable to bind Ca^{2+}.

These experiments have provided a starting point for studies on the roles of other transmembrane residues and on the arrangement of transmembrane helices.

E. PHOSPHOLAMBAN

β-adrenergic stimulation of cardiac muscle produces a marked positive inotropic effect. A large part of the effect can be attributed to a stimulation of the SR Ca^{2+} pump. A stimulated Ca^{2+} pump will increase the rate of Ca^{2+} resequestration that increases the rate of muscle relaxation. In addition, stimulated Ca^{2+} pumping will increase the SR Ca^{2+} load. More Ca^{2+} is then available for release resulting in an increase in contractile force (see Chapter 5).

The β-adrenergic stimulation of SR Ca^{2+} pump activity is due to protein kinase A (PKA)-dependent phosphorylation. However, the protein that is phosphorylated is not the Ca^{2+} pump itself, but another SR protein, phospholamban. Phospholamban is a small protein of 52 amino acids that forms homopentamers (Fig. 6). The first 29 amino acids from the N-terminus are located cytoplasmically and the C-terminal 23 amino acids are modeled to form an α-helix spanning the SR membrane. In the nonphosphorylated state, phospholamban associates with the Ca^{2+} pump and inhibits transport. PKA phosphorylates phospholamban on serine 16. After phosphorylation, phospholamban dissociates from the Ca^{2+} pump and inhibition is relieved. The major kinetic effect on the Ca^{2+} pump is to increase the apparent affinity for Ca^{2+} by about threefold.

Phospholamban is present in cardiac SR, but is completely absent in SR from fast-twitch skeletal muscle. Consistent with this is the lack of any effect of PKA on skeletal muscle SR Ca^{2+} pumping. The difference is not due to the presence of SERCA1 vs SERCA2. If phospholamban and the SERCA1 isoform of skeletal muscle are coexpressed in COS cells, PKA sensitivity can be induced in the skeletal muscle Ca^{2+} pump. Phospholamban is also phosphorylated by Ca^{2+}/calmodulin-dependent kinase on threonine 17. The effect of this phosphorylation is similar to the effect of phosphorylation at serine 16, although the physiological significance is not as clear.

Cross-linking experiments have defined the site of interaction of phospholamban on the Ca^{2+} pump. The site is near the aspartate of the Ca^{2+} pump that forms the high-energy phosphorylated intermediate in the catalytic cycle of the Ca^{2+} pump. This is an important region in Ca^{2+} pump function, and interaction at this site provides a molecular rationale for the inhibitory effects of phospholamban. A "knock out" mouse in which the phospholamban gene has been ablated has received much recent interest

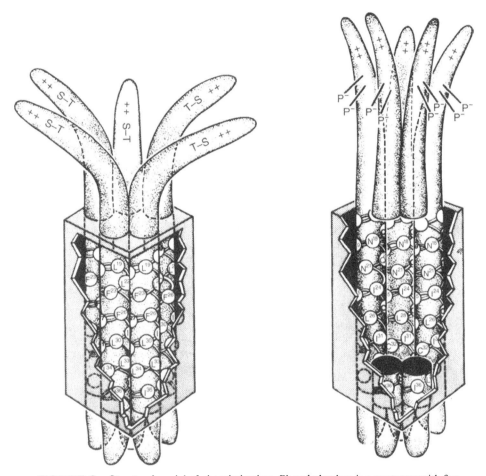

FIGURE 6   Structural model of phospholamban. Phospholamban is a pentamer with five identical subunits. Each subunit spans the SR membrane one time. The head groups in the cytoplasm interact with the Ca^{2+} pump to inhibit pump activity. Upon phosphorylation of a specific serine (S) or threonine (T), phospholamban undergoes a conformational change (right panel); the interaction with the Ca^{2+} pump is broken and inhibition is relieved. (Reprinted from Simmerman *et al.*, 1986, with permission from the American Society for Biochemistry and Molecular Biology.)

(Luo *et al.*, 1994). Generally, this genetic model provides extremely strong evidence for the model of phospholamban function previously presented. Hearts from phospholamban-deficient mice exhibited enhanced inotropy in the absence of a β-agonist. The myocardial performance of wild-type mice became similar to that of the phospholamban-deficient mice only after isoproterenol treatment. All data indicated that phospholamban is a key repressor of contractility and a major target for β-adrenergic agonists.

The laboratories of Tada and Jones have made many molecular advances in phospholamban research, although several others have also made important contributions. Phospholamban is reviewed in detail in Tada and Kadoma (1995).

In summary, the SR Ca^{2+} pump rapidly sequesters contractile Ca^{2+} following excitation. Regulation of the Ca^{2+} pump by phospholamban is a major pathway for controlling contractile strength in cardiac muscle. Progress has been made in determining the Ca^{2+} translocation pathway through the Ca^{2+} pump protein.

VII. Na^+-Ca^{2+} EXCHANGE

The Na^+-Ca^{2+} exchanger has a central role in cardiac E–C coupling. Some of the functions of the exchanger are well established, while others are controversial. In this section, we discuss the role of the Na^+-Ca^{2+} exchanger in cardiac cell physiology, but then focus on molecular advances in exchanger research. Relevant reviews include Reeves (1995), Hilgemann *et al.* (1996), Philipson (1996, 1997), and Khananshvili (1996).

A. CELL PHYSIOLOGY AND EXCITATION–CONTRACTION COUPLING

Na^+-Ca^{2+} exchange activity was first described in myocardium in 1968 (Reuter and Seitz, 1968) and in the squid giant axon in 1969 (Baker *et al.*, 1969). Some implications of this transport activity were immediately obvious. Baker *et al.* (1969), for example, recognized the potential role of the exchanger in explaining the positive inotropic effect of the drug digitalis.

Cardiac muscle contraction is initiated by Ca^{2+} influx through voltage-dependent Ca^{2+} channels. This Ca^{2+} current induces a larger release of Ca^{2+} from the SR. The component of Ca^{2+} that enters the cell with each contraction must subsequently be extruded to bring about muscle relaxation and to maintain Ca^{2+} homeostasis. It is clear from the work of several laboratories that the Na^+-Ca^{2+} exchanger is the primary Ca^{2+} extrusion mechanism in cardiac muscle. Thus, the exchanger has a key role in regulating myocardial contractility. Interventions that alter Na^+-Ca^{2+} exchange activity also modify contractility.

The exchanger catalyzes the countertransport of three Na^+ ions for one Ca^{2+} ion across the SL membrane. The exchanger effectively uses the energy stored in the outwardly directed Na^+ gradient to extrude cellular Ca^{2+}. Thus, Ca^{2+} extrusion is indirectly powered by the ATP-dependent Na^+ pump. With a stoichiometry of $3Na^+/1Ca^{2+}$, the exchanger is electrogenic. One positive charge is moved in the same direction as the Na^+ ions during each reaction cycle. Exchanger activity thus creates a depolarizing current during the Ca^{2+} extrusion process. Exchanger activity is also modulated by membrane potential. Negative membrane potentials will tend to accelerate exchanger-mediated Ca^{2+} efflux. The exchanger is reversible, however, and the net direction of Ca^{2+} movement is determined by the Na^+ and Ca^{2+} gradients and by membrane potential. Potentially, the exchanger could mediate net Ca^{2+} influx at the depolarizing voltages occurring during an action potential. This possibility is at the center of a controversy in the field of E–C coupling (see Chapter 5).

Although, as described, the exchanger is clearly the dominant Ca^{2+} efflux mechanism of cardiac muscle, some investigators additionally ascribe an important physiological role for the exchanger as a Ca^{2+} influx pathway. This hypothesis takes two forms: depolarization-induced Ca^{2+} influx and Na^+ current-induced Ca^{2+} influx. These possibilities are discussed briefly in turn.

During the plateau of the cardiac action potential, reverse exchange will at least transiently be favored due to depolarization. (The exchanger operating in the Ca^{2+} influx mode is often referred to as *reverse exchange*.) The question arises as to whether reverse exchange, like the Ca^{2+} current (I_{Ca}), can trigger SR Ca^{2+} release. One way to test this possibility is to study E–C coupling after blockade of I_{Ca} by various channel blockers. Indeed, under these conditions, a Ca^{2+} influx role for the exchanger can be clearly demonstrated. However, the conditions used in these experiments are often not completely physiological. Special attention must be paid to variables such as the internal Na^+ level and the SR Ca^{2+} load. In addition, the role of the exchanger may be different in the presence of a normal I_{Ca}. For example, I_{Ca} may induce a sufficiently high and rapid Ca^{2+} rise that the exchange will not function in the reverse mode (see Chapter 5). A representative article supporting the depolarization-induced reverse exchange hypothesis is by Levi *et al.* (1994).

A variation of this model is the Na^+ current-induced Ca^{2+} influx hypothesis first introduced by Leblanc and Hume (1990). In this case, Na^+ entering myocytes through Na^+ channels during the initial phase of the action potential causes a rise in Na^+ in a localized subsarcolemmal space. The elevated Na^+ level will then induce Ca^{2+} influx through the Na^+–Ca^{2+} exchanger. Support for this possibility has been reported by Lipp and Niggli (1994).

An aspect of these proposals is the possible need for a proximity of the exchanger and the Ca^{2+} release sites on the SR. The diadic cleft between the T tubule and the SR may be such a region of limited diffusion. These issues have been modeled extensively by Langer and Peskoff (1996) and are also addressed by Langer in this book (see Chapter 5). Sham *et al.* (1995) have provided evidence that Ca^{2+} delivered by I_{Ca} is much more effective than Ca^{2+} influx through the exchanger in inducing SR Ca^{2+} release. It was suggested that there is "privileged cross-signaling" between the dihydropyridine and ryanodine receptors for most efficient E–C coupling. Nevertheless, the role of the exchanger in Ca^{2+} influx remains an area of lively debate. Developments in the near future should help resolve this issue.

Although Na^+–Ca^{2+} exchange activity is especially high in cardiac muscle, activity is substantial in other tissues as well. In general, the physiological role of the exchanger in other tissues has not been as extensively studied as in heart tissue. Nevertheless, the exchanger is important in diverse processes such as synaptic transmission, renal Ca^{2+} reabsorption, and regulation of vascular tone.

B. MOLECULAR BIOLOGY

Isolation of the Na^+–Ca^{2+} exchanger by biochemical techniques led to cloning of the exchanger cDNA. Much new molecular information is now available.

1. Protein Isolation

The Na^+–Ca^{2+} exchanger is a low-abundance protein comprising only about 0.1% sarcolemmal protein. To isolate the exchanger, cardiac sarcolemmal membranes must first be isolated and the membranes solubilized with detergent. The solubilized proteins are then fractionated.

The only "marker" for the exchanger protein, however, is transport activity. Thus, to identify a protein fraction enriched in exchanger activity, each fraction must be reconstituted into liposomes to assay for exchange activity. The cardiac exchanger was isolated using this approach in 1988 (Philipson *et al.*, 1988). The mature exchanger protein has an apparent molecular weight of 120 kDa and is accompanied in the isolation by a proteolyzed active fragment of 70 kDa. Although only small quantities of exchanger could be isolated, polyclonal antibodies that became a valuable molecular tool were produced.

2. Cloning of Na^+–Ca^{2+} Exchangers

The Na^+–Ca^{2+} exchanger of cardiac SL (NCX1) was cloned by Nicoll *et al.* (1990). The cDNA coded for a protein of 970 amino acids. The N-terminal 32 amino acids represent a signal peptide that is cleaved from the protein during biosynthesis in the endoplasmic reticulum. A topological model of the Na^+–Ca^{2+} exchanger is shown in Fig. 7. As shown, the exchanger is modeled to have 11 transmembrane segments. The first group of five proposed α-helices is separated from the second group of six α-helices by a large intracellular loop. The intracellular loop is highly hydrophilic and is composed of 520 amino acids. N-linked glycosylation of unknown function occurs near the extracellular amino terminus, while the carboxyl terminus is modeled to be intracellular. The topological model is based largely on hydropathy analysis. Experimental support exists only for the extracellular

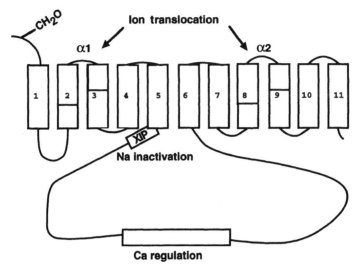

FIGURE 7 Topological model of the Na^+-Ca^{2+} exchanger. Shown are 11 putative trans-membrane segments. The intra- and extracellular media would be at the bottom and top, respectively. The large intracellular loop is involved in Na^+-dependent inactivation and the binding of regulatory Ca^{2+}. The $\alpha 1$ and $\alpha 2$ regions of the transmembrane segments are involved in ion translocation. A site of N-linked glycosylation is designated $-CH_2O$. XIP, exchanger inhibitory peptide.

location of the amino terminus and the intracellular location of the large hydrophilic loop. The cDNAs of two additional Na^+-Ca^{2+} exchangers have subsequently been cloned. These exchangers were both found in a rat brain library and are referred to as NCX2 and NCX3. The three NCXs are products of separate genes and are about 70% identical to each other at the amino acid level. NCX1 is expressed at high levels in the myocardium, but can be detected by Northern blot analysis in almost all tissues. In contrast, both NCX2 and NCX3 transcripts are present only in brain and skeletal muscle by Northern blot analysis. Detailed functional comparisons of the three isoforms have not been completed. NCX-type Na^+-Ca^{2+} exchangers have also been cloned from two invertebrates—*Drosophila* and *C. elegans.*

A distantly related $Na^+/Ca^{2+},K^+$ exchanger is present in high abundance in the outer segments of rod photoreceptors. This exchanger has been cloned (Reiländer *et al.,* 1992) and shows limited similarity to the NCX-type exchangers. Nucleotide sequences coding for other distantly related proteins are appearing in the data bases from large-scale sequencing projects. These sequences code for putative exchangers with as-yet unknown functions. Such sequences are present in human, *C. elegans,* yeast, and *E. coli* genomes.

3. Alternative Splicing

In addition to the presence of three distinct isoforms, the complexity of exchanger proteins is increased by alternative splicing. As characterized in several laboratories, a portion of the large intracellular loop is composed of different combinations of six small exons. At least 12 different splice variants of NCX1 have been described, and 3 splice variants of NCX3 have been detected. The splice variants are generally expressed in a tissue-specific manner, although the functional significance of alternative splicing is still unknown. Thus, there is a variety of Na^+-Ca^{2+} exchanger proteins. For those interested in mammalian cardiac muscle, however, the situation is relatively simple. Only one variant of one isoform (NCX1.1) is expressed in rat cardiac muscle. Most initial structure/function studies have used the cardiac form of the Na^+/Ca^{2+} exchanger.

C. REACTION MECHANISM

The Na^+-Ca^{2+} exchanger operates with a consecutive reaction mechanism; that is, the three Na^+ ions and the one Ca^{2+} ion are countertransported in separate reaction steps. For example, Na^+ ions bind at one surface, are translocated, and are released at the opposite surface. The binding sites are now available to bind one Ca^{2+} ion to initiate the reverse translocation process. Substantial evidence supports this general scheme, and much kinetic data are interpreted within this framework.

It also appears that the ion binding sites have a net negative charge, and some of this charge on the protein accompanies the Na^+ or Ca^{2+} ions through the potential field of the membrane. The nature of the protein charge movement will affect the electrogenicity of different steps in the reaction cycle. Na^+ translocation appears to carry more net charge than Ca^{2+} translocation. Recent noise and charge movement analysis indicates that the exchanger has a relatively high turnover rate ($5000\ s^{-1}$) (Hilgemann, 1996).

Detailed kinetics of the exchanger are likely to be quite complex. For example, kinetics may be affected by the order of binding of Na^+ ions and the different affinities of each Na^+ binding site. Likewise, multiple sites of binding may be involved as ions move through the protein. The site of Ca^{2+} binding will certainly not be equivalent to the binding sites of three Na^+ ions, and Ca^{2+} movement may follow a different pathway than Na^+ movement.

D. STRUCTURE–FUNCTION STUDIES

Biochemical analysis of the Na^+-Ca^{2+} exchanger was initiated by the development of the use of sarcolemmal vesicles to assay for exchange activity (Reeves and Sutko, 1979). Much kinetic and regulatory information was gathered and is reviewed in Reeves and Philipson (1989). The use of sarcolemmal vesicles has continued to be a productive approach. Nevertheless, this chapter focuses on recent advances brought about by the cloning of the Na^+-Ca^{2+} exchanger.

1. The Giant Excised Patch

The giant excised patch, as described by Hilgemann (1990), is the most important new technical advance for studying Na^+-Ca^{2+} exchange. Electrophysiology is a powerful tool for studying electrogenic reactions. However, the more common techniques were not optimal for molecular analysis of the exchanger. The small size of standard excised patches did not allow the small currents associated with transporters to be easily detected. With giant excised patches (about 30 μ diameter), the ionic environment of the Na^+-Ca^{2+} exchanger can be readily manipulated, and the resultant currents recorded. The technique has been especially efficacious for studying regulatory mechanisms (see p. 171), but has also been useful for studying the translocation mechanism. Giant patch technology has been applied both to the native exchanger in sarcolemmal membranes and to cloned exchangers expressed in *Xenopus* oocytes.

2. Ion Translocation

The large intracellular loop of the exchanger is not directly involved in ion translocation (see p. 171). The movement of ions is primarily the domain of the putative transmembrane segments of the exchanger. We have used site-directed mutagenesis to obtain initial information on regions and specific residues involved in ion translocation (Nicoll *et al.,* 1996). Our choice of candidate residues for mutation was aided by the discovery of the "α-repeats" in the exchanger sequence (Schwarz and Benzer, 1997).

The sequence of transmembrane segments 2 and 3 is similar to that of transmembrane segments 8 and 9. These regions have been named $\alpha 1$ and $\alpha 2$, respectively (see Fig. 7). This limited similarity between the first and second groups of transmembrane segments suggests that a gene duplication event was involved in exchanger evolution. The fact that intramolecular homology has been conserved only in the α-repeats further suggests that the α-repeats are especially important in exchanger function.

In an initial series of transmembrane segment mutations, mutations at 18 of 19 positions within the α-repeats resulted in exchanger protein with decreased or no activity. In contrast, mutations of only 3 of 11 residues within transmembrane segments, but outside of the α-repeats, had altered function. Based on this study, our hypothesis is that portions of transmembrane segments 2, 3, 8, and 9 form a portion of the ion translocation pathway.

3. Regulation

There is an interesting dichotomy of regulatory research on the Na^+–Ca^{2+} exchanger. On the one hand, a multitude of factors have been described that modulate exchange activity. On the other hand, it had been difficult to demonstrate that any of these modulatory mechanisms have physiological roles. We briefly review the older findings on this topic. Then, more recent experiments and possible physiological implications are discussed.

a. Vesicles

The ease of the vesicular assay system to monitor Na^+–Ca^{2+} exchange activity facilitated the discovery of different modulatory influences. For example, mild treatment of sarcolemmal vesicles with a proteinase markedly stimulates exchange activity. From more recent work, it appears that proteinases cleave the exchanger in the intracellular loop and relieve self-inhibition of exchange activity. Anionic lipid components (e.g., phosphatidylserine, unsaturated fatty acids) also stimulate vesicular exchange activity. These observations have been confirmed in giant excised patch experiments. The effects of membrane environment on exchange activity has been reviewed by Philipson (1990). Recent research has implicated exchanger–lipid interactions in cellular Ca^{2+} regulation (see p. 173).

Other regulatory influences on the exchanger have been described in vesicle experiments. These include redox modifications, pH, intravesicular Ca^{2+}, and EGTA. The vesicle studies are reviewed in Reeves and Philipson (1989).

b. Ca^{2+}-Dependent Regulation

Two regulatory mechanisms appear to be intrinsic to the exchanger protein: Ca^{2+}-dependent regulation and Na^+-dependent inactivation (see p. 172). Ca^{2+}-dependent exchanger of dialyzed squid axons (DiPolo, 1979), but it is now more readily studied in giant excised patches (Fig. 8). When outward Na^+–Ca^{2+} exchange currents are initiated by application of bath Na^+, full activation additionally requires the presence of micromolar levels of Ca^{2+} in the bath. This Ca^{2+} is not transported but binds to a high-affinity (K_D about 0.1–0.3 μM) regulatory site on the intracellular surface of the exchanger. Thus, the exchanger has distinct Ca^{2+} binding sites for transport and for regulation.

The physiological significance of Ca^{2+} regulation is unclear but at least two possibilities exist. First, rising intracellular Ca^{2+} will activate the exchanger to extrude Ca^{2+}. Second, at diastolic Ca^{2+} levels, Ca^{2+} will dissociate from the binding site, blocking further Ca^{2+} extrusion. This may prevent the exchanger from lowering Ca^{2+} to too low a level.

Some molecular insight into the Ca^{2+} regulatory process has been obtained. Initial deletion mutagenesis of the cardiac exchanger clone indicated

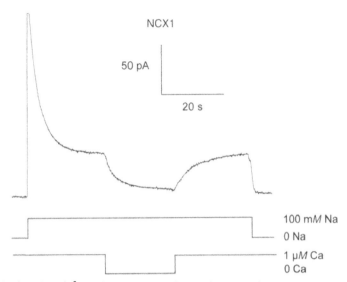

FIGURE 8 Na^+–Ca^{2+} exchange current. Shown is a typical outward exchange current measured in a giant patch excised from an oocyte expressing NCX1. Ca^{2+} is present in the pipette, and the bath composition at the intracellular surface is varied as shown in the bottom traces. Upon application of bath Na^+, outward exchange current is activated. A peak is rapidly achieved that partially inactivates (Na^+-dependent inactivation) to a steady-state plateau value. When regulatory Ca^{2+} is removed from the bath, exchange activity inactivates (Ca^{2+}-dependent regulation). (Courtesy of Dr. L. Hryshko.)

that the large intracellular loop was not required for transport but was essential for regulation by Ca^{2+}. The $^{45}Ca^{2+}$ overlay technique in combination with site-directed mutagenesis then localized a high-affinity Ca^{2+} binding site on the loop. The same mutations that decreased Ca^{2+} binding also affected Ca^{2+} regulation. Thus, the binding site for regulatory Ca^{2+} has been identified (see Fig. 7). (Levitsky *et al.*, 1994; Matsuoka *et al.*, 1995).

c. Na^+-Dependent Inactivation

When Na^+ is applied to the intracellular surface of a giant excised patch, the resultant outward exchange current rapidly peaks but then partially inactivates, over several seconds, to a steady-state level. Inactivation is a process more commonly associated with channels but is also inherent to the Na^+–Ca^{2+} exchanger (Hilgemann *et al.* 1992). The inactivated state can be modeled to arise from the Na^+-loaded exchanger molecule. That is, when three Na^+ ions bind at the intracellular surface, the ions can be translocated, or the exchanger can enter an inactivated state. In the terminology of Hilgemann, this is the I_1 state. In contrast, the inactivated state that forms in the absence of regulatory Ca^{2+} is known as I_2. The physiological significance of I_1 is unclear. Although I_1 and I_2 have been presented as independent phenomena, this is not the case. The conformational changes associated with I_1 and I_2 are likely to have some components in common.

d. Exchanger Inhibitory Peptide

A peptide with the same amino acid sequence as a portion of the intracellular loop of the exchanger is able to inhibit Na^+-Ca^{2+} exchange activity ($K_I \sim 1.5\,\mu M$) (Li et al., 1991). The peptide is known as exchanger inhibitory peptide (XIP) and the region of the exchanger with this sequence is known as the endogenous XIP region (see Fig. 7). It is postulated that the endogenous XIP region is involved in autoregulation of the exchanger. Consistent with this hypothesis, mutations within the XIP region can eliminate Na^+-dependent inactivation.

e. Phosphorylation

The squid Na^+-Ca^{2+} exchanger is strongly modulated by phosphorylation. Clear evidence for phosphorylation of the cardiac exchanger, however, has been lacking. Recent data now directly demonstrate that the cardiac exchanger can be phosphorylated by PKC (Iwamoto et al., 1996). Evaluation of the functional and physiological effects of this phosphorylation should be of much interest.

f. ATP and PIP$_2$

ATP stimulates the Na^+-Ca^{2+} exchanger of excised patches from cardiac myocytes. The stimulation appears to be unrelated to any kinase activity. It had previously been speculated that the ATP activated an aminophospholipid "flippase." The flippase would move anionic phospholipids into the inner leaflet of the sarcolemmal membrane and stimulate exchange activity. Hilgemann and Ball (1996) have recently presented compelling evidence that ATP induces synthesis of phosphotidylinositol 4,5 bisphosphate (PIP$_2$) and that the PIP$_2$ has a strong stimulatory effect on exchange activity. Dynamic regulation of levels of PI and PIP$_2$ may exert important influences on exchanger activity in vivo.

In summary, Na^+-Ca^{2+} exchange activity exerts a strong regulatory influence on myocardial Ca^{2+} homeostasis and on contractility. Exchange activity is altered by intracellular Na^+ and by the membrane potential. Other factors may exert complex regulatory influences.

VIII. OTHER TRANSPORTERS: Na^+-H^+ EXCHANGE

Ion transporters, such as the $Cl^--HCO_3^-$ exchanger, $Na^+-HCO_3^-$ cotransporter, and other Na^+-coupled transporters are also present in myocardial membranes. These transporters are undoubtedly important in myocardial function, but do not have key roles in E–C coupling and have not been discussed. However, another sarcolemmal transporter, the $Na^+–H^+$ exchanger, deserves a brief presentation. Although primarily involved in cellular pH regulation, the $Na^+–H^+$ exchanger can potentially move substantial amounts of Na^+ into myocardial cells with pathophysiological consequences. The general properties of the $Na^+–H^+$ exchanger are first presented to set the stage for the possible role of this exchanger in ischemia and reperfusion. More detail can be found in reviews by Fliegel and Fröhlich (1993), Fliegel and Dyck (1995), Counillon and Pouysségur (1995), and Terzic and Kurachi (1996).

At physiological pH, mechanisms other than the $Na^+–H^+$ exchanger maintain myocardial proton homeostasis. The activity of the $Na^+–H^+$ exchanger is negligible at pH 7.2. Only when internal pH falls does this exchanger become active. The $Na^+–H^+$ exchanger has a pH sensing mechanism that activates the exchanger following acidification. There is apparently a H^+ regulatory site on the $Na^+–H^+$ exchanger. The situation is analogous to that of the $Na^+–Ca^{2+}$ exchanger (see Section VII). The $Na^+–Ca^{2+}$ exchanger is activated by a rise in internal Ca^{2+} due to a Ca^{2+} regulatory site. Likewise, the $Na^+–H^+$ exchanger is turned on by a rise in internal H^+ to protect the cell from acidification. A maximal transport rate is achieved when pH falls by less than one pH unit. Interestingly, modulation of the $Na^+–H^+$ exchange activity is mediated through changes in the H^+ affinity of the regulatory site. For example, α_1-adrenoceptor agonists stimulate $Na^+–H^+$ exchange via this mechanism.

The $Na^+–H^+$ exchanger catalyzes the countertransport of one external Na^+ for one internal H^+. Although no energy source is used directly, ATP is consumed by the Na^+ pump to extrude the Na^+ back out of the cell. $Na^+–H^+$ exchanger can also use Li^+ as a substrate, in contrast to the $Na^+–Ca^{2+}$ exchanger, which cannot transport Li^+. Specific inhibitiors of the $Na^+–H^+$ exchanger are available. Amiloride derivatives, such as ethylisopropylamiloride, have IC_{50} values as low as 40 nM and are experimentally useful.

The $Na^+–H^+$ exchanger was first cloned in the laboratory of Pouysségur through elegant experiments involving complementation of exchanger-deficient cell lines (Sardet et al., 1989). Four isoforms have now been cloned and are known as NHE-1 to NHE-4. NHE-1 is a housekeeping gene expressed in all tissues and the isoform expressed in the myocardium. NHE-2, NHE-3, and NHE-4 have tissue-specific patterns of expression and are not present in the heart. The isoforms have different sensitivities to amiloride, with NHE-1 being the most sensitive.

Based on the cDNA sequence, the human NHE-1 consists of 815 amino acids and is modeled to have 12 transmembrane segments. There is a long (300 amino acids) hydrophilic C-terminal domain that is located intracellularly. The C-terminal domain is involved in regulation and contains phosphorylation sites and a calmodulin bind site. As mentioned previously, regulation is primarily mediated through changes in the proton affinity of the H^+ regulatory site. A distinctive feature of the $Na^+–H^+$ exchanger is the finding in several tissues of activation by all known mitogens to cause cell alkalinization. The molecular mechanism of this activation is unknown, and the role of the $Na^+–H^+$ exchanger as a mitogenic signal is controversial.

Various aspects of the role of the $Na^+–H^+$ exchanger during myocardial ischemia and reperfusion have been studied in several laboratories. An interesting model has evolved as follows: During ischemia, there is a large production of intracellular H^+. Due to activation of the $Na^+–H^+$ exchanger, there is a subsequent influx of Na^+. The Na^+ pump is unable to compensate quickly enough and internal Na^+ rises. When a large supply of external Ca^{2+} becomes available during perfusion, a large Ca^{2+} influx ensues due to $Na^+–Ca^{2+}$ exchange. This scenario undoubtedly occurs to some degree but the overall contribution of Ca^{2+} overload by this mechanism to myocardial damage is still unclear. The roles of the $Na^+–H^+$ and $Na^+–Ca^{2+}$ exchangers during pathophysiological situations is an active area of research.

IX. PERSPECTIVES

Physiological and molecular advances regarding the myocardial transporters will be forthcoming. Each of the transporters may be localized in microdomains near other transporters or channels. This may affect their activities with important functional consequences. Both new imaging and molecular biological approaches should provide new information on interacting molecules. Genetic techniques using transgenic mice may find unexpected roles for the different transporters. At the molecular level, the accumulation of new information and the application of new techniques should provide initial structural information and new insights on transport mechanisms.

REFERENCES

Baker, P.F., Blausteen, M.P., Hodgkin, A.L., and Steinhardt, R.A. (1969). The influence of calcium on sodium efflux in squid axons. *J. Physiol.* 200:431–458.

Bertorello, A.M., and Katz, A.I. (1995). Regulation of $Na^+–K^+$ pump activity: Pathways between receptors and effectors. *News Physiol. Sci.* 10:253–259.

Canessa, C., Jaisser, F., Horisberger, J.-D., and Rossier, B.C. (1994). Structure–function relationship of Na, K-ATPase: The digitalis receptor. *Curr. Top. Membr.* 41:71–85.

Carafoli, E. (1992). The Ca^{2+} pump of the plasma membrane. *J. Biol. chem.* 267:2115–2118.

Carafoli, E. (1994). Biogenesis: Plasma membrane calcium ATPase: 15 Years of work on the purified enzyme. *FASEB J.* 8:993–1002.

Caroni, P., and Carafoli, E. (1980). An ATP-dependent Ca^{2+}-pumping system in dog heart sarcolemma. *Nature* 283:765–767.

Chow, D.C., and Forte, J.G. (1995). Functional significance of the β-subunit for heterodimeric P-type ATPases. *J. Exp. Biol.* 198:1–17.

Clarke, D.M., Loo, T.W., Inesi, G., and MacLennan, D.H. (1989). Location of high affinity Ca^{2+}-binding sites within the predicted transmembrane domain of the sarcoplasmic reticulum Ca^{2+}-ATPase. *Nature* 339:476–478.

Clausen, T. (1996). Long- and short-term regulation of the $Na^+–K^+$ pump in skeletal muscle. *News Physiol. Sci.* 11:24–30.

Counillon, L., and Pouysségur, J. (1995). Structure–function studies and molecular regulation of the growth factor activatable sodium–hydrogen exchanger (NHE-1). *Cardiovasc. Res.* 29:147–154.

DiPolo, R. (1979). Calcium influx in internally dialyzed squid giant axons. *J. Gen. Physiol.* 73:91–113.

Falchetto, R., Vorherr, T., Brunner, J., and Carafoli, E. (1991). The plasma membrane Ca^{2+} pump contains a site that interacts with its calmodulin binding domain. *J. Biol. Chem.* 266:2930–2936.

Falchetto, R., Vorherr, T., and Carafoli, E. (1992). The calmodulin binding site of the plasma membrane Ca^{2+} pump interacts with the transduction domain of the enzyme. *Protein Sci.* 1:1613–1621.

Fambrough, D.M., Lemas, M.V., Takeyasu, K., Renaud, K.J., and Inman, E.M. (1994). Structural requirements for subunit assembly of the Na,K-ATPase. *Curr. Top. Membr.* 41:45–69.

Fliegel, L., and Dyck, J.R.B. (1995). Molecular biology of the cardiac sodium/hydrogen exchanger. *Cardiovasc. Res.* 29:155–159.

Fliegel, L., and Fröhlich, O. (1993). The Na^+/H^+ exchanger: An update on structure, regulation and cardiac physiology. *Biochem. J.* 296:273–285.

Gadsby, D.C., Rakowski, R.F., and DeWeer, P. (1993). Extracellular access to the Na,K pump: Pathway similar to ion channel. *Science* 260:100–103.

Hilgemann, D.W. (1990). Regulation and deregulation of cardiac $Na^+–Ca^{2+}$ exchange in giant excised sarcolemmal membrane patches. *Nature* 344:242–245.

Hilgemann, D.W. (1994). Channel-like function of the Na,K pump probed at microsecond resolution in giant membrane patches. *Science* 263:1429–1432.

Hilgemann, D.W. (1996). Unitary cardiac Na^+,Ca^{2+} exchange current magnitudes determined from channel-like noise and charge movements of ion transport. *Biophys. J.* 71:759–768.

Hilgemann, D.W., and Ball, R. (1996). Regulation of cardiac Na,Ca exchange and K_{ATP} potassium channels by the synthesis and hydrolysis of PIP_2 in giant membrane patches. *Science* 273:956–959.

Hilgemann, D.W., Matsuoka, S., Nagel, G.A., and Collins, A. (1992). Steady-state and dynamic properties of cardiac sodium–calcium exchange—Sodium-dependent inactivation. *J. Gen. Physiol.* 100:905–932.

Hilgemann, D.W., Philipson, K.D., and Vassort, G. (1996). Sodium–calcium exchange: Proceedings of the third international conference. *Ann. N.Y. Acad. Sci.* 779:1–593.

Hofmann, F., James, P., Vorherr, T., and Carafoli, E. (1993). The C-terminal domain of the plasma membrane Ca^{2+} pump contains three high affinity Ca^{2+} binding sites. *J. Biol. Chem.* 268:10252–10259.

Horisberger, J.-D. (1994). "The Na,K-ATPase: Structure–Function Relationship." R.G. Landes Company, Austin, TX.

Inesi, G., Lewis, D., Nikic, D., Hussain A., and Kirtley, M.E. (1992). Long-range intramolecular linked functions in the calcium transport ATPase. *In* "Advances in Enzymology and Related Areas of Molecular Biology" (Meister, A., ed.), Vol. 65, pp. 185–215. John Wiley & Sons, New York.

Iwamoto, T., Pan, Y., Wakabayshi, S., Imagawa, T., Yamanaka, H.I., and Shigekawa, M.

(1996). Phosphorylation-dependent regulation of cardiac Na^+/Ca^{2+} exchanger via protein kinase C. *J. Biol. Chem.* 271:13609–13615.

Kawakami, K., Noguchi, S., Noda, M., Takahashi, H., Ohta, T., Kawamura, M., Nojima, H., Nagano, K., Hirose, T., Inayama, S., Hayashida, H., Miyata, T., and Numa, S. (1985). Primary structure of the alpha-subunit of Torpedo californica ($Na^+ + K^+$)ATPase deduced from cDNA sequence. *Nature* 316:733–736.

Khananshvili, D. (1997). Structure, mechanism and regulation of the cardiac sarcolemma $Na^+–Ca^{2+}$ exchanger. *In* "Ion Pumps: Advances in Molecular and Cell Biology." JAI Press, Groton, CT.

Langer, G.A., and Peskoff, A. (1996). Calcium concentration and movement in the diadic cleft space of the cardiac ventricular cell. *Biophys. J.* 70:1169–1182.

Läuger, P. (1991). Electrogenic ion pumps. *In* "Distinguished Lecture Series of the Society of General Physiologists," Vol. 5, pp. 1–319. Sinauer Associates, Sunderland, MA.

Leblanc, N., and Hume, J.R. (1990). Sodium current induced release of calcium from cardiac sarcoplasmic reticulum. *Science* 248:372–376.

Levi, A.J., Spitzer, K.W., Kohmoto, O., and Bridge, J.H.B. (1994). Depolarization-induced Ca entry via Na–Ca exchange triggers SR release in guinea pig cardiac myocytes. *Am. J. Physiol.* 226:H1422–H1433.

Levitsky, D.O., Nicoll, D.A., and Philipson, K.D. (1994). Identification of the high affinity Ca^{2+}-binding domain of the cardiac $Na^+–Ca^{2+}$ exchanger. *J. Biol. Chem.* 269:22847–22852.

Li, Z., Nicoll, D.A., Collins, A., Hilgemann, D.W., Filoteo, A.G., Penniston, J.T., Weiss, J.N., Tomich, J.M., and Philipson, K.D. (1991). Identification of a peptide inhibitor of the cardiac sarcolemmal $Na^+–Ca^{2+}$ exchanger. *J. Biol. Chem.* 266:1014–1020.

Lingrel, J.B., and Kuntzweiler, T. (1994). Na^+,K^+-ATPase. *J. Biol. Chem.* 269:19659–19662.

Lingrel, J.B., Orlowski, J., Shull, M.M., and Price, E.M. (1990). Molecular genetics of Na,K-ATPase. *Prog. Nucleic Acid Res. Mol. Biol.* 38:37–83.

Lipp, P., and Niggli, E. (1994). Sodium current-induced calcium signals in isolated guinea-pig ventricular myocytes. *J. Physiol.* 474:439–446.

Lompré, A.-M., Anger, M., and Levitsky, D. (1994). Sarco(endo)plasmic reticulum calcium pumps in the cardiovascular system: Function and gene expression. *J. Mol. Cell. Cardiol.* 26:1109–1121.

Luo, W., Grupp, I.L., Harrer, J., Ponniah, S., Grupp, G., Duffy, J.J., Doetschman, T., and Kranias, E.G. (1994). Targeted ablation of the phospholamban gene is associated with markedly enhanced myocardial contractility and loss of beta-agonist stimulation. *Circ. Res.* 75:401–409.

Lytton, J., and MacLennan, D.H. (1992). Sarcoplasmic reticulum. *In* "The Heart and Cardiovascular System" (Fozzard, H.A. *et al.*, eds.), pp. 1203–1222. Raven Press, New York.

MacLennan, D.H. (1970). Purification and properties of an adenosine triphosphatase from sarcoplasmic reticulum. *J. Biol. Chem.* 245:4508–4518.

MacLennan, D.H., Brandl, C.J., Korczak, B., and Green, N.M. (1985). Amino-acid sequence of a $Ca^{2+} + Mg^{2+}$-dependent ATPase from rabbit muscle sarcoplasmic reticulum, deduced from its complementary DNA sequence. *Nature* 316:696–700.

Matsuoka, S., Nicoll, D.A., Hryshko, L.V., Levitsky, D.O., Weiss, J.N., and Philipson, K.D. (1995). Regulation of the cardiac $Na^+–Ca^{2+}$ exchanger by Ca^{2+}-mutational analysis of the Ca^{2+}-binding domain. *J. Gen. Physiol.* 105:403–420.

McDonough, A.A., and Farley, R.A. (1993). Regulation of Na,K-ATPase activity. *Curr. Sci. Nephrol. Hypertens.* 2:725–734.

McDonough, A.A., Azuma, K.K., Lescale-Matys, L., Tang, M.-J., Nakhoul, F., Hensley, C.B., and Komatsu, Y. (1992a). Physiologic rationale for multiple sodium pump isoforms. *Ann. N.Y. Acad. Sci.* 671:156–169.

McDonough, A.A., Hensley, C.B., and Azuma, K.K. (1992b). Differential regulation of sodium pump isoforms in heart. *Semin. Nephrol.* 12:49–55.

McDonough, A.A., Wang, J., and Farley, R.A. (1995). Significance of sodium pump isoforms in digitalis therapy. *J. Mol. Cell. Cardiol.* 27:1001–1009.

Medford, R.M. (1993). Digitalis and the Na^+,K^+-ATPase. *Heart Dis. Stroke.* 2:250–255.

Munzer, J.S., Daly, S.E., Jewell-Motz, E.A., Lingrel, J.B., and Blostein, R. (1994). Tissue-and isoform-specific kinetic behavior of the Na,K-ATPase. *J. Biol. Chem.* 269:16668–16676.

Nicoll, D.A., Longoni, S., and Philipson, K.D. (1990). Molecular cloning and functional expression of the cardiac sarcolemmal Na^+-Ca^{2+} exchanger. *Science* 250:562–565.

Nicoll, D.A., Hyrshko, L.V., Matsuoka, S., Frank, J.S., and Philipson, K.D. (1996). Mutation of amino acids residues in the putative transmembrane segments of the cardiac sarcolemmal Na^+-Ca^{2+} exchanger. *J. Biol. Chem.* 271:13385–13391.

Niggli, V., Penniston, J.T., and Carafoli, E. (1979). Purification of the $(Ca^{2+} + Mg^{2+})$-ATPase from human erythrocyte membranes using a calmodulin affinity column. *J. Biol. Chem.* 254:9955–9958.

Penniston, J.T., and Enyedi, A. (1997). Comparison of ATP-powered Ca^{2+} pumps. *In* "Ion Pumps. Advances in Molecular and Cell Biology." JAI Press, Greenwich, CT.

Philipson, K.D. (1990). The cardiac Na^+-Ca^{2+} exchanger—Dependence on membrane environment. *Cell Biol. Int. Rep.* 14:305–309.

Philipson, K.D. (1996). The Na^+-Ca^{2+} exchanger: Molecular aspects. *In* "Molecular Physiology and Pharmacology of Cardiac Ion Channels and Transporters" (Morad, M., ed.). pp. 435–446. Kluwer, Boston.

Philipson, K.D. (1997). Sodium–calcium exchange. *In* "Calcium as a Cellular Regulator" (Carafoli, E., and Klee, C., eds.). Oxford University Press, New York.

Philipson, K.D., Longoni, S., and Ward, R. (1988). Purification of the cardiac Na^+-Ca^{2+} exchange protein. *Biochim. biophys. Acta.* 945:298–306.

Reeves, J.P. (1995). Cardiac sodium–calcium exchange system. *In* "Physiology and Pathophysiology of the Heart" (Sperelakis, N., ed.), 3rd ed., pp. 309–318. Kluwer, Boston.

Reeves, J.P., and Philipson, K.D. (1989). Sodium–calcium exchanger activity in plasma membrane vesicles. *In* "Sodium–Calcium Exchange" (Noble, D., Reuter, N., and Allen, T.J.A., eds.), pp. 27–33. Oxford University Press, Oxford, England.

Reeves, J.P., and Sutko, J.L. (1979). Sodium–calcium ion exchange in cardiac membrane vesicles. *Proc. Natl. Acad. Sci. USA* 76:590–594.

Reiländer, H., Achilles, A., Friedel, U., Maul, G., Lottspeich, F., and Cook, N.J. (1992). Primary structure and functional expression of the Na/Ca,K-exchanger from bovine rod photoreceptors. *EMBO J.* 11:1689–1695.

Reuter, H., and Seitz, N. (1968). The dependence of calcium efflux from cardiac muscle on temperature and external ion composition. *J. Physiol.* 195:451–470.

Sardet, C., Franchi, A., and Pouysségur, J. (1989). Molecular cloning, primary structure, and expression of the human growth factor-activatable Na^+/H^+ antiporter. *Cell* 56:271–280.

Schatzmann, H.J. (1953). Herzglycoside als hemmstoffe für den aktiven kalium und natrium transport durch die erythrocyten-membran. *Helv. Physiol. Pharmacol. Acta* 11:346–354.

Schatzmann, H.J. (1966). ATP-dependent Ca^{2+}-extrusion from human red cells. *Experientia* 22:364–365.

Schwarz, E.M., and Benzer, S. (1997). Expression and evolution of Calx, a sodium–calcium exchanger of *Drosophila melanogaster*. Submitted for publication.

Sham, J.S.K., Cleeman, L., and Morad, M. (1995). Functional coupling of Ca^{2+} channels and ryanodine receptors in cardiac myocytes. *Proc. Natl. Acad. Sci. USA* 92:121–125.

Shamraj, O.I., and Lingrel, J.B. (1994). A putative fourth Na^+,K^+-ATPase α-subunit gene is expressed in testis. *Proc. Natl. Acad. Sci. USA* 91:12952–12956.

Shull, G.E., and Greeb, J. (1988). Molecular cloning of two isoforms of the plasma membrane Ca^{2+} transporting ATPase from rat brain. Structural and functional domains exhibit similarity to Na^+,K^+- and other cation transport ATPases. *J. Biol. Chem.* 263:8646–8657.

Shull, G.E., Schwartz, A., and Lingrel, J.B. (1985). Amino-acid sequence of the catalytic subunit of the $(Na^+ + K^+)$-ATPase deduced from a complementary DNA. *Nature* 316:691–695.

Simmerman, H.K.B., Collins, J.H., Theibert, J.L., Wegener, A.D., and Jones, L.R. (1986). Sequence analysis of phospholamban. Identification of phosphorylation sites and two structural domains. *J. Biol. Chem.* 261:13333–13341.

Skou, J.C. (1957). The influence of some cations on an adenosine triphosphatase from peripheral nerves. *Biochim. Biophys. Acta* 23:394–401.

Stauffer, T. P., Guerini, D., and Carafoli, E. (1995). Tissue distribution of the four gene products of the plasma membrane Ca^{2+} pump. *J. Biol. Chem.* 270:12184–12190.

Strehler, E.E. (1997). Sodium–calcium exchangers and calcium pumps. *In* "Principles of Medical Biology: Cell Chemistry and Physiology: Part III" Vol. 4, pp. 125–147. JAI Press, Greenwich, CT.

Tada, M., and Kadoma, M. (1995). Uptake of calcium by sarcoplasmic reticulum and its regulation and functional consequences. *In* "Physiology and Pathophysiology of the Heart" (Sperelakis, N., ed.), pp. 333–353. Kluwer Academic Publishers, Boston.

Terzic, A., and Kurachi, Y. (1996). Regulation of intracellular protons: Role of Na/H exchange in cardiac myocytes. *In* "Molecular Physiology and Pharmacology of Cardiac Ion Channels and Transporters" (Morad, M., Ebashi, S., Trautwein, W., and Kurachi, Y., eds.), pp. 555–561. Kluwer, Dordecht, The Netherlands.

Therien, A.G., Nestor, N.B., Ball, W.J., and Blostein, R. (1996). Tissue-specific versus isoform-specific differences in cation activation kinetics of the Na,K-ATPase. *J. Biol. Chem.* 271:7104–7112.

Toyoshima, C., Sasabe, C., and Stokes, D.L. (1993). Three dimensional cryo-electron microscopy of the calcium ion pump in the sarcoplasmic reticulum membrane. *Nature* 362:469–471.

Trumble, W.R., Sutko, J.L., and Reeves, J.P. (1980). ATP-dependent calcium transport in cardiac sarcolemmal membrane vesicles. *Life Sci.* 27:207–214.

Verma, A.K., Filoteo, A.G., Stanford, D.R., Wieben, E.D., Penniston, J.T., Strehler, E.E., Fischer, R., Heim, R., Vogel, G., Mathews, S., Strehler-Page, M.A., James, P., Vorherr, T., Krebs, J., and Carafoli, E. (1988). Complete primary structure of a human plasma membrane Ca^{2+} pump. *J. Biol. Chem.* 263:14152–14159.

Wang, K.K.W., Villalobo, A., and Roufogalis, B.D. (1992). The plasma membrane calcium pump: A multiregulated transporter. *Trends Cell Biol.* 2:46–52.

Wuddel, I., and Apell, H.-J. (1995). Electrogenicity of the sodium transport pathway in the Na,K-ATPase probed by charge-pulse experiments. *Biophys. J.* 69:909–921.

5

EXCITATION–CONTRACTION
COUPLING AND CALCIUM
COMPARTMENTATION

GLENN A. LANGER

I. INTRODUCTION

Cardiac tissue, when perfused with 1 mM [Ca]$_0$ contains approximately 2000 μmol Ca/kg wet weight of exchangeable Ca. So that maximal force is developed, no more than 80–100 μmol/kg wet weight needs to be released into the cell sarcoplasm. Therefore, 5% or less of total tissue Ca is required for cardiac muscle to develop full contractile force during a single beat. This is not to say, however, that the 95% of Ca not involved in a single contraction is of minor importance. Much of this Ca exchanges very rapidly both within and outside of the cell. Thus, the characteristics of distribution and compartmentation of the entire exchangeable pool determine the quantity and movement of the small fraction released into the sarcoplasm with each contraction.

This book focuses on cellular and subcellular processes. Before these processes are dealt with for Ca, it is important to review, briefly, the characteristics of that Ca that is outside of the cell. This enables the reader to place the cellular and subcellular Ca movements in quantitative context with respect to the functional tissue.

II. EXTRACELLULAR CALCIUM

With 1 mM [Ca] in the serum and normal coronary vascular blood flow of 1200 ml/kg/min, Ca passes through the capillaries at a rate of 20 μmol/s. The vascular Ca exchanges with Ca in the interstitial space. At

25% tissue volume, this space contains 250 μmol/kg wet weight in free solution. Kinetic compartmentation studies indicate that an amount at least equivalent to the free Ca content is bound within the interstitium. Therefore, the space contains approximately 500 μmol or one-fourth of the total exchangeable Ca in the tissue. This is a large Ca "sink" and is the immediate source of Ca for extracellular exchange. The Ca in the interstitium exchanges with vascular Ca with a half-time of about 60 seconds (Pierce *et al.*, 1987). This predicts that a change of [Ca] introduced in the arterial supply will produce a cellular contractile response with a half-time of 60 seconds. The response is, however, many times more rapid (Philipson and Langer, 1979). This is explained by the fact that as much as one-third of the surface of an individual cell is in virtual direct contact with the surface of a capillary with very little interstitium interposed (Frank and Langer, 1974). This provides, in effect, a "shunt" of the interstitium and permits the cell to respond within a few seconds to changes in the constituency of the blood perfusing the heart. Such a rapid response to humoral changes can be important for survival where large and immediate changes in cardiac output are required.

It has been known for well more than 100 years due to the work of Sidney Ringer that extracellular Ca is an absolute requirement for maintenance of cardiac contraction. Removal of extracellular Ca immediately following a full contraction will completely eliminate force development at the next electrical excitation (Rich *et al.*, 1988). This is in marked contrast to skeletal muscle in which large contractions persist for 15–20 minutes after complete removal of extracellular Ca (Armstrong *et al.*, 1972). From these results, it is clear that extracellular Ca plays a role in modulation of force development in the heart and that this is not the case for skeletal muscle. Voluntary skeletal muscle tissue has an extrinsic control mechanism by which force is modulated by recruitment of more or less motor units from its nerve supply. It only requires that a source of intracellular Ca be available for a full response when the fiber is stimulated via its motor nerve. If more force is required, the individual skeletal muscle cell does not need to develop more force; more force is developed by recruitment of more cells. Heart muscle is, however, an "all-or-none" muscle. In a normal heart, either all cells contract upon excitation or none contract. Since this is the case, force modulation must be at the level of the individual cell, and it is obvious that Ca, both extra- and intracellular, is the "linch pin" of the intrinsic control.

In general, the more rapid the basal ventricular beating rate, the greater is the contribution of Ca from intracellular stores [i.e., the sarcoplasmic reticulum (SR)] to force development (Bers, 1985; Rich *et al.*, 1988). The amphibian ventricle with heart rates in the 30 beats/min range or less and virtually no SR structure is entirely dependent on extracellular Ca for force development. At the other extreme, small rodents with heart rates in the 500 beats/min range are dependent on intracellular SR Ca stores for approximately 90% of their contractile-related Ca. The origin of contraction-related Ca is discussed in greater detail later in this chapter.

In summary, all cardiac muscle requires an immediate supply of extracellular Ca for contraction, regardless of the species. When intrinsic heart rate is low and force development is slow, all or almost all contraction-related Ca is derived from the cells' extracellular space. As intrinsic heart rate increases for the various species and the cardiac contraction cycle shortens, extracellular Ca is used as a "trigger" to promote the release of intracellular Ca stored in the SR. The release from the SR cisterns at the end of each sarcomere ensures that the activating Ca does not need to diffuse further than a distance of a half-sarcomere or about 1 μm. Therefore, the activating Ca can reach the myofilaments well within the 50–100-millisecond period required for support of contraction in a rapidly beating ventricle.

III. SUBCELLULAR CALCIUM COMPARTMENTATION

A. STEADY-STATE DISTRIBUTION

The resting ventricular cell has a free cytoplasmic calcium concentration of about 100 nM or perhaps a bit more. This is distributed in the nonmitochondrial cell water (40–50% of cell volume). Therefore, with approximately 75% of total tissue being cellular and 25% extracellular, the free Ca in the cell represents 30–40 nmol/kg wet tissue during diastole. Since exchangeable cellular Ca is about 1500 μmol/kg wet tissue (see previous section), the free Ca in the resting cell is a miniscule fraction (0.002–0.003%) of the cell's exchangeable pool. During systole the free [Ca] in the ventricular cell increases to a peak of 1.5–2.0 μM to produce near-maximum force development (Fabiato, 1983; Berlin and Konishi, 1993). To achieve this concentration, between 80 and 90 μmol/kg wet weight are released into the cell. This means that at peak systole less than 2% of the total Ca released remains free—all of the rest is buffered. It is obvious, then, that relatively small net changes in buffering could produce dramatic changes in free [Ca] in the cell and, therefore, marked changes in contractile function.

1. Sarcolemmal Calcium

The myocardial sarcolemma (SL) is a complex arrangement of phospholipids, proteins, and sugars. It carries a significant number of fixed negative charges, and these sites are capable of binding Ca. Evidence of Ca binding to the SL was suggested by early studies using the trivalent rare earth cation, lanthanum (La^{+++}) (Langer and Frank, 1972), in which localization of the electron-dense La^{+++} ion on the SL and glycocalyx of the cell coincided with displacement of 3 mmol Ca/kg dry weight from the cells (Langer *et al.*, 1990). Placed in context, this represented between 30 and 40% of the cell's total exchangeable Ca. Subsequent studies using isolated SL demonstrated that La could displace an amount of Ca from SL per se that accounted for about 70% of the amount displaced from whole, intact cells (Post and Langer, 1992a). Moreover, it was shown that if SR Ca was decreased using agents such as ryanodine or thapsigargin, the La displacement from whole cells was reduced by about 30%. This showed that at least part of the La-displaceable cellular Ca pool depended on a source of intracellular Ca, probably in the SR. Add to this the likelihood that La seems to be able to penetrate to at least the inner sarcolemmal surface (Peters *et al.*, 1989), and it supports the idea that a component of sarcolemmal-bound Ca may be in rapid equilibrium with a source of intracellular Ca. Studies on isolated cardiac SL provide strong support for a role of its inner lipid leaflet in Ca binding (Post *et al.*, 1988; Post and Langer 1992b).

The SL of the cardiac cell is, in its phospholipid constituency, remarkably similar to the membrane of the red blood cell. Phosphatidylcholine (PC) accounts for slightly more than 50% of the phospholipid and is about equally distributed between outer and inner monolayers. Sphingomyelin (Sph) accounts for about 14% with more than 90% in the outer monolayer. Neither of these phospholipids participate to any significant extent in Ca binding. Phosphatidylethanolamine (PE) represents 25% of total sarcolemmal phospholipid, and 75% is in the inner leaflet. Phosphatidylserine (PS) and phosphatidylinositol (PI) account for 7%, and they reside entirely within the inner monolayer. These zwitterionic (PE) and anionic (PS/PI) moieties account for all the membrane's phospholipid Ca binding.

Figure 1 illustrates apparent characteristics of the binding. The Scatchard analysis demonstrates two distinct classes of binding sites: (1) high-affinity binding (Kd $\sim$ 13 μM), saturating at $\sim$7 nmol Ca/mg sarcolemmal protein; and (2) low-affinity sites (Kd $\sim$ 1.1 mM), saturating at $\sim$85 nmol Ca/mg protein. The sarcolemmal membranes from which these data were obtained were present in the cells at a level of 44 g SL protein/kg dry weight cells (Post and Langer, 1992b). Using this conversion factor, the Ca bound to SL at saturation is 4.1 mmol/kg dry weight cells. At 1 mM [Ca]$_0$ (see Fig. 1), about 50% or 2 mmol Ca binding would be expected, which is in close agreement with the calculated membrane displacement by La from whole cells (0.7 $\times$ 3 mmol = 2.1 mmol).

Treatment of the SL with phospholipase C removes almost all Ca-binding phospholipid head groups, and such treatment completely eliminates the low-affinity binding sites, which account for more than 92% of the phospholipid-based Ca binding.

The level of sarcolemmal Ca binding reported here is the lowest in the literature, with other values for saturation ranging from 134 (Scarpa and Williamson, 1974) to 697 nmol/mg protein (Limas, 1977). However, all

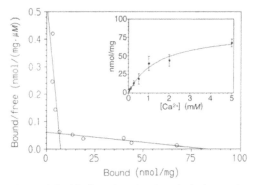

FIGURE 1 Sarcolemmal Ca binding. Scatchard analysis (saturation curve in the inset) of binding data obtained from "gas-dissected" sarcolemma from cultured cardiac cells. Two classes of binding sites are evident: one with a Kd = 13 μM and a capacity of 7 nmol/mg protein, and one with Kd = 1.1 mM and a capacity of 84 nmol/mg protein. (Reproduced from Post and Langer, 1992b, with permission.)

studies report both high-affinity sites (Kd 5–22 μM) and low-affinity sites (Kd 0.8–1.8 mM), with the low-affinity sites accounting for 80–97% of the total binding. The purity of the SL used and the method for measurement of Ca binding make it likely that the Post and Langer (1992) study represents the values most realistic for the intact cell.

From the preceding discussion, it is likely that La-displaceable Ca in the whole cell represents, for the most part, Ca bound to the SL and that most of this is bound to inner-leaflet phospholipids. The exchange rate of this sarcolemmal fraction was measured indirectly in adult rat ventricular cells. ^{45}Ca-labeled cells were isotopically washed out with ^{45}Ca-free solution at high perfusion rate (Langer *et al.*, 1990), and La was added after 2 seconds of washout. No evidence of ^{45}Ca displacement from the cells could be seen. It was demonstrated that, if the La-displaceable Ca had a $t_{1/2}$ of exchange of 1 second or longer, a displacement of ^{45}Ca would have been easily measured. Therefore, the sarcolemmal-bound Ca exchanges with a $t_{1/2}$ of less than 1 second, giving a flux rate of more than 1800 μmol/kg dry weight per second for this cellular Ca fraction. This rapid exchange rate explains why standard ^{45}Ca washout studies that employ "prewash" periods to remove extracellular ^{45}Ca consistently fail to demonstrate this large sarcolemmal-bound component of cellular Ca (Murphy *et al.*, 1987). The component can only be demonstrated with "on-line" isotopic techniques that obviate the need for an unrecorded extracellular wash period (Kuwata and Langer, 1989; Langer *et al.*, 1990).

In summary, Ca bound to the SL accounts for 35–40% of the cell's exchangeable Ca (500–600 μmol/kg wet tissue). It exchanges with the extracellular space very rapidly, with a $t_{1/2}$ of less than 1 second. It appears that a significant fraction is in rapid equilibrium with intracellular Ca, probably in the closely juxtaposed lateral cistern for the SR. Most of the Ca is bound to zwitteronic (PE) or anionic (PS, PI) phospholipid, 80% of which is located in the inner leaflet of the SL. The role of this Ca and its binding sites are discussed later in this chapter.

2. Sarcoplasmic Reticulum Calcium

The Ca content of the SR varies greatly according to the method used for its measurement. Levitsky *et al.* (1981) measured maximum Ca binding (ATP-dependent Ca uptake without addition of oxalate as a precipitating agent in the vesicles) in highly purified SR at about 150 μmol Ca/kg wet weight guinea pig ventricle. Bridge (1986) measured, by rapid cooling technique to induce SR Ca release, a rest-induced loss of Ca of 258 μmol/kg wet weight rabbit ventricle, which was assumed to be derived from the SR. Lewartowski *et al.* (1984), using a ^{45}Ca isotopic technique, found a very large net loss of 1200 μmol/kg wet weight after 4 minutes of rest following steady stimulation at 60 beats/min in guinea pig ventricle. This is 4.7 times the value found by Bridge. Bassani *et al.* (1995) used 8 mM [Ca]$_0$ to "highly" load the SR of ferret ventricular myocytes. They measured steady-state SR content by measuring the Ca released by application of 10 mM caffeine. The "highly" loaded SR contained 95 μmol/liter accessible cell water (38 μmol/kg wet wt) of which 56 μmol/liter (22 μmol/kg wet wt) was released upon stimulation. Other values in the literature place SR content between 100 and 300 μmol Ca/kg wet weight (Dani *et al.*, 1979; Kawai and Konishi, 1994).

Therefore, there is a range of SR content reported between 38 and 1200 μmol/kg wet weight. What is a reasonable value? Many of the lower values assume that caffeine induces a complete release of all Ca stored in the SR, and the work quoted is that of Weber (1968) and Weber and Herz (1968). These studies used SR isolated from frog or rabbit skeletal muscle. Under various concentrations of ATP, Mg, and caffeine, the maximum caffeine-induced Ca loss was in no case greater than 50% and in many cases in the 35% range. Therefore, the studies in which caffeine was used as the Ca-releasing agent could underestimate total SR Ca by at least 50%. The isotopic technique as used by Lewartowski *et al.* (1984) could very well greatly overestimate SR content since contributions to isotope loss could be derived from other regions of the whole tissue used in this study.

Our study used an isotopic technique but in isolated rat ventricular cells perfused at 170 ml/min to diminish isotopic reflux during washout. The technique greatly improves the accuracy of kinetic analysis of cellular Ca compartmentation with no chance of contribution of isotope from noncellular tissue elements (Langer *et al.*, 1990). ^{45}Ca washout of these cells clearly demonstrates three compartments (Fig. 2A).

A fourth La displaceable compartment with a $t_{1/2}$ of less than 1 second was discussed previously and is believed to be based primarily at sarcolemmal sites. The washout shows a "slow" compartment with $t_{1/2}$ for exchange of 3.6 minutes. This most likely represents mitochondrial Ca and is discussed later. The remaining two compartments (Fig. 2) are intermediate in their exchange rates with $t_{1/2}$ of 3.5 and 19 seconds. Importantly, the amount of ^{45}Ca released by caffeine applied at various times during washout, when

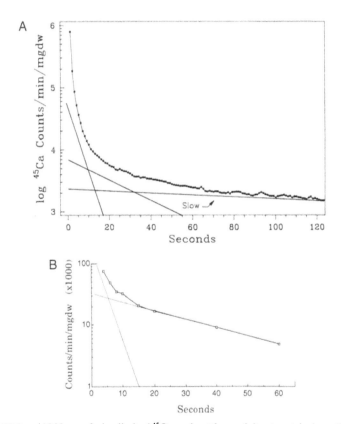

FIGURE 2   (A) Nonperfusion limited ^{45}Ca washout from adult rat ventricular cells. Longer washouts defined the "slow" compartment, and its extrapolation is shown. Its contribution is subtracted and the curve peeled by computer between 2 and 120 s to yield the two steeper exponential components shown. A more rapid component representing a La^{+++}-displaceable compartment ($t_{1/2} < 1$ s) is not shown. The two components shown are, then, intermediate in exchange rate between the La-displaceable and "slow" compartments and have $t_{1/2}$ of 3.5 and 19 s. (B) Nonperfusion limited ^{45}Ca washouts in which each point represents the mean ^{45}Ca release upon addition of caffeine at each point in time. Each point is the mean of at least six caffeine additions. Peeling of the curve yields two exponential components with $t_{1/2}$ of 2 and 19 s or very similar to the intermediate components defined in (A). (Reproduced from Langer *et al.*, 1990, with permission.)

plotted semilogarithmically (Fig. 2B), showed two components with $t_{1/2}$ very similar to the "intermediate" components—2 and 19 seconds. This strongly suggests that the origin of the intermediate components is the SR. The total Ca in the intermediate compartments was 393 μmol/kg wet weight tissue when the ventricular cell is perfused with 1 mM [Ca]$_0$ (Langer et al., 1990). This is 2 to 3 times the level reported in the studies using caffeine-releasable Ca as a measure of SR content and is therefore consistent with the Weber studies on the ability of caffeine to release SR Ca. In our study, caffeine was capable of releasing only 21% of the intermediate compartments' content.

Williamson et al. (1983), using a null point titration technique and induction of SR release by the ionophore A23187 in permeabilized cells, reported a Ca content of 2.5 nmol/mg dry cell weight of rat ventricular cells. This converts to 460 μmol/kg wet tissue, a value similar to that measured isotopically.

It would seem, taking all factors into consideration, that an SR content of 400 μmol/kg wet weight tissue is reasonable under physiological conditions. This represents 25–30% of the cell's exchangeable Ca. The biphasic kinetics of the putative SR Ca ($t_{1/2}$ of 3.5 and 19 s) is of interest in the light of the structural diversity of the SR. Jorgensen et al. (1985) found that calsequestrin (the Ca-binding protein responsible for storage of large amounts of Ca) is present only in the lumen of peripheral (under surface SL) SR, in the interior junctional (abutting on the T tubules) SR, and in the lumen of the corbular (distributed in the interfibrillar spaces of the I band regions) SR. Calsequestrin is absent from the lumen of the network (primary Ca pumping site) SR. The ratio of calsequestrin content was ~40:60%, corbular to junctional. If SR Ca content is proportional to calsequestrin content, then 60% of SR Ca would be released from the SR in the space directly under the SL (i.e., have direct access to transsarcolemmal exchange systems). In contrast, 40% of the content would be released within the interfibrillar spaces and be required to diffuse to sarcolemmal sites so that it can be exchanged. It is of interest that the rapidly exchangeable ($t_{1/2} = 3.5$ s) component of the putative SR compartment contains 60% of the Ca, whereas the slowly exchangeable component ($t_{1/2} = 19$ s) contains 40% (Langer et al., 1990). This is consistent with the corbular:junctional distribution of Ca found by Jorgensen et al. (1985).

If the value of 400 μmol/kg wet weight approximates the SR content under physiological conditions, then maximum contraction would be achieved by release of 20–25% of the total content of the SR. It should be emphasized that, dependent on the inotropic state of the cell, Ca in the SR would be expected to vary over a large range. For example, our study (Langer et al., 1995) measures a ninefold increase in Ca flux through the SR when [Ca]$_0$ is increased from 0.5 to 4.0 mM.

In summary, Ca in the SR represents 25–30% of the cell's exchangeable

Ca, with the caveat that the value assumes that caffeine releases only 30–50% of total SR Ca. Ca exchange in isolated cells measured under conditions of minimal isotopic reflux indicate that SR Ca is distributed within two kinetically distinct compartments with exchange $t_{1/2}$ of ~4 and 20 seconds. It is suggested that these compartments may represent junctional and corbular SR Ca, respectively. SR Ca content can be expected to vary widely, dependent on the cell's contractile state and extent of Ca loading.

3. Mitochondrial Calcium

The mitochondria occupy 30–35% of the cell volume and, as we see later in this chapter, are active participants in the cell's Ca exchange. The mitochondrion is a large structure, easily identifiable in frozen tissue and relatively easily sampled by an electron probe. Three such studies (Walsh and Tormey, 1988; Moravec and Bond, 1992; Isenberg et al., 1993) of mammalian ventricular cells find the mitochondrial Ca content in the resting cell with $[Ca]_0$ between 1 and 2 mM to be in the range of 500–550 μmol/ kg dry mass mitochondria. This converts to 60–70 μmol/kg wet tissue and would account for only 4–5% of the cell's exchangeable Ca. However, mitochondrial content varies dramatically with activity and metabolic state of the cell. The maximal capacity of the mitochondria was calculated by Fry et al. (1984) to be an astounding 17 mmol/kg wet weight when the cells were permeabilized to permit free Ca entry. Isenberg et al. (1993) measured Ca in the mitochondria 40 milliseconds after excitation and found a six fold increase as compared to the resting state or about 300 μmol/kg wet weight tissue. Wolska and Lewartowski (1991) found that mitochondrial content of guinea pig ventricular muscle strips that were measured by CCCP (mitochondrial uncoupler) release increased by 2.5 times when the strips were stimulated as compared to when at rest.

As in the case with the SR, our on-line studies using ^{45}Ca kinetic measurements give relatively high values for mitochondrial Ca content. The kinetically defined "slow" compartment, with $t_{1/2}$ equaling 3.6 minutes, accounts for all of the remaining exchangeable Ca in the cell after the rapidly exchangeable La-displaceable and the intermediate compartments. The Ca content of the slow compartment can be specifically augmented by the addition of a proton donor (e.g., $-H_2PO_4$) to the perfusing solution. When the acid phosphate enters the cell followed by loss of proton to the alkaline matrix of the mitochondria (Lehninger, 1974), the resulting excess anion provides the milieu for accumulation of Ca as the phosphate salt. This sequence is specific for the mitochondria. The Ca accumulation attendant to proton donation in these cells is halted by inhibition of mitochondrial respiration (e.g., by warfarin or antimycin A) (Langer et al., 1990). These agents prevent the donated protons from passing along the proton transport chain, thereby preventing excess anion accumulation and the enhanced Ca uptake. The proton-augmented uptake exchanges with a $t_{1/2}$ of 3.3 minutes virtually identical to the "slow" compartment $t_{1/2}$ of 3.6 minutes. This is strong evidence for the mitochondrial origin of the kinetically defined "slow" compartment.

The "slow" compartment content of rat ventricular cells in 1 mM [Ca]$_0$ is 370 μmol/kg wet cell or 300 μmol/wet tissue. This value compares well with that found by Williamson *et al.* (1983) for rat ventricular myocytes using the null point titration technique and induction of Ca release by FCCP, an uncoupler of mitochondrial respiration in permeabilized cells. Williamson *et al.* found 4.8 nmol Ca per milligram of mitochondrial protein in cells perfused with 1 mM [Ca]$_0$. Using 77 milligrams of mitochondrial protein per gram wet weight tissue (Scarpa and Graziotti, 1973), the mito-chondrial Ca content is 370 μmol/kg wet tissue, a value similar to that measured with isotopic technique. If these higher values are realistic then the mitochondria account for 20–25% of the exchangeable cell Ca.

It is, perhaps, not surprising that reported steady-state levels of Ca in the mitochondria vary widely. It is very clear that the mitochondria respond to increased cellular Ca loads and concentration. Lukcs and Kapus (1987) measured the mitochondrial Ca content by atomic absorption after they measured the free matrix [Ca]$_m$ with fura-2 dye in isolated mitochondria. They found that a rise in [Ca]$_m$ from 100 to 1000 μM caused a rise in Ca content from 1 to 6 μmol/mg protein. This corresponds to an increase of mitochondrial content (based on 77 mg mitochondrial protein/g wet tissue) from 77 to 460 μmol/kg wet tissue. The [Ca]$_m$ is within the range required for regulation of intramitochondrial dehydrogenases (Denton and McCor-mack, 1985; Hansford, 1985), which play a major role in the control of intramitochondrial energy metabolism. In addition, the level of proton donation will have a major effect on mitochondrial content. The addition of 10 mM NaH$_2$PO$_4$ to perfusate produced more than a 900 μmol/kg wet increase in Ca in rat ventricular cells localized to the mitochondria (Langer *et al.*, 1990) (Fig. 3). This increase was largely eliminated by warfarin inhibi-tion of mitochondrial respiration. Therefore, as with the SR, the mitochon-drial content will vary greatly dependent on the inotropic and energetic state of the cell.

In summary, as with the SR, mitochondrial Ca content varies greatly with the functional state of the cell. After long-term rest, content may be less than 100 μmol/kg wet weight tissue. Under physiological perfusion and activity, levels may rise into the 300–400 μmol range and approach 1000 μmol under the condition of high-proton donation. In response to pathological conditions, loading levels can exceed 10 mmol/kg wet weight. Mitochondrial Ca content responds to the free [Ca] concentration in the cytosol and thereby adjusts mitochondrial metabolic activity to cellular energy demand. The exchange of mitochondrial Ca with the extracellular

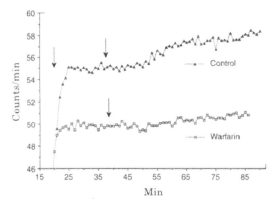

FIGURE 3 On-line monitoring of ^{45}Ca uptake in two groups of adult rat ventricular cells. Cells are introduced into the counting chamber at first arrow. In the top curve, 10 mM NaH$_2$ PO$_4$ is added as a proton donor to the ^{45}Ca labeling at the second arrow. There follows a progressive net gain in ^{45}Ca representing 4.1 mmol Ca/kg dry weight cells (900 μmol/kg wet wt) in this experiment. In subsequent washout, the net gain was localized entirely in the "slow" compartment (see Fig. 2A). In the bottom curve, 10 mM NaH$_2$ PO$_4$ plus 10^{-5} M warfarin was added at the arrow. The net gain of Ca was reduced to 1 mmol Ca/kg dry weight by the warfarin-induced inhibition of mitochondrial respiration. (Reproduced from Langer *et al.*, 1990, with permission.)

space has a $t_{1/2}$ between 3 and 4 minutes—much slower than sarcolemmal ($t_{1/2} < 1$ s) and SR ($t_{1/2} \sim 4$ and 20 s, biphasic) exchange.

4. Nuclear Calcium

Other than the myofilaments, the remaining organelle in the cell is the nucleus. In the rat myocyte, its content measured by electron microprobe was 1.0 mmol/kg dry weight nucleus (Miller and Tormey, 1993). Nuclear volume is about 2.5% cell volume (Legato, 1979). Thus, nuclear Ca is about 20 μmol/kg wet tissue—a very small component of tissue Ca. Its content does, however, seem to be in equilibrium with cytosolic [Ca]. A dye study of nuclear [Ca] in neonatal rat hearts (Minamikawa *et al.*, 1995) found that nuclear [Ca] followed cytosolic [Ca] with a lag of 100–200 milliseconds. The increase in the nucleus was initiated at the edge of the nucleus where cytosolic [Ca] was high and spread inward to the nuclear center. With peak cytosolic [Ca] levels induced by caffeine peak nuclear [Ca] attained levels greater than 1000 nM (Minamikawa *et al.*, 1995). Current evidence indicates that Ca in combination with calmodulin regulates a number of nuclear functions including DNA synthesis and repair, gene transcription, phosphorylation and dephosphorylation of nuclear proteins, control of an intranuclear contractile system, chromatin condensation, and programmed cell death (Bachs *et al.*, 1992).

In summary, nuclear Ca content is 1% or less of total tissue, but its variation with cytosolic levels seems to be of major importance in the control of nuclear DNA and genetic processes.

5. Cytosolic Calcium

If we add the contributions of extracellular ($\sim$500 μmol), sarcolemmal ($\sim$600 μmol), sarcoplasmic reticular ($\sim$400 μmol), and mitochondrial ($\sim$300 μmol) Ca they account for about 90% of the total exchangeable tissue Ca. There remains the Ca bound to the myofilaments and to molecules such as calmodulin. Pierce *et al.* (1985) measured the passive Ca-binding capacity of a crude tissue homogenate at 72 μmol Ca/kg wet tissue when free [Ca] was 1 μM. Berlin *et al.* (1994) found maximum passive binding to be 123 μmmol/liter cell H_2O, which converts to about 90 μmol Ca/kg wet tissue. These values are consistent with the concept that, to achieve a level of 1–2 μM free [Ca], 80–90 μmol has to be released to the cytosol.

In summary, cytosolic Ca binding adds about 4% to the total exchangeable Ca of the tissue. Given the approximation required, virtually all exchangeable Ca is accounted for among the compartments discussed.

6. "Inexchangeable" Calcium

Analysis of total cellular Ca from ashed cells (Langer *et al.*, 1990) gives a value that is consistently higher than the total derived from asymptotic isotopic measurements. The difference represents "inexchangeable" or extremely slowly exchangeable Ca in the cell and amounts to about 275 μmol/kg wet tissue (Langer *et al.*, 1990). Williamson *et al.* (1983) found a "nonreleasable form" of Ca of about 175 μmol/kg wet weight. The origin or origins of the inexchangeable Ca are unknown.

Finally, it should be emphasized that all studies except those using the "on-line" isotopic technique (Langer *et al.*, 1990) report cellular Ca levels that do not include the La-displaceable component based in the SL. This is because, with standard isotopic labeling and extracellular washing technique, the labeled La-displaceable Ca is lost before cellular Ca is measured. As reviewed previously, this represents about 600 μmol Ca/kg wet tissue or about 30% of the wet tissue's exchangeable Ca. As is discussed later in this chapter, this component probably plays an important role in Ca-related processes in the cell.

B. CALCIUM FLUX AT THE ORGANELLES

1. Sarcolemma

Currently, there are three known systems in the SL responsible for moving Ca. These are the Ca channels (see Chapter 3), the Na–Ca exchangers, and the sarcolemmal Ca pump (see Chapter 4). Measurements in rat ventricular cells in which Ca current was integrated over the course of an excitation indicated a single-cell maximum Ca influx of 18×10^{-17} mols (Wang et al., 1993). This converts to about 9 μmol/kg wet weight tissue per excitation. Therefore, the Na/Ca exchangers have to remove 9 μmol/kg so that a steady-state cellular Ca level is maintained. The rat ventricle normally beats at rates greater than 300 beats/min. At a rate of 300, the channel/exchange-mediated flux would approximate 50 μmol/kg wet wt/s with, perhaps, a small efflux via the sarcolemmal ATPase. In the section in which sarcolemmal Ca is discussed (Section III,A,1), we indicate that ^{45}Ca exchange studies indicated a sarcolemmal-based exchange of more than 1800 μmol/kg dry wt/s. This converts to more than 414 μmol/kg wet wt/s or an exchange greater than nine times that of the channel/exchanger-mediated flux of the rat ventricle. The difference is almost certainly due to "exchange diffusion" as first suggested by Ussing (1948). This is the process in which "uphill" transport of one ion is coupled to the "downhill" transport of another ion of the same species. This process can only be observed with the use of isotopic tracers. Thermodynamically, the energy expended is nil. Ussing (1948) emphasized that the carrier involved in exchange diffusion, in this case, Ca–Ca exchange, does not need to be an independent molecule. Indeed, Slaughter et al. (1983) clearly demonstrated in cardiac sarcolemmal vesicles that the Na–Ca exchanger carries a large component of Ca–Ca exchange in this preparation. Under their in vitro conditions they found a $t_{1/2}$ of a few seconds for Ca equilibrium exchange "as would be expected for the process of sarcolemmal 'exchange diffusion' " and, consistent with the very rapid exchange of the sarcolemmal bound, La-displaceable fraction of adult rat ventricular cells. Routes other than Na–Ca exchange, as yet unidentified, may also be involved.

In summary, the sarcolemmal Na–Ca exchanger accounts for the major fraction of active Ca transport from the cell. Of necessity, this matches influx through the sarcolemmal Ca channels so that steady-state intracellular Ca levels are maintained. In addition to active transport, a component of nonenergy-dependent Ca–Ca exchange takes place across the SL. This is represented by the La-displaceable fraction and may represent more than 90% of total sarcolemmal exchange.

2. Sarcoplasmic Reticulum

Sipido and Wier (1991) measured SR Ca flux in guinea pig ventricular myocytes. They calculated the flux by measurement of $[Ca]_i$ transients and Ca currents and by assuming characteristics of Ca-binding ligands. Measurements were made with the cells in 1 mM $[Ca]_0$. They found a large range of flux values from 2.7 to 9.5 mmol/s/liter accessible cell water flux rate associated with a depolarization. A midrange value is 6.1, which converts to 2.4 mmol/s/kg wet weight. If we assume that SR release occurs over a 20-

ms period, and the release is linear over time, then a single depolarization would release 48 μmol Ca/kg wet weight (0.020 s $\times$ 2400 μmol). Such release would produce about 55% of maximum contractile force (Fabiato, 1983). If SR content is about 400 μmol/kg wet, then 55% maximum force would be attained with release of about 12% of the total SR compartment (48/400). In contrast, in ferret myocytes at 2 mM [Ca]$_0$, Bassani et al. (1995) measured an SR content (by caffeine release) of 36 μmol/kg wet with a release of 13 μmol, or 36%, of the content. This amount of release would generate no more than 5% maximum force. Since 2 mM [Ca]$_0$ supports significant force in all mammalian hearts, the measured SR release must be artifactually low in this study.

If we assume that the values obtained by Sipido and Wier (1991) are physiologically representative then a ventricle beating at a rate of 80 beats/ min and generating about 50% maximum force would release about 4000 μmol Ca/kg wet weight from the SR over the period of 1 minute. As discussed later in this chapter, when Na–Ca exchange-mediated Ca flux is evaluated, about 10% of Ca released by the SR with each beat is predicted to exit the cell via Na–Ca exchange with 90% going to the cytosol and returning to the SR in an intracellular cycle. Thus for a release of 50 μmol/ kg wet/beat, 5 μmol would exit and 45 μmol would recycle. Therefore, it would be predicted that an isotopically labeled SR content of 400 μmol/ kg wet weight would nearly completely exchange in about 1 minute at a heart rate of 80 beats/min. The mean exchange rate of the ventricular cell's intermediate compartment proposed to represent the SR indicates that SR turnover would be 96% complete within 1 minute. The similarity of the predicted turnover times may be fortuitous since the beat rate of the cells during isotopic washout was unknown. The example does, however, indicate the magnitude of SR flux, its exchange with extracellular Ca, and the obvious influence of heart rate on this exchange.

The previous discussion deals with SR Ca flux in the beating cell. It is generally agreed that in the absence of beating there is a continuing exponential decay of Ca in the SR due to a continuing "leak." This leak is removed from the cell, in mammalian hearts other than the rat, via Na–Ca exchange (Bassani and Bers, 1994). In rabbit ventricular myocytes, the rate constant for Ca loss measured about 0.009/s ($t_{1/2} = 77$ s) (Bassani and Bers, 1994) and in the guinea pig, 0.029/s ($t_{1/2} = 24$ s) (Terracciano et al., 1995). With an SR content of 400 μmol/kg wet weight the initial net loss from the SR would then occur at a rate of 4–12 μmol/kg wet wt/s and decay to steady state in 2–3 minutes. The resting steady-state SR efflux was measured at about 0.15 μmol/kg wet wt/s (Bassani and Bers, 1995) or only about 0.3% of the flux associated with a single contraction capable of 55% maximal force development. It is this resting leak that is believed to generate so-called elementary Ca release events recorded as the response of intracellular Ca dyes called "Ca sparks" (Cheng et al., 1993).

In summary, a ventricle beating 80 times per minute and developing ~55% maximal force would have an SR flux of about 4000 μmol/kg wet wt/min. About 90% of this flux is directed to the cytoplasm and the myofilaments to be pumped back into the SR and, therefore, recycled. About 10% will exit the cell via Na–Ca exchange and account for most of the active transport across the SL. This flux (about 7 μmol/kg wet wt/beat) is easily handled by the Na–Ca exchangers. It represents about 2% of sarcolemmal Ca exchange, the remaining 98% being "energy neutral" exchange diffusion. Upon rest, the SR loses Ca at an initial rate of 4–12 μmol/kg wet/s. This falls to a steady-state turnover rate of about 0.15 μmol/s over a period of 2–3 minutes.

3. Mitochondria

Ca influx to the mitochondria is via the so-called "Ca uniporter" (Crompton, 1990). In the presence of a finite mitochondrial membrane permeability, Ca will enter passively down the very large (-160 to -200 mV) electrochemical gradient. For each Ca that enters, two H^+ are extruded via H^+ pumps of the respiratory chain and this provides the charge compensation. Ca efflux from the mitochondria is via an electroneutral Na–Ca exchanger (i.e., two Na^+ for one Ca^{2+}). It has, therefore, a different stoichiometry than the sarcolemmal (i.e., three Na^+ for one Ca^{2+}) exchanger.

As is the case with the other organelles, there is a range of flux values for the mitochondria reported in the literature. Crompton (1990) estimated a minimum value for V_{max} of 10 nmol Ca/mg mitochondrial protein per second at 25. We have measured a Q_{10} for mitochondrial exchange at 1.4 (Marengo *et al.*, 1997). Converting to wet weight and 37°, the V_{max} is about 1.2 μmol Ca/kg wet wt/s. The "slow compartment," which is assumed to be mitochondrial, exchanges at 1.6 μmol/kg wet/s under nonperfusion-limited conditions. These values are in reasonable agreement. The low flux rate indicates that mitochondrial Ca would respond slowly to extracellular Ca modifications. Similarly, Moravec and Bond (1991) demonstrated, using electron microprobe technique, that the mitochondria are not involved in intracellular contraction-producing fluxes on a beat-to-beat time frame. The flux rate in the range of 1.5 μmol/s is grossly insufficient for activation of the contractile elements. Comparatively, Na–Ca exchange has a V_{max} of about 100 μmol/kg wet wt/s (Hilgemann *et al.*, 1991), and Levitsky *et al.*, (1981) found an energy-dependent Ca uptake of 25 μmol/kg wet wt/ 100 ms, or 250 μmol/s, for the SR. Dependent on the level of mitochondrial loading, higher flux values for mitochondria have been reported (e.g., 7 μmol/kg wet wt/s) (Fry *et al.*, 1984). However, it is apparent that mitochondrial Ca turnover is not involved in the beat-to-beat regulation of excitation–contraction (E–C) coupling.

The relatively small turnover is, however, important in the feedback between myocardial energy production and demand. Leisey *et al.* (1993)

found that mitochondrial matrix Ca increased to steady state after an increase in extra mitochondrial average Ca concentration with a $t_{1/2} \sim 2$ min. They concluded that this was a "mechanism of signal transduction from the cytosol to the mitochondrial matrix." The work of Denton and McCormack (1980) and Hansford (1985) has demonstrated that Ca-sensitive dehydrogenases [pyruvate dehydrogenase (PDH), NAD-isocitrate dehydrogenase (NAD-ICDH), and 2-oxyglutarate dehydrogenase (OGDH)] catalyze an irreversible oxidative decarboxylation and are the main sites of respiratory CO_2 production. They are important regulatory enzymes of mitochondrial oxidative metabolism and all are regulated by free [Ca] in the range found within the mitochondria. Kd values for PDH and OGDH are 0.4–0.8 μM and for NAD-ICDH about 10-fold higher (Rutter, 1990). This could allow Ca to regulate mitochondrial oxidative metabolism in a graded manner over a wide range of [Ca] (McCormack and Denton, 1994). All increases in contractile state involve increases in the cytosolic Ca transient. Crompton (1990) has modeled such increased transients and shows that they are likely to produce quasi-steady state increases in mitochondrial free [Ca] as experimentally demonstrated by Leisey et al. (1993). The feedback is, then, obvious: Increased contractile state with increased energy demand → increased beat-to-beat Ca → increased steady-state mitochondrial matrix [Ca] → increased dehydrogenase activation → increased oxidative metabolism → increased energy supply. It should be noted that matrix [Ca] is not the sole signal for control of mitochondrial energy metabolism. ATP-consuming processes lead to a decrease in the so-called phosphorylation potential, ATP/ADP + P_i. Within the mitochondria, a decreased ATP:ADP ratio will cause an increased flow of electrons along the respiratory chain and increased O_2 consumption. This is a stimulation of state 3 mitochondrial respiration and would not require an increase in mitochondrial matrix [Ca] (McCormack and Denton, 1994).

In summary, mitochondrial Ca flux will vary with Ca load but under basal conditions is in the 1–2 μmol/kg wet wt/s range with a $t_{1/2}$ of exchange with the extracellular space of 3–4 minutes. Matrix [Ca] will vary proportionally with cytoplasmic [Ca] with quasi-steady state achieved in about 2 minutes. This relation provides the signal transduction by which the mitochondrial dehydrogenases regulate their activity to control oxidative metabolism and thereby match energy output to demand.

4. Myofilaments

Cardiac troponin C (TnC) has a single specific site for Ca binding (Thompson et al., 1990). It has a concentration of 2.8 μmol/kg wet weight of tissue (Fabiato, 1983) with a Kd (Ca) = 1.0×10^{-6} M (Holroyde et al., 1980). The "operative flux" at the myofilaments is determined by the binding and unbinding of Ca to TnC. The amount bound in a single cardiac cycle is characterized by Michaelis–Menten kinetics and is therefore dependent on the level of free [Ca] in the cytoplasm. For example, according to Fabiato (1983) $[Ca]_i$ of 2.7×10^{-6} M gives 50% maximal contractile force. To achieve this free $[Ca]_i$ level, approximately 46 μmol/kg wet weight needs to be released to the cytosol. This release will be bound primarily to three types of sites—TnC, calmodulin, and SR binding sites— according to their content and affinities. If the binding reaches equilibrium, TnC will bind 20.4, calmodulin 7.0, and SR 13.7 μmol/kg wet weight of the 46 μmol released. These are clearly approximate values since they require extrapolation from *in vitro* binding measurements to the *in vivo* state and the change in environment could make for significant variation. Nevertheless, the calculation indicates that, at 50% maximal force, about 20 μmol will bind to TnC with each beat. Thus a ventricle beating at 80 beats/min would have a "myofilament exchange" of 1600 μmol/min of Ca. Since the total SR exchange at this rate and with 50% force development is about 4000 μmol, 40% is destined for TnC binding sites to participate in the last step in E–C coupling.

In summary, there is a single site on TnC to which Ca binds repetitively with each beat. The exchange of Ca at this site represents 35–40% of the SR flux, with the ventricle beating at 80 beats/min with 50% maximal force.

Table 1 summarizes the Ca content within, and the fluxes from, the individual organelles for cells at basal levels of function and physiological $[Ca]_0$.

IV. CALCIUM MOVEMENT IN SUBSARCOLEMMAL DIADIC CLEFT

The diadic cleft space is the region between the junctional sarcoplasmic reticulum (jSR) membrane and the inner leaflet of the transverse (T) tubular sarcolemmal membrane. As results accumulate from various laboratories, it may not be an exaggeration to describe the diadic cleft as the "control center" for cardiac contraction.

A. CLEFT SPACE STRUCTURE

Figure 4 presents, in schematic form, the important elements of the cleft space (see Chapter 1 for electromicroscopic definition of the region). The lateral cisterns of the jSR directly appose about 50% of the T tubular membrane (Page, 1978). The space between the outer cisternal membrane and the inner leaflet of the T tubule SL is spanned by the so-called "feet" or ryanodine receptors. These are tetrads with a central channel opening into the cistern and with side channels opening into the cleft space (Radermacher *et al.*, 1994). They are the sites of Ca release from the jSR. The "feet" are 12 nm high and 29 nm square. Estimates as to their number in

TABLE 1 Organelle Calcium Content and Fluxes[a]

Organelle or site	Content (μmol/kg wet wt)	Flux (μmol/kg wet wt/s)		Comments
		Intracell cycle	Intracell–extracell	
Sarcolemma	500–600		>300	(1) 75% bound to inner SL leaflet phospholipid (2) 90% of flux is "exchange diffusion" (3) La^{+++} displaceable
Sarcoplasmic reticulum	~400	~50 (Heart rate ~80/min, 50% maximum force)	~7 (via Na–Ca exchange)	(1) Biphasic intracell–extracell kinetics (2) Resting flux ~ 4–12 μmol/s declining to ~0.15 μmol/s steady state (3) Content not totally caffeine releasable
Mitochondria	100–300 (Dependent on contractile activity)	~1.5	~1.5	(1) Intracell–extracell exchange has $t_{1/2} = 3$–4 min (2) Ca controls dehydrogenase activity
Myofilaments	20	~25	—[b]	(1) Binding to single TnC site
Nucleus	20	?	?	(1) Nuclear [Ca] equilibrates with cytosol in 100–200 ms
Nonmyofilamentous cytosol (calmodulin, phosphocreatine, ATP)	~70	—	—	
"Inexchangeable"	175–275	—	—	(1) $t_{1/2}$ of exchange is many hours or days

[a] Values apply to cells in 1 mM [Ca]$_0$.

T TUBULE

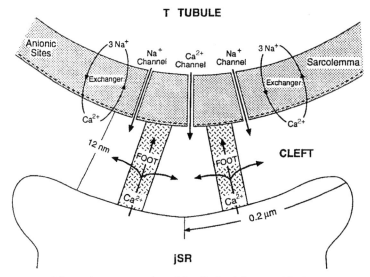

FIGURE 4   Schematic representation of the diadic cleft space. The important constituents relative to Ca exchange are represented. The diadic cleft regions represent apposition of junctional sarcoplasmic reticulum (jSR) to inner leaflet of the sarcolemma (SL) in transverse (T) tubules. Similar cleft structure is probably present at the surface membrane as well (Sun *et al.*, 1995). Note particularly the "feet" spanning the cleft, the anionic sites at the inner sarcolemmal leaflet, the proposed entry of Na^+ and Ca^{2+} channels into the cleft, and the proposed presence of Na^+–Ca^{2+} exchangers at the cleft. Schematic is not to scale. (Reproduced from Langer and Peskoff, 1996, with permission.)

a cleft vary, but Wibo *et al.* (1991) indicate feet density to be at 765 ft/μm^2 of junctional area. This would place 100 feet structures in a cleft of 200 nm radius. Wibo *et al.* (1991) also estimated the density of Ca channels in the cleft at 84 μm^2 or a ratio of 9 feet to 1 channel. This would suggest an arrangement as depicted in Fig. 5. One hundred feet within a cleft would

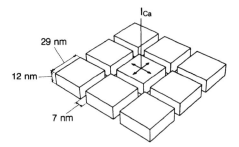

FIGURE 5   Schematic representation proposed for the relation of Ca L channel to the "feet" structures in the diadic cleft. Ratio of number of feet to a channel is from Wibo *et al.* (1991) at 9:1. The feet and their spacing are to scale and, at this ratio, demonstrate their high density in the space. (Reproduced from Langer and Peskoff, 1996, with permission.)

occupy two-thirds of the cleft volume and all 9 feet would be less than 50 nm from the centrally placed Ca channel. As noted previously, the ratio of feet to channels may vary as well as cleft dimension but the structural arrangement presented (see Figs. 4 and 5) is probably a good approximation and serves as the basis for a comprehensive model of Ca concentration and movement in the diadic cleft space (Langer and Peskoff, 1996).

B. CALCIUM BINDING

The single most important factor in the determination of Ca concentration and movement in the space is the presence of anionic phospholipid binding sites on the inner sarcolemmal leaflet (see Fig. 4). The distribution of phospholipid in cardiac SL was first described by Post *et al.* (1988) followed by characterization of Ca-binding properties of the SL (Post and Langer, 1992b). Seventy-five percent of Ca binding of the SL is attributable to PS, PI, and PE located within the inner leaflet. These sites have low affinity for Ca (Kd = 1.1 mM), but are abundant—about 900 μmol/kg wet weight. Their presence has the effect of markedly delaying the diffusion of Ca from the cleft. Figure 6 illustrates Ca concentration in the cleft as a

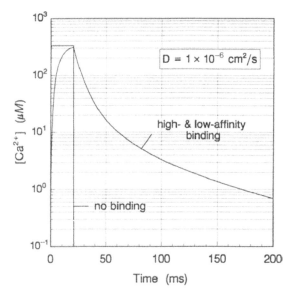

FIGURE 6   The effect of inner sarcolemmal leaflet anionic Ca-binding sites on Ca diffusion within the cleft. The Ca concentration is averaged over the volume of the cleft space with diffusion rate within the space set at approximately 20% of free aqueous diffusion for Ca (D = 1 $\times$ 10^{-6} cm^2/s). Ca is released from the "feet" for 20 ms and then ceases. Although there is little effect on the concentration level reached during release, note the striking slowing effect on [Ca] decline with the anionic sites present.

function of time after a 20-ms release from the feet into the cleft. The amount released is sufficient to produce near-maximum force (2 × 10^{-19} mol/individual cleft or 70 μmol/kg wet tissue or about 20% of the SR content). First, note the "no binding" plot. This represents the condition in which, in the model, all inner sarcolemmal anionic sites have been removed. Diastolic [Ca] is set at 100 nM. As Ca is released from the jSR feet, [Ca] increases almost instantaneously to more than 300 μM and remains at this level as the 20-millisecond release continues. Upon cessation of release, the [Ca] returns to the diastolic level in less than 1 millisecond. This is because diffusion within the cleft is very rapid even when the value for diffusion within the cleft space has been reduced to 1 × 10^{-6} cm^2/s (see inset), or to about 20% that of free diffusion in aqueous solution. Now note the dramatic effect of adding the anionic sites in the quantity and with the characteristics as experimentally measured (Post and Langer, 1992b). [Ca] increases more slowly but to a level just below the "no binding" plot. Upon cessation of release, as Ca diffusion is slowed by the sarcolemmal binding, [Ca] falls at a remarkably low rate. It is still more than 3 μM or 30 times the diastolic level 80 milliseconds after release has ceased.

In summary, the presence of predominantly low-affinity Ca binding sites at the inner sarcolemmal leaflet in the diadic cleft delays diffusion of Ca from the region and, thereby, causes [Ca] to remain many fold higher than in the bulk cytoplasm for more than 100 milliseconds. As will become evident, this has significant implications for a number of elements of the E–C coupling process.

C. "TRIGGER" CALCIUM

The cleft model allows the reader to explore an important question, "Does the Ca that serves as the trigger for Ca-induced release of Ca from the feet come from the Ca channel or from 'reverse' Na–Ca exchange?" There are arguments on both sides of the question (Leblanc and Hume, 1990; Levi et al., 1994; Lipp and Niggli, 1994; Sham et al., 1995; Cannell et al., 1994).

In Fig. 7, a Ca channel at the cleft center and in the center of a 9-foot array (see Fig. 5) is modeled to pass a current of 0.3 pA with a rectangular pulse lasting 1 millisecond. Note the curve designated "50 nm" (third curve from top). This depicts the [Ca] at a distance of 50 nm from the channel opening and, with the arrangement depicted in Fig. 5, would include all 9 feet in the array. Despite a 500-μs delay before [Ca] starts to rise above the diastolic level of 100 nM, it reaches 1 μM in the next 500 μs and reaches 10 μM in the next 300 μs or within 800 μs from the time the channel initially opens. Therefore, all 9 feet within the domain of the Ca channel will "see" a minimum 1 μM concentration at a maximum time of 1 ms. According to Fabiato (1985), this concentration

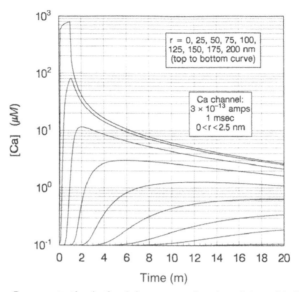

FIGURE 7 Ca concentration in the cleft space as a function of time with Ca influx giving a 0.3 pA maximum current with a square pulse of 1 ms duration. The Ca channel is placed at the cleft center (0 nm), and the [Ca] profiles are plotted at radial intervals of 25 nm for a 200-nm radius cleft. Note the 50-nm curve. This radius would encompass all nine "feet" in an array as depicted in Fig. 5. Note that [Ca] reaches more than 10 μM approximately 1 ms after it starts to rise.

increase within this time will release an amount of Ca sufficient to activate approximately 50% of the maximum force.

We now compare the [Ca] profile in the cleft secondary to channel entry with the profile seondary to Ca entry via "reverse" Na–Ca exchange (Fig. 8). We used the data of Matsuoka and Hilgemann (1992) in our computation to estimate the exchanger current at the cleft as the cell depolarizes to +20 mV and remains at this level for the initial 10 ms of the action potential. We had previously computed the [Na] profile in the cleft so that the effect of [Na]$_i$ on the Ca movement via the exchangers could be evaluated. On the basis of Hilgemann *et al.* (1991), who estimated sarcolemmal exchanger density, and Frank *et al.* (1992), who provided evidence for preferential T tubule localization, we assumed an exchanger density of 800/μm^2 cleft membrane. This places 100 exchangers within a single cleft. Under these conditions of the model we find that [Ca] in the cleft increases from the 0.1 μM diastolic level to about 0.5 μM over 10 ms. This will produce about 20% activation of contractile force (Fabiato, 1985). Note that entry via the channel produced twice the [Ca] in 10% of the time (see Fig. 7).

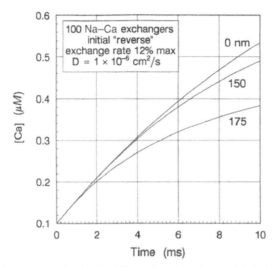

FIGURE 8   Ca concentration in the cleft as a function of time with Ca influx via "reverse" Na–Ca exchange. Model places 100 Na–Ca exchangers in the cleft and Vm is clamped at +20 mV. At this voltage and with the calculated $[Na]_i$ and $[Ca]_i$, the exchangers will be operating at a maximum of 12% V_{max} (Matsuoka and Hilgemann, 1992). Compare the [Ca] with Fig. 7 in which Ca influx is via the channel. Note that with "reverse" Na–Ca exchange about 9 ms are required to reach 0.5 μM [Ca] at the cleft center (0 nm). In Fig. 7, [Ca] is twice this level within 1 ms. (Reproduced from Langer and Peskoff, 1996, with permission.)

Although Ca entry via the channel is fully capable of providing a strong trigger stimulus over a domain of nine receptors (see Fig. 5), according to the model, this stimulus could be augmented by a chain reaction in which the cluster of SR channels are mutually coupled by their own Ca release, as suggested by Stern (1992). Although this mechanism might be redundant dependent on frequency of Ca channel openings in a cleft, a locally regenerative release could significantly amplify the relatively weak signal based on "reversed" Na–Ca exchange.

In summary, it is clear that if the model is a reasonable approximation of the diadic region Ca entry via its channel provides a much more efficient trigger stimulus than does entry via "reverse" Na–Ca exchange. Nevertheless, in the absence of Ca current or under conditions where it may be delayed or reduced, reversed exchange is capable of providing a triggering stimulus.

D. SODIUM–CALCIUM EXCHANGE

Dependent on the amount of Ca released from the jSR cytosolic [Ca] reaches peak levels between 1 and 2 μM for 30–40 ms falling to mean levels of less than 1 μM for the remaining 150–200 ms of the [Ca] transient (Berlin and Konishi, 1993). A reasonable value for [Ca] to produce half-maximal stimulation of the Na–Ca exchanger is 5 μM (Kd Ca) (Hilgemann *et al.*, 1992). Therefore, the levels of cytosolic [Ca] over the course of a cardiac contraction cycle will result in much less than 50% maximal Na–Ca exchanger activity. If all Na–Ca exchange takes place over the entire sarcolemmal surface with no specialized cleft structure to facilitate the exchange, and the exchangers have the characteristics as defined by Hilgemann *et al.* (1992) and Matsuoka and Hilgemann (1992), it would require ~1 second for the exchangers to expel the amount of Ca that had entered via the Ca channels (Langer and Peskoff, 1996). Therefore, if the cell beat rate is more than 60 times per minute, a progressive increase in intracellular Ca would occur. Such obviously cannot be the case in the functional cell. If, however, the exchangers are preferentially distributed at the cleft sites, as suggested, the cell will have little difficulty in the maintenance of steady-state intracellular Ca levels up to beating rates of 300 or more.

Figure 9 illustrates the [Ca] at various radial locations of the cleft versus time after a 20-ms release of Ca from feet throughout the cleft. The amount released is sufficient to produce near maximal contractile force

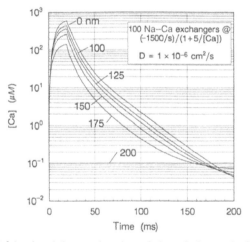

FIGURE 9   [Ca] in the cleft as a function of time during and after uniform release throughout the cleft space from the "feet." The release is an amount that will produce near maximal force (2×10^{-19} mol/cleft or ~70 μmol Ca/kg wet wt). During the 200-ms period plotted, Vm is assumed to be −60 mV, a reasonable mean for the rat ventricular action potential. This sets the velocity of the Na–Ca exchangers at 30% V_{max} or 1500 per second (Matsuoka and Hilgemann, 1992). Kd (Ca) is taken as 5 μM (Hilgemann *et al.*, 1992). The velocity then changes with instantaneous [Ca] according to Michaelis–Menten kinetics (see inset). Diastolic [Ca] is set at 0.1 μM. [Ca] from cleft center (0 nm) to periphery (200 nm) is indicated. The model predicts continued expulsion of Ca so that [Ca] slides less than 0.1 μM at about −170 ms. In reality, this will not occur as diastolic steady state is reestablished.

$(2 \times 10^{-19}$ mol/cleft/beat or $\sim 70\ \mu$mol/kg wet weight tissue). The numbers on the curves represent radial distances from the cleft center. It is striking that [Ca] reaches 600 μM at cleft center and more than 100 μM at cleft periphery. After release from the jSR ceases at 20 ms, [Ca] begins to decrease, but relatively slowly due to the sarcolemmal binding sites as shown in Fig. 6. Note that throughout most of the cleft [Ca] is still more than 1 μM at 100 ms and an order of magnitude higher at 50 ms. Therefore, the Na–Ca exchangers in the cleft are exposed to high [Ca] for a relatively prolonged period and therefore Ca efflux is enhanced. (Note that the [Ca] falls more rapidly in Fig. 9 as compared to Fig. 6. This is because [Ca] is affected only by diffusion as modeled in Fig. 6, whereas efflux from the cleft via Na–Ca exchange is added in Fig. 9).

It was suggested previously that the presence of inner sarcolemmal leaflet Ca binding sites, predominantly phospholipid, has significant effects. The model certainly predicts such for Na–Ca exchange.

Figure 10 illustrates the effect of progressive removal of the inner leaflet sites on Ca efflux from the cleft via Na–Ca exchange. The Ca release is over a period of 20 milliseconds as in Fig. 9. Because [Ca] rises much higher than 100 μM during release regardless of whether binding sites are present, efflux is not affected during this period. However, after release ceases, the effect is obvious. With no binding sites (0% curve), only 32% of total required efflux occurs within 200 milliseconds. At the rate of efflux indicated it would require more than 3 seconds to exchange the amount of Ca required

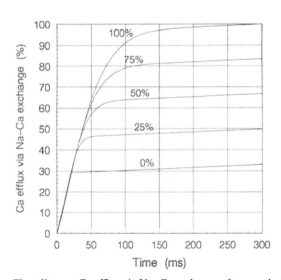

FIGURE 10   The effect on Ca efflux via Na–Ca exchange of removal of inner leaflet SL anionic binding sites. jSR Ca release is over 20 ms. Exchange is set at 100% for a full complement of anionic sites. The predicted effect of progressive removal is clearly shown. (Reproduced from Langer and Peskoff, 1996, with permission.)

to maintain steady state. Therefore, beat rates of more than 20 beats/min would cause progressive increase in [Ca]ᵢ if all Ca efflux occurred at the diadic cleft sites.

Experimental confirmation of the effect of anionic site removal has recently been demonstrated (Wang *et al.,* 1996). It was shown that the local anesthetic, dibucaine, competitively displaces Ca specifically from anionic phospholipid head groups in intact myocardial cells. Dibucaine, in effect, neutralizes the anionic sites, which are tantamount to their removal. Such treatment decreased Na–Ca exchange-mediated flux by almost 60%.

In summary, the cell's Na–Ca exchange requires that it be localized, in significant part, at the diadic cleft regions of the cell so that it may be able to match Ca influx and maintain steady-state levels of intracellular Ca. The cleft environment, particularly its inner sarcolemmal anionic sites, is responsible for increase of [Ca], after jSR release, by 2 to 3 orders of magnitude above bulk cytoplasmic level for a significant fraction (50–100 ms) of the contraction cycle. This optimizes Ca efflux via the exchangers.

E. CHARACTERISTICS OF CALCIUM IN THE CLEFT

The profile of Ca in the cleft and moving out of the cleft during and following a 20-ms release of 70 μmol Ca/kg wet weight is shown in Fig. 11.

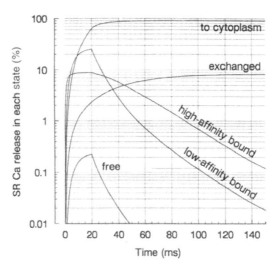

FIGURE 11   A profile of Ca in the diadic cleft space as predicted by the model. The percentage of Ca existing in various states is plotted vs time beginning with Ca release from the jSR. At any instant after release is completed (>20 ms) addition of all fractions adds to 100%. Note that at completion of release two-thirds of the Ca destined for the cytoplasm has exited the cleft but that three-fourths of Ca destined to exit via Na–Ca exchange (exchanged fraction) is still in the space. Of this latter fraction, 99% is bound and <1% is free. (Reproduced from Langer and Peskoff, 1996, by permission.)

This represents the total number of Ca ions free, bound, and exchanged via Na–Ca exchange and diffused to the cytoplasm as a function of time following the onset of jSR release. The amounts have been normalized with respect to the total jSR release so that at any instant after release is completed (t > 20 ms) addition of all fractions adds to 100% of the amount released. Note that 91% moves to the cytoplasm for contractile activation. The remaining 9% or between 6–7 μmol/kg wet weight leaves the cell via Na–Ca exchange. This is in agreement with Bassani *et al.* (1994) who found that about 7% of the Ca released during a contraction was transported by Na–Ca exchange. It should be noted, however, that as jSR release decreases, the fraction exchanged increases quite steeply. For example, with release of 30 μmol Ca/kg wet weight, the fraction released to the cytoplasm falls to 80% with 20% leaving the cell via the exchangers. The steep, nonlinear increase in exchanger fraction is predicted by hyperbolic Michaelis–Menten kinetics for exchanger operation with saturation of the exchangers in the cleft even at low levels of jSR release. In any event, Ca movement out of the cell, as predicted by the cleft model, is in the range of 5–8 μmol/kg wet weight dependent on the magnitude of jSR release. This efflux of Ca needs to match the influx of Ca through the channels with each beat if steady-state intracellular [Ca] is to be maintained. Wang *et al.* (1993) found a range of 11–18 $\times$ 10^{-17} mol/cell Ca entry via the channels in the rat ventricular cell as contractile shortening increased from 40 to 350% (100% at $[Ca]_0$ = 1 mM). The Ca entry converts to 5.3–8.7 μmol/kg wet weight— remarkably similar to the Na–Ca exchange mediated Ca efflux as calculated from the model.

Returning to Fig. 11, note that free Ca in the cleft never exceeds 0.2% of the amount released and that, by 50 milliseconds (30 ms after termination of release), 98% of the Ca that is destined for the cytoplasm has left the cleft. At this point all of the remaining Ca to be transported out of the cell via Na–Ca exchange is derived from the low- and high-affinity binding sites.

It is appropriate to pose the question, "How does the Na–Ca exchange system 'know' the amount of Ca that enters via the channel so that it can adjust efflux to match?" An early clue comes from the Callewaert *et al.* (1989) study in which it was shown that release of Ca from the SR produced a transient inward current that represented efflux of Ca via Na–Ca exchange. This implied a coupling between SR and Na–Ca exchange in ventricular cells. This idea was given further support when Janiak *et al.* (1996) found that interventions that stopped Ca flux through the SR (longitudinal SR Ca uptake to jSR to release from the "feet") greatly decreased the rate of outward Ca transport via Na–Ca exchange. The results were consistent with the proposal that Ca released from the SR interacts with the cell's Na–Ca exchangers most probably within the cleft regions. Lewartowski *et al.* (1997) examined the magnitude and time course of the so-called tail currents generated by Na–Ca exchange. It was found that the amplitude of the tail currents was higher during the rising phase (first 50 ms) of the cytoplasmic Ca transient (recorded with Indo-1 dye) than during the descending phase, although the Ca transient was lower. This dissociation between Na–Ca current and bulk cytoplasmic [Ca] was abolished by thapsigargin, a drug that blocks SR CaATPase and prevents SR Ca uptake. The results strongly suggested that, over the initial 50 milliseconds after jSR Ca release, the Na–Ca exchangers are exposed to a Ca concentration higher than that in the bulk sarcoplasm. It is proposed that this occurs because the Ca released from the jSR first enters the cleft and stimulates Na–Ca exchange before it enters the general cytoplasm to contribute to the Ca transient as recorded by the Indo-1. All of these results support the concept that movement of Ca through the jSR is tightly coupled to Na–Ca exchangers in the cleft.

If such is the case then SR Ca content should be proportional to Na–Ca exchange-mediated Ca flux. Langer et al. (1995) used an on-line isotopic technique in which Na–Ca-mediated Ca exchange could be compared with SR content. Treatment of the cells with thapsigargin reduced SR content to 29% of control and Na–Ca-mediated flux to 30% control; caffeine treatment reduced SR content to 38% and flux to 36%. Elevation of SR content always increased flux though not in 1:1 proportion. All results, taken in toto, support the contention that the SR content is a major determinant of Ca efflux from the cell via Na–Ca exchange. It is also generally accepted that the magnitude of Ca influx via Ca channels is reflected, in the steady state, in the SR Ca content. This is, after all, the accepted mechanism for the classic force–frequency or staircase relation (Bouchard and Bose, 1989; Borzak et al., 1991). Frampton et al. (1991) showed that, when stimulation rate was increased, the Ca load of the SR increased as well. It seems likely, therefore, that the channel flux is a major determinant of SR content and the latter, in turn, is a determinant of the amount of Ca "fed" to the exchanger thereby providing the essential feedback in the matching of efflux to influx.

In summary, the conditions proposed for the cleft model account for Ca movements consistent with physiological requirements. Specifically, Ca entry via Ca channels is matched by Ca efflux via Na–Ca exchange. This is accomplished by virtue of the links among jSR content, channel influx, and jSR release, with the release directly into the cleft region where the cell's Na–Ca exchange is controlled.

V. CALCIUM IN THE CYTOPLASM

As Ca enters the general cytoplasm through noncleft sarcolemmal channels and from the clefts, it is extensively buffered. The major buffers are (1) calmodulin, (2) troponin C, (3) SR binding and uptake, (4) SL binding,

and (5) phosphocreatine. It is the combination of these buffers, with dif-
fering affinities and capacities, that accounts for the fact that at peak $[Ca]_i$
only 1–2% of the Ca entering the cytosol is free. Berlin *et al.* (1994), using
the Ca indicator Indo-1, estimated (after subtraction for the dye itself)
cytosolic buffering at 123 ± 18 μmol/liter cell H_2O with a "lumped"
Kd (Ca) = 0.96 ± 0.18 μM. The data confirmed that during a rapid rise
of $[Ca]_i$ from 0.1 to 1.0 μM approximately 98% of the Ca entering the
cytosol is buffered. This study excluded SR uptake, which contributes more
to late buffering.

Direct measurement of the magnitude and time course of the cytosolic
[Ca] transient is measured by one of many available Ca indicators. The
dyes are either directly injected into isolated cells in their free form or
allowed to enter in a lipophilic esterified form. After entry the ester is
cleaved by intracellular enzyme activity. As might be expected, the amplitude of the transient (therefore, the estimate of free $[Ca]_i$) and its time
course are dependent on the dye selected. A dye with high affinity will
saturate relatively early and release its Ca relatively slowly. This will decrease the peak amplitude and delay decline of the signal. Figure 12 simu-
lates the response of a high-affinity dye (fura-2) and a low-affinity dye
(furaptra) to an identical influx of Ca to the cytosol (Berlin and Konishi,
1993). Note that the peak [Ca] recorded by fura-2 ($\sim$500 nM) is less than
the peak ($\sim$1600 nM) recorded by furaptra. The estimated Ca involved in
the transients is, however, not that different (see $[Ca^{2+}]$ TOT for each
transient). Berlin and Kohnishi (1993) explain that with furaptra most of
the Ca is bound to intrinsic Ca buffers (e.g., troponin C and SR Ca ATPase).
In contrast, with fura-2 with much higher affinity, the Ca indicator itself is
a major buffer "binding similar, if not greater, amounts of Ca than troponin
or SR ATPase." Note also that the time course of the decline in the signal
is much delayed with fura-2. In addition, other factors such as distribution
of the indicator in the cell, extent of deesterification if an ester is used and
differences between *in vitro* and *in vivo* calibration of the indicator can
make reliable quantification of the Ca signal difficult. All of this not with-
standing, the Ca dyes have given additional insight into cellular Ca movements.

A. CALCIUM "SPARKS" AND "WAVES"

It has been proposed by Cannell *et al.* (1995) and Lopez-Lopez *et al.*
(1995) that the whole-cell Ca transient (see Fig. 12) represents a summation
of a large number of microscopic SR Ca release events called Ca sparks.
Sparks may occur spontaneously but their occurrence is greatly increased
by triggering from Ca entering via the L-type Ca channels. A spark has
been defined by Santana *et al.* (1996) as follows: (1) a peak $[Ca]_i$ at least
50 nmol/liter cell water greater than $[Ca]_i$ in the neighboring region;

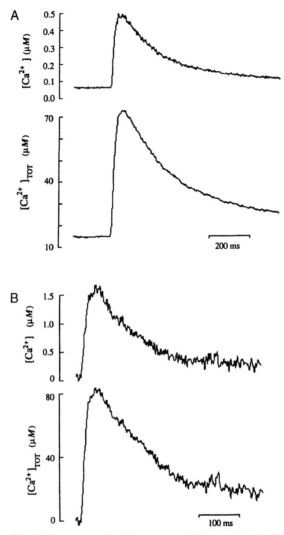

FIGURE 12   Simulations comparing the amount of Ca involved, [Ca]$_{TOT}$, in [Ca] transients measured in cells loaded with a high-affinity (fura-2) and a low-affinity (furaptra) Ca-sensitive dye. Fura-2 (A) records a peak [Ca] transient of about 0.5 μM when [Ca]$_{TOT}$ is slightly more than 70 μ M. Furaptra (B) records a peak more than three times higher when [Ca]$_{TOT}$ is only slightly higher. Note also the much more rapid decline of the [Ca] transient as measured by the low-affinity furaptra. (Reproduced from Berlin and Konishi, 1993, with permission.)

(2) time to peak concentration between 2 and 20 ms with a half-time of decay between 10 and 40 ms; and (3) width of the spark (measured as full width across the spark at half its maximum amplitude) at least 0.5 μm, but not more than 3 μm. These sparks are measured as indicator fluorescence changes over one plane of a cell by confocal microscopy. Spatial resolution of the technique will not allow definition of a spark within a single cleft. However, Shacklock et al. (1995) using dual-channel confocal laser scanning microscopy have shown that 85% of all Ca sparks evoked by electrical stimulation occurred within 0.5 μm of a T tubule, and 30% occurred within 0.2 μm of a T tubule. A histogram plotting number of sparks versus the distance of spark origin from the nearest T tubule gave a normal distribution with extremes of 1.0 μm to the left and 1.0 μm to the right of the nearest T tubule with the center at 0. This is consistent with placing the origin of the sparks at the diadic clefts (see diadic cleft model).

Sparks of low magnitude (200–300 nM) do not give rise to propagating transients of increased Ca concentration over the whole cell or a significant part of the cell. This is the condition seen in the resting state. It would seem that a so-called "macrospark" (peak in the 500-nM range) is necessary for the initiation of a propagating transient (Cheng et al., 1993). This requires release from the feet or ryanodine receptors and is produced by Ca channel triggering in the diadic cleft space. Further evidence that the feet are the origin of the initiating sparks is the demonstration that high-dose ryanodine (which closes the feet channels) eliminates sparks in most cells (Cheng et al., 1993).

If the cleft model is realistic, it would be expected that the onset of an electrically invoked Ca transient would be dependent on the time course of Ca movement to the cytoplasm from the cleft space. Figure 11 predicts that the transient should appear within 10–20 ms of electrical stimulation and be of full magnitude within 50–60 ms at which time all of the Ca destined to leave the cleft for the cytoplasm has done so. The measurement of cellular Ca transients by Cannell et al. (1994) confirms the prediction.

If the jSR cistern and its associated cleft is associated with two one-half sarcomeres as assumed and supported by structural data, then the maximum distance over which Ca must diffuse axially after it leaves the cleft space is about 1 μm (sarcomere). In the course of this diffusion it is, of course, affected by all the buffer sites that will diminish free [Ca]. The release from the feet of 70 μmol Ca/kg wet weight (2×10^{-19} mol/cleft) over 20 ms will cause [Ca] at the exit from the cleft to rise to more than 100 μM at end-release. As it diffuses, the [Ca] falls to less than 1 μM over the half-sarcomere within 100 ms (Peskoff, unpublished data, 1997).

We have, up to this point, ignored another source of cytoplasmic Ca— the so-called corbular SR (cSR) (see Chapter 1). As with the jSR, the cSR extends from the longitudinal or network SR tubules. It does not, however, interface with the SL. It is most densely distributed in the interfibrillar spaces near the center of the I-band of the sarcomere. The cSR appears as vesicular-like pouches extending from the longitudinal SR (Dolber and Sommer, 1984) and Jorgensen *et al.* (1985) have found that it contains calsequestrin (a Ca-binding protein) as does jSR. Most important, it has been demonstrated that cSR contains Ca release channels/ryanodine receptors (feet) (Jorgensen *et al.,* 1993) and, therefore, is capable of releasing Ca if activated by an appropriate trigger stimulus. Jorgensen *et al.* (1985) reported that as much as 40–50% of the total calsequestrin in the cells was localized to the cSR, which implies that a similar fraction of Ca could be stored in and released from these organelles. The model presented does not include consideration of the cSR as a source of activating Ca, nor does any other model of which we are aware. Although cSR has been structurally defined there is no functional data available that can be fit into the E–C coupling scheme. Jorgensen *et al.* (1993) speculate that Ca is initially released from jSR and that this release, "in turn, possibly by a regenerative mechanism, induces further release via ryanodine receptors from cSR located further away from the SL and jSR." Given the interfibrillar location, Ca release from cSR would augment release from the clefts. One might expect, if such augmentation were taking place, that a nonuniformity in the [Ca]$_i$ transient at the subsarcomeric level might be evident. The spatial resolution required to examine this possibility is not possible with present techniques.

In summary, most of the Ca released to the cytoplasm exits from the clefts after its release from the jSR feet. It is also likely that a significant fraction is derived from the cSR. The Ca is heavily buffered by diverse binding sites and by reuptake by the longitudinal SR so that only 1–2% remains "free." The "free" Ca gives rise to a Ca transient measured by a Ca indicator dye. Quantitation of the transient varies with the dye used and must be interpreted with care. The whole-cell Ca transient most likely represents a summation of a large number of microscopic "release events" (as Ca exits from the individual clefts) called Ca "sparks." A large spark (macrospark) initiates a Ca transient following triggered release of Ca in the cleft. The cleft model predicts that a transient would commence 10–20 ms after electrical excitation and reach its peak within 50–60 ms. The form of experimentally measured Ca transients fits the prediction.

VI. DRUGS AND CALCIUM

There are two major sites of drug action affecting Ca in the heart—the SL and the SR. The SL Ca channels are sites at which drugs act to produce significant effects in cardiac muscle, the so-called T and L channels (see Chapter 3). The T channels are most prominent in cells capable of spontaneous diastolic depolarization or pacemaking activity. They contribute to the inward current that carries sinus node and atrioventricular node cells to the threshold for L channel current, which is responsible for the action potential "spike" in these cells.

A. CATECHOLAMINES

The L channel is the site at which most of cellular Ca influx occurs. The influx is in the form of a $\sim$1 ms rectangular pulse giving a peak current of $\sim$0.3 pA/channel (Rose et al., 1992). The focal points for control of this current are adenylate cyclase, which is activated by the sarcolemmal β_1 receptors via stimulatory G proteins. Catecholamine interacts primarily via β_1 stimulation to activate adenylate cyclase to form cyclic-AMP, which activates protein kinase A (PKA), which mediates phosphorylation of the L channel. Phosphorylation increases the probability that a channel is available to open if and when its threshold is reached. β_1-selective catecholamines (e.g., isoproterenol) are capable of increasing I_{Ca} by more than threefold over basal levels (Trautwein et al., 1987) and contributes to their markedly positive inotropic effect.

The inotropically effective catecholamines are not only associated with increased force and increased rate of force development, but they also are associated with an increased rate of relaxation. This rapid relaxation is primarily due to the effect of the catechols on the rate of Ca pumping by the longitudinal SR. Elevated levels of a cyclic-AMP stimulate a protein kinase-dependent phosphorylation of phospholamban (see Chapter 4), which diminishes phospholamban's inhibitory effect on the SR Ca pump and, therefore, increases Ca uptake from the cytoplasm. Another system adds to the augmentation of SR Ca pumping. As indicated previously, the catechols augment Ca influx through the channels thereby increasing "trigger" Ca leading to increased Ca release to the cytoplasm and positive inotropy. The increased cytoplasmic [Ca] activates a Ca-calmodulin protein kinase leading to additional phospholamban phosphorylation and enhanced Ca uptake. Therefore, β_1 activation by catechols increases Ca entry via the Ca channels in the SL and, at the same time, stimulates its uptake by the SR. Another cyclic-AMP dependent system activated by catechol β_1 stimulation is the phosphorylation of troponin I (TNI) (Robertson et al., 1982). This lowers TNC's affinity for Ca, which would also be expected to hasten the process of relaxation. The net result of the β_1 stimulation by catecholamines is increased force with increased rate of development and increased rate of relaxation.

In summary, the final common pathway following β_1 receptor activation by catecholamines is phosphorylation. Phosphorylation of the L channels increases Ca influx across the SL; phosphorylation of phospholamban and TNI increases SR Ca uptake and decreases TNC's affinity for Ca, respectively. This combination produces a "yin–yang" effect—increased force and rate of force development on the one hand and increased rate of relaxation on the other hand. This results in a larger and faster contraction. It should be kept in mind that β_1-activating catechols have not only an inotropic but also a chronotopic effect. The latter is via their effect on diastolic currents in pacemaker cells leading to increased heart rate (see Chapter 3). So that stroke volume is maximized and coronary flow preserved as heart rate increases, it is necessary to preserve the diastolic period during which the ventricles fill to the greatest extent possible. The more rapid systole produced by the catechols does just that.

B. CALCIUM BLOCKERS

This group of drugs acts by directly binding to the channel itself, blocking it progressively as dosage is increased. There are three classes of these compounds: (1) benzeneacetonitrile (verapamil), (2) dihydropyridine (nifedipine), and (3) benzothiazepine (diltiazem). The order of potency for blocking the L channel of cardiac tissue is nifedipine > diltiazem ≥ verapamil (Li and Sperelakis, 1983). These agents are relatively specific in that their effect is entirely attributable to blockade of the slow inward Ca current, thereby eliminating the "trigger" Ca.

The drugs, to varying degree (nifedipine the least), are frequency dependent (i.e., they bind to a greater extent when the channel is activated). Therefore, their effect is greater the higher the heart rate. They are capable of blocking the channels to the degree that the "trigger" Ca falls to a level that is insufficient to induce jSR release and, therefore, contractile force would fall to zero. In this context, we noted (see "Trigger Ca") that in the absence of Ca current (as produced by high-dose nifedipine) "reversed" Na–Ca exchange might be expected to provide the "trigger" stimulus, though less efficiently. Levi *et al.* (1994) found that a depolarization to +10 mV in the presence of 20 μM nifedipine, which completely blocked I_{Ca}, still produced a contraction 43% of control.

Blockade of Ca channels obviously reduces Ca influx and should result in a proportional reduction in Na–Ca exchange-mediated Ca efflux as a new steady state is reached. A study by Marengo *et al.* (1997) showed that 1 μM nifedipine reduced efflux via Na–Ca exchange by approximately 40%. It is of interest that a nifedipine analogue, Bay K 8644, binds differently to the "L" channel and, rather than blocking, enhances its opening. As expected it produces significant positive inotropy (Schramm *et al.*, 1983). It would be expected that Bay K 8644 would induce, in the steady state, an increase in Na–Ca exchange-mediated Ca efflux. An appropriate 30% increase has been demonstrated (Wang *et al.*, 1997).

In summary, Ca channel antagonists and agonists affect channel current by direct action on the channel structure and not through cyclic AMP as an intermediate, as with the catecholamines. Current evidence indicates that Na–Ca exchange appropriately adjusts to the changes in channel flux so as to maintain steady-state intracellular Ca.

C. CARDIAC GLYCOSIDES

In the first edition of this text, we summarized concepts of the mechanism of action of the digitalis glycosides as of 1974. At that time, there was controversy as to whether the well-known and very specific inhibition of the sarcolemmal Na-KATPase (Na–K pump) by glycosides was the basis for the increased cellular Ca uptake associated with glycoside-induced inotropy (Langer and Serena 1970; Lee and Klaus, 1971). We had previously shown that increased cellular Ca uptake occurred coincident with a period of increasing cellular [Na] following an abrupt increase in beat rate. The period during which $[Na]_i$ was rising we called the "Na pump lag" (Langer, 1965) and proposed (Langer, 1971) that the increased $[Na]_i$ was coupled to the increase in cellular Ca through the newly discovered Na–Ca "membrane carrier system" (Baker *et al.,* 1969; Reuter and Seitz, 1968). We explained the classic Bowditch force–frequency inotropic response on the basis of this sequence (i.e., Na pump lag → increased $[Na]_i$ → increased net Ca uptake via effect on Na–Ca exchange). The "Na pump lag" hypothesis has now come to be generally accepted (Langer, 1983; Harrison and Boyett, 1995). We noted that cardiac glycosides produced the same series of effects on Na and Ca exchange as occurred due to a rapid increase in frequency of contraction (Langer and Serena, 1970). Increase in frequency is associated with a "lag" in the ability of the Na pump to increase efflux to match increased influx. Glycoside inhibits the pump such that efflux lags behind influx at unchanged frequency. Both result in accumulation of intracellular Na and subsequent increase of intracellular Ca via the effect on Na–Ca exchange. Figure 13 illustrates the sequence. Prior to digitalis, intracellular Na is maintained at a relatively low level by the Na^+–K^+ pump, and the [Na](out)/[Na](in) gradient is relatively high. This favors a high Na influx (arrow 1) and, coupled to Ca (arrow 2), a relatively high Ca efflux (arrow 3). With glycoside inhibition of the Na^+–K^+ pump, intracellular Na increases to an elevated steady-state level, reducing the [Na]out/[Na]in gradient. This decreases Na influx (arrow 1) and, therefore, Ca efflux (arrow 3) via Na^+–Ca^{2+} exchange. Thus, intracellular Ca rises to a new steady-state level and is, presumably, stored in the SR to be released in increased amounts to produce positive inotropic response to digitalis administration. This basic concept of Na–Ca relations was presented by Repke (1964) to explain the action of digitalis at the same time the concept of Na pump lag was being presented to explain the Bowditch force–frequency response (Langer, 1965).

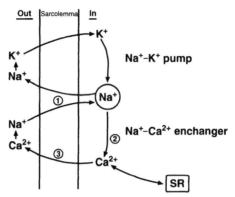

FIGURE 13 The relationship between the Na^+-K^+ pump and the Na^+-Ca^{2+} exchanger. Failure of the Na^+-K^+ pump to remove Na influx either due to "Na pump lag" or digitalis administration causes intracellular Na^+ (encircled) to rise. This, in turn, partially "reverses" the Na^+-Ca^{2+} exchanger with reduction of Ca efflux (arrow 3) and consequent increase of intracellular Ca leading to increase in SR Ca content and positive inotropy.

The presumption that the increased cellular Ca is stored in the SR is supported by kinetic studies of Ca distribution after glycoside administration (Langer *et al.*, 1995). Administration of ouabain to tissue culture cells increased measured steady-state flux through the SR by more than twofold. The same study showed indirectly that [Na] concentration in the cleft space was significantly increased in the presence of ouabain. This is consistent with the sequence illustrated in Fig. 13.

In summary, the mechanism of glycoside action in the production of increased contractile force is now understood and strongly supported by numerous experimental studies. The increase of $[Na]_i$ by the glycosides is the crux of their action. Increased $[Na]_i$ reduces Na_0 to Na_i flux, which reduces Ca_i to Ca_o flux via the Na–Ca exchangers resulting in increased cellular Ca. This last is the basis for the glycoside-induced inotropy.

D. METHYLXANTHINES

This group includes caffeine, theophylline, and theobromine. It is problematic whether therapeutic doses of these compounds produce drug levels that produce any significant effects on cellular Ca movements *in vivo*. These drugs have, however, multiple actions *in vitro,* all of which impinge on some aspect of Ca movement in the cell. These include inhibition of phosphodiesterase activity, stimulation of Ca release from the SR, adenosine receptor blockade, and increased myofilament sensitivity to Ca^{2+} (Akera, 1990). Caffeine, in millimolor concentrations, has proven to be the most useful with respect to definition of the role of the SR in intracellular Ca movement.

Rousseau and Meissner (1989), using canine myocardial SR vesicles, showed that millimolar concentrations of caffeine increase the SR channel open time without changing its unit conductance. They indicated that caffeine- and Ca-induced activation of single Ca release channels were similar.

Callewaert *et al.* (1989) used caffeine as a specific releasing agent for SR Ca and demonstrated that the transient inward current generated by this release represented efflux of Ca via the Na–Ca exchangers. Once again this demonstrates the close coupling between SR Ca and the Ca component transported from the cell by the exchangers. This close coupling has been proposed to occur within the diadic cleft (see Fig. 13). Frampton *et al.* (1991) used caffeine to demonstrate that the increased Ca influx that occurred when stimulation rate was increased was associated with an increased SR Ca content. This is also consistent with the couplings illustrated in Fig. 13. Increased frequency elevates steady-state intracellular Na (encircled in Fig. 13), which decreases Ca efflux (arrow 3). This increases intracellular Ca, which is, as demonstrated by Frampton *et al.* (1991), stored in the SR to participate in the increased force development typical of the force–frequency response (Bowditch "staircase").

It should be reemphasized that, although caffeine is an excellent probe for Ca in the SR, it is likely that the drug does not release all of this Ca (Weber, 1968; Weber and Herz, 1968).

In summary, the methylxanthines have multiple sites of action that affect Ca movements in the cell. Caffeine has proved to be the most useful of the group due to its ability to specifically release a portion of SR Ca.

E. RYANODINE

This is a drug with a fascinating history (Jenden and Fairhurst, 1969). It was first isolated and crystallized from ground stem wood and root of the *Ryania* speciosa Vahl, a plant native to Trinidad. It was first used as an insecticide. Its negative inotropic effects on the heart of mammalian species were first reported more than 40 years ago (Furchgott and de Gubareff, 1956; Hillyard and Procita, 1956). It was of interest that the ventricles of frog and turtle were not significantly affected by the drug (Jenden and Fairhurst, 1969; Furchgott, personal communication). Given the paucity of SR in amphibian hearts, this might have served as an early clue to the site of the drug's action.

A series of elegant studies (Inui *et al.*, 1987, 1988) using tritiated ryanodine clearly identified the jSR feet structures as the ryanodine receptor sites and defined two classes of binding—high (Kd $\sim$ 8 nM) and low (Kd $\sim$ 1 μM) affinity. They showed that ryanodine binding to the high-affinity sites locks the Ca release channels in the open state, whereas binding to the low-affinity sites closes the channels. Low doses would, then, be predicted to cause a continuous leak through the feet with continuous cycling through the cleft. Most would exit the cleft to the cytoplasm to be taken up again by the longitudinal SR in a continuous cycle. However, as Ca entered the cleft, some would be transported out of the cell by the Na–Ca exchangers and this would result in progressive depletion of SR Ca. The results of Janiak *et al.* (1996) support this sequence. In contrast, high-dose ryanodine would be expected to block the release of Ca into the cleft with retention in the SR. Both low and high doses result in reduction of developed force.

Ryanodine has been useful in further exploration of the cleft model (see Fig. 4). The model indicates that, in the steady state, Ca in the jSR is in equilibrium with the inner sarcolemmal leaflet binding sites within the cleft space. If such is the case, it would be predicted that low-dose ryanodine would deplete the jSR and decrease sarcolemmal binding proportionally. Using the "gas-dissection" technique for instantaneous sarcolemmal isolation from intact cells, both jSR content and sarcolemmal binding were reduced by low-dose ryanodine as expected (Langer *et al.*, 1995). By contrast, high-dose ryanodine left jSR content unchanged, but decreased sarcolemmal binding by 95%. This is exactly as predicted by the model (Rich *et al.*, 1988; Bers, 1989), since Ca access to the cleft space is denied in the presence of high-dose ryanodine.

There is a marked species difference in response to ryanodine. Rat ventricle is very sensitive, showing 80–90% decline of contractile force, while rabbit ventricle is much less affected (10–15% decline). Although peak force development is maintained, the rate of force development is significantly decreased by ryanodine in the rabbit (Janiak *et al.*, 1996). The responses are consistent with the greater dependence of the rat on SR Ca release for its force development. In the rabbit, transsarcolemmal Ca channel flux is sufficient to produce near normal force but at a much slowed rate in the absence of a large component of triggered SR release.

In summary, ryanodine is a specific SR probe owing to its specific binding to the SR feet. The drug maintains the feet in an open state at low concentration but closes them at high concentration. Use of the drug has emphasized species differences in E–C coupling and supported the concept of the cleft's role in control of Ca movements in the cell.

F. THAPSIGARGIN

Thapsigargin is a sesquiterpene lactone that was found, in isolated rat liver microsomes to induce a rapid release of stored Ca (Thustrup *et al.*, 1990). Sagara and Inesi (1991) next found that, in vesicular fragments of longitudinal SR from skeletal muscle at concentration as low as 10^{-10} M, ATP-dependent Ca uptake was inhibited. At somewhat lower sensitivity, a similar, specific effect was found for cardiac SR (Kijima *et al.*, 1991). Janczewski and Lakatta (1993) documented that thapsigargin reduced SR Ca content by blocking its uptake in contrast to low-dose ryanodine, which increased its leak rate. As expected, the drug markedly reduces contractile amplitude and fluorescently measured Ca transients.

Since its primary effect is to decrease Ca uptake and deplete the SR, the cleft model predicts that Na–Ca mediated efflux should be diminished by thapsigargin. One μM thapsigargin reduced this flux by 70%. In addition, jSR content and sarcolemmal binding (see Fig. 4) should be reduced proportionately by the drug. SR content decreased by 68%; sarcolemmal binding by 71% (Langer *et al.*, 1995).

The use of ryanodine in combination with thapsigargin has proved useful in the further definition of intracellular Ca movement (Janiak *et al.*, 1996). In single ventricular cells from guinea pig treated with low-dose ryanodine, the amplitude of electrically stimulated fluorescently measured Ca transients was reduced by at least 50%, and the initial rapid phase (assumed to be SR derived) of the transients was slowed (Fig. 14). Responses of these cells to brief superfusions with 15 mM caffeine were decreased, which confirmed that SR Ca stores were, indeed, reduced. At this point, the cells were treated with 2×10^{-7} M thapsigargin and the amplitude of the Ca transients *increased* significantly as shown in Fig. 14. The most likely interpretation of this result is as follows: In the presence of low-dose ryanodine, SR Ca uptake continues but because of the ryanodine-induced leak it is released into the cleft space where a fraction is diverted out of the cell via Na–Ca exchange before it can diffuse to the cytoplasm and activate the myofilaments. Inhibition of SR uptake by thapsigargin removes the SR from the pathway and diverts more Ca to the cytoplasm, thus increasing the fluorescently measured transient.

In summary, thapsigargin, through its ability to inhibit SR Ca ATPase, specifically blocks SR Ca uptake. Its use as a specific probe has helped to further define Ca movement through the SR → cleft → cytoplasm cycle.

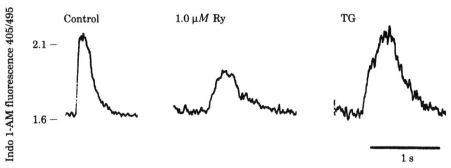

FIGURE 14   Indo 1-AM fluorescence in a single guinea pig ventricular myocyte. Cell was continuously electrically stimulated at 30/min. From left to right: Control, after 30 min superfusion with 1 μM ryanodine (Ry) and after 2×10^{-7} M thapsigargin (TG) in the continued presence of ryanodine. (Reproduced from Janiak *et al.*, 1996, with permission.)

The drugs discussed in this section are those most commonly used clini- cally and/or in the laboratory. There are other agents (e.g., phosphodiester- ase inhibitors, ionophores), but these are little used and would not have provided further insight into Ca compartmentation and movement. A sum- mary of the drugs discussed in this section is presented in Table 2.

VII. CALCIUM MOVEMENT IN THE CELL

Figure 15 summarizes, in schematic form, Ca movements during the course of a single contraction cycle. The numbers indicate the chronological sequence of events, some of which are occurring simultaneously:

1. Ca enters the cleft space via the L channel. The magnitude of this entry, once the channel's voltage threshold has been reached, is controlled by the extent to which the channel is phosphorylated. This is controlled, in turn, by a cyclic AMP-dependent protein kinase. The maximum Ca influx/excitation via the channels is in the range of 10 μmol/kg wet weight ventricle. This is, by itself, insufficient to produce force development. It is, however, capable of raising [Ca] in the cleft space to 1 μM in less than 1 ms and triggering subsequent release.

2. Ca-induced Ca release from the jSR via the feet into the diadic cleft space. The magnitude of the release is determined by the concentration of Ca in the cistern and the amount of trigger Ca and the rate at which it arrives at the "feet" structures. [Ca] in the space increases to more than 100 μM during release and then, largely because of the presence of inner sarcolemmal leaflet phospholipid Ca binding, diffuses relatively slowly from the cleft. ([Ca] is still 1–10 μM 80 ms after release has ceased.)

3. Ca leaves the cleft by two routes. It diffuses to the myofilaments to bind to TnC and activate contraction with a small fraction taken up by the mitochondria to play a role in the regulation of energy production via the Ca-sensitive dehydrogenases. The remainder of the Ca released through the feet, approximately 8–10%, will be transported out of the cell via Na– Ca exchangers. This transport is optimized by the maintenance of [Ca] level in the cleft for a period sufficient to stimulate the exchangers with Kd(Ca) ~5–6 μM. Efflux via Na–Ca exchange is adjusted to match influx via the Ca channels so as to maintain steady-state intracellular Ca levels.

4. Ca reuptake via the CaATPase (Ca pump) in the longitudinal SR. The small efflux from the mitochondria seems also to be taken up by the SR. This Ca diffuses to the jSR to be bound to calsequestrin for rerelease upon subsequent activation of the Ca channels in the feet. During diastole the high-affinity [Kd(Ca) ~ 0.5 μM)], low-velocity sarcolemmal Ca pump and continuing low-level Na–Ca exchange maintain [Ca]$_i$ at ~100 nM.

TABLE 2 Drugs Affecting Cardiac Ca^{2+} Movement

Drug	Effect on Ca^{2+}	Mechanism of action	Functional effect
β_1 Catecholamines	Increases L channel flux Increases longitudinal SR uptake Decreases TnC affinity	Phosphorylation of L channel[a] Phosphorylation of phospho-lamban[a] Phosphorylation of TnC[a]	Increased force Increased contraction and relaxation velocity
Calcium channel antagonists	Decreased L channel flux	Direct action on channel structure	Decreased force Depress supraventricular pacemaker activity
Cardiac glycosides	Increased cellular Ca^{2+}	Inhibition of Na–K ATPase → increased Na_i → Decreased Ca efflux via Na–Ca exchange → increased SR Ca storage	Increased contractility
Methylxanthine	Transient increased cellular Ca followed by cellular depletion	Inhibition of phosphodiesterase Adenosine receptor blockade Inhibition of Ca uptake, stimulation of Ca release by SR Increased myofilament sensitivity to Ca	Increased force dependent on dose and time
Ryanodine	Low dose—depletes SR Ca stores High dose—blocks SR release	Low dose—feet channels open High dose—feet channels closed	Decreased contractile force
Thapsigargin	Depletes cellular Ca	Inhibits SR Ca ATPase	Decreased contractile force

[a] Cyclic AMP-mediated.

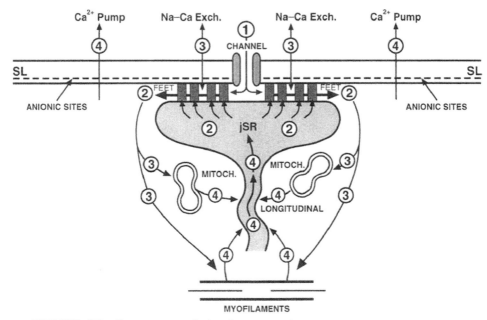

FIGURE 15 Ca movements during the course of a contraction cycle. The numbers indicate the chronological sequence beginning with membrane depolarization and Ca entry through the channel ① and ending with Ca pumping by the longitudinal SR and sarcolemmal Ca pump ④. See text for description.

VIII. CALCIUM IN THE METABOLICALLY COMPROMISED CELL

It is well established that inhibition of the heart's metabolism, whether by ischemia-reperfusion, anoxia, reoxygenation, or chemically induced inhibition, leads to an increase of intracellular Ca (Bourdillon and Poole-Wilson, 1982; Miyata *et al.*, 1992; Barrigon *et al.*, 1996; Hohl and Altschuld, 1991). This is discussed in Chapter 7, but in this chapter, it is appropriate to emphasize certain aspects of Ca movement and compartmentation as the cell is progressively compromised.

A. AFFECTED SYSTEMS

1. Calcium Leak Channels

Studies from a number of laboratories indicate that the increased uptake of Ca by the cell may be due to a number of factors. Coulombe *et al.* (1989) identified "leak" channels specific to Ca that are conductive at resting potential (-80 to -90 mV) (see Chapter 3). Wang *et al.* (1995) showed that the activity of these channels was increased markedly over a period of a few minutes of metabolic inhibition. The channel conductance was increased about 10-fold. Given the low level of conductance in the control state, this large increase probably accounts for a relatively small fraction of net Ca uptake. However, it has an early onset and may be an important initial response to metabolic inhibition.

2. Exchangers

The major factor, with respect to the determination of Ca uptake under metabolic inhibition, is almost certainly the level of Na_i. This level, in turn, is largely determined by the level of activity of the Na–H exchangers, first described by Frelin *et al.* (1984) in cardiac cells. When intracellular pH is lowered, the Na–H exchange becomes the major pH_i regulating system. As there is a net H^+ movement outward, there is a net Na^+ movement inward. $[H^+]_i$ is generated from two sources: (1) as aerobic metabolism declines, glycolytic metabolism will generate lactic acid as long as substrate (e.g., glycogen) is available and pH_i remains high enough not to inhibit glycolytic enzymes; and (2) the generation of protons as a by-product of ATP hydrolysis (Eisner *et al.*, 1989). It has been calculated that the hydrolysis of 1 mol ATP (to ADP + P_i) will generate 0.8 mol of H^+ (Wilkie, 1979), a significant source of protons. The importance of operation of the Na–H exchangers is demonstrated by the abolition of the reperfusion rise in Na_i by treatment with amiloride, a Na–H exchange inhibitor (Tani and Neely, 1989).

Another factor that will contribute to increased Na_i is failure of the Na–K pump. Although pump activity is of the highest priority in hierarchy of cellular function and can function for 15–20 minutes anaerobically (Kleber, 1983; Weiss and Shine, 1982), it does fail and will begin to contribute to increased Na_i.

An elevation of Na_i has predictable effects on Ca_i (see previous). As Na_i rises due to increased H^+ out/Na^+ in via the Na–H exchangers and failure of the Na–K pump the Na–Ca exchangers will operate less in the net Ca efflux mode. Ca will rise in the cell. This assumes that Na–Ca exchange remains operative for an extended period in the metabolically compromised cell. Barrigon *et al.* (1996) showed that the Na–Ca exchangers remain capable of Ca transport for at least 40 minutes after ATP levels had decreased to 10% of preinhibition levels in cultured heart cells. In contrast, Satoh *et al.* (1995) found in metabolically inhibited isolated guinea pig myocytes that, although $[Na]_i$ increased threefold, $[Ca]_i$ as measured by fluo 3 flourescence did not increase. This was interpreted to indicate an inhibition of Na–Ca exchange. However, such may not be the case. Free $[Ca]_i$ might well be maintained at control levels in the presence of increased Ca stores (e.g., in the SR), which continue to exchange via an operative Na–Ca exchange system. Barrigon *et al.* (1996) clearly showed a metabolically induced increased cell Ca uptake in the presence of continuing Na–Ca exchange and, as discussed in the next section, evidence of increased SR and mitochondrial Ca. Moreover, net Ca uptake did not begin to increase in the Barrigon *et al.* (1996) study until 25–30 minutes of metabolic inhibition at which time Satoh *et al.* (1995) commenced energy repletion and then found a large increase in $[Ca]_i$.

In summary, as the cardiac cell becomes metabolically compromised, intracellular Ca increases. A small fraction of this increase may be attributed to enhanced leak via specific Ca leak channels. The major fraction, however, is based on the sequence: intracellular acidosis with $[H^+]_i$ increase → activation of Na–H exchange with increase $[Na]_i$ → increase of intracellular Ca secondary to net uptake via Na–Ca exchange.

B. CALCIUM COMPARTMENTATION

Figure 16 (Barrigon *et al.*, 1996) shows the course of net Ca uptake in cultured heart cells exposed to 1 mM iodoacetic acid (IAA) plus 10 mM deoxyglucose (2-DOG) at 0 time. In these cells, ATP has decreased to 10% control after 10 minutes. Note that cellular Ca increases by 4 mmol/kg dry weight (about 700 μmol/kg wet wt) after 50 minutes of metabolic inhibition. It had been previously shown that there is no intracellular enzyme release (lactate dehydrogenase) up to 60 minutes of metabolic inhibition (Post *et al.*, 1993). Therefore, the uptake is not due to a nonspecific leak of Ca into the cell.

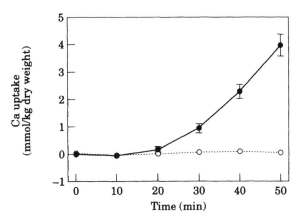

FIGURE 16   Effect of metabolic inhibition on net Ca uptake by cultured myocardium. After ^{45}Ca labeling to asymptote, metabolic inhibition (1 mM iodoacetic acid and 10 mM deoxyglucose) was instituted at 0 time in plot indicated by closed circles. After a 20-min delay, Ca uptake increases progressively. Its distribution is discussed in the text. Open circles indicate maintenance of asymptotic level under control (no metabolic inhibition) conditions. (Reproduced from Barrigon *et al.*, 1996, with permission.)

Barrigon *et al.* (1996) examined the distribution of the additional 4 mmol of Ca taken up by the cells. Eleven percent, or 0.44 mmol, exchanged with the characteristics of the "slow compartment," which, as discussed in Section III,A,3, places it in the mitochondria. Eighty percent, or 3.21 mmol, of the increase exchanged much more rapidly via Na–Ca exchange and turned over within less than 2 minutes. This exchange rate is consistent with an SR origin and the proposal that a component of SR Ca exits the cell via Na–Ca exchangers in the cleft space (see Section IV,E). Therefore, 3.65 of the 4 mmol Ca gain or more than 90% remains exchangeable. The proportion that remains reversible could not be directly ascertained since IAA metabolic inhibition cannot be readily reversed. Since most remains exchangeable, however, it would be predicted that most could be cleared from the cell if normal metabolism were to be reestablished. The results indicate that, in these cultured cells, the SR serves as the predominant buffer for Ca during the early stages of metabolic compromise with the mitochondria beginning to contribute. An earlier study (Post *et al.*, 1993) demonstrated much greater mitochondrial participation (70% of the uptake localized to the mitochondria). It seems that the SR/mitochondrial distribution of Ca is variable during the earlier stages of metabolic inhibition, but it is clear that both organelles participate in the buffering of the increased cellular Ca load.

In summary, metabolic inhibition causes a net increase of cellular Ca. An initial small leak via specific Ca channels may contribute to early net uptake. The dominant mechanism is based on increased Na_i. The Na_i gain occurs via Na–H exchange stimulated by intracellular acidosis. The elevated Na_i, in turn, induces a net Ca uptake via its effect on Na–Ca exchange, which remains operative in the face of ATP depletion. Both the SR and mitochondria serve as buffers for the increased Ca uptake.

C. SARCOLEMMAL INTEGRITY

As metabolic inhibition is prolonged in the cultured cell model beyond 60 minutes, LDH begins to leak from the cells. This is indicative of structural changes in the sarcolemmal membrane and may indicate the start of irreversibility. Prior to LDH release a specific change in sarcolemmal phospholipid asymmetry begins (Post *et al.*, 1993; Musters *et al.*, 1993). PE is normally distributed such that 75% is in the inner sarcolemmal leaflet (see Section III,A,1). As ATP declines and intracellular Ca starts to rise there is a migration of PE outward so that its outer monolayer concentration doubles. There are probably two mechanisms responsible for the maintenance of the normal asymmetrical phospholipid distribution: (1) an interaction of the head groups of the amino-containing phospholipids with the underlying cytoskeleton (Haest and Deuticke, 1982); and (2) an ATP-dependent process that translocates amino-containing phospholipids from the outer to inner monolayer, the so-called "flippase" or translocase (Seigneuret and DeVeaux, 1984). It seems that with the decline in ATP there may be a break in cytoskeleton connections and an inhibition of the translocase. The increase in Ca_i might also be involved in the inhibition of translocase activity, since Ca has been shown to inhibit the enzyme (Bitbol *et al.*, 1987). These changes would allow the outward movement of PE. As PE moves out of its normal environment at the inner leaflet and away from PS, the bilayer stabilizing effect of PS on PE is lost. An expression of PE's nonbilayer behavior leads to uncontrolled fusion events such as formation of multilamellar vesicles (MLVs) leading to sarcolemmal destruction (Musters *et al.*, 1991) and uncontrolled, irreversible Ca entry.

In summary, as metabolic inhibition progresses and ATP declines and Ca_i rises, PE migrates from inner to outer sarcolemmal leaflet. This leads to destabilization of the membrane structure with progression to an irreversible path to cell death marked by uncontrolled Ca entry and enzyme leakage.

IX. THE FUTURE

As the reader scans the content of the first edition of *The Mammalian Myocardium* (1974), it is striking that study of subcellular cardiac function was in its infancy and molecular studies were not yet born. In the intervening years between the first and the present second edition, our increased understanding of subcellular and molecular function has been truly spectacular. This is certainly true for the area of E–C coupling and Ca compartmentation. In 1974, if the word *compartmentalization* was used in reference to cellular Ca, one was roundly criticized for the concept. Currently, it has become obvious that Ca is compartmentalized not only at the organelle level but also at the suborganelle and even molecular level. In fact, it is the breakdown of this compartmentation that frequently leads to cellular malfunction and death.

Myocardial study over the last quarter century has been, as with most other biological studies, reductionist in nature or to quote the popular phrase, "learning more and more about less and less." It now seems appropriate, as we enter the twenty-first century, that we begin to reintegrate. We cannot simply extrapolate from the isolated molecular or organellar study and assume that *in vivo* function can be accurately predicted. It is now the challenge to test our molecular findings and models in the functional cell and tissue to see if, indeed, they make sense. An elegant example of meeting this challenge is the development of the murine model for study of the relation of genetic mechanisms to the physiology of ventricular overload (Rockman *et al.*, 1993). Similar models are needed in all areas including E–C coupling and will, no doubt, be forthcoming. As these develop, we will continue to fill in the gaps that remain in our knowledge of the processes involved in myocardial contractile control. As Stanley Schultz (Chairman, Department of Integrative Biology, University of Texas Medical School, Houston) states, "Everybody tries to work for first principles. However, eventually, you need to construct the whole system, look at how it functions together. ... We're just now developing the tools to approach these questions" (quoted in *The Scientist,* June 8, 1992).

REFERENCES

Akera, T. (1990). Pharmacological agents and myocardial calcium. *In* "Calcium and the Heart" (Langer, G.A., ed.), pp. 299–331. Raven Press, New York.

Armstrong, C.M., Bezanilla, F.M., and Horowicz, P. (1972). Twitches in the presence of ethyleneglycolbis (aminoethylether)-N,N1 tetraacetic acid. *Bioch. Biophys. Acta* 267: 605–608.

Bachs, O., Agell, N., and Carafoli, E. (1992). Calcium and calmodulin function in the cell nucleus. *Biochim. Biophys. Acta* 1113:259–270.

Baker, P.F., Blaustein, M.D., Hodgkin, A.L., and Steinhardt, R.A. (1969). The influence of calcium on sodium efflux in squid axons. *J. Physiol. (London)* 200:431–458.

Barrigon, S., Wang, S.-Y., Ji, X., and Langer, G.A. (1996). Characterization of the calcium overload in cultured neonatal cardiomyocytes under metabolic inhibition. *J. Mol. Cell. Cardiol.* 28:1329–1337.

Bassani, R.A., and Bers, D.M. (1994). Na–Ca exchange is required for rest-decay but not for rest potentiation of twitches in rabbit and rat ventricular myocytes. *J. Mol. Cell. Cardiol.* 26:1335–1347.

Bassani, R.A., and Bers, D.M. (1995). Rate of diastolic Ca release from the sarcoplasmic reticulum of intact rabbit and rat ventricular myocytes. *Biophys. J.* 68:2015–2022.

Bassani, J.W.M., Bassani, R.A., and Bers, D.M. (1994). Relaxation in rabbit and rat cardiac cells: Species-dependent differences in cellular mechanisms. *J. Physiol. (London)* 476: 279–293.

Bassani, J.W.M., Yuan, W., and Bers, D.M. (1995). Functional SR Ca release is regulated by trigger Ca and SR Ca content in cardiac myocytes. *Am. J. Physiol.* 268:C1313–C1319.

Berlin, J., and Konishi, M. (1993). Ca^{2+} transients in cardiac myocytes measured with high and low affinity Ca^{2+} indicators. *Biophys. J.* 65:1632–1647.

Berlin, J.R., Bassani, J.W.M., and Bers, D.M. (1994). Intrinsic cytosolic calcium buffering properties of single rat cardiomyocytes. *Biophys. J.* 67:1775–1787.

Bers, D.M. (1985). Ca influx and sarcoplasmic reticulum Ca release in cardiac muscle activation during post-rest recovery. *Am. J. Physiol.* 248:H366–H381.

Bitbol, M., Fellman, P., Zachawski, A., and Deveaux, P.F. (1987). Ion regulation of phosphatidylserine and phosphatidyl-ethemolamine outside-inside translocation in human erythrocytes. *Bioch. Biophys. Acta* 904:268–282.

Borzak, S., Murphy, S., and Marsh, J.D. (1991). Mechanisms of rate staircase in rat ventricular cells. *Am. J. Physiol.* 260:H884–H892.

Bouchard, R.A., and Bose, D. (1989). Analysis of the interval–force relationship in rat and canine ventricular myocardium. *Am. J. Physiol.* 257:H2036–H2047.

Bourdillon, P.D., and Poole-Wilson, P.A. (1982). The effects of verapanil, quiescence and cardioplegia on calcium exchange and mechanical function in ischemic rabbit myocardium. *Circ. Res.* 50:360–368.

Bridge, J.H.B. (1986). Relationships between the sarcoplasmic reticulum and sarcolemmal calcium transport revealed by rapidly cooling rabbit ventricular muscle. *J. Gen. Physiol.* 88:437–473.

Callewaert, G., Cleeman, L., and Morad, M. (1989). Caffeine-induced Ca^{2+} release activates Ca^{2+} extrusion via $Na^{+}-Ca^{2+}$ exchanger in cardiac myocytes. *Am. J. Physiol.* 257:C147–C152.

Cannell, M.B., Cheng, H., and Lederer, W.J. (1994). Spatial nonuniformities in [Ca]$_i$ during excitation–contraction coupling in cardiac myocytes. *Biophys. J.* 67:1942–1956.

Cannell, M.B., Cheng, H., and Lederer, W.J. (1995). The control of calcium release in heart muscle. *Science* 268:1045–1049.

Cheng, H., Lederer, W.J., and Cannell, M.B. (1993). Calcium sparks: Elementary events underlying excitation–contraction coupling in heart muscle. *Science* 262:740–744.

Coulombe, A., Lefevre, I.A., Baro, I., and Coraboeuf, E. (1989). Barium and calcium-permeable channels open at negative membrane potentials in rat ventricular myocytes. *J. Membr. Biol.* III:57–67.

Crompton, M. (1990). The role of Ca^{2+} in function and dysfunction of heart mitochondria. *In* "Calcium and the Heart" (Langer, G.A., ed.), pp. 167–198. Raven Press, New York.

Dani, A.M., Cittadini, A., and Inesi, G. (1979). Calcium transport and contractile activity in dissociated mammalian heart cells. *Am. J. Physiol.* 237:C147–C155.

Denton, R.M., and McCormack, J.G. (1980). On the role of the calcium transport cycle in heart and other mammalian mitochondria. *FEBS Lett.* 119:1–8.

Denton, R.M., and McCormack, J.G. (1985). Ca^{2+} transport by mammalian mitochondria and its role in hormone action. *Am. J. Physiol.* 249:E543–E554.

Dolber, P.C., and Sommer, J.R. (1984). Corbular sarcoplasmic reticulum of rabbit cardiac muscle. *J. Ultrastruc. Res.* 87:190–196.

Eisner, D.A., Nichols, C.G., O'Neill, S.C., Smith, G.L., and Valdeolmillos, M. (1989). The effects of metabolic inhibition on intracellular calcium and pH in isolated rat ventricular cells. *J. Physiol. (London)* 411:398–418.

Fabiato, A. (1983). Calcium-induced release of calcium from the cardiac sarcoplasmic reticulum. *Am. J. Physiol.* 245:C1–C14.

Fabiato, A. (1985). Time and calcium dependence of activation and inactivation of calcium-induced release of calcium from the sarcoplasmic reticulum of a skinned cardiac Purkinje cell. *J. Gen. Physiol.* 85:247–289.

Frampton, J.E., Orchard, C.H., and Boyett, M.R. (1991). Diastolic, systolic and sarcoplasmic reticulum $[Ca^{2+}]$ during inotropic interventions in isolated rat myocytes. *J. Physiol. (London)* 437:351–375.

Frank, J.S., and Langer, G.A. (1974). The myocardial interstitium: Its structure and its role in ionic exchange. *J. Cell Biol.* 60:586–601.

Frank, J.S., Mottino, G., Reid, D., Molday, R.S., and Philipson, K.D. (1992). Distribution of the $Na^{+}-Ca^{2+}$ exchange protein in mammalian cardiac myocytes: An immuno-fluorescence and immunocolloidal gold-labeling study. *J. Cell Biol.* 117:337–345.

Frelin, C., Vigne, P., and Lazdunski, M. (1984). The role of the Na^{+}/H^{+} exchange system in cardiac cells in relation to the control of the internal Na^{+} concentration. *J. Biol. Chem.* 259:8880–8885.

Fry, C.H., Powell, T., Twist, V.W., and Ward, P.T. (1984). Net calcium exchange in adult rat ventricular myocytes: An assessment of mitochondrial accumulating capacity. *Proc. Royal Soc. B* 223:223–238.

Furchgott, R.F., and de Gubareff, T. (1956). Depression of contractile force of isolated auricles by ryanodine. *Fed. Proc.* 15:425.

Haest, C.W.M., and Deuticke, B. (1982). Interaction between membrane cytoskeletal protein and the intrinsic domain of the erythrocyte membrane. *Bioch. Biophys. Acta* 694:33–352.

Hansford, R.G. (1985). Relation between mitochondrial calcium transport and control of energy metabolism. *Rev. Physiol. Biochem. Pharmacol.* 102:1–72.

Harrison, S.M., and Boyett, M.R. (1995). The role of the Na^+–Ca^{2+} exchanger in the rate-dependent increase in contraction in guinea-pig ventricular myocytes. *J. Physiol.* (*London*) 482:555–566.

Hilgemann, D., Nicoll, D.A., and Philipson, K.D. (1991). Charge movement during Na^+ translocation by native and cloned cardiac Na^+/Ca^{2+} exchanger. *Nature* 352:715–718.

Hilgemann, D.W., Collins, A., and Matsuoka, S. (1992). Steady state and dynamic properties of cardiac sodium–calcium exchange: Secondary modulation of cytoplasmic calcium and ATP. *J. Gen. Physiol.* 100:933–961.

Hillyard, I.W., and Procita, L. (1956). Action of ryanodine on isolated kitten auricle. *Fed. Proc.* 15:438.

Hohl, C.M., and Altschuld, R.A. (1991). Response of isolated adult canine cardiac myocytes to prolonged hypoxia and reoxygenation. *Am. J. Physiol.* 260:C383–C391.

Holroyde, M.J., Robertson, S.P., Johnson, J.D., Solaro, R.J., and Potter, J.D. (1980). The calcium and magnesium binding sites on cardiac troponin and their role in the regulation of myofibrillar adenosine triphosphatase. *J. Biol. Chem.* 255:11688–11693.

Inui, M., Saito, A., and Fleischer, S. (1987). Purification of the ryanodine receptor and identity with feet structures of junctional terminal cisternae of sarcoplasmic reticulum from fast skeletal muscle. *J. Biol. Chem.* 262:1740–1747.

Inui, M., Wang, S., Saito, A., and Fleischer, S. (1988). Characterization of junctional and longitudinal sarcoplasmic reticulum from heart muscle. *J. Biol. Chem.* 263:10843–10850.

Isenberg, G., Hans, S., Schiefer, A., and Gallitelli, M.-F. (1993). Changes in mitochondrial calcium concentration during the cardiac contraction cycle. *Cardiovasc. Res.* 27:1800–1809.

Janczewski, A.M., and Lakatta, E.G. (1993). Thapsigargin inhibits Ca^{2+} uptake and Ca^{2+} depletes sarcoplasmic reticulum in intact cardiac myocytes. *Am. J. Physiol.* 256:H517–H522.

Janiak, R., Lewartowski, B., and Langer, G.A. (1996). Functional coupling between sarcoplasmic reticulum and Na–Ca exchange in single myocytes of guinea-pig and rat heart. *J. Mol. Cell. Cardiol.* 28:253–264.

Jenden, D.J., and Fairhurst, A.S. (1969). The pharmacology of ryanodine. *Pharmacol. Rev.* 21:1–25.

Jorgensen, A.O., Shen, C.-Y., and Campbell, K.P. (1985). Ultrastructural localization of calsequestrin in adult rat atrial and ventricular muscle cells. *J. Cell Biol.* 101:257–268.

Jorgensen, A.O., Shen, AC-Y., Arnold, W., McPherson, P.S., and Campbell, K.P. (1993). The Ca^{2+}-release channel/ryanodine receptor is localized in junctional and corbular sarcoplasmic reticulum in cardiac muscle. *J. Cell Biol.* 120:969–980.

Kawai, M., and Konishi, M. (1994). Measurement of sarcoplasmic reticulum calcium content in skinned mammalian cardiac muscle. *Cell Calcium* 16:123–136.

Kijima, Y., Ogunbunmi, E., and Fleischer, S. (1991). Drug action of thapsigargin on the Ca^{2+} pump protein of sarcoplasmic reticulum. *J. Biol. Chem.* 266:22912–22918.

Kleber, A.G. (1983). Resting membrane potential, extracellular potassium activity, and intracellular sodium activity during acute global ischemia in isolated perfused guinea pig hearts. *Circ. Res.* 52:442–450.

Kuwata, J.H., and Langer, G.A. (1989). Rapid, non-perfusion limited calcium exchange in cultured neonatal myocardial cells. *J. Mol. Cell. Cardiol.* 21:1195–1208.

Langer, G.A. (1965). Calcium exchange in dog ventricular muscle: Relation to frequency of contraction and maintenance of contractility. *Circ. Res.* 17:78–90.

Langer, G.A. (1971). The intrinsic control of myocardial contraction-ionic factors. *New Engl. J. Med.* 285:1065–1071.

Langer, G.A. (1983). The "sodium pump lag" revisited. *J. Mol. Cell. Cardiol.* 15:647–651.

Langer, G.A., and Frank, J.S. (1972). Lanthanum in heart cell culture. Effect on calcium exchange correlated with its localization. *J. Cell Biol.* 54:441–455.

Langer, G.A., and Peskoff, A. (1996). Calcium concentration and movement in the diadic cleft space of the cardiac ventricular cell. *Biophys. J.* 70:1169–1182.

Langer, G.A., and Serena, S.D. (1970). Effects of strophanthidin upon contraction and ionic exchange in rabbit ventricular myocardium: Relation to control of active state. *J. Mol. Cell. Cardiol.* 1:65–90.

Langer, G.A., Rich, T.L., and Orner, F.B. (1990). Ca exchange under non-perfusion-limited conditions in rat ventricular cells: Identification of subcellular compartments. *Am. J. Physiol.* 259:H592–H602.

Langer, G.A., Wang, S.Y., and Rich, T.L. (1995). Localization of the Na–Ca exchange-dependent Ca compartment in cultured neonatal rat heart cells. *Am. J. Physiol.* 268:C119–C126.

Leblanc, N., and Hume, J.R. (1990). Sodium current-induced release of calcium from sarcoplasmic reticulum. *Science* 248:372–376.

Lee, K.S., and Klaus, W. (1971). The subcellular basis for the mechanism of inotropic action of cardiac glycosides. *Pharmacol. Rev.* 23:193–261.

Legato, M. (1979). Cellular mechanisms of normal growth in mammalian heart. II. A quantitative and qualitative comparison between right and left ventricular myocytes in the dog from birth to five months of age. *Circ. Res.* 44:263–279.

Lehninger, A. (1974). Role of phosphate and other proton-donating anions in respiration-coupled transport of Ca^{2+} by mitochondria. *Proc. Natl. Acad. Sci. USA* 71:1520–1524.

Leisey, J.R., Grotyohann, L.W., Scott, D.A., and Scaduto, R.C., Jr. (1993). Regulation of mitochondrial calcium by average extramitochondrial calcium. *Am. J. Physiol.* 265:H1203–H1208.

Levi, A.J., Spitzer, K.W., Kohmoto, O., and Bridge, J.H.B. (1994). Depolarization-induced Ca entry via Na–Ca exchange triggers SR release in guinea pig cardiac myocytes. *Am. J. Physiol.* 266H422–H433.

Levitsky, D.O., Benevolewsky, D.S., Levchenko, T.S., Smirnov, V.N., and Chazov, E.I. (1981). Calcium binding rate and capacity of cardiac sarcoplasmic reticulum. *J. Mol. Cell. Cardiol.* 13:785–796.

Lewartowski, B., Pytkowski, B., and Janczewski, A. (1984). Calcium fraction correlating with contractile force of ventricular muscle of guinea pig heart. *Pfluger's Arch.* 401:198–203.

Lewartowski, B., Janiak, R., Wang, S.Y., and Langer, G.A. (1997). The effect of sarcoplasmic reticulum calcium release into the diadic region on Na–Ca exchange current in single myocytes of guinea-pig and rat heart. In preparation.

Li, T., and Sperelakis, N. (1983). Calcium antagonist blockade of slow action potentials in cultured chick heart cells. *Can. J. Physiol. Pharmacol.* 61:957–966.

Limas, C.J. (1977). Calcium-binding sites in rat myocardial sarcolemma. *Arch. Biochem. Biophys.* 179:302–309.

Lipp, P., and Niggli, E. (1994). Sodium current-induced calcium signals in isolated guinea pig ventricular myocytes. *J. Physiol. (London)* 474:439–446.

López-López, J.R., Shacklock, P.S., Balke, C.W., and Wier, W.G. (1995). Local calcium transients triggered by single L-type calcium channel currents in cardiac cells. *Science* 268:1042–1045.

Lukcs, G.L., and Kapus, A. (1987). Measurement of the matrix free Ca^{2+} concentration in heart mitochondria by entrapped fura-2 and quin 2. *Biochem. J.* 248:609–613.

Marengo, F.D., Wang, S.Y., and Langer, G.A. (1997). The effects of temperature upon calcium exchange in intact cultured cardiac myocytes. *Cell Calcium.* In press.

Matsuoka, S., and Hilgemann, D. (1992). Steady-state and dynamic properties of cardiac sodium–calcium exchange. Ion and voltage dependencies of the transport cycle. *J. Gen. Physiol.* 100:963–1001.

McCormack, J.G., and Denton, R.M. (1994). Signal transduction by intramitochondrial Ca^{2+} in mammalian energy metabolism. *News Physiol. Sci.* 9:71–76.

Miller, T.W., and Tormey, J. McD. (1993). Calcium displacement by lanthanum in subcellular components of rat ventricular myocytes: Characterization by electron probe microanalysis. *Cardiovasc. Res.* 27:2106–2112.

Minamikawa, T., Takahashi, A., and Fujita, S. (1995). Differences in features of calcium transients between the nucleus and the cytosol in cultured heart muscle cells: Analyzed by confocal microscopy. *Cell Calcium* 17:165–176.

Miyata, H., Lakatta, E.G., Stern, M.D., and Silverman, H.S. (1992). Relation of mitochondrial and cytosolic free calcium to cardiac myocyte recovery after exposure to anoxia. *Circ. Res.* 71:605–613.

Moravec, C.S., and Bond, M. (1991). Calcium is released from the junctional sarcoplasmic reticulum during cardiac muscle contraction. *Am. J. Physiol.* 260:H989–H997.

Moravec, C.S., and Bond, M. (1992). Effect of inotropic stimulation on mitochondrial calcium in cardiac muscle. *J. Biol. Chem.* 267:5310–5316.

Murphy, J.G., Smith, T.W., and Marsh, J.D. (1987). Calcium flux measurements during hypoxia in cultured heart cells. *J. Mol. Cell. Cardiol.* 19:271–279.

Musters, R.J.P., Post, J.A., and Verkleif, A.J. (1991). The isolated neonatal rat-cardiomyocyte used in an *in vitro* model for "ischemia." I. A morphological study. *Bioch. Biophys. Acta* 1091:270–277.

Musters, R.J.P., Olten, E., Biegelmann, E., Bijvelt, J., Keijzer, J.J.H., Post, J.A., OpdenKamp, J.A.F., and Verkleij, A.J. (1993). Loss of asymmetric distribution of sarcolemmal phosphati-dylethanolamine during simulated ischemia in the isolated neonatal rat cardiomyocyte. *Circ. Res.* 73:514–523.

Page, E. (1978). Quantitative ultrastructural analysis in cardiac membrane physiology. *Am. J. Physiol.* 4:C147–C158.

Peters, G.A., Kohmoto, O., and Barry, W.H. (1989). Detection of La^{3+} influx in ventricular cells by indo-1 fluorescence. *Am. J. Physiol.* 256:C351–C357.

Philipson, K.P., and Langer, G.A. (1979). Sarcolemmal-bound calcium and contractility in the mammalian myocardium. *J. Mol. Cell. Cardiol.* 11:857–875.

Pierce, G.N., Rich, T.L., and Langer, G.A. (1987). Trans-sarcolemmal Ca^{2+} movements associated with contraction of the rabbit right ventricular wall. *Circ. Res.* 61:805–814.

Post, J.A., Langer, G.A., Opden Kamp, J.A.F., and Verkleij, A.J. *et al.* (1988). Phospholipid asymmetry in cardiac sarcolemma. Analysis of intact cells and "gas-dissected" membranes. *Biochim. Biophys. Acta* 943:256–266.

Post, J.A., and Langer, G.A. (1992a). Cellular origin of the rapidly exchangeable calcium pool in the cultured neonatal rat heart cell. *Cell Calcium* 13:627–634.

Post, J.A., and Langer, G.A. (1992b). Sarcolemmal calcium binding sites in heart. I. Molecular origin in "gas-dissected" sarcolemma. *J. Membr. Biol.* 129:49–57.

Post, J.A., Clague, J.R., and Langer, G.A. (1993). Sarcolemmal phospholipid asymmetry and Ca fluxes on metabolic inhibition of neonatal rat heart cells. *Am. J. Physiol.* 265:H461–H468.

Radermacher, M., Rao, V., Grassuci, J., Frank, J., Timerman, A.P., Fleischer, S., and Wagenknecht, T. (1994). Cryo-electron microscopy and three-dimensional reconstruction of the calcium release channel/ryanodine receptor from skeletal muscle. *J. Cell Biol.* 127:411–423.

Repke, K. (1964). Über den Biochemischen Winkingsmoders von Digitalis. *Klin. Wochenschr.* 42:157–165.

Reuter, H., and Seitz, N. (1968). The dependence of calcium efflux from cardiac muscle on temperature and external ion composition. *J. Physiol. (London)* 195:45–70.

Rich, T.L., Langer, G.A., and Klassen, M.G. (1988). Two components of coupling calcium in single ventricular cell of rabbits and rats. *Am. J. Physiol.* 254:H937–H946.

Robertson, S.P., Johnson, J.D., Holroyde, M.J., Kranias, E.G., Potter, J.D., and Solaro, R.J. (1982). The effect of troponin I phosphorylation on the Ca^{2+}-binding properties of the Ca^{2+}-regulatory site of bovine cardiac troponin, *J. Biol. Chem.* 257:260–263.

Rockman, H.A., Knowlton, K.U., Ross, J.R., Jr., and Chien, K.R. (1993). *In vivo* murine cardiac hypertrophy: A novel model to identify genetic signaling mechanisms that activate an adaptive physiologic response. *Circulation* 87:14–21.

Rose, W.C., Balke, C.W., Wier, W.G., and Marban, E. (1992). Macroscopic and unitary properties of physiological ion flux through L-type Ca^{2+} channels in guinea pig heart cells. *J. Physiol.* (*London*) 456:267–284.

Rousseau, E., and Meissner, G. (1989). Single cardiac sarcoplasmic reticulum Ca^{2+}-release channel: Activation by caffeine. *Am. J. Physiol.* 256:H328–H333.

Rutter, G.A. (1990). Ca^{2+} binding to citrate cycle dehydrogenases. *Int. J. Biochem.* 22:1081–1088.

Sagara, Y., and Inesi, G. (1991). Inhibition of the sarcoplasmic reticulum Ca^{2+} transport ATPase by thapsigargin at subnanomolar concentrations. *J. Biol. Chem.* 266:13503–13506.

Santana, L.F., Cheng, H., Gomez, A.M., Cannell, M.B., and Lederer, W.J. (1996). Relation between the sarcolemmal Ca^{2+} current and Ca^{2+} sparks and local control theories for cardiac excitation–contraction coupling. *Circ. Res.* 78:166–171.

Satoh, H., Hayashi, H., Katoh, H., Terada, H., and Kobayashi, A. (1995). Na^+/H^+ and Na^+/Ca^{2+} exchange in regulation of $[Na^+]_i$ and $[Ca^{2+}]_i$ during metabolic inhibition. *Am. J. Physiol.* 268:H1239–H1248.

Scarpa, A., and Williamson, J.R. (1974). Calcium binding and calcium transport by subcellular fractions of heart. *In* "Calcium Binding Proteins" (Drabikowski, W., Strzelecka-Gabaozewska, H., and Carafoli, E., eds.), pp. 547–585. Elsevier, Amsterdam, The Netherlands.

Scarpa, A., and Graziotti, P. (1973). Mechanisms for intracellular calcium regulation in heart. I. Stopped-flow measurements of Ca^{++} uptake by cardiac mitochondria. *J. Gen. Physiol.* 62:756–772.

Schramm, M., Thomas, G., Towart, L., and Franckowiak, G. (1983). Novel dihydropyridines with positive inotropic action through activation of Ca^{2+} channels. *Nature* 303:535–537.

Seigneuret, M., and DeVeaux, P.F. (1984). ATP dependent asymmetric distribution of spin labeled phospholipids in the erythrocyte membrane: Relation to shape changes. *Proc. Natl. Acad. Sci. USA* 81:3751–3755.

Shacklock, P.S., Wier, W.G., and Balke, C.W. (1995). Local Ca^{2+} transients (Ca^{2+} sparks) originate at transverse tubules in rat heart cells. *J. Physiol.* (*London*) 487:601–608.

Schultz, S. (1992, June 8). *The Scientist.*

Sham, J.S.K., Cleeman, L., and Morad, M. (1995). Functional coupling of Ca channels and ryanodine receptors in cardiac myocytes. *Proc. Natl. Acad. Sci. USA* 92:121–125.

Sipido, K.R., and Wier, W.G. (1991). Flux of Ca^{2+} across the sarcoplasmic reticulum of guinea pig cardiac cells during excitation–contraction coupling. *J. Physiol.* 435:605–630.

Slaughter, R.S., Sutko, J.L., and Reeves, J.P. (1983). Equilibrium calcium–calcium exchange in cardiac sarcolemmal vesicles. *J. Biol. Chem.* 258:3183–3190.

Stern, M.D. (1992). Theory of excitation–contraction coupling in cardiac muscle. *Biophys. J.* 63:497–517.

Sun, X.-H., Protasi, F., Takahashi, M., Takeshima, H., Ferguson, D.G., and Franzini-Armstrong, C. (1995). Molecular architecture of membranes involved in excitation–contraction coupling of cardiac muscle. *J. Cell Biol.* 129:659–671.

Tani, M., and Neely, J.R. (1989). Role of intracellular Na^+ in Ca^{2+} overload and depressed recovery of ventricular function of reperfused ischemic rat hearts: Possible involvement of H^+/Na^+ and Na^+/Ca^{2+} exchange. *Circ. Res.* 65:1045–1056.

Terracciano, C.M.N., Naqvi, R.U., and MacLeod, K.T. (1995). Effects of rest interval on the release of calcium from the sarcoplasmic reticulum in isolated guinea pig ventricular myocytes. *Circ. Res.* 77:354–360.

Thompson, R.B., Warber, K.D., and Potter, J.D. (1990). Calcium at the myofilaments. *In* "Calcium and the Heart" (Langer, G., ed.), pp. 167–198. Raven Press, New York.

Thustrup, J.P., Cullen, J., Drobek, B., Hanley, M.R., and Davson, A.P. (1990). Thapsigargin, a tumor promoter, discharges intracellular Ca^{2+} stores by specific inhibition of the endoplasmic reticulum Ca^{2+}-ATPase. *Proc. Natl. Acad. Sci. USA* 87:2466–2470.

Trautwein, W., Cavali, A., Flockerzi, V., Hofmann, F., and Pelzer, D. (1987). Modulation of calcium channel function by phosphorylation in guinea pig ventricular cells and phospholipid bilayer membranes. *Circ. Res.* 61(suppl. I):117–123.

Ussing, H.H. (1948). The use of tracers in the study of active ion transport across animal membranes. *Cold Spring Harbor Symp. Quant. Biol.* 13:193–200.

Walsh, L.G., and Tormey, J.McD. (1988). Subcellular electrolyte shifts during *in vitro* myocardial ischemia and reperfusion. *Am. J. Physiol.* 255:H917–H928.

Wang, S.-Y., Winka, L., and Langer, G.A. (1993). Role of calcium current and sarcoplasmic reticulum calcium release in control of myocardial contraction in rat and rabbit myocytes. *J. Mol. Cell Cardiol.* 25:1339–1347.

Wang, S.-Y., Clague, J.R., and Langer, G.A. (1995). Increase in calcium leak channel activity by metabolic inhibition or hydrogen peroxide in rat ventricular myocytes and its inhibition by polycation. *J. Mol. Cell. Cardiol.* 27:211–222.

Wang, S.-Y., Peskoff, A., and Langer, G.A. (1996). Inner sarcolemmal leaflet Ca^{2+} binding: Its role in cardiac Na–Ca exchange. *Biophys. J.* 70:2266–2274.

Wang, S.-Y., Dong, L., and Langer, G.A. (1997). Matching Ca efflux and influx to maintain steady-state levels in cultured cardiac cells. Flux control in the subsarcolemmal cleft. *J. Mol. Cell. Cardiol.* In press.

Weber, A. (1968). The mechanism of action of caffeine on sarcoplasmic reticulum. *J. Gen. Physiol.* 52:760–772.

Weber, A., and Herz, R. (1968). The relationship between caffeine contracture of intact muscle and the effect of caffeine on reticulum. *J. Gen. Physiol.* 52:750–759.

Weiss, J., and Shine, K.I. (1982). Extracellular K^+ accumulation during myocardial ischemia in isolated rabbit heart. *Am. J. Physiol.* 268:H1239–H1248.

Wibo, M., Bravo, G., and Godfraind, T. (1991). Postnatal maturation of excitation–contraction coupling in rat ventricle in relation to the subcellular localization and surface density of 1,4-hydropyryidine and ryanodine receptors. *Circ. Res.* 68:662–673.

Wilkie, D.R. (1979). Generation of protons by metabolic processes other than glycolysis in muscle cells. *J. Mol. Cell. Cardiol.* 11:325–330.

Williamson, J.R., Williams, R., Coll, K.E., and Thomas, A.P. (1983). Cytosolic free Ca^{2+} concentration and intracellular calcium distribution of Ca^{2+}-tolerant isolated heart cells. *J. Biol. Chem.* 258:13411–13414.

Wolska, B.M., and Lewartowski, B. (1991). Calcium in the *in situ* mitochondria of rested and stimulated myocardium. *J. Mol. Cell Cardiol.* 23:217–226.

6

MECHANICS AND FORCE
PRODUCTION

KENNETH P. ROOS

I. INTRODUCTION

The continuous mechanical pumping action of the heart is essential for the sustenance of life by providing an adequate blood supply throughout the entire organism. This is achieved in the mammalian myocardium by the coordinated contraction–relaxation cycling in the billions of cells that constitute the atrial and ventricular chambers of the heart. Mammalian ventricles are formed from a complex three-dimensional (3-D) syncytium of cells whose orientations and vectors of force vary continuously across the thickness of the chamber wall (Streeter *et al.*, 1969). This leads to difficulties in the identification and interpretation of the underlying cellular and molecular mechanisms that determine mechanical function (Brady, 1974, 1991b; Brady *et al.*, 1981). Yet taken at the most simplistic, whole organ level, the combined force development and shortening of cells elicits a rise in chamber pressure and a decrease in chamber volume, which ejects blood into the circulatory system. Thus, in the intact heart, the clinically measurable parameters of pressure and volume are related, but clearly not identical, to the parameters of force, length, velocity, and time that are classically used to characterize single skeletal fibers or papillary and trabecular cardiac muscle function.

In the previous edition of this book, Allan Brady (1974) evaluated and modeled these classical force, length, velocity, and time relations primarily from papillary muscle data. These and whole heart studies by others have historically provided the basis for our understanding of myocardial mechanics. Today, critical new information continues to be obtained from these types of organ and multicellular preparations, particularly in relation to the role of the collagenous extracellular matrix, 3-D diastolic and systolic properties, and pathophysiological changes in heart function (Conrad *et al.*, 1995; de Tombe and ter Keurs, 1992; MacKenna *et al.*, 1994, 1996; Perreault *et al.*, 1990; ter Keurs *et al.*, 1980a, 1980b; Van Leuven *et al.*, 1994; Weber *et al.*, 1994). However, interpretation of data from these multicellular strip (papillary and trabecula) and whole heart preparations is necessarily limited by their complicated structure, attachment-induced artifact, and unpredictable nonuniformities in space and time (Brady, 1991b; Donald *et al.*, 1980; Holmes *et al.*, 1995; Krueger and Pollack, 1975; Omens *et al.*, 1996; Pollack and Krueger, 1976; Rodriguez *et al.*, 1993; Streeter *et al.*, 1969). Significant efforts have been made to circumvent these issues through mathematical modeling, length clamping, and perturbation analyses (Brady *et al.*, 1981; Guccione *et al.*, 1993; Little *et al.*, 1995; Rodriguez *et al.*, 1992; ter Keurs *et al.*, 1980a, 1980b). Although the importance of these classical preparations has not diminished, cellular and molecular approaches developed since the mid-1970s provide a simplified system where less ambiguous results permit a better evaluation of the fundamental mechanisms of myocardial mechanics.

Following the theme of this book, this chapter deals almost exclusively with the cellular and subcellular mechanical function associated with adult mammalian ventricular cells (Fig. 1). When isolated from the collagenous extracellular matrix, the ventricular myocyte remains an excellent physiological model of myocardial function (Brady, 1984, 1991b; Fabiato and Fabiato, 1975, 1976, 1978; Frank, 1989; Frank *et al.*, 1986; Granzier and Irving, 1995; Kent *et al.*, 1989; Krueger, 1988; Lieberman *et al.*, 1987; McDon-

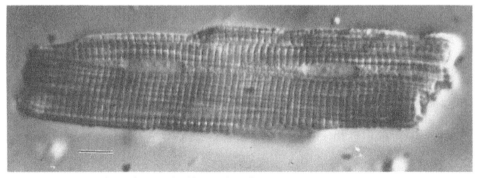

FIGURE 1   Photomicrograph of isolated cardiac myocyte. This micrograph is of a cardiac myocyte isolated from a Sprague-Dawley (SD) rat heart by collagenase treatment. It is bathed in relaxing solution and clearly shows the A–I-band cross-striation pattern and two nuclei. Domains of sarcomeres are longitudinally bisected by the nuclei and by strips of mitochondria (not resolved). Average sarcomere periodicity is 1.89 μm. Calibration bar = 10 μm. (Reproduced with permission from Roos and Brady, 1989.)

ald *et al.*, 1995; White *et al.*, 1995) or dysfunction (Brooksby *et al.*, 1992; Delbridge *et al.*, 1996; Kobayashi *et al.*, 1995; Mann *et al.*, 1991; Spinale *et al.*, 1992; Zile *et al.*, 1995). Despite significant progress in other areas of study, only modest success has been made until recently in evaluating the mechanical properties of myocytes and their molecular components. This is due primarily to the difficulty in assessing mechanical function from small structures that are not easily attached to transducers or visualized. The ventricular myocyte and its sarcomeres consist of the membrane-based regulatory systems (see Chapters 3, 4, and 5), the cytoskeleton (see Chapter 1), and the filamentous regulatory and contractile proteins. This chapter focuses on the mechanical function associated with the cytoskeletal, regulatory, and contractile proteins in adult mammalian cells. Section II commences with a brief review of the structural basis of mechanical function in the cell and its sarcomeres. Section III examines force development, and Section IV examines contractile regulation at the myofilament and molecular levels. Section V concludes with a review of passive and active mechanical properties in isolated cardiac myocytes. Additional discussion is included, when appropriate, on the relationship between normal cellular mechanical function and that of cells from hypertrophied heart.

II. CELLULAR STRUCTURE—MECHANICAL ASPECTS

As previously described in Chapter 1, the myocardium is a highly differentiated and complex structure. In the intact ventricular myocardium, the individual myocytes are surrounded by collagenous extracellular matrix, fibroblasts, and the vasculature (MacKenna *et al.*, 1994; Weber *et al.*, 1994). Whether mechanically disaggregated (DeClerck *et al.*, 1984; Hofmann and Moss, 1992; Sweitzer and Moss, 1990, 1993) or isolated using standard Langendorff enzymatic (collagenase) digestion methods (Brady *et al.*, 1979; Delbridge *et al.*, 1996; Krueger *et al.*, 1980; Lieberman *et al.*, 1987; Palmer *et al.*, 1996; Roos and Brady, 1989), myocytes are released from an extracellular matrix that can modulate the cell's contractile function in unknown ways. Yet these rodlike cells retain their sarcolemmal, glycocalyx, cytoskeletal, contractile filament, and intercalated disk integrity (Fig. 2), and are tolerant of physiological levels of calcium in their saline bathing solutions (Brady *et al.*, 1979; Frank, 1989; Frank *et al.*, 1986; Lieberman *et al.*, 1987). This permits mechanical studies to be performed with suitable optical and mechanical measurement methods without confounding extracellular collagen and cell orientation problems (Brady, 1984, 1990, 1991a, 1991b; Delbridge and Roos, 1997; Fabiato, 1981; Garnier, 1994; Kent *et al.*, 1989; Krueger, 1988; Mukherjee *et al.*, 1993; Palmer *et al.*, 1996; Roos and Taylor, 1993; White *et al.*, 1995).

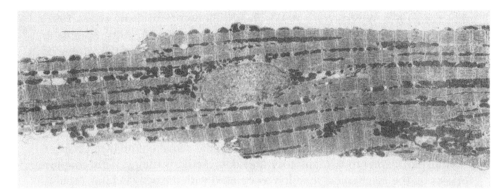

FIGURE 2   Electronmicrograph of isolated cardiac myocyte. This thin section micrograph illustrates the central portion of a fixed isolated rat cardiac myocyte. The organization of the cell's sarcomere fields, mitochondrial strips, sarcolemmal membrane, and a nucleus are clearly depicted. The sarcomere striation pattern is generally in register, although there is a small amount of lateral shifting (shear) between some sarcomere fields. Note the clustering of the mitochondria adjacent to the nucleus where the sarcomeres weave around the nucleus in and out of the plane of the section. Calibration bar = 5 μm. (Reprinted with permission from Frank, 1989. Copyright CRC Press, Boca Raton, Florida, © 1989.)

In addition, mechanical studies can be performed not only on cells with intact membrane systems but also on "skinned" cells or on molecular-based assay systems. Cells whose membrane systems are "skinned" or disrupted either mechanically or with detergent treatment offer an alternative system for mechanical measurement (Araujo and Walker, 1994, 1996; Fabiato and Fabiato, 1975, 1976, 1978; Granzier and Irving, 1995; Hofmann and Moss, 1992; Palmer *et al.*, 1996; Siri *et al.*, 1991; Sweitzer and Moss, 1990, 1993). The levels and time course of activating calcium can be modulated independently from the normal membrane-based regulatory systems described elsewhere in this book. Whether the membrane is skinned or intact, the cardiac myocyte retains the essential structural entities of the contractile and cytoskeletal proteins that characterize mechanical behavior at the cellular and sarcomeric levels. At the molecular level, the development of the *in vitro* motility assay system has permitted the evaluation of mechanical function in isolated motor proteins from skeletal and cardiac muscle (Harris *et al.*, 1994; Kron and Spudich, 1985; VanBuren *et al.*, 1994, 1995; Warshaw, 1996; Yanagida *et al.*, 1984). These cellular, sarcomeric, and molecular mechanical assessment approaches are discussed later in this chapter.

A. CELL MORPHOLOGY

Although mammalian ventricular myocytes (see Fig. 1) vary somewhat in size and shape, typical cells from normal adult hearts are about 120 μm in length and have an ellipsoidal (~2:1 aspect ratio) cross-section 10–35 μm in width (Gerdes et al., 1994; Sorenson et al., 1985). The electron micrograph in Fig. 2 shows the salient ultrastructural features of an isolated rat cardiac myocyte. The dominant features of the myocyte are the fields of longitudinally oriented sarcomeres that weave around the nuclei and mitochondria along the length of the cell (Frank, 1989; Sommer and Johnson, 1979). From a mechanical point of view, a key structural consideration is that the sarcomeres generally run parallel to each other, thus providing a linear force vector. In rat heart, Page (1978) reported that the sarcomeric myofilaments make up about 55% of the cell volume as compared to more than 80% in skeletal fibers (Eisenberg, 1983). The mitochondria run in longitudinal strips between the sarcomeric fields and are especially dense near the nuclei (see Fig. 2). Mitochondria make up 36% of the rat cardiac cell volume, far more than the 5% in the less energetically demanding skeletal muscle. The balance of the cell consists of the nuclei, SR, cytoskeleton, and cytosol. These values can be altered somewhat in other species (Blumberg et al., 1995) and significantly so in hypertrophied heart (Vulpis et al., 1995).

B. CONTRACTILE APPARATUS

Electron micrographs (e.g., Fig. 2) clearly show that the interdigitating thick and thin contractile filaments are regularly ordered in both mammalian skeletal and cardiac sarcomeres. This gives the A- and I-band striated appearance to cardiac myocytes when viewed under the optical microscope (see Fig. 1). Figure 3 shows a representation of the organization of the sarcomere and its constituent filament systems. Longitudinally centered in the sarcomere, the thick filaments are about 1.6 μm in length (A-band length) and 15 nm in diameter. They are anchored at the M-band with the cytoskeletal proteins myomesin and M-protein (Obermann et al., 1995), and stabilized longitudinally by titin (Horowits and Podolsky, 1987; Keller, 1995; Maruyama, 1994; Trinick, 1994, 1996; K. Wang et al., 1993). Thick filaments are formed primarily from the large (480 kDa, ~170 nm long) filamentous protein myosin, which composes about 43% of the total myofibrillar protein in skeletal muscle (Warrick and Spudich, 1987; Yates and Greaser, 1983). Myosin "heads" protrude regularly at a 14.3-nm spacing from the thick filament in a hexagonal arrangement (Squire, 1981; Squire et al., 1994) to form cross-bridges with the thin filament during contraction (Fig. 3). As discussed in Section III, muscle myosin has been sequenced and the cross-bridge 3-D crystal structure determined (Cheney et al., 1993; Rayment et al., 1993a). Additional components of the thick filament include the structural/regulatory C and X proteins (Gautel et al., 1995; Hofmann and Lange, 1994; Weisberg and Winegrad, 1996). The regulatory function of C-protein is discussed in Section IV.

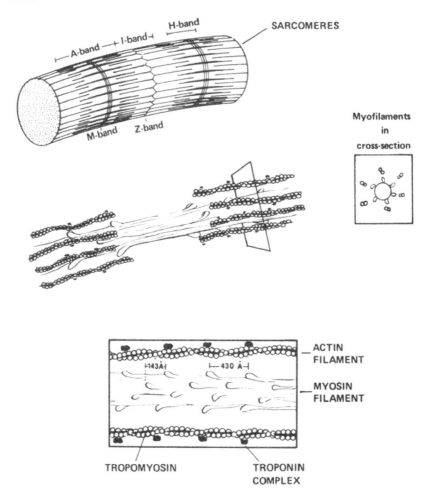

FIGURE 3   Sarcomere and myofilament organization. This drawing illustrates the myofilament arrangement in the sarcomere. Each sarcomere (top) is composed of interdigitating
thick and thin filaments (middle) whose optical properties form the A- and I-bands of striated
muscle visible under the light microscope. The H-band corresponds to the central portion of
the A-band in which the thin filaments have not interdigitated. The M-band is at the center
of the A-band and the Z-band in the center of the I-band. The thin filaments and the myosin
cross-bridges are hexagonally arranged (right and bottom) about the thick filament. The
troponin–tropomyosin regulatory proteins are distributed along the thin filament (bottom).
(Reproduced with permission from Lieber, 1992.)

The thin filaments extend about 1.1 μm toward the center of the sarco-mere from each Z-line and are about 10 nm in diameter in cardiac muscle (see Fig. 3). The thin filaments are arranged and stabilized at the Z-disk lattice by the cytoskeletal proteins, predominantly α-actinin (Goldstein *et al.*, 1989; Isobe *et al.*, 1988). Although composed of many different proteins, the thin filament itself consists primarily of the ubiquitous globular G-actin protein. G-actin (43 kDa) composes about 22% of the total myofibrillar protein in skeletal muscle (Yates and Greaser, 1983) and consists of 375 amino acids that have been sequenced. Its 3-D structure (5.5 nm in diame-ter) has also been determined (Holmes *et al.*, 1990) and divided into four structural domains. One domain contains the myosin binding site for force transmission during contraction, and the others contain regulatory and structural binding sites. G-actin polymerizes into fibrous F-actin under physiological conditions to form the thin filament backbone as a double-stranded super helix with a 38-nm half-turn repeat.

In addition to providing the myosin binding site, the thin filament also consists of the troponin–tropomyosin complex that allosterically regulates cross-bridge function (see Section IV) and thus force development in car-diac and skeletal muscle (Chalovich, 1992; H. Huxley, 1996; Tobacman, 1996). Tropomyosin filaments (68 kDa, 40-nm filaments) are helically ar-ranged in proximity to the myosin binding sites along the F-actin helices. The globular cardiac troponins (Tn-C, 18.4 kDa; Tn-I, 23.8 kDa; Tn-T, 38 kDa) are distributed along the length of the thin filament in association with both the actin and tropolyosin (see Fig. 3).

The sarcomeric organization of these interdigitating thick and thin fila-ments (see Fig. 3) formed the original basis for the sliding filament hypothe-sis of muscular contraction that is applicable to both cardiac and skeletal muscle systems (Gordon *et al.*, 1966; A.F. Huxley, 1957, 1988, 1995; A.F. Huxley and Niedergerke, 1954; H.E. Huxley, 1957, 1996; H.E. Huxley and Hanson, 1954). For further information regarding the structural configura-tion of the sarcomere, its contractile filaments, and their regulatory protein constituents, see Amos (1985), H.E. Huxley (1996), Lieber (1992), Squire (1981), and Squire *et al.* (1994).

C. CYTOSKELETON

The functional components of the cardiac myocyte are interconnected by a noncontractile filamentous network—the cytoskeleton. Although not normally seen in micrographs due to their small size relative to the contrac-tile filaments and membrane systems, cytoskeletal components comprise more than 10% of the myocyte's total protein and play an essential mechani-cal role in contractile function. Kuan Wang (1985) has conveniently divided this network into the exo- and endosarcomeric lattices. These are schematized by Price (1991) for a generalized striated muscle and are shown here in Fig. 4.

1. Exosarcomeric Lattice

The sarcomeres, sarcolemma, mitochondria, nuclei, and other organelles are interconnected in the cell by an exosarcomeric lattice (Milner *et al.*, 1996; Price, 1984, 1991; K. Wang, 1985; K. Wang and Ramirez-Mitchell, 1983; Watkins *et al.*, 1987; Watson *et al.*, 1996). In cardiac myocytes, this consists largely of a desmin-based intermediate filament network that runs longitudinally, transversely, and circumferentially in the cell (see Fig. 4). Desmin is nearly 2% of a cardiac myocyte's total protein and is generally believed to be responsible for the localization and orientation of structures within the cell, particularly during growth (Milner *et al.*, 1996; Price, 1984; Watkins *et al.*, 1987). Mechanically, the transversely oriented desmin filaments couple the sarcomeres together at the Z-line to maintain striation registration. However, the fields of sarcomeres are often separated by large strips of mitochondria in cardiac muscle (see Fig. 2), which implies a longer desmin tethering and possible mechanical shear of sarcomere registration during contraction. Data also suggest that the exosarcomeric intermediate filament lattice may contribute to a small portion of the passive length–tension relation and the internal restoring forces as discussed in Section V (Brady, 1991a, 1991b; Brady and Farnsworth, 1986; Granzier and Irving, 1995; Roos and Brady, 1989). Desmin message and protein levels in heart have also been reported to increase in response to artificially induced stretch (Watson *et al.*, 1996) or hypertrophy (Collins *et al.*, 1996), reinforcing the important structural and organizational role of the intermediate filaments during growth and development.

Microtubules (not shown in Fig. 4) are only about 0.01% of the total protein in adult myocytes and are also distributed in the extracellular matrix in heart (Watkins *et al.*, 1987). As important organizing structures during cell growth, microtubules have recently been suggested as a source of contractile dysfunction in hypertrophied feline right ventricle (Tagawa *et al.*, 1996; Zile *et al.*, 1994). But similar studies on the spontaneously hypertensive rat (SHR) (Roos, unpublished observations, 1997) and aortic-banded guinea pig heart (Collins *et al.*, 1996) have not indicated a significant passive or active mechanical role for microtubules in hypertrophy or failure. The mechanical role of the exosarcomeric lattice in cardiac cells is addressed in Section V.

2. Endosarcomeric Lattice

Within the sarcomere itself, there are several noncontractile cytoskeletal proteins with potential mechanical roles (see Fig. 4). Titin (also called connectin) is about 8% of total muscle protein and is the largest (~3 MDa)

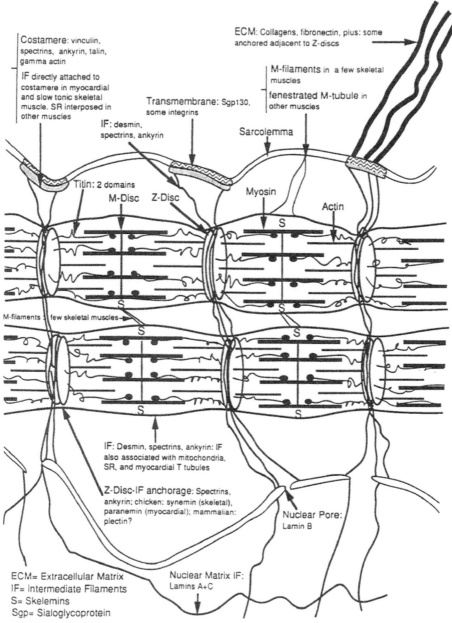

Costamere: vinculin, spectrins, ankyrin, talin, gamma actin

IF directly attached to costamere in myocardial and slow tonic skeletal muscle. SR interposed in other muscles

IF: desmin, spectrins, ankyrin

ECM: Collagens, fibronectin, plus: some anchored adjacent to Z-discs

M-filaments in a few skeletal muscles

Transmembrane: Sgp130, some integrins

fenestrated M-tubule in other muscles

Sarcolemma

Titin: 2 domains

M-Disc   Z-Disc

Myosin

Actin

S

M-filaments in few skeletal muscles

S

S

S

S

S

S

S

IF: Desmin, spectrins, ankyrin: IF also associated with mitochondria, SR, and myocardial T tubules

Z-Disc-IF anchorage: Spectrins, ankyrin; chicken: synemin (skeletal), paranemin (myocardial); mammalian: plectin?

Nuclear Pore: Lamin B

ECM= Extracellular Matrix
IF= Intermediate Filaments
S= Skelemins
Sgp= Sialoglycoprotein

Nuclear Matrix IF: Lamins A+C

FIGURE 4   Cytoskeletal lattices in striated muscle. The endosarcomeric and exosarcomeric cytoskeletal lattices are illustrated in a generic model of striated muscle. Of particular importance for cardiac mechanics are the transverse exosarcomeric desmin filaments interconnecting the Z-discs and the longitudinal endosarcomeric titin filaments. See text and reference for details. (Reproduced with permission from Price, 1991.)

single protein described to date (Furst and Gautel, 1995; Granzier *et al.*, 1996; Horowits and Podolsky, 1987; Jin, 1995; Keller, 1995; Labeit and Kolmerer, 1995; Linke *et al.*, 1996; Maruyama, 1994; Maruyama *et al.*, 1976; Price, 1991; Trinick, 1994; K. Wang *et al.*, 1984, 1991, 1993; S.-M. Wang and Greaser, 1985). Figure 5 shows the current model for cardiac titin and its relationship to the sarcomere's structure (Trinick, 1996). Titin spans the entire half sarcomere running from the M-line to the Z-line (not shown in Fig. 4) and is divided into several domains and zones (see Fig. 5). At the Z-line, titin's N-terminus may be associated with and is probably anchored to α-actinin (Keller, 1995). At the M-line, titin's C-terminus is associated with and probably anchored to both M-protein and myomesin (Obermann *et al.*, 1955; Vinkemeier *et al.*, 1993). This structural configuration strongly suggests that titin has a mechanical role in the passive behavior of striated muscle (Brady, 1991b; Gautel and Goulding, 1996; Granzier and Irving, 1995; Granzier *et al.*, 1996; Horowits and Podolsky, 1987; Houmeida *et al.*, 1995; Improta *et al.*, 1996; Linke *et al.*, 1996; Matsubara and Maruyama, 1977; Politou *et al.*, 1995; Trinick, 1996; K. Wang *et al.*, 1991, 1993).

Labeit and Kolmerer (1995) recently sequenced human cardiac titin in segments and found it to have 26,926 residues and a predicted molecular mass of 2993 kDa. However, they demonstrated that titin can apparently exist in many isoforms of varying protein lengths in striated muscle. This giant molecule is formed as a modular configuration of 112 immunoglobulin-like (Ig) and 132 fibronectin III-like (FN3) domains of 100 amino acids each. Near the A–I-band junction, there is an interesting segment, 70% of

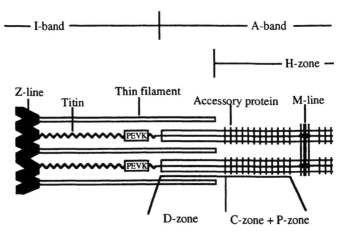

FIGURE 5   Titin structure and localization in the sarcomere. This drawing illustrates molecular titin spanning one-half of the sarcomere from the Z-line to the M-line. The PEVK and the zigzag portions in the I-band represent possible elastic regions of the molecule. The C, D, and P zones may provide a scaffold for the thick filament. See text for details. (Reproduced with permission from Trinick, 1996.)

which is composed largely of just four amino acids: proline (P), glutamic acid (E), valine (V), and lysine (K). This so-called PEVK segment is structurally distinct from the multiple repeating Ig or FN3 domains and can have significant molecular plasticity. In particular, the PEVK segment appears to be lengthened with additional immunoglobulin domains added in series to the I-band segments of skeletal muscle titin as compared to cardiac isoforms (Labeit and Kolmerer, 1995; Linke et al., 1996).

Titin has multiple functional roles in striated muscle. First, titin may serve as a protein template for the thick filaments during myofibrilagenesis (Trinick, 1994). Second, titin may stabilize the thick filament via an association with C-protein (Freiburg and Gautel, 1996; Furst and Gautel, 1995; Gautel et al., 1995; Houmeida et al., 1995). Third, because titin is an elastic molecule that spans the entire half sarcomere from the M- to the Z-lines (Gautel and Goulding, 1996; Politou et al., 1995; K. Wang, 1985), it most likely serves a mechanical role in mediating the passive length–tension relation in striated muscle (Brady, 1991b; Granzier and Irving, 1995; Granzier et al., 1996; Horowits and Podolsky, 1987; Improta et al., 1996; Keller, 1995; Labeit and Kolmerer, 1995; Linke et al., 1996; Palmer et al., 1996; K. Wang et al., 1993). Finally, altered expression and molecular changes have been reported in titin in association with remodeling in hypertrophied and failing heart (Ausma et al., 1995; Collins et al., 1996; Hein et al., 1994; Morano et al., 1994). The mechanical role of titin in cardiac muscle is further discussed in Section V.

In addition to titin, α-actinin (not shown in Fig. 4) is a significant component of the endosarcomeric cytoskeleton. It is about 2% of the total myofibrillar protein and forms the Z-disk lattice (Isobe et al., 1988; Yates and Greaser, 1983). This lattice structurally organizes the thin filament array and is the transmitter of longitudinal force between adjacent sarcomeres. In addition, Goldstein et al. (1989) have observed conformational changes in this lattice during contraction in cardiac muscle. This suggests that the lattice may also function as a radial spring modulating the passive characteristics, such as the restoring force in the cells. Finally, the newly described Z-line and thin filament associated protein nebulette (related to the skeletal muscle protein nebulin) may have some unspecified structural and mechanical roles in cardiac muscle (Moncman and Wang, 1995).

In summary, the myocyte is a highly differentiated structure that contains the force-generating contractile apparatus of thick and thin filaments. The myofilament-containing sarcomeres and other noncontractile structures of the myocyte are organized by the cytoskeleton. The interplay between the force generators and the passive elastic characteristics of some cytoskeletal components determine the overall mechanical function of the cell. The cardiac myocyte and its molecular constituents provide a simplified model for the evaluation of cardiac mechanics that cannot be achieved with whole heart or multicellular preparations.

III. MYOFILAMENTS AS FORCE
GENERATORS—MOLECULAR MOTORS

Since the formation of the sliding filament hypothesis in the mid-1950s (A.F. Huxley, 1957; A.F. Huxley and Niedergerke, 1954; H.E. Huxley, 1957; H.E. Huxley and Hanson, 1954), considerable effort has been made to elucidate the molecular events that cause the filaments to slide past one another and produce sarcomeric shortening. Despite substantial progress, there has not yet been a complete reconciliation of all the structural, mechanical, and biochemical data to present a complete explanation of how a muscle converts the chemical energy from ATP hydrolysis into mechanical work. Until recently, work concentrated on myofilament structure (Amos, 1985; H.E. Huxley, 1996; Squire, 1981), muscle fiber mechanics (Ford et al., 1977; Gordon et al., 1966; A.F. Huxley, 1988, 1995; A.F. Huxley and Simmons, 1971), energetics (Gibbs and Barclay, 1995; Homsher, 1987), and solution biochemistry (Brenner and Eisenberg, 1987; Eisenberg et al., 1980; Lymn and Taylor, 1971). These data suggested a model in which the myosin cross-bridge cyclically attaches to and detaches from actin (like an oar) to propel the myofilaments past one another using ATP as an energy source (Hibberd and Trentham, 1986). Despite this generally accepted paradigm, it has not been clear how the myosin molecule changes its conformation, how far this conformational change will displace a cross-bridge, or how much force a cross-bridge can produce given sufficient ATP as an energy source.

Considerable progress has been made since the mid-1980s to address these questions by elucidating the molecular structures and mechanisms underlying all forms of motility in living organisms. In particular, this has been facilitated in striated muscle by two recent technical advances: (1) the determination of the 3-D molecular structures of the actomyosin complex and other related molecular motors, and (2) the ability to measure forces and displacements from the interaction of single cross-bridges instead of ensembles of cross-bridges working asynchronously. Although primarily utilizing skeletal, smooth, or microtubule-based motor proteins, data derived from studies using these two approaches are of such a fundamental nature that they generally apply to cardiac function at the cellular and organ levels. Any cardiac specific differences in motor function are discussed when applicable. In the balance of this section, these structural and functional advances are described and related to the overall cross-bridge cycle, sliding filament hypothesis, and cardiac function.

A. GENERAL FEATURES OF MOLECULAR MOTORS

Molecular motor proteins have been divided into the kinesin, dynein, or myosin super-families based on their molecular sequences and functional environments (Hackney, 1966; Johnson, 1985; Sweeney and Holzbaur, 1996; Vallee and Sheetz, 1996; Warrick and Spudich, 1987). Although the overall protein structures of kinesin, dynein, and myosin are tailored to their function, all three of these families have the common dual features of being molecular force generators and ATPases. Despite differences in their gross structures and chemomechanical transduction cycles, the most critical aspects of the actual molecular force-producing mechanisms and structures are amazingly similar (Cope *et al.*, 1996; Hackney, 1996; Howard, 1995; Johnson, 1985; Kull *et al.*, 1996; Rayment, 1996a; Romberg and Vale, 1993; Sellers, 1996; Warrick and Spudich, 1987). A complete discussion of the structure and function of all the distinct forms of molecular motor proteins [which may approach 100 (Spudich, 1994)] is beyond the scope of this chapter. Thus, after a brief description of kinesin and dynein, this section concentrates on molecular structure and mechanical function of myosin II and its interaction with actin.

1. Kinesin and Dynein

Both the kinesin and dynein motor protein families are involved in microtubule-based intracellular transport, chromosome movement, ciliary and flagellar motion, and some forms of cell motility. Kinesin is a two-headed heterotetramer formed from two heavy chains (each 120 kDa) and two light chains (each 62 kDa). Although it has little sequence homology, kinesin's 3-D structure resembles a shortened myosin molecule with two heads and a tail (see Fig. 6). Its mechanism of interaction with tubulin is now fairly well understood as it produces movement of substrates toward the plus (growing) end on microtubules (Hoenger *et al.*, 1995; Kikkawa *et al.*, 1995; Sweeney and Holzbaur, 1996). Dynein is a single-, double-, or triple-headed motor protein of variable size ($\sim$500–1900 kDa) whose various structural forms and mechanisms of action are under intense investigation (Johnson, 1985; Sweeney and Holzbaur, 1996; Vallee and Sheetz, 1996; Z. Wang *et al.*, 1995). Dynein appears to be involved in retrograde movement toward the minus end on microtubules and organelles. In heart, both kinesin and dynein are involved in microtubule's role in cell proliferation during fetal growth and development and with hypertrophic remodeling (see Chapter 2).

2. Myosin

The myosin super-family has at least 11 distinct forms of varying sizes, structures, and functions (Cheney *et al.*, 1993; Rayment, 1996b; Sweeney and Holzbaur, 1996; Warrick and Spudich, 1987). Myosin II (hereafter referred to as myosin) is the form associated with muscle motility and has numerous isoforms of its own (McNally *et al.*, 1989). In striated muscle, myosin aggregates to form the bipolar thick filament (see Fig. 3, Section II).

Myosin is derived from three gene products, which code for a heavy chain (200 kDa each) and two light chains (20 kDa each). The hexameric structure of each muscle myosin molecule consists of two copies of each chain as shown in Fig. 6. The heavy chain forms a long α-helical rod segment and two globular S1 segments. The ~150-nm long rod is subdivided into two segments: (1) the 90–100-nm long light meromyosin (LMM) section, which forms the thick filament backbone; and (2) the 50–60-nm long S2 rod segment. The S2 segment is probably hinged away from the thick filament to permit a greater mobility of the S1 head to attach to actin during the cross-bridge cycle (see Fig. 9, Section IV). Indeed, Margossian *et al.* (1991) observed that filament sliding ceased when antibodies were bound to the LMM-S2 hinge region to restrict its mobility. Although the S2 is not the force-generating element of myosin, it appears to act as a damped spring like element transmitting S1-generated displacements to the LMM of the thick filament under isotonic conditions or storing S1-generated forces under isometric conditions (Ford *et al.*, 1977; A.F. Huxley, 1988, 1995; A.F. Huxley and Simmons, 1971). C-protein, which binds to both LMM and the S2 near this hinge region, is involved in the regulation of cross-bridge function, as also described in Section IV (Gautel *et al.*, 1995, Hofmann and Lange, 1994; Weisberg and Winegrad, 1996).

The 17–20-nm long S1 segment of the heavy chain consists of a slender neck region and globular head motor domain (see Fig. 6). One essential light chain (ELC or LC1 or "a" light chain) and one regulatory light chain (RLC or LC2 or "p" light chain) are bound to each neck segment of the

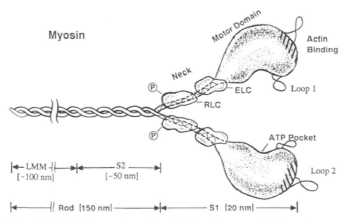

FIGURE 6   Model of the myosin molecule. The entire myosin II structure consists of the long α-helical rod segment and two globular head S1 segments. The rod portion of the molecule is divided into the light meromyosin (LMM) and S2 segments. Each S1 segment contains the motor domain and a neck region that binds the regulatory (RLC) and essential (ELC) light chains. Each motor domain of the S1 has an ATP enzymatic pocket (cleft), two loops, and an actin binding site. See text for details. (Adapted and reproduced with permission from Warshaw, 1996.)

myosin heads. As detailed in Section IV, the RLC and ELC light chains are essential for full force production in striated muscle, and they regulate cardiac contractility (Levine *et al.*, 1996; Lowey and Trybus, 1995; Lowey *et al.*, 1993a, 1993b; Morano and Ruegg, 1986; Morano *et al.*, 1995a; Opie, 1995; Patel *et al.*, 1996; Sweeney *et al.*, 1993; VanBuren *et al.*, 1994). The globular head region of each S1 (the motor domain) contains actin binding and ATP hydrolytic sites along with two surface loops. These two loops (see Fig. 6) are not well conserved between myosin isoforms and appear to be involved in the regulation of ATPase kinetics (loop 1) and motor velocity (loop 2) (Bobkov *et al.*, 1996; Spudich, 1994).

Cardiac, skeletal, and smooth muscle myosin each have their own distinct isoforms with characteristic actomyosin ATPase rates in solution that correlate with their maximal velocity of shortening (Marston and Taylor, 1980; Yamashita *et al.*, 1992). In cardiac myosin, there are two heavy chain isoforms designated α and β. These combine to form three different hexameric myosins called V1, V2, and V3, whose classification is based on their electrophoretic mobility, ATPase rates, and unloaded velocities of shortening (Harris *et al.*, 1994; Hoh *et al.*, 1977; McNally *et al.*, 1989; Pagani and Julian, 1984). V1 myosin, which has the fastest ATPase and shortening rates, is formed from two copies of the α-type of heavy chain isoform. The slow V3 myosin has two copies of the β-type of heavy chain. Despite these functional differences, the α- and β-isoforms of the heavy chain have a 93% sequence homology (McNally *et al.*, 1989). The V2 myosin has intermediate function and consists of one α heavy chain and one β heavy chain. The relative proportions of these V1, V2, and V3 cardiac myosins vary between species with, for example, normal adult rabbit heart containing about 70% of V3 myosin, whereas normal rat heart has an age varying percentage ranging from about 60 to 100% of V1 myosin (Hoh *et al.*, 1977). Furthermore, cardiac hypertrophy and its associated dysfunction are often associated with isoform changes or mutations in the myosin heavy chain. For example, the high proportion of the fast V1 myosin normally expressed in rodent models is replaced by the slower V3 myosin when the heart responds to a hypertrophic stimulus (Dool *et al.*, 1995; Hoh *et al.*, 1977; Tanamura *et al.*, 1993). Thus, these cardiac heavy chain myosin isoforms are particularly important in the regulation of contractile function between different species and in the adaptive response to cardiomyopathies (Cuda *et al.*, 1993; Lankford *et al.*, 1995; Morano *et al.*, 1995b; Morkin, 1993; Yamashita *et al.*, 1992). Similar alterations in the structure and function of the RLC and ELC light chains are discussed in Section IV.

B. MYOSIN MOTOR STRUCTURE

Muscle myosins from many sources have been fully sequenced for some time, but their primary molecular structure gives little insight into the actual chemomechanical transduction mechanism (Cheney *et al.*, 1993; A.F.

Huxley, 1988; McNally *et al.*, 1989; Warrick and Spudich, 1987). Virtually all these efforts have utilized frog, chicken, or rabbit skeletal muscle preparations as the basis for these structural and mechanical studies. Efforts to identify the molecular site of force development and estimate the myosin step size (the working stroke or displacement of the cross-bridge) have, until recently, depended on both structural and mechanical evidence from skeletal muscle fibers (Burton, 1992; Cooke, 1995; A.F. Huxley, 1995; H.E. Huxley, 1996; Squire *et al.*, 1994). Classical analysis of frog skeletal muscle single fibers suggested that the cross-bridge working stroke cross-bridge displacement was on the order of 14 nm/half sarcomere during maximal contraction (Ford *et al.*, 1977; Huxley and Simmons, 1971). However, these data are confounded by variations in individual myofilament lengths (Sosa *et al.*, 1994) and an unknown degree of myofilament compliance during contraction (Higuchi *et al.*, 1995; Mijailovich *et al.*, 1996).

Regardless of these considerations, the gross molecular shape of the S1 myosin head and neck derived from micrographic and diffraction data appear to conform to the mechanical data from fibers. Assuming that a rigid 17–20-nm long S1 myosin head and neck can swivel about the S2 segment, a maximal working stroke of 40 nm could be achieved with a full 180° rotation (Fig. 6). Although there has been considerable controversy on this issue (Burton, 1992; Higuchi and Goldman, 1995; Warshaw, 1996), the real cross-bridge displacement under low-velocity conditions is likely less and closer to the mechanical data estimates. This is due to S2 elasticity, the physical constraints preventing full head rotation within the myofilament environment, and the molecular location of the conformational change somewhere within the S1 (see Fig. 7 insert).

The recent publication of the S1 3-D structure from chicken skeletal muscle myosin has greatly advanced our understanding of the molecular mechanism of contraction (Rayment *et al.*, 1993a). When combined with the Å level molecular structure of F-actin (Holmes *et al.*, 1990) in an unregulated (no troponin–tropomyosin) thin filament, the entire actomyosin cross-bridge has been modeled (Rayment *et al.*, 1993b). This 3-D model of the S1 segment's 850 amino acids has fostered additional studies permitting the localization of molecular sites associated with the protein's light chains, ATPase activity, and actin binding, which are all critical for motor function. Furthermore, these structural sites and their orientations during cross-bridge cycling suggest possible regions in the S1 domain for the conformational changes necessary to elicit motor function (Banos *et al.*, 1996; Holmes *et al.*, 1990, 1995; Kinose *et al.*, 1996; Park *et al.*, 1996; Rayment, 1996b; Rayment and Holden, 1994; Rayment *et al.*, 1996; Ruppel *et al.*, 1994; Spudich, 1994).

Figure 7 is a model of the actomyosin cross-bridge cycle that will be described in more detail later in this section. The insert in Fig. 7 (lower right corner) shows three conformational configurations of the S1 myosin head hypothesized to occur during this cycle (Ostap *et al.*, 1995). These

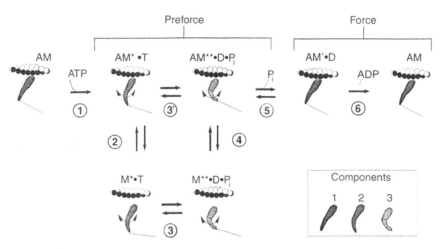

FIGURE 7 Cross-bridge cycle and myosin conformations. A six-step model of the actomyosin cross-bridge cycle is illustrated. The inset shows the three conformational configurations of the myosin S1 corresponding to an ordered rigor-like (1) and two disordered, nonforce-producing (2 and 3) states. See text and reference for details. (Adapted and reproduced with permission from Ostap *et al.*, 1995.)

conformational shapes are based on both structural and optical data. Specifically, the structural data have indicated that the actin binding site is near the end of the S1, and the ATP binding pocket (or enzymatic cleft) is located about 5–6 nm away (see Fig. 6). Optical studies have shown that the actin binding site does not rotate (see Fig. 7) once the myosin is bound (Cooke, 1995; Ostap *et al.*, 1995; Zhao *et al.*, 1995). Thus, Rayment (Rayment *et al.*, 1993b; Rayment and Holden, 1994) proposed that the enzymatic cleft was a logical site for a conformational change in S1 because it is the location of the cyclic ATP hydrolysis and phosphate release. Also, this site is the farthest away from the S1–S2 junction, which would leave 11–12 nm of the S1 neck region available as a lever arm for displacement (see Fig. 9, Section IV). However, this does not appear to be the case. Other optical studies have indicated that there is no orientation change in probes labeling the ATP pocket portion of the S1 during the cross-bridge cycle (Cooke, 1995; Ostap *et al.*, 1995; Zhao *et al.*, 1995). Thus, any conformational change or changes must occur elsewhere between the enzymatic site and the S1–S2 junction. Finally, it is unlikely that a conformational change in any one location could change sufficiently (20–30°) to achieve the cross-bridge lever displacement suggested by the mechanical data due to the high energetic cost (Cooke, 1995; Ford *et al.*, 1977).

More recent optical and structural data suggest S1 locations other than the ATP pocket and acting binding site might be involved in the step motion of the myosin motor. Park *et al.* (1996) have reported the closure during hydrolysis of another cleft region located opposite the ATP pocket.

Kinose *et al.* (1996) have also suggested a nearby region is essential for motor activity. In mutational manipulation of *Dictyostelium* myosin, Ruppel and Spudich (1995) sequentially reduced the length of the S1 heavy chain neck region in two segments to effectively reduce the lever arm length (see Fig. 6). Starting at the S1–S2 junction, they deleted the first 4 nm or so of the myosin neck along with the RLC light chain and observed a 50% reduction in shortening velocities using the *in vitro* motility assay method (see following discussion). A second 4-nm deletion along with the ELC further reduced the velocity to only 10% of normal. Thus, a majority of the lever arm length and rotation reside in the S1 neck region and not near the actin binding site. Other studies using optical probes (fluorescent polarization and EPR spectroscopy) or flash photolysis have suggested that conformational changes also occur further from the actin binding site at or near the RLC associated neck region (Allen *et al.*, 1996; Irving *et al.*, 1995; Ling *et al.*, 1996; Ostap *et al.*, 1995). The involvement of the RLC portion of the S1 neck in the motor's conformational change also makes sense in that these light chains are known to modify cross-bridge function (Lowey *et al.*, 1993a, 1993b; VanBuren *et al.*, 1994). These data strongly suggest that the light chain-associated neck region is also involved in cross-bridge motion as much or more than the enzymatic region (see Fig. 6). Finally, Banos *et al.* (1996) have modeled the actomyosin interaction and suggest that small changes in multiple sites are sufficient to elicit motor function consistent with the mechanical data. As a whole, these data from skeletal muscle suggest that there are multiple sites or a continuum of bending in the S1 between the RLC and the ATP pocket sufficient to permit adequate displacement of the myosin head on each cross-bridge cycle (see Fig. 7).

Finally, in cardiac muscle myosins, the sequence differences between α and β heavy chain myosin isoforms and the mutations identified on the β heavy chain in hypertrophic cardiomyopathy strongly suggest that they are responsible for alterations in motor function typical of this group of diseases. Most of the 7% sequence differences between the α and β heavy chain isoforms cluster within the critical actin binding. ATP binding pocket, and light chain-associated neck domains of the molecule (McNally *et al.*, 1989; VanBuren *et al.*, 1995). For example, the two surface loops shown in Fig. 6 contain many of these sequence alterations and are involved in the regulation of ATPase kinetics and motor velocity as previously described (Bobkov *et al.*, 1996; Spudich, 1994). Similarly, a variety of single-point mutations found in human hypertrophic cardiomyopathy were associated with these three critical domains, but they were not equally dysfunctional (Fananapazir and Epstein, 1994; Lankford *et al.*, 1995; Rayment *et al.*, 1995). Mutations closely correlated with sudden death in humans were found to be located in the actin binding and essential light chain neck regions, and these severely limit motor function. A relatively less severe mutation was found to be located in the enzymatic ATP pocket site and had no impact on isolated motor function. Thus, the molecular level structural data with subsequent optical and pathophysiological assessment experiments have identified some of the crucial binding, enzymatic, and potential conformational change sites involved in contractile function and dysfunction.

C. ASSESSMENT OF ACTOMYOSIN MECHANICAL FUNCTION

Parallel with the determination of the actomyosin crystal structure, the *in vitro* motility assay was developed to more directly assess myofilament and cross-bridge function. Motility assays can be classified into three related methods illustrated in Fig. 8, which have provided measures of myofilament velocities (V_{max}), cross-bridge force (F_{avg}, F_{uni}), and myosin step size (d_{uni}). As opposed to mechanical measurements from muscle fibers or cells that evaluate system function from an ensemble of molecules working asynchronously (see Section V), these new approaches aim to reveal the function of the myofilaments and individual molecules. These molecular function data can then be correlated with the structural data to provide a better understanding of cross-bridge function.

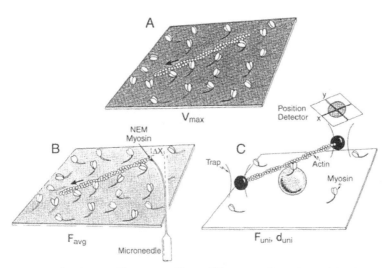

FIGURE 8 *In vitro* motility assays. Three different configurations of *in vitro* motility assays are illustrated. Panel A diagrams the basic assay system where myosin is plated on a nitrocellulose-coated coverslip upon which movement of fluorescently labeled actin filaments can be observed under various conditions to determine their maximal velocities of motion V_{max}. Panel B shows a variant of this basic assay where a microneedle is attached to the end of an actin filament to directly assay force (F_{avg}). Panel C shows an optical trap motility assay system where a few or single myosins are placed on a bead. By holding both ends of the actin filament in a laser trap, both unitary force (F_{uni}) and displacement (d_{uni}) can be estimated. See text and reference for details. (Reproduced with permission from Warshaw, 1996.)

1. *In Vitro* Motility Assay

The basic motility assay was developed first with a *Nitella*-based moving actin cable system (Sheetz and Spudich, 1983) and later with a simplified myosin-coated surface assay (Kron and Spudich, 1986; Yanagida *et al.*, 1984) as shown in Fig. 8A. Here, highly purified myosin is plated on nitrocellulose-coated cover slips, upon which fluorescently labeled F-actin filaments are allowed to settle. With the addition of sufficient levels of ATP, these unregulated actin filaments move unidirectionally, and their velocities (V_{max}) can be determined by video microscopic analysis (Homsher *et al.*, 1992; Warrick *et al.*, 1993). Under normal motility assay conditions, the external load, the number of cross-bridges cycling at any one time, and the percentage of the cycle time that they are firmly attached to the actin filament (the duty cycle) are unknown. However, the basic motility assay system permits comparative measurements of maximal unloaded filament velocities (V_{max}) between different myosin substrates under controlled conditions.

In early work, it was quickly determined that only the 17-nm long S1 portion of the myosin molecule and ATP were necessary to demonstrate motility (Toyoshima *et al.*, 1987). Although it was observed that the actin fiber velocity was less with the S1 as compared to more full length versions of myosin, these data directly demonstrated that the S1 contains all the regions necessary for enzymatic activity and force generation. Subsequent work using fast skeletal muscle myosin showed that the actin filaments moved consistently in the 5–6-μm/s range as long as the amounts of plated myosin and ATP were adequate (Sellers and Kachar, 1990; Sellers *et al.*, 1993; Warrick *et al.*, 1993). In comparison, the unloaded velocity with rabbit cardiac V1 myosin measured with a *Nitella*-based system was about 0.6 μm/s, one-tenth the skeletal value (Yamashita *et al.*, 1992). These filament velocities were found to be consistent with unloaded fiber shortening data from skeletal muscle under the same biochemical and mechanical conditions (Homsher *et al.*, 1992).

Because the force–velocity relationship is a key feature of striated muscle mechanics (Brady, 1974, 1984; A.F. Huxley, 1988, 1995), considerable effort has been made to correlate function at the cross-bridge level to that of whole muscle. The motility assay best evaluates unloaded velocities of unregulated actin filaments on plated myosin. Therefore, it has been extensively used to compare the mammalian cardiac (V1 and V3), skeletal, and smooth muscle myosins at various ATP concentrations. These data have clearly shown that measured differences in a muscle mechanical motility can be largely attributed to their myosin isoform and correlated to their ATPase activity (Harris *et al.*, 1994; Sata *et al.*, 1993; Yamashita *et al.*, 1992). Specifically, Yamashita *et al.* (1992) observed that rabbit cardiac V1 myosin had faster filament velocities and ATPase activities than rabbit cardiac V3 myosin. Sata *et al.* (1993) varied the relative proportions of the V1 and V3 myosins and reported nonlinear ATPase and velocity responses. Harris *et al.* (1994) observed no differences between the velocities of actin filaments derived from cardiac and skeletal sources when assayed with the same myosin isoform. As a whole, these studies suggest that it is the myosin, not the actin, isoforms that determine filament sliding velocity. Finally, the average force per cross-bridge (F_{avg}) was estimated from a mathematical model to be the greatest in phosphorylated smooth muscle myosin followed by progressively lower levels in skeletal, V3 cardiac, and V1 cardiac myosins (Harris *et al.*, 1994). The combination of these motility data suggest that rabbit cardiac V1 myosin appears to be a faster, but less forceful, isoform than the V3 myosin. This is consistent with force–velocity curves derived from whole muscle preparations.

To obtain data over a wider range of the force–velocity relationship, several attempts have been made to alter the basic unloaded system. For example, when the amount of plated skeletal muscle myosin is reduced to very low levels, filament sliding ceases. Uyeda *et al.* (1991) determined the threshold plating and ATP levels above which unloaded filament sliding occurs and observed that the motion appeared as series of unitary events. From these very low velocity data, they estimated that the myosin step size (d_{uni}) was on the order of 5–20 nm/ATP hydrolyzed with no external load. In another study, the force of a single kinesin molecule (F_{uni}) was estimated to be 4.2 pN by externally loading a fully plated system with a viscous solution until sliding stalled (Hunt *et al.*, 1994). Similar force estimates have yet to be made with a myosin-based system with the basic motility assay. These data have expanded the range of the basic motility assay to permit more than single-point comparisons between myosin isoforms along the force–velocity curve.

Other motility assay data have evaluated the regulatory roles of the thick and thin filaments (see Section IV) or the functional changes associated with myosin mutations. Ruppel and Spudich (1995) used the motility assay to characterize the effect of shortening the neck region of the S1 as previously described. Sata *et al.* (1995) measured the sliding characteristics of reconstituted thin filaments on cardiac myosin and evaluated their calcium sensitivity. Cuda *et al.* (1993) and Lankford *et al.* (1995) demonstrated that the single-point mutations on the β-myosin heavy chain associated with hypertrophic cardiomyopathy are directly responsible for significant reductions in actin filament sliding velocities as compared with normal. Although sensitive to multiple experimental factors (Homsher *et al.*, 1992), the basic *in vitro* motility assay technique can provide a reasonable index of muscle function, which is particularly useful for the comparison among species, myosin isoforms, myosin mutations, and the biochemical environment (Cuda *et al.*, 1993; Harris *et al.*, 1994; Homsher *et al.*, 1996; Lankford *et al.*, 1995; Lin *et al.*, 1996; Ruppel *et al.*, 1994; Sata *et al.*, 1993, 1995; Spudich, 1994; Warshaw, 1996; Yamashita *et al.*, 1992).

2. Microneedle Force Assay

Kishino and Yanagida (1988) modified this basic motility assay system by sticking glass microneedles on the ends of the actin filaments (Fig. 8B). By measuring the displacement of a calibrated needle pulled by the filament on the assay plate, forces (F_{avg}) can be estimated under various experimental conditions (Ishijima et al., 1996; VanBuren et al., 1995; Yanagida and Ishijima, 1995). This approach thus provides direct information about force generation under high load, no (or low) velocity conditions. But the absolute values for the force per cross-bridge (F_{uni}) or the myosin step size (d_{uni}) depend on estimating the actual number of cross-bridges attached at any one time. This introduced an unknown amount of error, especially in some of the early measurements, leading to widely varying results (Burton, 1992; Higuchi and Goldman, 1995; Warshaw, 1996). Despite this problem, the microneedle assay can generate force–velocity data over a wider range of values.

Using the microneedle method, VanBuren et al. (1995) confirmed their previous estimates from rabbit heart with the basic motility assay (Harris et al., 1994) and estimated that the V1 form of cardiac myosin produces only one-half the cross-bridge force of the V3 form while demonstrating three times the velocity. Ishijima et al. (1996) reduced the myosin concentration to very low levels and oriented the filaments attached to a microneedle in the same direction as the myosin heads (as in the sarcomere) to achieve maximal mechanical function. Using perturbation and noise analysis, they reported single skeletal actomyosin force (F_{uni}) to be in the 5–6-pN range for about 50 ms at high loads and its displacement (d_{uni}) at about 20 nm at near zero load.

3. Optical Trapping

The third motility assay method (Fig. 8C) utilizes the trapping power of high numerical aperture laser light to hold or move polystyrene beads, which are in turn attached to actin filaments or coated with a molecular motor (Block, 1990). The laser beam can be moved about the chamber to position the beads, and its trapping force can be modulated with its intensity. This method was first applied to single beads coated with kinesin motors whose displacement was measured using a interferometer (Block, 1990; Svoboda et al., 1993). They observed about an 8-nm step size (d_{uni}) for single kinesin molecules, which are smaller in molecular size than myosin.

Optical trapping has now been applied to the direct measurement of unitary force (F_{uni}) and steps (d_{uni}) from single actomysin interactions using dual beam traps to control both ends of an actin filament (Finer et al., 1994; Miyata et al., 1994; Molloy et al., 1995; Simmons et al., 1996). As shown in Fig. 8C, an actin filament is trapped at each end and placed over the myosin-plated bead. Finer et al. (1994) pioneered this approach for myosin-based systems and reported an average step size of 11 nm and a peak force of

3–4 pN when the velocity approached zero. Miyata *et al.* (1994) observed 7-nm steps at high trapping forces and up to 30-nm steps at low trapping forces. Using a similar system with a different noise analysis, Molloy *et al.* (1995) reported a smaller step size of 4 nm and 1.7 pN of force at high trap forces. Despite these differences in absolute values, these groups confirmed that the S1 is the exclusive force producing entity in muscle. Finally, Sugiura *et al.* (1996) observed no differences in the force-generating capacity between the cardiac V1 and V3 myosin isoforms of rat heart as measured from a trap system. They attribute the discrepancy between their results and those of Harris *et al.* (1994) and VanBuren *et al.* (1995) on experimental or species differences. But the values of Sugiura *et al* (1996) were an order of magnitude lower than those obtained with the microneedle method.

Although data from some of the earlier studies were varied (Burton, 1992; Warshaw, 1996), most recent efforts using all three of these motility assay methods provide relatively consistent molecular-based estimates of 1.7–6-pN levels of maximal unitary force at zero velocity (F_{uni}) and 4–20-nm displacements (d_{uni}) in the actomyosin motors. Table 1 summarizes some of the large number of force and step size values reported in the literature using a variety of methods and protocols. In general, these data are within an order of magnitude of estimates based on muscle fiber experiments. Note that the step size displacement values can be quite variable depending on the load. Higuchi and Goldman (1995) have reported up

TABLE 1   Motility Assay Data

Motor type	Method	Step size (nm) (d_{uni})	Force/CB (pN) (F_{uni})	Source
Skeletal	Fiber mechanics	14	—[a]	Huxley and Simmons, 1971; Ford *et al.*, 1977
Skeletal	Basic assay	5–20	—	Uyeda *et al.*, 1991
Kinesin	Basic assay	—	4.2 (max)	Hunt *et al.*, 1994
Skeletal	Microneedle	—	0.2 (min)	Kishino and Yanagida, 1988
Skeletal	Microneedle	20	5–6 (max)	Ishijima *et al.*, 1996
Cardiac	Microneedle	—	0.15 (V1) 0.3 (V3)	VanBuren *et al.*, 1995
Kinesin	Trap	8	—	Svoboda *et al.*, 1993
Skeletal	Trap	11	3–4 (max)	Finer *et al.*, 1994
Skeletal	Trap	7	—	Miyata *et al.*, 1994
Skeletal	Trap	4	1.7 (max)	Molloy *et al.*, 1995
Cardiac	Trap	—	0.03 (avg)	Sugiura *et al.*, 1996

[a]—, No data obtained.

to 190 nm displacements per ATP utilized at high velocities in fibers. Displacements exceeding 20 nm have also been reported in motility assay data under unloaded, high-velocity conditions (Burton, 1992; Ishijima *et al.*, 1996; Miyata *et al.*, 1994; Yanagida and Ishijima, 1995). These motility assay approaches are still under development (Simmons *et al.*, 1996) and will undoubtedly provide more controlled and precise values in the near future.

D. ACTOMYOSIN CROSS-BRIDGE CYCLE

When combined with the large amount of biochemical and mechanical data from muscle regarding cross-bridge kinetics (Brenner and Eisenberg, 1987; Brenner *et al.*, 1995; Dantzig *et al.*, 1992; Eisenberg *et al.*, 1980; Ford *et al.*, 1977; Higuchi and Goldman, 1995; Homsher and Millar, 1990; A.F. Huxley, 1988, 1995; A.F. Huxley and Simmons, 1971; Kawai *et al.*, 1993; Lymn and Taylor, 1971; Millar and Homsher, 1992; Regnier *et al.*, 1995; Smith and Geeves, 1995), these new molecular-based structural and functional data permit an improved, but still incomplete, understanding of the cross-bridge mechanism. Solution biochemistry laid the kinetic basis for the actomyosin interaction with a simple four-step process proposed initially by Lymn and Taylor (1971). Reconciliation with the new structural data and other optical or mechanical data from transient length perturbation analyses and flash photolysis of caged compounds (see Section V) on glycerinated rabbit psoas skeletal muscle fibers has generated a number of schemes with at least six steps (Brenner and Eisenberg, 1987; Brenner *et al.*, 1995; Dantzig *et al.*, 1992; Geeves and Conibear, 1995; Homsher and Millar, 1990; Kawai *et al.*, 1993; Millar and Homsher, 1992; Regnier *et al.*, 1995). These additional steps are due to the identification of at least three basic binding states for the actomyosin complex, which are commonly called the strongly bound, weakly bound, and unbound cross-bridge states. Figure 7 illustrates the generally accepted simplified model of the cross-bridge cycle combined with an illustrated interpretation of the skeletal myosin S1 conformation based on the molecular orientation of electron spin resonance labels (Ostap *et al.*, 1995). This cycle is always assumed to be operational (unregulated) with sufficient ATP and actomyosin substrates to permit function. The regulation of this cycle is discussed in Section IV. Although a comprehensive discussion of the basis for each step in this unregulated cross-bridge cycle for striated muscle is beyond the scope of this chapter, a brief sketch of the current configuration based on skeletal muscle studies completes this section.

Starting at the upper left corner of the cross-bridge cycle in Fig. 7, the myosin S1 is believed to be in a relatively straight and unvarying molecular configuration diagrammed as component 1. It is strongly bound to actin in a rigor-like "AM" state. The bound S1 myosin is rotated about 45° relative to the thin filament straining the S2 segment, thereby transferring force to the thick filament. The myosin rod S2 segment, illustrated here as a line at a right angle to the myosin S1, is believed to have some elasticity to permit the storage of energy as in a spring. Following ATP binding (step ①) at the enzymatic nucleotide binding site, the myosin S1 changes conformation to a slightly bent and more disordered configuration (component 2) whose neck regions can rotate with time (arrows). Remember that the actin binding site on myosin S1 does not rotate when bound. The nucleotide binding in this AM*•T state not only changes the overall S1 shape, but also permits the S1 to release itself from actin at its binding site (step ②), thus relieving any strain on the S2 segment. ATP hydrolysis of the M*•T product to the M**•D•P_i product (step ③) elicits a further conformational bending in the S1 to the configuration of component 3. This conformation is also free to rotate in the neck domain (arrows) and can now rebind weakly to the same or another actin (step ④), thus resulting in the AM**•D•P_i state. At this point in the cycle, the S2 segment is not stretched and no force is being generated between the thick and thin filaments. This weakly bound preforce state is rapidly changed to a strongly bound AM'•D state (step ⑤) in a multistep, force- generating process. This weak to strong binding is accompanied by the major conformational change or "power stroke" of the cross-bridge (from 3 to 1), which straightens the myosin S1 and presumably stretches the elastic S2 segment (see Fig. 9, Section IV). If the thick and thin filaments are not constrained, they can now slide past one another to shorten the sarcomere. If the myofilaments are constrained, isometric force is generated and is likely stored in the S2. Also in association with the power stroke, phosphate is released as part of step ⑤. Evidence strongly suggests that the power stroke precedes the release of Pi (Dantzig *et al.*, 1992; Kawai *et al.*, 1993; Millar and Homsher, 1992; Smith and Geeves, 1995). Following the multistep phosphate release and power stroke, ADP is irreversibly released and the cycle returns to the rigor-like AM strong binding state (step ⑥).

Although the biochemical steps and their kinetics are well established in solution, the exact conformational changes associated with these steps shown in Fig. 7 are not fully established. As previously described, it is not entirely clear in which part or parts of the S1 the conformational changes occur and by what extent. Nor have all the steps characterizing the rapid power stroke been firmly established and reconciled with other forms of data. Furthermore, no account has been taken of any regulatory mechanisms (next section), which could alter the kinetics of the cycle. In addition, Smith and Geeves (1995) have proposed that time varying strain dependence of protein–protein interactions might alter the conditions sufficiently to require many additional steps to form a complete model applicable to a range of loads. Finally, in cardiac muscle, Kawai *et al.* (1993) suggested a similar scheme based on mechanical perturbation analysis, but with kinetic rate constants an order of magnitude slower than those derived from rabbit psoas muscle models. These data correspond to the previously described order of magnitude difference between cardiac and skeletal actomyosin velocity data obtained from motility assays. Although incomplete, this reconciliation of the biochemical, structural, and mechanical data does provide a reasonable framework for the chemomechanical transduction of ATP energy into force and motion. For a more detailed analysis of the cross-bridge cycle, see Brenner and Eisenberg (1987), Brenner et al. (1995), Dantzig et al. (1992), Eisenberg et al. (1980), Kawai et al. (1993), Lymn and Taylor (1971), Millar and Homsher (1992), Regnier et al. (1995), or Smith and Geeves (1995).

In summary, the characterization of the molecular events leading to force production and displacement in muscle has progressed substantially since the 1950s since the formation of the sliding filament hypothesis. This has been made possible combining data derived from solution biochemistry, cardiac and skeletal muscle mechanics, x-ray diffraction, optical localization, and most recently the molecular level structural and functional analyses. The molecular structure of the actomyosin complex has revealed the 3-D localizations of the regulatory, actin, and nucleotide binding sites. Combined with optical probe data, potential sites for the skeletal myosin S1 conformational changes associated with force development can be suggested. A variety of motility assays have provided a molecular level assessment of muscle function that have estimated single actomyosin maximal force in the 2–6-pN range and unloaded step size displacements of 4–20 nm. Studies evaluating various normal (V1 and V3) and mutated (cardiomyopathic) cardiac myosin heavy chain isoforms have directly correlated mechanical function of the muscle tissue to the molecular level. As a whole, these data strongly suggest that force in cardiac and skeletal muscle is generated by molecular changes in the S1 portion of the myosin molecule, which are associated with the cyclic hydrolysis of ATP and its interaction with actin to elicit myofilament sliding and, thus, sarcomere shortening.

IV. MYOFILAMENTS AS REGULATORS OF FORCE GENERATION

The previous discussion of force generation in the myofilaments has largely ignored aspects of contractile regulation by calcium, ATP, and other cytosolic constituents that are the focus of the majority of this volume. Virtually all the studies described in Section III evaluated biochemical kinetics, structures, and force development from unregulated molecular entities. For that discussion, we assumed that there was no inhibition of actomyosin binding and unlimited supplies of ATP (see Fig. 7). If that were the case, the actomyosin motors would be continuously switched on and heart would be in contracture, unable to pump blood. It is the goal of this section to now introduce the myofilament-based mechanisms that cyclically switch on and off the ensemble of cross-bridges in each sarcomere to elicit a contraction and modulate overall heart performance.

As detailed in the previous chapters in this volume, the calcium concentration surrounding the myofilaments is exquisitely varied in a cyclic manner following each membrane depolarization. This acts as an ionic switch turning "on" and "off" each contraction. Because muscle contraction requires both actin and myosin, it is not surprising that both these and other proteins bound to the thick and thin filaments are involved in the beat-to-beat and long-term regulation of force development in heart. Figure 9 summarizes the myofilament constituents that regulate the contraction in cardiac muscle. These include actin, tropomyosin, and troponin along the thin filament, and myosin and C-protein along the thick filament. Figure 9 also shows the metal ion (calcium and magnesium) and phosphorylation (P) binding sites involved in contractile regulation.

Primary contractile activation resides on the thin filament in cardiac muscle and has two regulatory mechanisms: the calcium binding troponin–tropomyosin complex and the facilitation of activation by the binding of myosin to actin itself. The initial interaction between the actin and myosin is tightly controlled by the calcium ion concentration. The ionic switch that responds to the cyclic changes in calcium concentration resides in the

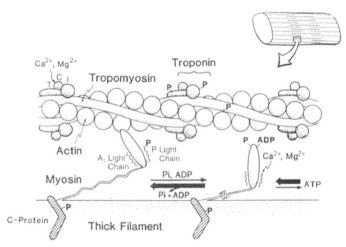

FIGURE 9 Thick and thin filament regulatory structure. The ultrastructure of the thick and thin filaments of cardiac muscle is depicted in the systolic (bound, left cross-bridge) and diastolic (unbound, right cross-bridge) states. The troponin–tropomyosin complex on the thin filament has regulatory sites that bind calcium or magnesium (Ca^{2+}, Mg^{2+}) and other sites that can be phosphorylated (P). Myosin light chains and C-protein on the thick filament also have phosphorylation sites (P) that regulate contractile function. (Reproduced with permission from Solaro *et al.*, 1993.)

troponin–tropomyosin complex on the thin filament (Chalovich, 1992; Fuchs, 1995; Solaro and Van Eyk, 1996; Tobacman, 1996; Zot and Potter, 1987). When the calcium concentration is low ($<1 \mu M$), cross-bridge cycling is allosterically inhibited by the physical blocking of the actomyosin binding sites with tropomyosin. When the calcium concentration rises into the micromolar range, it binds to the troponin C (TnC) subunit in a highly cooperative manner eliciting a cascade of conformational changes in the troponin–tropomyosin complex. These changes expose the myosin binding site on actin sufficiently to permit strongly bound cross-bridge formation and force generation. Full force development actually requires a second thin filament-based mechanism in which additional calcium binding cooperativity is conferred on the TnC by actomyosin binding of the cross-bridge itself (Butters et al., 1993; Chalovich, 1992; Fraser and Marston, 1995; Fuchs, 1995; Hofmann and Fuchs, 1988; Holmes, 1995; Lehrer, 1994; Metzger, 1995; Solaro and Van Eyk, 1996; Swartz et al., 1996; Tobacman, 1996).

These two mechanisms work together in striated muscle to cooperatively regulate contraction as characterized by the steep sigmoid force–pCa relationship (Fig. 10) generated from a skinned myocyte isolated from rat heart; pCa is the negative log of the calcium concentration. The shape, position, and steepness of this relationship indicate the overall calcium sensitivity and degree of cooperativity of the system. Force–pCa data, such as those in Fig. 10, are easily fit by the classical Hill equation (Hill, 1910) to obtain numerical indices of sensitivity and cooperativity. Myofilament calcium sensitivity is characterized by the horizontal position of the curve as indi-

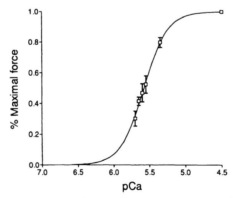

FIGURE 10   Force–pCa relationship. Force data obtained from eight skinned WKY rat cardiac myocytes activated with various levels of calcium was averaged and fitted with the Hill equation (Hill, 1910). These data give a sigmoid relationship between the normalized force and the inverse log of the calcium concentration (pCa). The average maximal force developed from these cells at pCa = 4.5 ($32 \mu M$) was 22.3 mN/mm^2. After curve fitting, the Hill coefficient (n_H) = 2.62 ± 0.14 and the pCa_{50} = 5.58 ± 0.01. See text for details. (Original data courtesy of Dr. Roy E. Palmer.)

cated by the pCa_{50} value; pCa_{50} is the value at which the preparation develops 50% maximal force. If the pCa_{50} increases to a higher numerical value (curve shifts to the left), the myofilaments are more sensitive to calcium. If the pCa_{50} decreases to a lower value (curve shifts to the right), the myofilaments are less sensitive to calcium. The degree of cooperativity is characterized by the Hill coefficient (n_H) where a greater value indicates increased cooperativity (Hill, 1910). In Fig. 10, the pCa_{50} is 5.58 (2.6 μM) and the n_H is 2.62, which represent typical values for cardiac muscle under these experimental conditions. This terminology is used throughout the remainder of this chapter in reference to contractile function and its regulation.

Under normal conditions, the myofilaments in heart are bathed in a free calcium concentration of a few μM (pCa 5.5–5.8) at the peak of systole, which is only part way up the force–pCa curve shown in Fig. 10. However, cardiac muscle has the ability to modulate the force of contraction via several mechanisms associated with the concept of contractility (Opie, 1995; Solaro et al., 1993). This is quite different than the mechanism in skeletal muscle where the developed force is largely graded by the differential recruitment of motor units under direct neural control. In the heart, the pressure developed in a ventricular contraction is generally modulated either by altering myofilament calcium sensitivity, the myosin ATPase rate, or the myofilament-mediated mechanisms associated with end diastolic volume (preload). These myofilament-based regulatory mechanisms reside on both the thick and thin filaments in the sarcomere and operate on a beat-to-beat and longer term basis.

On a beat-to-beat basis, the sarcomere length (preload)-dependent calcium sensitivity is a critical factor in regulating force along the ascending limb of the Starling relationship where the heart normally functions (Allen and Kentish, 1985, 1988; Fuchs, 1995; Gulati et al., 1991; Moss et al., 1991; Patterson and Starling, 1914). On a longer term basis, β-adrenergic stimulation leads to positive inotropic and chronotropic effects in the heart, which are mediated through the cAMP-dependent protein kinase (PKA) system. This cascade leads to increased phosphorylation of calcium channels, phospholamban on the SR, and myofibrillar proteins such as troponin I, myosin light chains, and C-protein. The membrane- or channel-associated mechanisms (Chapters 3–5, 7) can alter the amount of calcium available on each beat. In this case, higher calcium levels would produce more force by moving up the force–pCa relationship toward the plateau of maximal force generation (see Fig. 10). Alternatively, other myofilament-based regulatory mechanisms are essential for the normal long-term modulation of contractility in the heart. In this case, the force–pCa relationship would shift to the right or left to alter myofilament sensitivity to a given level of calcium. For example, the myosin ATPase rate is modulated by the phosphorylation (P) of myosin light chain and C-protein myofilament subunits (Gautel et al.,

1995; Hofmann and Lange, 1994; Levine *et al.,* 1996; Lowey and Trybus, 1995; Morano and Ruegg, 1986; Opie, 1995; Patel *et al.,* 1996; Sweeney *et al.,* 1993).

Because both these beat-to-beat and longer term regulatory processes are highly complex and interactive, their detailed description is beyond the scope of this volume. Thus, the following discussion only provides a sketch of each these regulatory mechanisms. Refer to Chalovich (1992), Fuchs (1995), Metzger (1995), Solaro and Van Eyk (1996), Tobacman (1996), and Zot and Potter (1987) for more detailed reviews of myofilament regulation. The mechanical pumping function of the heart is subsequently discussed in terms of these regulatory mechanisms in Section V.

A. THIN FILAMENT REGULATION—TROPONIN AND TROPOMYOSIN

Contraction is initiated in striated muscle by the arrival of calcium from the SR and other membrane sources (Chapters 3–5) to elicit a conformational change in the troponin–tropomyosin complex. Although all the critical components have been sequenced and many of their 3-D structures were obtained *in vitro,* the complex structural changes of this system *in vivo* have not been clearly worked out. However, the general scheme involves the highly cooperative interaction of actin, troponin, and tropomyosin subunits that have a 7 : 1 : 1 stoichiometry (see Fig. 9). Thus, there is a functional unit consisting of seven G-actins along the thin filament for each troponin and tropomyosin. The high degree of cooperativity seen in the force–pCa relationship (see Fig. 10) must be due to long-range effects of the troponin on the tropomyosin, which spans all seven actins and other significant longer range protein–protein interactions along the thin filament (Chalovich, 1992; Metzger, 1995; Solaro and Van Eyk, 1996; Tobacman, 1996; Zot and Potter, 1987). The troponin itself is divided into three separate functional subunits that bind the calcium [troponin C (TnC)], inhibit contraction when no calcium is present [troponin I (TnI)], and bind the other subunits to the tropomyosin [troponin T (TnT)]. Enlarged schematic representations of the troponin–tropomyosin complex along the thin filament are shown in longitudinal (Fig. 11) and cross-sectional (Fig. 12) views during diastole and systole. Figures 11 and 12, as well as Fig. 9, serve as a basis for the remainder of the discussion on thin filament regulation.

1. Troponin C

Troponin C (TnC) is the calcium binding protein subunit that initiates the sequence of conformational changes on the thin filament. TnC is a dumbbell-shaped 18.4 kDa (in cardiac) molecule bound to the troponin I (TnI) subunit that contains four metal ion binding sites (See Figs. 11 and 12). The N-terminus of the molecule (the N-domain) contains two sites

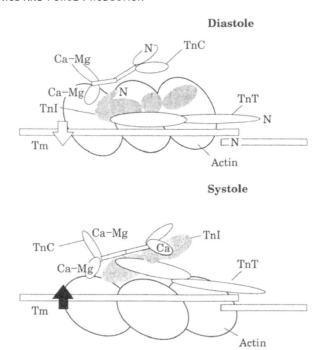

FIGURE 11   Regulatory functional unit of the thin filament. This drawing illustrates (side view) the hypothesized structure and positions of the functional regulatory units along a three actin stretch of the thin filament during diastole (top) and systole (bottom). "N" represents the N-terminal end of the tropomyosin and troponin molecules. The three active binding sites (II, III, and IV) on the cardiac troponin C (TnC) are shown with site II (N-terminal end) filled with calcium during systole. Binding site I is inactive in cardiac TnC. The conformational change in TnC is transmitted through the troponin I (shaded, TnI) and troponin T (TnT) to shift the tropomyosin (Tm) rods up and down (arrows) (see text for details). (Reproduced with permission from Solaro and Van Eyk, 1996.)

designated I and II, whereas the C-terminus has the two remaining sites, which are designated III and IV. Sites III and IV have relatively high affinities for both calcium and magnesium and, therefore, are always filled (probably with magnesium) under physiological conditions during both contraction and relaxation (Chalovich, 1992; Dong and Cheung, 1996; Solaro and Van Eyk, 1996; Tobacman, 1996; Zot and Potter, 1987). In fast skeletal TnC, both sites I and II in the N-domain are active and preferably bind calcium at a relatively low affinity. But cardiac and slow skeletal TnC differ from the fast skeletal isoform in that site I is inactive due to several amino acid substitutions in the binding area. Furthermore, mutational analysis has shown that a functional site II cannot be replaced by site I and is therefore essential for normal contractile function (Gulati *et al.*, 1992; Sweeney *et al.*, 1990). Thus, only a single calcium ion reversibly binds at low-affinity site II in each cardiac TnC to elicit contraction and relaxation.

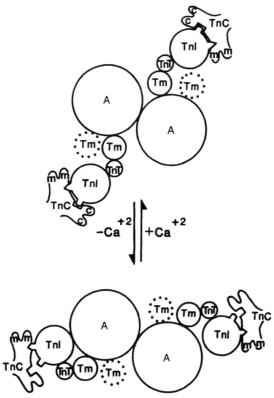

FIGURE 12   Thin filament regulation. This drawing illustrates (end view) the hypothesized structure and positions of the troponin–tropomyosin regulatory structures of the thin filament during systole (top) and diastole (bottom). Under each condition, two actin filaments (A) are shown with the tropomyosin (Tm) and troponin (TnC, TnI, TnT) complex. Comparison of the Tm positions (solid circle vs dotted circle) suggests the allosteric blocking of the actomyosin binding site. (Reprinted by permission of the publisher from J. M. Chalovich, *Pharmacol. Ther.* 55, pp. 95–148. Copyright 1992 by Elsevier Science Inc.)

This important difference is discussed later in terms of myofilament cooperativity and length-dependent calcium sensitivity.

Although 3-D structures of skeletal and cardiac TnC have not been reported with calcium bound to the low affinity sites, there is strong mutational, NMR, and optical evidence that a major conformational change opens up the N-domain (see Fig. 11) when calcium is bound to site II (Babu *et al.*, 1993; Dong and Cheung, 1996; Rao *et al.*, 1995; Tobacman, 1996). The molecular alteration in cardiac TnC may be different than that of the fast skeletal isoform because of the site I alterations and single site II calcium binding (Rao *et al.*, 1995). TnC appears to be highly conserved, and there is no evidence for isoform transitions in TnC from normal fetal to adult heart (Schiaffino et al., 1993).

2. Troponin I

Troponin I (TnI) is the protein subunit that inhibits muscle contraction in the absence of calcium. TnI is a 23.8 kDa (in cardiac) globular molecule binds to TnC, TnT, and actin (see Figs. 11 and 12). The conformational changes from TnC are transmitted to the TnI to release its inhibitory binding to actin in the on state when calcium is present (Chalovich, 1992; Tobacman, 1996; Zot and Potter, 1987). There are skeletal and cardiac isoforms of TnI that often manifest different functional characteristics. The skeletal isoform of TnI strongly inhibits actomyosin ATPase even without tropomyosin (Tm) present; the cardiac isoform's inhibition is less strong. The stability of the TnI–TnC complex is sensitive to calcium, which suggests that their interface is the likely site of signal transmission from TnC to the rest of the thin filament. Furthermore, mechanical and mutagenic analyses indicate that TnI has a regulatory role via PKA-induced phosphorylation (see Fig. 9) or decreased pH leading to reduced contractile force (Guo et al., 1994; Hofmann and Lange, 1994; Zhang et al., 1995). The expression of the cardiac TnI isoform is altered during the transition from fetal to adult development and as a result of hypertrophy (Gulati et al., 1994; Schiaffino et al., 1993).

3. Troponin T

Troponin T (TnT) is the protein subunit that binds the TnI at one site and Tm at two sites. TnT is a 38 kDa (in cardiac) rodlike molecule that spans the length of several G-actins and holds the regulatory system together (see Figs. 11 and 12). Similar to TnI, TnT can inhibit actomyosin ATPase in the presence of Tm and has multiple isoforms that can be altered during fetal development (Chalovich, 1992; Schiaffino et al., 1993; Tobacman, 1996). TnT is a key regulatory subunit whose isoforms can alter the force–pCa functional relationship (Nassar et al., 1991) and are implicated in certain forms of hypertrophic cardiomyopathy (Anderson et al., 1995; Gulati et al., 1994; Lin et al., 1996; Malhotra, 1994; Saba et al., 1996).

4. Tropomyosin

Although actin is essential as the site of force transmission and troponin as the primary site of regulation, no regulation would occur without the direct structural blocking action of the tropomyosin (Tm) (Chalovich, 1992; Holmes, 1995; Tobacman, 1996; Zot and Potter, 1987). It is a dimeric α-helical coiled coil (68 kDa) about 42 nm in length that binds to F-actin and troponin (see Figs. 11 and 12) (Hitchcock-DeGregori and Varnell, 1990). Tm rods overlap at each end, binding them together to facilitate longer range cooperative interaction as detailed later. There are two iso-forms (α and β) coded on separate genes, but the α-isoform is predominant in large mammalian adult heart (Censullo and Cheung, 1994; Malhotra, 1994; Tobacman, 1996). Smaller mammals express a mixture of isoforms, but there appears to be no significant functional difference. During diastole, Tm filaments are held away from the groove of the F-actin helix while they lie close to the groove during systole (see Figs. 11 and 12). Sequential Tm filaments overlap to form a continuous strip along the entire length of the thin filament. It appears that Tm has considerable transverse but little longitudinal flexibility (Censullo and Cheung, 1994). It is likely that this transverse flexibility along with tropomyosin's binding kinetics are key factors in the nature of its regulatory blocking role along the thin filament.

As with the myosin isoforms, pressure overload hypertrophy appears to trigger an α- to β-isoform shift in Tm in rat heart (Malhotra, 1994). Because the β-isoform is also expressed during development (Schiaffino *et al.*, 1993), this is another example of the fetal program being expressed with hypertro-phy. In addition, single-point mutations in α-tropomyosin have been directly linked to human hypertrophic cardiomyopathy (Watkins *et al.*, 1995). These pathophysiological alterations further emphasize the critical functional im-portance of Tm in heart function.

5. Regulation by the Complex

X-ray and optical diffraction studies (Amos, 1985; H.E. Huxley, 1996; Squire, 1981) clearly indicate that Tm shifts laterally out of the F-actin helical groove during contraction without any gross change in actin ordering (Fig. 12). Although there is some lateral flexibility of the Tm filaments (Censullo and Cheung, 1994), it is generally believed that the modulation of calcium on TnC triggers a cascade of conformational events in the Tn subunits that shift the Tm rods closer to or away from the actomyosin binding site (see Figs. 11 and 12). When calcium binds to TnC, all the interactions among the three Tn subunits become stronger, whereas those to Tm and actin become weaker. Thus, the Tm rod moves toward its preferred actin binding site near the thin filament groove. This pulls the unbound TnI away from the actin, exposing the actomyosin binding site (Chalovich, 1992; Fuchs, 1995; Tobacman, 1996; Zot and Potter, 1987). Although the precise conformational changes of each unit are not fully worked out, this general scheme permits the allosteric blocking and un-blocking of the actomyosin binding site as a function of calcium concentra-tion. Thus, the troponin–tropomyosin complex provides a direct mechanism to initiate force development following membrane depolarization.

B. THIN FILAMENT REGULATION—ACTOMYOSIN

Numerous biochemical and mechanical studies have determined that the degree of cooperativity conferred by the troponin–tropomyosin complex is insufficient to explain all of the observed cross-bridge force development (Chalovich, 1992; Fuchs, 1995; Holmes, 1995; Lehrer, 1994; Metzger, 1995; Solaro and Van Eyk, 1996; Swartz *et al.*, 1996; Tobacman, 1996). Specifically, steady-state force measurements from skinned muscle in a range of calcium concentrations give a very steep force–pCa relationship (see Fig. 10) with Hill coefficients (n_H) in the 4–8 range for fast skeletal muscle and in the 2–5 range for cardiac preparations (Allen and Kentish, 1985, 1988; Moss *et al.*, 1986; Palmer *et al.*, 1996; Sweitzer and Moss, 1990). Because there are only one (cardiac) or two (fast skeletal) active calcium binding sites on TnC, the high Hill coefficients suggest additional cooperativity from other protein–protein interactions.

Bremel and Weber (1972) reported that skeletal fiber contraction could occur in the absence of calcium under low ATP and ionic strength conditions. Under these low ATP conditions, rigor cross-bridges could form long enough to displace the tropomyosin and permit binding from additional cross-bridges. They further suggested that the binding of cross-bridges along one functional group (1 Tm, 7 actins long) would facilitate the actomyosin binding along adjacent groups. This implies that a bound cross-bridge physically pushes the tropomyosin away from its inhibiting position along a fairly long stretch of the thin filament (see Figs. 11 and 12).

Subsequent studies have confirmed their findings under more physiological conditions and in cardiac muscle (Butters *et al.*, 1993; Chalovich, 1992; Fraser and Marston, 1995; Fuchs, 1995; Hancock *et al.*, 1993; Homsher *et al.*, 1996; Metzger, 1995; Palmer and Kentish, 1994; Sata *et al.*, 1993, 1995; Solaro and Van Eyk, 1996; Swartz *et al.*, 1996; Tobacman, 1996). Butters *et al.* (1993) reported that the skeletal troponin–tropomyosin complex promoted conformational changes within the actin filament. Hancock *et al.* (1993) demonstrated that cardiac muscle force is regulated by the kinetics of additional actomyosin binding interactions. Both Fraser and Marston (1995) and Sata *et al.* (1995) suggested that actin–tropomyosin filaments were regulated as a single unit based on data from motility assays that showed very steep unloaded velocity–pCa relationships. Palmer and Kentish (1994) examined the effect of pH and phosphate on calcium sensitivity and concluded that TnC only partly accounts for the observed function. Thus, kinetic mechanics associated with cross-bridge attachment and cycling must also be included. Metzger (1995) recently compared the effects of myosin binding-induced activation between cardiac and skeletal muscle systems. He found that cardiac muscle was more sensitive to activation from cross-bridge binding in the absence of calcium than skeletal muscle and that this difference disappeared when TnC was extracted. Homsher *et al.* (1996) studied calcium regulation on regulated cardiac thin filaments with a motility assay. Their data suggested that this regulation correlated to the number of cross-bridges interacting with the thin filaments but not the ATPase or unloaded filament velocity rates. Finally by evaluating the rigor binding characteristics of fluorescently tagged myosin S1, Swartz *et al.* (1996) reported that even high levels of calcium alone do not activate the thin filament along its entire length.

As a whole, these data suggest that cardiac muscle has a greater proportion of activation from cross-bridge binding than does skeletal muscle. The facilitation of activation by myosin binding also relates to the previously discussed (Section III) issue of the effect of strain or load on the kinetics of the cross-bridge actomyosin cycle (Geeves and Conibear, 1995; Millar and Homsher, 1992; Smith and Geeves, 1995). Thus, after calcium initiates the contraction by conformationally altering TnC, subsequent cross-bridge attachment serves as a feedback mechanism to increase the TnC's calcium sensitivity. Therefore, both calcium activation via the troponin–tropomyosin complex and cross-bridge binding facilitation are clearly important and necessary components of contractile regulation, especially in cardiac muscle.

C. THICK FILAMENT REGULATION

Although the primary site for contractile regulation in striated muscle resides in the thin filaments, there are also critical regulatory mechanisms associated with the thick filament proteins. These are particularly significant in the heart where the overall contractility of the system must be altered on demand to meet the changing needs of the organism. In addition to the heart's ability to modulate the levels of calcium through extrinsic mechanisms (Chapters 3–5, 7), the myofilament sensitivity to calcium can also be altered by PKA-mediated phosphorylation of myosin light chains or C-protein.

1. Light Chains

The essential (ELC or LC1) and regulatory (RLC or LC2) light chains are bound to and stabilize the neck region of myosin S1 (see Fig. 6, Section III). The presence of the ELC, closest to the motor domain, is required for full contractile function *in vitro* or *in vivo* (Trybus, 1994). ELC removal reduced isometric force to 50% of control (VanBuren *et al.*, 1994) and the motility assay velocity by an order of magnitude (Lowey *et al.*, 1993b). Morano *et al.* (1995a) observed interactions between the ELC and the C-terminus of actin, which modulated force in the human heart at the same level of calcium. Interestingly, all the reductions in function occur without any significant reduction in myosin ATPase activity under physiological conditions. Thus, the ELC does not appear to have any regulatory function via phosphorylation or other pathways under normal physiological conditions, but it is critical for normal function.

However, the RLC is phosphorylatable and clearly modulates contractile function. In smooth and invertebrate muscle that do not have thin filament regulation (troponin is absent), these light chains are the primary regulators of contraction (Trybus, 1994). In striated muscle, RLC is similarly phosphorylated by myosin light chain kinase (MLCK) and calmodulin to modulate myosin's sensitivity to calcium, particularly at low myoplasmic concentrations (Levine *et al.*, 1996; Morano and Ruegg, 1986; Patel *et al.*, 1996; Sweeney *et al.*, 1993). RLC phosphorylation shifts the force–pCa relationship (see Fig. 10) leftward, thus increasing the force development at a given calcium concentration. Sweeney *et al.* (1993) suggested that the RLC phosphorylation tended to move the S1 farther away from the thick filament backbone to a more advantageous position for actomyosin binding. This type of structural change following activation has been confirmed in the RLC by optical methods (Allen *et al.*, 1996) and in the thick filament by x-ray diffraction studies (Levine *et al.*, 1996). Margossian *et al.* (1992) linked a reduction in myosin function in cardiomyopathy to protease-mediated cleavage and loss of RLC. Finally, Patel *et al.* (1996) reported that the RLC modulates calcium-dependent force development in skeletal muscle. Thus, the phosphorylation of RLC provides mechanisms using alternative pathways that can modulate force generation without changing myoplasmic calcium levels in cardiac muscle (Levine *et al.*, 1996; Lowey and Trybus, 1995; Margossian *et al.*, 1992; Morano and Ruegg, 1986; Opie, 1995; Patel *et al.*, 1996; Sweeney *et al.*, 1993).

2. C-Protein

C-protein is a myosin binding protein distributed along the thick filament backbone of striated muscle and binds to both LMM and S2 (see Fig. 9). The cardiac isoform has recently been sequenced and found to be 137 kDa, which is smaller than the skeletal form (Gautel *et al.*, 1995). It serves as both an organizer and a regulator of contraction by direct protein–protein interactions with myosin and titin (Freiburg and Gautel, 1996; Houmeida *et al.*, 1995). Hofmann *et al.* (1991) observed a reversible increase in tension and a decrease in cooperativity at low myoplasmic calcium concentrations with the partial extraction of C-protein from cardiac myocytes. Phosphorylation of C-protein by PKA exposure significantly decreased the isometric force generation in skinned cardiac myocytes, but it did not reduce their maximal unloaded shortening velocity (Hofmann and Lange, 1994). Molecular analysis has identified four PKA- and calmodulin-regulated phosphorylation sites on C-protein (Gautel *et al.*, 1995). Optical diffraction studies suggest that phosphorylation of C-protein extends the cross-bridges from the thick filament backbone and alters their orientation (Weisberg and Winegrad, 1996). These changes could alter the cross-bridge kinetics to modify force production, much like the similar changes resulting from RLC phosphorylation. Analysis has further linked C-protein to human chromosome 11, which is the site of familial hypertrophic cardiomyopathy (Gautel *et al.*, 1995). Thus, C-protein is clearly involved in the PKA-associated modulation of myosin cross-bridge function and possibly in cardiomy-opathy.

D. LENGTH-DEPENDENT CALCIUM SENSITIVITY

The length-dependent modulation of calcium sensitivity is an intrinsic regulatory feature of heart function associated with the myofilaments (Allen and Kentish, 1985, 1988; Fuchs, 1995; Kentish *et al.*, 1986; Solaro *et al.*, 1993). It has long been realized that an increase in preload will increase the force of contraction by moving up the ascending limb of the Starling relation (Patterson and Starling, 1914). This important intrinsic control mechanism, which prevents overfilling or underfilling of the heart on a beat-to-beat basis, is determined by the steep ascending segment of the cardiac force–length relation (see Fig. 14, Section V). Figure 13 shows how the steepness of the ascending limb of the force–length relationship (bold line) might be altered by calcium sensitivity and sarcomere length in cardiac muscle. The light lines show a family of hypothetical force–length curves at different calcium levels where the shape of each curve is based on cross-bridge dynamics expected from myofilament overlap alone. During a real cardiac contraction, the amount of force generated at a new end diastolic sarcomere length (preload) would depend on both the change in myofila-

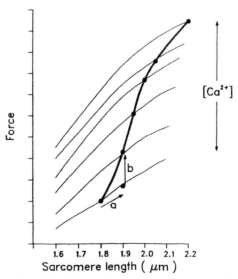

FIGURE 13   Ascending limb of the cardiac force–length relationship. This diagram shows how the shape of the ascending limb of the cardiac force–length relationship is due in part to modulation of force at a given sarcomere length by the calcium concentration. See text for details. (Reproduced with permission from Fuchs, 1995.)

ment overlap (arrow "a") and the change in calcium sensitivity due to enhanced cooperative interactions at that length (arrow "b"). This length-dependent calcium sensitivity (see Fig. 13) is not significant in skeletal muscle where the ascending limb is less steep (Fuchs, 1995). Thus, length-dependent alterations in calcium sensitivity could be accounted for by either an increase in TnC affinity for calcium or by an increase in the number of cross-bridges generating force at a given level of calcium saturation.

To assess the role of TnC in this issue, Hofmann and Fuchs (1988) examined the length-dependent calcium binding characteristics in skinned cardiac muscle and clearly demonstrated that TnC calcium affinity was length sensitive. Because skeletal muscle does not exhibit this degree of length sensitivity, functional differences between fast skeletal and cardiac TnC isoforms have been evaluated by mechanical, biochemical, and extraction/resubstitution experiments on skinned fibers (Fuchs, 1995; Gulati *et al.*, 1991, 1992; McDonald *et al.*, 1995; Moss, 1992; Moss *et al.*, 1986, 1991). Extraction/resubstitution experiments exchanged skeletal and cardiac isoforms of TnC in skinned fibers to identify their relative contributions to length-dependent sensitivity (Fuchs, 1995; Gulati *et al.*, 1991; Moss, 1992; Moss *et al.*, 1991). The results were initially conflicting and appeared to depend on the specific experimental protocols. Recent transgenic mouse data have confirmed that cardiac muscle has a greater length-dependent calcium sensitivity than skeletal muscle. But these changes are not dependent on the specific troponin isoform because cardiac and skeletal TnCs had equal sensitivity when in the cardiac muscle environment (McDonald *et al.*, 1995).

One additional problem with skinned cell studies is that the myofilament lattice of muscle swells when the membrane systems are removed. This may be the reason Kentish *et al.* (1986) and Gao *et al.* (1994) noted differences in myofilament calcium sensitivity between intact and skinned rat cardiac muscle. As the myofilament lattice spacing increases from either swelling in skinned muscle or due to the isovolumetric behavior of intact cells at shorter sarcomere lengths, the myosin S1 head lies farther from its binding site on actin. Thus, the amount of developed force will decline at a given myofilament calcium concentration. The degree of swelling in skinned muscle is also likely species and sarcomere length dependent. Thus, it is not surprising that McDonald and Moss (1995) reported a reduction in length-dependent calcium sensitivity in skinned cardiac fibers along the ascending limb of the force–length relation. They modulated the myofilament lattice spacing by changing length or with osmotic recompression. Finally, data from Fuchs and Wang (1996) suggested that interfilament spacing, not the TnC isoform, is the primary determinant of the length-dependent calcium sensitivity characteristic of cardiac muscle. Thus, it appears that length-dependent calcium sensitivity in cardiac muscle (see Fig. 13) is a fundamental feature of myofibrillar lattice spacing changes that occur along the ascending limb of the force–length relationship where cardiac muscle normally functions. These changes in myofibrillar organization modulate the calcium binding characteristics of the thin filament and are not solely dependent on unique cardiac regulatory protein isoforms.

In summary, the myofilaments are not only the generators of force, but major regulators of force in cardiac muscle. Thin filament regulation by the troponin–tropomyosin complex provides the ionic switch that turns "on" and "off" each contraction. Calcium binding to the single low-affinity site on cardiac troponin C elicits a cascade of conformational changes that move the tropomyosin rod away from the actomyosin binding site. This highly cooperative conformational process permits the myosin S1 of the thick filament to bind strongly to actin on the thin filament. Then repetitive attachment and detachment linked to ATP hydrolysis can occur as described in the previous section. The overall force sensitivity to calcium is modulated by a number of myofilament-mediated factors, including actomyosin binding, inotropic stimulation, and sarcomere length (preload). Full activation in cardiac muscle also requires the long-range cooperativity conferred by the actomyosin binding itself. After calcium initiates the contraction by conformationally altering TnC, subsequent cross-bridge attachment serves as a feedback mechanism to increase the TnC's calcium sensitivity and further activate the muscle. Therefore, both calcium activation via the troponin–tropomyosin complex and cross-bridge binding are necessary components of contractile regulation, especially in cardiac muscle. Although the thin filaments are the primary regulators, the thick filaments can also modulate long-term changes in cardiac muscle contractility by the PKA-mediated phosphorylation of myosin light chains and C-protein. This mechanism moves the myosin S1 head closer to the actin to facilitate strong binding and cross-bridge cycling. Length-dependent calcium sensitivity appears to be the result of changes in myofilament spacing, which affect the calcium binding kinetics to troponin. The specific isoform of troponin itself does not appear to be a critical factor. When combined with myocyte structure and cross-bridge cycling, these myofilament regulatory mechanisms provide the foundation for the evaluation of mechanical function at the cellular level.

V. CELLULAR FUNCTION—SYSTOLIC AND DIASTOLIC PROPERTIES OF CELLS

The overall systolic and diastolic function of the heart is determined by the time varying interplay between multiple cellular, extracellular, viscoelastic, and geometric factors. At the cellular level, interpretation of the system's mechanical function is simplified by removing the extracellular matrix and geometric factors that may alter recorded values by unknown or unmeasurable amounts. The syncytial geometric organization of heart is necessary to optimize its pumping function (Streeter *et al.*, 1969). However, this complex geometry confounds the direct interpretation of the pressure–volume–time relationships measured in the whole heart in terms of the force, length, and time parameters that characterize the mechanical function from the contractile apparatus in the cell. Although considerable effort has been made to evaluate cardiac function with more linear multicellular preparations, such as trabeculae and papillary muscles (Brady, 1984; Brady *et al.*, 1981; de Tombe and ter Keurs, 1992; Krueger and Pollack, 1975; ter Keurs *et al.*, 1980a, 1980b), these preparations still manifest the length- and time-dependent properties derived from attachment artifact or extracellular factors that modify systolic and diastolic function in unknown ways. Therefore, the characterization of myocardial mechanics from isolated cells can provide a clearer picture of the cellular contributions and thus the underlying mechanisms that determine overall heart function.

Traditionally, mechanical function of striated muscle preparations has been evaluated with respect to both isometric force development and isotonic shortening behavior. For both, it is necessary to carefully determine the changes in either force or length with time while the other parameter is temporarily held constant. For the examination of passive and active force–length relationships, length is held constant at a series of values while the maximal force is determined. The force–length relationships are shown for a variety of skeletal and cardiac preparations in Fig. 14. Here the active (*a*) and passive (*f*) force–length relationships are redrawn (dashed lines) from data for frog skeletal muscle (Gordon *et al.*, 1966). In comparison, the active cardiac force–length relationship (solid line *b*) is much steeper along the ascending limb (1.60–2.10 μm) than that for skeletal muscle (dashed line *a*). The passive force–length relationship in rat papillary muscle (solid line *c*) is greater than that for isolated rat fibers (solid line *d*), which is greater than that for isolated dog fibers (solid line *e*). All the passive force relationships are greater than the passive skeletal muscle relationship (line *f*). The isolated cells have about the same compliance (shape of the curve) as the whole heart, but the curve is shifted right to longer sarcomere lengths (Brady, 1991a; Granzier and Irving, 1995; MacKenna *et al.*, 1994). Furthermore, the absolute passive force in isolated cardiac cells is species dependent with the dog cells being less stiff than the rat cells (Brady, 1984, 1990, and 1991b; Fabiato and Fabiato, 1978). These intrinsic cardiac force–length characteristics are an important aspect of the classic "Starling relationship," which relates preload (end diastolic sarcomere length) to stroke volume (sarcomere shortening) along the working range of the heart (Patterson and Starling, 1914). Thus, it is essential to characterize cellular mechanics in terms of force, length, and time to better understand the steeper ascending limb of the active curve (*b*) and the alterations in the passive curves (c–f) shown in Fig. 14. Similarly, the measurement of isotonic

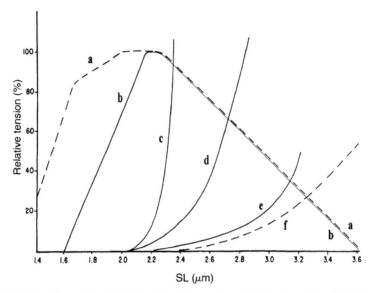

FIGURE 14   Force–length relationship. The passive and active force–length relationships are drawn on the same axes from both cardiac and skeletal muscle preparations. Dashed line *a* is the maximally activated force from frog skeletal muscle at a range of sarcomere lengths from 1.4 to 3.6 μm. Solid line *b* is activated rat papillary muscle data from 1.6 to 3.6 μm sarcomere lengths. Solid lines *c*, *d*, and *e* are the resting tension relationships respectively for whole rat papillary muscle, a skinned rat ventricular fiber, and a skinned dog ventricular fiber. The dashed line *f* is the resting tension relationship from a skinned rat skeletal muscle. These data were compiled by Brady (1984) from a number of studies. See text and reference for details. (Adapted with permission from Brady, 1984.)

contraction data to generate force–velocity relationships provides some insight into cross-bridge function under various loading and inotropic conditions. Combined, these force–length–velocity–time measures permit the evaluation of a full range of mechanical function at the cellular level that can be related to the molecular entities that generate and regulate force development in the whole organ.

The remainder of this chapter outlines our current understanding of mechanical function in isolated cardiac myocytes. Despite extensive efforts, only modest progress has been achieved in evaluating the full spectrum of mechanical properties from cells. Although the isolated cardiac myocyte is a structurally simpler system in which to evaluate mechanical function than is the whole heart, it is far from problem free. Its small size and lack of natural attachment sites make the recording of these parameters technically difficult. Thus, in the following section, the methodology used to evaluate length and force in heart cells is briefly outlined. Subsequently, the majority of this chapter details the systolic and diastolic mechanical data obtained from cardiac myocytes isolated from normal and hypertrophied heart.

A. METHODS OF MEASURING FORCE AND LENGTH

The unambiguous determination of changes in length and force with time is essential for the complete characterization of mechanical function in cardiac muscle at the cellular level. In the whole heart, it is relatively easy to measure chamber volumes and pressures (Conrad *et al.*, 1995; MacKenna *et al.*, 1994; Van Leuven *et al.*, 1994). In multicellular preparations such as papillary and trabecular muscles, the determination of muscle length and force is somewhat more difficult and subject to attachment artifact, but it is generally routine (de Tombe and ter Keurs, 1992; Donald *et al.*, 1980; Krueger and Pollack, 1975; Pollack and Krueger, 1976; ter Keurs *et al.*, 1980a, 1980b). In the smaller isolated cardiac myocyte preparation, a variety of methods have been developed to assess force and length that have been recently reviewed by Delbridge and Roos (1997), Garnier (1994), and Palmer *et al.* (1996). Although the full spectrum of mechanical analyses is not as yet possible in isolated myocytes, substantial success has been achieved, particularly with the assessment of unloaded shortening in contracting cells and steady-state force from skinned cells. These methods are now briefly described and serve as a basis for the subsequent discussion of mechanical data.

1. Measurement of Cell and Sarcomere Length

A variety of optical techniques have been developed to measure sarcomere or cell length at rest or during contraction. The majority of these efforts have been applied to the evaluation of unloaded shortening of calcium-tolerant myocytes subject to various inotropic, electrophysiological, or mechanical interventions. But recently some methods have been adapted for experiments that also measure force from cells attached to transducers (see following discussion). In general, a measure of length is obtained from isolated cardiac myocytes at high spatial and temporal resolution by directly projecting the cell image with a microscope objective onto a detector. A variety of analysis methods have been employed that monitor either one or both the cell ends or the sarcomere striation pattern to obtain a measure of length.

The simplest methods monitor one or both cell ends with single photodiode, single line, or area array (video) detectors (Boyett *et al.*, 1988; Delbridge *et al.*, 1989; Harding *et al.*, 1988; Krueger, 1988; Spurgeon *et al.*, 1990; Steadman *et al.*, 1988). Cell length is either inferred from the movement of one cell end or directly calculated from the positional difference beween the two ends. The advantage of these cell end detection systems is their simplicity; the major disadvantage is that they are subject to optical and geometric errors and do not directly measure sarcomere length (Delbridge and Roos, 1997). These systems are in wide use for the study of unloaded contractile characteristics in myocytes during inotropic or electrophysiological interventions (Brooksby *et al.*, 1992; Delbridge *et al.*, 1996; Duthinh and Houser, 1988; Kobayashi *et al.*, 1995; Mukherjee *et al.*, 1993; Rich *et al.*, 1988; Sollott and Lakatta, 1994; Spinale *et al.*, 1992; Vescovo *et al.*, 1989; Zile *et al.*, 1995).

The determination of sarcomere length provides a more direct measure of myofilament overlap than does cell length. Thus, many labs have developed either light diffraction or direct striation pattern imaging methods to determine sarcomere periodicity from whole cells, specified cell regions, or even individual sarcomeres. Light diffraction patterns provide the average striation spacing in the region of the muscle or cell illuminated (Leung, 1983a; Roos and Leung, 1987) and have been applied to studies monitoring cardiac muscle sarcomere dynamics in multicellular preparations (Krueger and Pollack, 1975; Pollack and Krueger, 1976; ter Keurs *et al.*, 1980a, 1980b) and myocytes (Haworth *et al.*, 1987; Kent *et al.*, 1989; Krueger, 1988; Krueger *et al.*, 1980; Leung, 1983b; Mann *et al.*, 1991; Niggli and Lederer, 1991; Urabe *et al.*, 1993; Wussling *et al.*, 1987). The sarcomere length of cardiac myocytes has been determined by directly imaging the cell's striation pattern at high resolution. Sarcomere length has been determined either by Fourier analysis to provide average values (Krueger, 1988; Krueger and Denton, 1992; Krueger *et al.*, 1992; Siri *et al.*, 1991) or by direct determination of striation positions to provide both discrete and average values (DeClerck *et al.*, 1984; Gannier *et al.*, 1993; Linke *et al.*, 1993; Lundblad *et al.*, 1986; Mukherjee *et al.*, 1993; Nassar *et al.*, 1987; Palmer *et al.*, 1996; Roos, 1986a, 1986b, 1987; Roos and Brady, 1982, 1989, 1990; Roos and Taylor, 1989, 1993; Roos *et al.*, 1982; Z.L. Wang *et al.*, 1996; Zile *et al.*, 1995). These direct striation pattern monitoring approaches provide a more direct assessment of sarcomere dynamics than the cell length approaches, but they are technically more difficult to implement and often require considerable off-line analysis.

Finally, it is assumed that all these optical monitoring approaches provide an accurate index of myofilament overlap, which permits the correlation of cross-bridge function to cellular and whole organ contractile mechanics. However, the cell's shape and optical characteristics are not ideal and certain monitoring techniques may not accurately represent sarcomere dynamics and, thus, cross-bridge function. The Cardiovascular Research Laboratory at UCLA School of Medicine (hereafter referred to as "this laboratory") has recently rigorously analyzed cell and sarcomere contractile dynamics from the same contractions and confirmed that cell length can be an adequate index of sarcomere dynamics under carefully controlled experimental conditions (Delbridge and Roos, 1997). Thus, numerous methods have been utilized by various investigators to evaluate changes in unloaded cardiac myocyte shortening in response to inotropic or electro-physiological interventions. A more complete description of these methods, their advantages, and potential errors for cell or sarcomere length determination are presented in Delbridge and Roos (1997).

2. Methods of Measuring Force and Stiffness

The second essential parameter that must be determined to fully characterize the mechanical function of heart cells over time is force. The measurement of unloaded velocities of shortening by monitoring cell or sarcomere length may serve as an index of contracility over a limited range (Brady, 1991b; Delbridge and Roos, 1997), but it does not provide a complete measure of mechanical function. As with the measurement of cell and sarcomere length, several different approaches have been utilized to determine force and/or stiffness from isolated cardiac myocytes with and without intact membrane systems. Force is the direct determination of the amount of tension produced by a preparation, whereas stiffness is the relative change in force from an induced step change in length (static stiffness) or oscillatory changes in length at various frequencies (oscillatory stiffness). Stiffness modulus normalizes for the cross-sectional area, which permits comparison between preparations of different size (Brady, 1990, 1991a, 1991b; Brady and Farnsworth, 1986).

Isolated amphibian atrial and ventricular cells were among the first whose force and stiffness characteristics were evaluated because their long slender shape permitted attachment to mechanical transducers (Cecchi et al., 1993; Copelas et al., 1987; Tarr et al., 1979; Tung, 1986). The more rectangular shape of the mammalian cardiac myocyte, its membrane fragility, and its lack of natural attachment sites are some of a large number of technical problems that have inhibited the widespread evaluation of their active or passive forces. These cell and other transducer considerations have limited the evaluation of mammalian cardiac cells using current technologies to relatively low levels of force in myocytes with intact sarcolemma or to forces obtained from chemically activated skinned myocytes.

The mechanical or chemical (detergent) removal of the sarcolemmal membrane from cardiac myocytes permits their attachment to force transducers and length displacement devices without the concern of sarcolemmal membrane fragility. These so-called "skinned cells" can be maintained in a relaxing solution (pCa > 7) or activated to develop steady-state forces with a range of calcium containing activation solutions (see Section IV and following discussion). Direct force and stiffness has been obtained from skinned cardiac cells by gripping their ends with microtools (DeClerck et al., 1977; Fabiato and Fabiato, 1975, 1976, 1978; Fabiato, 1981; Fish et al., 1984), by gluing their ends to glass probes with silicone or other types of adhesives (Araujo and Walker, 1994, 1996; Granzier and Irving, 1995; Granzier et al., 1996; Heyder et al., 1995; Hofmann and Lange, 1994; Hofmann and Moss, 1992; Hofmann *et al.*, 1991; Linke *et al.*, 1993, 1994, 1996; Metzger, 1995; Strang *et al.*, 1994; Strang and Moss, 1995; Sweitzer and Moss, 1990, 1993; Vannier *et al.*, 1996), or by sucking the cell ends into single- or double-barreled pipettes coated with glue (Brady, 1990, 1991a, 1991b; Brady and Farnsworth, 1986; Brady *et al.*, 1979; Palmer *et al.*, 1996; Roos and Brady, 1989, 1990).

The last method, using vacuum suction pipettes, was developed in this laboratory by Allan J. Brady (Brady *et al.*, 1979). Figure 15 illustrates a chemically skinned rat myocyte attached between a pair of double-barreled micropipettes that were custom fabricated and coated with a thin layer of adhesive. The outer pipette serves as a docking port to align the cell end perpendicular against the inner pipette as the cell is pulled into the system with vacuum suction. Unlike other attachment approaches, the intercalated disk of the cell rests squarely against the inner pipette to transmit force as uniformly as possible. After allowing the glue to set, the vacuum is released leaving the cell firmly attached to the pipette without any induced nonuniformity of sarcomere length (see Fig. 3 in Palmer *et al.*, 1996). A very small portion of the cell is actually inside the pipette; therefore, 85–90% of the cell length is clear of the attachment system and available for functional measurement. This system has been used to obtain active and passive force and stiffness measurements (see following discussion) with simultaneous sarcomere length monitoring (Palmer *et al.*, 1996; Roos and Taylor, 1993). A fully submersible force transducer and attachment system on a single integrated circuit chip is under development (Lin *et al*, 1996).

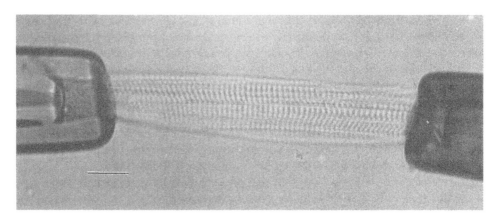

FIGURE 15   Myocyte attached to pipettes. This photomicrograph illustrates a rat cardiac myocyte attached to concentric double-barreled pipettes. Less than eight sarcomeres are drawn into the pipettes at each end with the intercalated disc resting against the inner pipette. There is no detectable sarcomere length nonuniformity between the pipettes induced by this attachment method. Calibration bar = 20 μm. (Reproduced with permission from Palmer *et al.*, 1996.)

The measurement of force or stiffness from calcium-tolerant cardiac myocytes with intact sarcolemma (unskinned) is technically more difficult than measurements from skinned cells due to membrane fragility. As with skinned cells, intact cells have been attached to force transducers with vacuum suction single-barreled micropipettes (Brady *et al.*, 1979) and micro-tools (Fabiato, 1981). Both of these approaches were abandoned because of their low success rate and and technical difficulty due to membrane fragility. More recently, several investigators have attempted to measure forces from unskinned cells glued to glass needles, plates, or carbon fibers (Bluhm *et al.*, 1995; Copelas *et al.*, 1987; Gannier *et al.*, 1993; Garnier, 1994; Shepherd *et al.*, 1990; White *et al.*, 1995). All of these glue and carbon fiber approaches have produced new information about cellular mechanical function (see following discussion), but they also have one or more technical problems that restrict their measurements to low force and potentially nonuniform sarcomere length conditions. Because cell attachment is techni-cally difficult, several alternative approaches have been employed to assess cellular mechanics by modifying the external load on the cell by either embedding freshly isolated cells in agarose or fibrin gels to induce an external load (Sollott and Lakatta, 1994; Zile *et al.*, 1994), by osmotically stressing cells with hypo- and hypertonic saline solutions (Roos, 1986a), or by externally loading cells with highly viscous methylcellulose containing saline solutions (Kent *et al.*, 1989; Mann *et al.*, 1991; Urabe *et al.*, 1993).

Thus, skinned and unskinned cardiac myocytes attached to force trans-ducers by a variety of methods have provided considerable, but incomplete, new mechanical function data (see following discussion). These data have largely been limited due to the immense technical challenge of assessing the force from the small cells without induced artifact and noise. For a more detailed analysis of the methods utilized for the attachment and measurement of mammalian cardiac myocyte force and stiffness, see the recent reviews by Garnier (1994) and Palmer *et al.* (1996).

B. PASSIVE PROPERTIES OF CARDIAC MYOCYTES

The passive properties of the heart are important arbiters of both its systolic and diastolic function. During diastole the ventricles refill (preload) adequately to eject sufficient blood as demanded by the organism. The volume of blood that preloads the heart is determined by the balance between the venous filling pressure and the stiffness of the ventricular chamber. Without any passive resistance to filling, the heart could overfill during diastole and move down the descending limb of the force–length relationship (see right side of curves a and b in Fig. 14). Under these conditions, the heart would produce insufficient force to eject the blood against the afterload of the systemic circulation, and the ventricle would progressively accumulate more blood with each beat. However, if the chamber is too stiff, it will not preload sufficiently and move down the ascending limb of the force–length relationship (see left side of curve b in Fig. 14), again producing inadequate force. Thus, the passive stiffness of the ventricular chambers is a critical factor in determining the preload and the subsequent systolic function of the heart via the classic Starling mechanism (Patterson and Starling, 1914). Alterations in the passive properties of the heart following hypertrophy can therefore alter systolic and diastolic function leading to failure (Collins et al., 1996; Conrad et al., 1995; Spinale et al., 1992; Zile et al., 1994).

Considerable evidence suggests that the passive properties influencing myocardial mechanics of normal heart are determined to a large extent by the cardiac myocytes themselves (Brady, 1990, 1991a, 1991b; Brady and Farnsworth, 1986; Granzier and Irving, 1995; Helmes et al., 1996; Roos and Brady, 1989). At the cellular level, evaluation of diastolic mechanical function is presumably simplified by removing the passive extracellular matrix and geometric factors. The geometric factors are not easily evaluated in the whole heart (Guccione et al., 1993; Little et al., 1995; Omens et al., 1996; Rodriquez et al., 1992; Streeter et al., 1969) but are somewhat simplified in the linear strip multicellular preparations of papillary or trabecular muscle (Brady, 1974, 1984, 1990, 1991a, 1991b; Brady et al., 1981; de Tombe and ter Keurs, 1992). Some insight into the extracellular matrix's contribution has been obtained from the direct comparison of the passive properties between multicellular and cellular preparations. As previously shown in Fig. 14, the passive force–length relationships are different in various skeletal and cardiac preparations. Thus, the assessment of the cellular component of diastolic stiffness is essential to our understanding of mechanical performance in the whole heart. The passive mechanical properties of skinned and unskinned cardiac myocytes have been obtained in a number of laboratories using the various cell attachment and force measurement methods described previously. These studies are now described in terms of the internal forces that modulate cellular elongation or reextension and are related to the elastic cytoskeletal structures that mediate these properties.

1. Passive Force–Length Relationship

When stretched above slack length, the passive force in skinned rat cardiac myocytes increases exponentially with sarcomere length up to about 2.5 μm (see curve d, Fig. 14). But sarcomere length dependence of this relationship is clearly species and preparation dependent (see curves c–f, Fig. 14). There have been few studies that have directly compared the passive properties between preparations (see Table 3 in Brady, 1991a, for synopsis). Differences between species were reported by Fabiato and Fabiato (1978) who found that the passive stiffness sequentially increased in mechanically skinned frog, dog, and rat heart cells. Similarly, Brady (1991a) reported that the stiffness modulus increased sequentially in chemically skinned guinea pig, rabbit, and rat cells. Fish *et al.* (1984) evaluated the differences in passive properties between skeletal and cardiac muscle cells in the hamster. They found that the passive length tension curves from the two preparations had the same shape, but that the cardiac curve was shifted to the left implying greater stiffness at any given length. Finally, Granzier *et al.* (1996) observed that skinned rat cardiac myocytes were substantially stiffer than similarly prepared rabbit skeletal muscle.

To determine the relative roles of cellular and extracellular factors, Kentish *et al.* (1986) reported only a small drop in passive stress on the chemical skinning of RV rat trabeculae, particularly at sarcomere lengths <2.2 μm, but greater differences at extended sarcomere lengths (Fig. 4 in Kentish *et al.*, 1986). They attributed their findings to the possible release of extracellular elastic components from the cells and the osmotic swelling characteristics of skinned cells. MacKenna *et al.* (1994) collagenase treated whole rat hearts and measured their diastolic pressure–volume characteristics. They reported (Fig. 6 in MacKenna *et al.*, 1994) no difference in the ventricular compliance (the shape of the curve) but a shift to the right indicating an increase in ventricular volume and thus sarcomere length. Finally, Brady (1990, 1991a) and Granzier and Irving (1995) both reported a proportionally higher passive stiffness in multicellular versus cellular preparations. But the stiffness in the majority of the working range of the heart (sarcomere lengths <2.20 μm) was primarily of cellular origin. Taken together, these data suggest that the extracellular matrix may only modify the preferred diastolic sarcomere length but not significantly contribute to the passive mechanical properties of the normal heart in its working range.

To obtain passive force and oscillatory stiffness relationships from isolated cardiac myocytes, the skinned cells are attached to mechanical transducers with the double-barreled pipette method (see Fig. 15) and stretched to extended lengths above slack length. Figure 16 illustrates a typical incremental step stretch protocol applied to a rat myocyte as used in this laboratory (Palmer *et al.*, 1996). The direct force (lower) and 100 Hz oscillatory stiffness (upper) traces are qualitatively the same and show significant increases as the cell is stretched. In addition, both traces demonstrate a stress relaxation that persists for minutes following the step stretch. Force or stiffness values for a given sarcomere length are obtained from the curves following the stress relaxation just prior to the next stretch. These can then be plotted against sarcomere length to produce a sigmoid relationship as shown in Fig. 17 for two different rat strains. Regardless of whether absolute passive force or stiffness modulus is evaluated in terms of length, a roughly exponential increase is exhibited up to sarcomere lengths of about 2.4–2.5 μm. Beyond that length, the data tend to plateau. The structural sources that determine the shape of these curves in Fig. 17 are discussed later in this chapter.

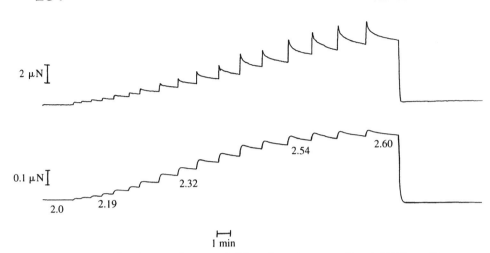

FIGURE 16   Passive stretch protocol. Direct force (top record) and 100 Hz oscillatory stiffness (bottom record) are displayed during an incremental stretch protocol on a Wistar-Kyoto (WKY) cell. Each upstroke in force and stiffness represents a step increase in cell length. The force and stiffness traces transiently increase, then exhibit a stress relaxation at each increment. The numbers below the stiffness trace indicate average sarcomere length for that step. See text for further details. (Reproduced with permission from Palmer *et al.*, 1996.)

Similar results have been obtained in other laboratories from skinned rat myocytes (Granzier and Irving, 1995; Granzier *et al.*, 1996) and myofibrils (Linke *et al.*, 1994, 1996). Although these labs used different preparations, stretch protocols, or attachment methods, their passive tension data were qualitatively and quantitatively the same as ours up to about 2.5 μm sarco-

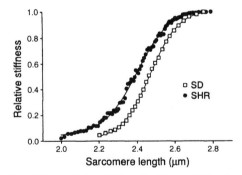

FIGURE 17   Passive stiffness relationship in cells. The normalized passive stiffness–length relationship is plotted from two skinned cardiac myocytes. One cell is from a hypertrophied spontaneously hypertensive rat (SHR: ●) heart and the other from a normotensive control Sprague-Dawley (SD: □) rat heart. Both curves have a sigmoid shape with the majority of the rise in the working range of the heart prior to a 2.5 μm sarcomere length. The data plateau thereafter with additional stretch. The hypertrophied SHR cell is clearly stiffer at all sarcomere lengths than the normal SD cell. (Original data courtesy Dr. Roy E. Palmer.)

mere lengths. However, at extended lengths, the force did not plateau nearly to the extent seen in skeletal muscle preparations (K. Wang *et al.*, 1991, 1993) and the cardiac data found in this laboratory (see Fig. 17). This is probably due to the continuous slow stretch protocol used by Granzier and Irving (1995), which may retain a portion of the stress relaxation forces, or to the fragility of the myofibrillar preparation used by Linke *et al.*, (1994, 1996). Interestingly, Linke *et al.*, (1994) noted considerable stress relaxation in myofibrils, indicating that its structural source is sarcomeric in nature. Thus, cells and their myofibrillar components exhibit a significant resistance to stretch.

Figure 17 plots relative oscillatory stiffness data averaged from control Sprague-Dawley (SD) LV myocytes and hypertrophied spontaneously hypertensive rat (SHR) LV myocytes. The SHR model of hypertrophy parallels human pressure overload hypertrophy in terms of increased systolic pressures and diastolic stiffness (Conrad *et al.*, 1995). Although there is cell-to-cell variability (Palmer *et al.*, 1996), SHR myocytes are clearly stiffer than control on average. Furthermore, the absolute maximal stiffness is three- to four-fold greater in the SHR than in the SD at all sarcomere lengths between the minimum measurable value and the plateau. For example, we find SD cells have a stiffness modulus of about 2–3 mN/mm^2 as compared to the 9–12 mN/mm^2 found in the SHR cells at a sarcomere length of 2.3 μm. Absolute and normalized data (not shown) for the Wistar-Kyoto (WKY) rat (genetic control for SHR) fall in-between at all lengths. These data suggest that the cells may make a significant contribution to the increased diastolic stiffness characteristic of hypertrophied heart, which has traditionally been attributed to the increased fibrosis of the extracellular matrix (Weber *et al.*, 1994).

2. Restoring Forces in Cells

Cardiac myocytes not only exhibit an inherent resistance to passive stretch, but they resist passive shortening. When the heart and its myocytes relax, they rapidly return to their preferred size and shape. The rate of myocyte relaxation is on the same order as its rate of unloaded shortening (see following discussion). Attempts to passively shorten isolated myocytes by pushing on their ends or retarding their relaxation causes the cells to buckle (Helmes *et al.*, 1996; Krueger *et al.*, 1980, 1992). Thus, there is a significant restoring force inherent to the cells that is different from the force mediating extension.

Logically, this restoring force must be considerably less than the force of externally unloaded shortening. Fabiato and Fabiato (1976) estimated the force of reextension in a mechanically skinned rat myocyte to be about 4% of its maximal active force by measuring how hard the cell pushed against a transducer during relaxation. Similar values were obtained from direct force measurements on small trabecular preparations (ter Keurs *et al.*, 1980b), from velocities of reextension following photolytic release of ATP on rigor cells (Niggli and Lederer, 1991), and from relengthening behavior in asynchronous contractions of cells (Krueger *et al.*, 1992). However, these values may underestimate the restoring force by an unknown amount because of viscoelastic effects and residual cross-bridge activity during relaxation. Recently, Helmes *et al.* (1996) estimated the restoring force to be on the order of 6–10% of the maximal active force in skinned rat myocytes based on relaxation dynamics of rigor shortened cells. Furthermore, they suggested that this force was about 70% of the total restoring force in the heart based on comparisons to trabeculae. Finally, they demonstrated that the cytoskeletal protein titin modulated the restoring force in their preparation. Thus, restoring forces are a significant cellular entity that modulate the relaxation dynamics in the heart.

3. Structural Source of Passive Elasticity in Cells

Both the passive resistance to elongation and the restoring forces expressed in cardiac myocytes have structural sources that reside in one or more components of the cytoskeleton (see Fig. 4). Although there may be a small amount of baseline cross-bridge activity during diastole (Brenner and Eisenberg, 1987; Eisenberg *et al.*, 1980; Kawai *et al.*, 1993; Smith and Geeves, 1995; Sollott *et al.*, 1996), comparative studies of intact and skinned cells suggest that it has no significant effect on the force–length relationship or restoring forces (Brady, 1991a, 1991b; Brady and Farnsworth, 1986; Granzier and Irving, 1995; Helmes *et al.*, 1996; Roos and Brady, 1989). As previously outlined (Section II), the cytoskeleton of the myocyte consists of a number of exosarcomeric and endosarcomeric structures that comprise more than 10% of the cell's total protein (Price, 1991; K. Wang, 1985). The primary structural candidates contributing to passive forces in cardiac myocytes are the longitudinally oriented titin (also called connectin) protein associated with the myofilaments (see Fig. 5) and the transversely oriented desmin and α-actinin proteins in the Z-disk (see Fig. 4). Microtubules do not appear to have a significant role in the passive properties of normal adult heart (Collins *et al.*, 1996; Granzier and Irving, 1995; Watkins *et al.*, 1987; Zile *et al.*, 1994).

A balance of forces between these longitudinal and radial cytoskeletal structures determines the cell's preferred shape and size. Osmotic stress studies (Roos *et al.*, 1982; Roos, 1986a) have provided some clues as to the relative importance of these structures. During hypertonic stress, unskinned cells shrank proportionally in both the radial and longitudinal directions, indicating approximately equal contributions of these structures to the restoring forces following compression. However, hypotonic swelling did not elicit any cell elongation; all of the cell's volume increase was radial. Thus, the longitudinal constraining structures such as titin are proportionally more stiff than the radial structures associated with the intermediate filaments or the α-actinin of the Z-disk.

Considerable effort has been made to identify the structural source or sources of the passive resistance to stretch in cardiac myocytes. By far the most productive approach has been to selectively extract structures from the cell and measure the changes in passive properties over a range of lengths. Brady and collaborators (Brady, 1990; Brady and Farnsworth, 1986; Roos and Brady, 1989) pioneered this selective extraction approach. By attaching unskinned cells to their transducer system, they were able to follow the changes in force and stiffness at each extraction step in the same cell. Chemical skinning with Trition x-100 had minimal effect on the passive properties. The small decline in passive stiffness sometimes observed on skinning at any length is most likely due to the internal rearrangement of the myofilament lattice that swells in skinned cells. Although sarcolemmal-associated structures are removed during this process, they do not appear to have any significant effect on the cell's passive properties. Following chemical skinning, cells were exposed to various concentrations of high salt solutions to selectively extract the myofilament apparatus. They found that the passive stiffness significantly declined to about 20% of control following thick filament extraction and to about 10% of control following thin filament extraction. These data strongly suggested that some proteins associated with the myofilament apparatus, such as titin, determines the majority of passive stiffness–length relationship.

These initial extraction studies were later confirmed over extended lengths by Granzier and Irving (1995) who additionally assessed the relative roles of titin, intermediate filaments, collagen, and microtubules in cardiac myocytes (the latter two structures have already been discussed). By comparing results from high salt extraction experiments to those obtained following mild trypsin digestion, which degrades titin, they directly confirmed that titin is the structural source that is responsible for the majority of the resting tension in the working sarcomere length range of the heart. Intermediate filaments appear to have little contribution to the resting tension except at nonphysiological sarcomere lenths greater than 3.0 μm.

Based on this cardiac extraction information and studies from skeletal muscle (Horowits and Podolsky, 1987; Improta et al., 1996; Linke et al., 1996; K. Wang et al., 1991), the most obvious structural candidate for the resting tension in cardiac myocytes is the megadalton protein titin that runs from the M-line to the Z-disk (Fig. 5) Horowits and Podolsky (1987) demonstrated that titin was essential to maintain the positional stability of the A-band during tetanic contractions in skeletal muscle fibers. In addition to its localization in the sarcomere, molecular titin is known to be elastic (Maruyama, 1994; Maruyama et al., 1976; Matsubara and Maruyama, 1977; Trinick, 1994, 1996; K. Wang et al., 1984, 1991, 1993). The molecular structure of titin is now known (Furst and Gautel, 1995; Gautel and Goulding, 1996; Jin, 1995; Labeit and Kolmerer, 1995; Linke *et al.*, 1996; Politou *et al.*, 1995) and is consistent with elasticity in the I-band region of the protein. These elastic properties are due to the folded PEVK and repeating immuno-globulin (Ig) domains (the pleated portion of the I-band in Fig. 5). Finally, titin expression is altered in hypertrophied, failing, and hibernating heart suggesting its important mechanical role (Ausma *et al.*, 1995; Collins *et al.*, 1996; Hein *et al.*, 1994).

Prior to this complete molecular analysis of titin, K. Wang *et al.* (1991, 1993) suggested that titin functioned as a dual molecular spring and pro-posed the segmental-extension model to explain passive force–length rela-tionships in striated muscle. This model proposed that the I-band segment of titin was elastic and could be stretched like a spring over the functional range of the muscle. On further extension past a "yield point," the portion of the titin molecule associated with the thick filament would be irreversibly pulled off or damaged (see Fig. 8 in K. Wang *et al.*, 1993). At the yield point, the passive force–length relationship starts to plateau (as in Fig. 17). This segmental extension model is based on a large body of evidence from combined immunocytochemical and mechanical analyses in skeletal muscle fibers.

Specifically, K. Wang and colleagues observed the displacement of anti-body epitopes distributed along the titin molecule. Consistent with the molecular data, they observed that the I-band region was elastic and the A-band region inelastic during stretch. These data were correlated to the passive force measured at each length, which plateaued at extended lengths. Repetitive stretch and release cycles to increasingly longer lengths past the initial yield point produced an irreversible reduction in force and a shift of the passive force–length relationship to the right. Furthermore, the specific sarcomere length at which this initial yield point occurred directly correlated to the molecular weight of titin in specific fiber types of skeletal muscle; the lower the molecular weight, the shorter the sarcomere length at which the yield occurred. This finding was highly suggestive that a lower molecular weight titin molecule was shorter in the I-band region and, therefore, became taut at shorter sarcomere lengths shifting the passive force–length relationship to the left.

Subsequent molecular structure and immunocytochemical studies of titin (Furst and Gautel, 1995; Gautel and Goulding, 1996; Labeit and Kolmerer, 1995; Linke *et al.*, 1996; Politou *et al.*, 1995; Trinick, 1996) confirmed the basic premise of Wang's segmental-extension model (K. Wang *et al.*, 1991, 1993). However, these new studies suggest a more complicated elastic sys-tem than originally envisioned. For example, Labeit and Kolmerer (1995) have suggested that there are multiple isoforms of titin that can be expressed even in the same muscle type. Indeed, in the PEVK and repeating Ig domains, segments are inserted in or deleted from titin depending on the isoform or muscle type. Thus, a shorter titin I-band segment would provide a shorter molecular tether that would shift the passive force–length relationship to the left. Although still unresolved, the dual stage spring action may reside only in the I-band region with differential elasticity of the PEVK and repeating Ig domains. Linke *et al.* (1996) carefully tracked several antibody epitopes in the I-band on stretch of skeletal and cardiac myofibrills and reported that the stretch and molecular elasticity is nonuniform in the I-band. Based on these data, Linke *et al.* (1996) have proposed a complicated sequence of molecular unfolding and elasticity with passive stretch, suggesting that the PEVK segment may determine the tissue-specific stiffness, whereas the Ig domain in the I-band sets the slack length and yield point for that muscle. Qualitatively similar results were reported by Granzier *et al.* (1996).

K. Wang *et al.* (1991) suggested that the steeper passive force–length relationship in cardiac muscle was due to the lower molecular weight and presumably shorter titin molecule. The passive force–length relationships for mammalian skeletal muscle fibers start to rise at sarcomere lengths of 2.8–3.6 μm and have yield points in the 3.8–4.8-μm range. The data from cardiac cells and myofibrils fit Wang's hypothesis. The cardiac passive force–length relationship starts to rise in the 2.0–2.2-μm range in data from this (see Fig. 17; Palmer *et al.*, 1996) and other laboratories (Granzier and Irving, 1995; Linke *et al.*, 1996). The yield point is less well defined. Granzier and Irving (1995) and Linke *et al.* (1996) report cardiac yield points in the 2.9–3.0-μm range based on the onset of the tension plateau. Fig. 18 illustrates a WKY myocyte subjected to a multiple stretch–relaxation protocol in this laboratory. The first two stretches ($\bigcirc$ and $\blacksquare$) to 2.4 μm and back did not alter the oscillatory stiffness modulus–length relationship. However, the third stretch ($\triangle$) to 2.8 μm significantly reduced the stiffness in the subsequent fourth stretch ($\blacktriangledown$). Based on this and other cell data, we suggest that the yield point for rat myocytes is in the 2.7–2.8-μm range. This value is slightly less than that reported elsewhere and may reflect differences in the preparation, attachment methods, or stretch protocols.

The structural source of the restoring forces is more difficult to identify than that related to passive stretch (Brady 1984, 1991b). Possible structural sources could involve a combination of several entities, including the compression or stretch of longitudinal or circumferential components or the stretch of radial components in the Z- or M-lines. When a cell contracts, its isovolumetric nature means that the cell gets wider as it shortens. This strains any circumferential and radial structures that could store forces generated during the shortening to later restore the cell to its normal shape during relaxation. Indeed, Goldstein *et al.* (1989) noted considerable structural changes in the Z-band lattice during contraction. Although these changes in the α-actinin and desmin Z lattice could be solely from contractile forces of the thin filaments, some could be from radial and circumferential

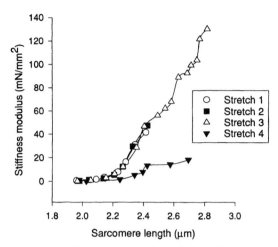

FIGURE 18   Yield point with passive stretch. An isolated Wistar-Kyoto (WKY) rat cell is repeatedly stretched and relaxed to increasingly extended lengths. The oscillatory stiffness modulus–length relationships of first and second stretches (○ and ■) to 2.4 μm are superimposible. The third and fourth stretches (∇ and ▼) are to about 2.8 μm, but only the third stretch superimposes upon the first and second. At the fourth stretch, the cell has lost considerable stiffness, having surpassed its yield point. See text for details. (Original data courtesy Dr. Roy E. Palmer.)

stretching. Interestingly, Fabiato and Fabiato (1978) reported that detergent-treated cells did not relengthen as did their mechanically skinned cells. This finding suggests the importance of interfibrillar structures such as the intermediate filaments that might be released by membrane disruption. These structures are different from the longitudinal structures, such as titin, which resist stretch and are not affected by detergents. In terms of longitudinal compressive forces, there is no obvious cytoskeletal component that could act as a compressed spring. However, Helmes *et al.* (1996) recently suggested that titin could be involved in restoring the cell's diastolic length following shortening. However, these data were derived from detergent skinned cells that were not isovolumetric. As a whole, current data do not permit a clear identification of a structure or structures that produce the restoring force in cardiac myocytes.

C. SARCOMERE LENGTH DYNAMICS AND UNIFORMITY IN ISOLATED CARDIAC MYOCYTES

As previously discussed, sarcomere or cell length is an essential parameter in the assessment of mechanical function in cardiac myocytes. Both systolic and diastolic function are sensitive to changes in sarcomere length, which indicate the degree of myofilament interdigitation in the myocyte.

Changes in sarcomere length at the end of diastole (preload) modulates the stroke volume of the heart by moving up or down the ascending limb of the active force–length (Starling) relationship (see left side of curve b, Fig. 14). In addition, the sarcomere length during the contraction modulates the calcium sensitivity of the troponin–tropomyosin regulatory system to steepen this part of the ascending limb (see Section IV, Figs. 13 and 14).

In an isolated cell, free of the extracellular matrix, the degree of myofilament interdigitation or sarcomere length at rest is determined by a balance of passive internal restoring forces, which tend to elongate the cell, and internal compressive forces, which tend to shorten the cell. During a cell contraction with no external load, the forces developed by the cross-bridges temporarily alter the balance of passive forces to shorten the cell against the restoring force. In the whole heart or in multicellular preparations, the extracellular matrix may modulate the sarcomere length to an unknown degree by altering the balance of forces with an external factor. In this case, the cross-bridges must work against both the internal and external loads to shorten the cell. During relaxation, as the cross-bridges detach, elongation occurs as the cell returns to its preferred diastolic length, once again balancing all the internal and external opposing forces.

In this section, sarcomere length dynamics and uniformity data are described from resting and contracting isolated cardiac myocytes under various experimental conditions. Differences in contractile performance between normal and hypertrophied heart are highlighted. A discussion of the validity of using unloaded shortening as a measure of myocardial contractility concludes the section.

1. Sarcomere Length Uniformity in Cells at Rest

The average sarcomere periodicity reported in the literature for untethered mammalian cardiac myocytes at rest is typically in the 1.80–1.95-μm range (DeClerck et al., 1984; Gannier et al., 1993; Haworth et al., 1987; Kent et al., 1989; Krueger, 1988; Krueger et al., 1980, 1992; Leung, 1983b; Lundblad et al., 1986; Mann et al., 1991; Mukherjee et al., 1993; Nassar et al., 1987; Roos, 1986a, 1986b, 1987; Roos and Brady, 1982, 1989, 1990; Roos and Taylor, 1993; Roos et al., 1982; Siri et al., 1991; Sollott et al., 1996; Z.L. Wang et al., 1996; Wussling et al., 1987). Although there is cell-to-cell variability, this range of averaged lengths is consistent across the spectrum of mammalian species; location in the heart (e.g., RV vs LV, or endocardium vs epicardium); and relatively independent of calcium concentration, within the physiological range, in unskinned cells. Sollott et al. (1996) reported a small (~1%) elongation of unskinned rat myocytes during diastole when exposed to 6 mM BDM (2,3-Butanedione Monoxime). Because BDM is a potent inhibitor of force production in striated muscle, these data suggest that residual cross-bridge interaction during diastole plays a minor but measurable role in the determination of resting sarcomere length. On average, there is a small increase in sarcomere length (from 1.85 to 1.89 μm) observed upon chemical skinning in a population of rat myocytes (Roos and Brady, 1989). This is presumably due to small changes in the balance of forces with a reduction in cross-bridge interaction in relaxing solution and the effect of membrane disruption with its subsequent cell swelling. In the whole heart or in multicellular preparations, the average resting sarcomere length is generally reported to fall within a longer range of 2.0–2.4 μm (de Tombe and ter Keurs, 1992; Kentish et al., 1986; Krueger and Pollack, 1975; ter Keurs et al., 1980a, 1980b). In this case, the extracellular matrix or external loading conditions (preload) probably add an external factor that alters the balance of forces to extend the cell. The sarcomere length at which maximal force is developed (L_o) is in the 2.2–2.3-μm range. Thus, the isolated cardiac myocyte at rest expresses a relatively narrow range of average sarcomere lengths that are midway along the working range of the ascending limb of the force–length relationship.

Up to this point, the isolated cardiac myocyte has been viewed as a single unit whose sarcomere length is uniform throughout its volume. Thus, average values from a single or a few myocytes have been considered as representative of the whole population. This may not be the case. It is clear that there are regional differences in stress and strain at the level of the whole heart (Guccione et al., 1993; Holmes et al., 1995; Omens et al., 1996; Rodriquez et al., 1993). When the ventricles contract, there is considerable slippage and shear between the cell layers. How might these global changes affect myocyte mechanics at the subcellular and sarcomere levels?

Even though there is no difference, on average, between populations of cells derived from the RV versus the LV or the endocardium versus the epicardium, there is cell-to-cell variability. Within a population of cells from the same ventricle, there will be a spectrum of average sarcomere lengths obtained from cell to cell (Roos et al., 1982). In addition, there can be cell-to-cell variation in the passive (Palmer and Roos, unpublished observations, 1997) and active (see following discussion) properties from cell to cell. Thus, these cell-to-cell differences in averaged sarcomere length and force characteristics may reflect the local environment of a particular cell within the heart.

The structural organization of cardiac myocytes (see Fig. 2; Sommer and Johnson, 1979) is suggestive of cellular subdivisions that could operate with some degree of independence. The longitudinal strips of mitochondria and nuclei separate fields or domains of sarcomeres. Sarcomeres within a single domain are probably tightly coupled by desmin intermediate filaments at the Z-discs (see Fig. 4; Price, 1991), but the separations between domains are spanned by longer desmin tethers, which may not couple the sarcomeres between domains as tightly. Thus, each domain could function semi-independently in terms of sarcomere length, passive force, and contractile dynamics.

There have been a few reports evaluating the sarcomere length uniformity within different regions or domains of resting cardiac myocytes. DeClerck *et al.* (1984) reported cell-to-cell sarcomere length variability, but found similar Gaussian-like distributions of sarcomere length within small windows of the same rat cells. Krueger and Denton (1992) also observed Gaussian distributions of sarcomere lengths within whole guinea pig cells. However, higher resolution analysis revealed evidence for discrete populations of sarcomeres within a resting single cell. This confirms earlier data from this laboratory from one-, two-, and three-dimensionally reconstructed resting rat myocytes, which clearly showed discrete domains of sarcomeres with significantly different periodicities (Roos, 1986b, 1987; Roos and Brady, 1982; Roos *et al.*, 1982). High resolution one-dimensional line scans along the length of cells indicated longitudinal uniformity of sarcomere length (Roos and Brady, 1982; Roos *et al.*, 1982). Measured differences in individual sarcomere lengths were small and randomly distributed along these representative line scans. Later, Roos (1986b, 1987) fully analyzed all the sarcomere lengths within cell volumes and found that each cell could be divided into two, three, or more domains that ran along the full length of the myocyte. Domain boundaries correlated to longitudinally oriented noncontractile zones of mitochondria and nuclei. Sarcomere striation periodicity could be different between domains transversely across the cell, but it was always highly uniform within a domain along the length of the cell. Although the differencs in resting sarcomere length between domains could be small in some cases ($\sim$0.03 μm), they are significant in terms of the numbers of sarcomeres examined and the steepness of the ascending limb of the active force–length relation at these lengths (see curve b, Fig. 14).

This domain/field organization implies that the cell is not necessarily a single functional unit. During contraction, these domains could slip and shear just as cell layers do in the whole heart. Thus, the preferred average sarcomere length measured in isolated cardiac myocytes falls within a consistent range of values. But the field/domain organization of cardiac myocytes permits a variable degree of diastolic length nonuniformity within cells that might be expected to alter cellular mechanics.

2. Sarcomere Dynamics in Normal and Hypertrophied Cells

The contractile behavior of unloaded mammalian cardiac myocytes in response to electrical stimuli has been extensively examined with a variety of cell and sarcomere monitoring methods (see previous discussion). A large majority of these studies have examined qualitative changes in contractile function in response to specific inotropic or electrophysiological interventions (Boyett *et al.*, 1988; Delbridge *et al.*, 1989; Duthinh and Houser, 1988; Harding *et al.*, 1988; Haworth *et al.*, 1987; Krueger, 1988; Mukherjee *et al.*, 1993; Rich *et al.*, 1988; Spinale *et al.*, 1992; Spurgeon *et al.*, 1990; Steadman *et al.,* 1988; Zile *et al.,* 1995; and many others previously described in this book). A complete description of this large body of work is beyond the scope of this chapter; therefore, this section concentrates on the more limited body of work that describes sarcomere dynamics and uniformity in myocytes isolated from normal and hypertrophied heart.

Among the pioneers in the field, Krueger *et al.* (1980) utilized light diffraction and cine photomicrography methods to evaluate the shortening and relaxation behavior of untethered rat ventricular myocytes under physiological conditions. Sequences of paced electrical stimuli on these rat cells elicited an initial negative staircase (in extent and duration of shortening) followed by steady, superimposible shortening curves after the second or third beat. Sarcomere shortening behavior was also calcium concentration and stimulus rate dependent. Average sarcomere spacing appeared to be uniform throughout the contraction, although there were some unexplained shortening-related changes in the diffraction pattern light intensity. All of these factors are consistent with shortening behavior from multicellular preparations under zero load conditions and confirmed that the isolated myocyte was indeed a reasonable model of myocardial mechanics under these conditions.

Subsequent studies from this and other laboratories confirmed these findings in the rat and other preparations under various conditions. Using single line CCD imaging, Roos *et al.* (1982) reported no difference in the standard deviation of sarcomere lengths between resting and contracted rat myocytes as measured from the same representative linear segment of the cell. Using the same line image methods, Roos and Brady (1982) characterized the contractile dynamics from a number of representative groups of sarcomeres and noted that the peak rates of shortening and relaxation were of the same order of magnitude. Furthermore, the contraction occurred synchronously in all the individual sarcomeres along the entire length of these cells at room temperature, at least within the 13-ms intervals resolved by this system.

Other laboratories have also characterized sarcomere dynamics from cardiac myocytes using a variety of imaging and analysis methods (see previous discussion). DeClerck *et al.* (1984) analyzed regional sarcomere shortening behavior with a video method and also reported uniform sarcomere shortening behavior in isolated cells within the 20-ms temporal resolution limit of their system. Nassar *et al.* (1987) compared the ultrastructure and sarcomere dynamics between neonatal and adult rabbit myocytes with their 16.7-ms resolution video method. They reported faster, more vigorous contractions in the adult cells, which were attributed to the developmental differentiation of the SR and T tubules. Wussling *et al.* (1987) characterized the contractile properties and force–interval relationships of guinea pig and mouse myocytes with a 10-ms temporal resolution light diffraction system. Kent *et al.* (1989) characterized the contractile function of feline RV myocytes at 37° under viscous loading conditions with their 1-ms temporal resolution diffraction system. By bathing cells in saline solutions of increasing viscosity, they reported that both the rate and the extent of shortening varied inversely with viscosity, which presumably modulated the external load. In a variation on this theme, Sollott and Lakatta (1994) embedded fresh cells in fibrin and monitored their auxotonic contractile function and calcium dynamics at a series of stretched lengths. These cells shortened less in magnitude, shortened less rapidly, and had a shorter duration of Indo-1 calcium transients following a stretch. This increased viscosity or stretch applies an unknown but probably constant amount of external force to the cells during a contraction. In general, these data demonstrate reduced levels of sarcomere shortening with increasing load, which would be expected from the hyperbolic force–velocity relationship.

There has been considerable interest in evaluating the differences in contractile mechanics in cells isolated from normal and hypertrophied heart in an effort to identify the source of the pathophysiological changes associated with this disease process. The gross changes in contractile function at the whole heart and multicellular levels are fairly well understood during the development of hypertrophy and failure in most models (Brady, 1984; Conrad et al., 1995; Spinale et al., 1992; Zile et al., 1995). The characterization of isolated myocyte contraction assists in the separation of the cellular from the extracellular changes that occur in these pathophysiological models. However, a rigorous comparison requires that both the single cell and whole heart studies are performed under similar conditions. Although this is rarely done in the same study, some representative studies using various imaging methods include the contractile characterizations of pressure overload (arterially constricted) RV rat (Vescovo et al., 1989), RV feline (Mann et al., 1991), LV guinea pig (Siri et al., 1991), and volume overloaded (AV fistula) RV feline myocytes (Urabe et al., 1993). Most of these studies reported diminished contractile responses as compared to control, which correlate, in general, to diminished whole heart function under similar conditions.

Of particular interest to this laboratory are the mechanical changes that occur in the spontaneously hypertensive rat (SHR) relative to its Wistar-Kyoto (WKY) control. This polygenomic model of pressure overload hypertrophy is quite different from arterially constricted or hormonally induced models. In the SHR, blood pressure rises progressively to an elevated level (~200 mm/Hg) during the sixth to the tenth week of age. The heart is maintained in a hypertrophied state without failure for months. During this nonfailing, hypertrophied stage (e.g., ~3–16 months), the SHR heart develops more force, has a faster rate of tension development, has a longer time to peak tension, and has a similar rate of relaxation as compared to WKY hearts (Conrad et al., 1995). Only later, at 20 months of age or more, does the heart decompensate and fail.

Several reports in the literature (Brooksby *et al.*, 1992; Delbridge *et al.*, 1996; Kobayashi *et al.*, 1995) demonstrate that isolated myocytes from young (10–30 weeks old) SHR and WKY hearts have similar changes in contractile function as compared to the compensated 12-month-old hearts described by Conrad *et al.* (1995). In general, nonfailing but hypertrophied SHR myocytes shortened faster and to a greater extent than WKY myocytes under the same unloaded conditions with variable changes in their time to peak tension. These differences were independent of stimulus frequency and calcium concentrations greater than or equal to 1.5 mM (Brooksby *et al.*, 1992). At 1.0 mM calcium concentrations or less, the contractile response of these young adult SHR myocytes diminished greatly to a level below that of the WKY (Fig. 8 in Brooksby *et al.*, 1992; Delbridge *et al.*, 1996). Thus, the contractile function of nonfailing, hypertrophied SHR myocytes is more sensitive to external calcium concentration changes in the 0.5–1.5 mM range than control WKY cells. Kobayashi *et al.* (1995) reported changes of the same proportion in whole heart and myocyte mechanics, thus confirming that unloaded myocyte contractile function reflects that of the whole organ.

Figure 19 illustrates the steady state contractile time courses at 2.0 mM calcium and at 31° obtained in this laboratory from an SHR and a WKY myocyte isolated from 4-month-old animals. The sarcomere length is determined every 4 ms from an entire 2-D focal plane image of the cell. There are striking differences in the contractile time courses in these two cells. Although they have similar initial maximal rates of unloaded shortening, the SHR myocyte contracts to a greater extent and for a longer duration. Despite the usual cell-to-cell variability in diastolic sarcomere length, these

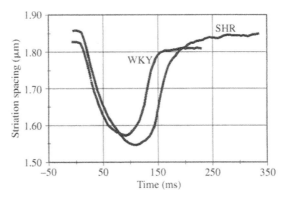

FIGURE 19   Sarcomere dynamics in normal and hypertrophied cells. The contractile time courses of average sarcomere length from a spontaneously hypertensive rat (SHR) cell and a Wistar-Kyoto rat (WKY) cell are plotted on the same time scale. In these rested state contractions in 2.0 mM Ca^{2+} at 31°C, the SHR cell shortens to a greater extent and for a longer period of time. The stimulus is at time = 0 ms. (Original data from K. P. Roos.)

results are typical of a population of cells (n = 20) under the same conditions, which demonstrated a 16% increase in the magnitude of shortening and a 24% increase in the time to peak tension. The maximal rates of shortening and relaxation were not significantly different in this population under these conditions. Although obtained at a slightly lower temperature, these data are consistent with the myocyte and whole heart data obtained at 37° in that these nonfailing hypertrophied hearts have an enhanced contractile response at physiological levels of calcium. As a whole, the variations in myocyte contractile data from different models of hypertrophied heart demonstrate the complicated nature of these pathophysiological processes.

3. Sarcomere Uniformity in Contracting Cells

All the previous studies assessing the contractile characteristics of myocytes either assumed cell length as an index of sarcomere length, measured all the sarcomeres in a focal plane section of a cell image, or measured small representative line segments of sarcomeres. There have been very few studies in contracting cells evaluating the potential sarcomere length differences between domains as described earlier for resting cells (Roos, 1986b, 1987). Although interpretation of light diffraction pattern fine structure is difficult, Leung (1983a, 1983b) suggested that there were discrete groups of sarcomeres whose dynamics could be followed during contractions. Roos and Taylor (1989) imaged 25-μm square segments of contracting rat myocytes at high speed and reported that all the sarcomeres within this segment contracted and relaxed synchronously. However, they noted that differences in absolute sarcomere periodicity were maintained between adjacent domains during a contraction. That is, if a domain had shorter diastolic sarcomere length relative to an adjacent domain, it remained proportionally shorter throughout a contraction. Krueger and Denton (1992) confirmed these results by simultaneously windowing dual sarcomere segments in contracting cells. They also observed synchronous contractile dynamics with a 1-ms time resolution, but they found sufficiently different sarcomere lengths in each windowed segment to consider them as separate populations. Recent work in this laboratory (Roos, unpublished observations, 1997) characterized the sarcomere dynamics in multiple domains from entire cells with a 4-ms time resolution. As with the smaller samples of sarcomeres, these data demonstrate consistent and uniform sarcomere dynamics within each domain as well as domains with different resting sarcomere lengths that are maintained during contraction. At this time, there is insufficient data to determine whether the extensive remodeling accompanying hypertrophy alters this domain behavior.

These subcellular and domain studies of sarcomere dynamics and uniformity reveal two important insights into mechanical function of the cell and heart. First, synchronous sarcomere dynamics imply that the cell contracts as a functional unit in terms of time. Even if there are cellular nonuniformities in calcium kinetics, these do not seem to alter cell shortening within a 1-ms time scale. It is unclear whether this temporal uniformity in shortening is due to the cooperative regulation at the cross-bridge level (Section IV) or from a tight longitudinal coupling of sarcomeres by the cytoskeleton (Section II). The second insight is that there is uniformity of sarcomere length within domains but nonuniformity among some domains. Because isolated myocytes are released from the extracellular matrix, the functional significance of these sarcomere length differences between domains remains unclear. At a minimum, these differences in sarcomere length strain reflect localized differences in the balance of cytoskeletal elastic structures that may arise from nonuniform stresses imposed by the whole heart during contraction. If these differences in sarcomere length occur to a great extent in cells in the intact heart, the steep ascending limb of the force–length relationship (see curve b, Fig. 14) would imply different levels of developed force within regions of the cell. But isolated myocytes rarely exhibit nonuniform motions that might be expected from this difference. Thus, the cell can be considered as a single contracting unit in terms of time but not necessarily in terms of mechanical stress and strain.

4. Cell Shortening as an Index of Contractility

Most of the previously described studies characterize the contractile dynamics from isolated myocytes that are not externally loaded. With the internal restoring force load being no more than 4–10% of the maximal active force (Fabiato and Fabiato, 1976; Helmes *et al.*, 1996; Niggli and Lederer, 1991; ter Keurs *et al.*, 1980b), the rate and magnitude of cell shortening is essentially maximal. It is generally assumed that the time course and magnitude of unloaded cell shortening is an index of the cell's contractility, indicative of the underlying inotropic status of the myocardium. However, cell shortening alone does not provide a complete evaluation of a cell's inotropic status (Allen and Kentish, 1985; Brady, 1991b; Delbridge and Roos, 1997). In reality, myocyte force development and shortening are related, but they are not equivalent indices of contractile dynamics. Thus, a complete measure of contractility must include force along with length. As discussed previously, this is technically very difficult in calcium-tolerant myocytes and has yet to be achieved over a full range of inotropic states (see the next section).

Because shortening cannot be directly related to cross-bridge function, the force developed during contraction is unknown in externally unloaded myocytes. No data or models have been developed that establish a predictable relationship between contractility and unloaded shortening (Brady, 1974, 1984, 1991b). As detailed by Brady (1991b), variables that can confound shortening and force relationships include the unknown effects of thin filament interactions, the internal load, and length- or time-dependent activation or relaxation factors. Some of these factors were discussed previously (Sections II and IV).

There are two significant components to the internal load. These are the static load and a viscous component that depends on the velocity of shortening. Thus, the amount of force necessary to shorten the cell may be different from that necessary to reextend the cell (Brady, 1991b; Krueger, 1988; Krueger and Denton, 1992). The static internal load estimated in the 4–10% range is like a spring with its level of force dependent on sarcomere length. Therefore, direct measures of sarcomere length, not just cell length, are important when evaluating contractile mechanics. Also, it should be noted that these estimates of static load are based on comparisons to maximal force. If a cell were only contracting at 40% maximal force (an activation level common to normal heart function), the internal load could be 25% of the force ($2.5 \times 10\%$).

In addition to the static internal load, the length- and time-dependent viscous component is important to consider. At low shortening velocities, the viscous damping is low, and a high velocities, it is large. The viscous damping could be as much as 70% of the total resistive load during rapid shortening (Niggli and Lederer, 1991). Thus, an increase in the unloaded shortening velocity measured following the application of a positive inotropic agent could significantly underestimate the true increase in cross-bridge force development by an unknown amount. The greater the inotropic change and the greater the velocity of shortening, the greater the error. This problem is not limited to isolated cells in that de Tombe and ter Keurs (1992) reported that the viscous component limited the maximal velocity of shortening in their trabecular preparations. Thus, at low levels of baseline inotropy, a small increase in contractility can elicit a relatively large change in apparent cell shortening. At relatively high levels of baseline inotropy, it would take a large additional increase in contractility to elicit even a small change in cell shortening behavior.

Despite these considerations, the qualitative consistency of data between multicellular and cellular studies suggest that comparative measurements of contractility are possible from unloaded shortening experiments under very carefully controlled conditions. The limited amount of data from cells lightly loaded by viscous solutions, gels, or compliant transducer systems (see next discussion, and Bluhm et al., 1995; Gannier et al., 1993; Garnier, 1994; Kent et al., 1989; Sollott and Lakatta, 1994; White et al., 1995) suggest that limited changes in load do not significantly alter the cell's inotropic status. For a more detailed evaluation of this issue, see Brady (1991b) and Delbridge and Roos (1997).

D. ACTIVE PROPERTIES OF CARDIAC MYOCYTES

The final parameter necessary to fully assess the mechanical function of isolated cardiac myocytes is the force developed during systole. Sufficient pressure must be developed in the ventricular chambers of the heart to eject blood into the circulatory system against an afterload. This pressure is determined in large part from the summation of systolic forces generated in each cell. The active force in a myocyte is, in turn, derived from the summation of the individual forces generated by cycling cross bridges along the myofilaments (Section III). The actual amount of cellular force or developed pressure is modulated by multiple structural and regulatory factors, some of which can vary with time or diastolic sarcomere length (Sections II and IV).

To characterize myocardial function in terms of interpretable force–length or force–velocity relationships, direct force and stiffness measurements have been obtained from a variety of preparations using the cell attachment methods described previously. In the first part of this section, data from steady-state measurements on skinned cells are described. These represent the majority of force studies that have been particularly useful in evaluating regulatory proteins and the length-dependent changes associated with activation of cardiac muscle (Section IV). Second, recent experiments examining transient mechanical responses from cells following rapid changes in length or substrate concentrations are discussed. These transient studies help correlate cellular function to cross-bridge kinetics (Section III). Finally, the few studies measuring forces from calcium-tolerant myocytes (intact sarcolemma) are described in relation to other mechanical function studies. All of these studies are related to systolic and diastolic function from normal and hypertrophied heart when applicable.

1. Steady-State Force Measurements from Skinned Cells

Direct force has been measured in a number of laboratories by attaching skinned cardiac myocytes to mechanical transducers as described previously. In addition to the assessment of a skinned cell's passive properties in relaxing solution, the cross-bridges in myocytes can be activated to develop steady-state levels of force by directly bathing the myofilaments in calcium-containing activation solutions. Figure 20 illustrates a typical activation experiment on a rat myocyte exposed to a series of activation solutions containing carefully controlled levels of calcium (pCa = 4.5–7.0) interspersed at each step with relaxing solution (pCa > 7.0). Each activation solution elicts a specific level of force according to the number of cross-bridges in a force generating state (according to the thin filament regulatory scheme described in Section IV). As before (see Fig. 16), the lower trace is the cell stiffness, which is demodulated from the direct force shown in the upper trace. Force–pCa relationships (see Fig. 10, Section IV) are derived by plotting the developed force for each level of calcium. As previously described (Section IV), rightward shifts in this curve indicate decreased myofilament calcium sensitivity, whereas leftward shifts indicate increased myofilament calcium sensitivity. Similarly, a steeper sigmoid

pCa 4.5   pCa 5.7   pCa 5.65   pCa 5.6   pCa 4.5

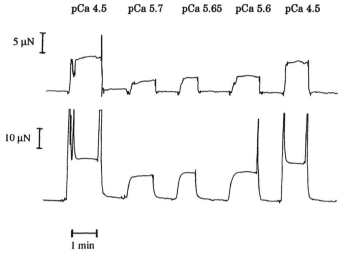

FIGURE 20   Activation protocol from skinned myocyte. Direct force (top record) and 100 Hz oscillatory stiffness (bottom record) are displayed during an activation protocol on a skinned Wistar-Kyoto (WKY) cell. The cell is initially bathed in relaxing solution (pCa < 7), which is rapidly exchanged for a series of activation solutions containing a range of calcium concentrations as indicated by the pCa value at the top. Between each activation, the cell is returned to the relaxing solution. There was no significant change in the force from the beginning to the end of the experiment at full activation. Note the high signal-to-noise ratio and the steady baseline. (Modified with permission from Palmer *et al.*, 1996.)

shape indicates an increased system cooperativity that can be characterized by the Hill coefficient, n_H (Hill, 1910).

The use of force–pCa relationships to assess the active mechanical function in skinned isolated cardiac myocytes was pioneered by A. Fabiato and colleagues (Fabiato, 1981; Fabiato and Fabiato, 1975, 1978). Using their elegant microtool attachment and solution exchange system, Fabiato and Fabiato (1975) generated and compared force–pCa relationships before and after mechanically skinned cells were detergent treated to remove their sarcoplasmic reticulum (SR). Their data were qualitatively similar in shape to those derived from skeletal muscle. Furthermore, they observed no significant difference in the myofilament-generated force with or without the SR present, which indicates that the developed force is directly related to the calcium concentration at the myofilaments regardless of its source. Interestingly, Fabiato and Fabiato (1975, 1978, 1981) observed cyclic contractions in mechanically skinned cells bathed in highly EGTA-buffered calcium solutions. These oscillations in force were attributed to a calcium-induced release of calcium by the SR, which has become a central component of cardiac excitation–contraction (E–C) coupling as discussed in Chapters 3–5. Finally, by plotting the maximal force developed in a series of force–pCa curves generated at different sarcomere lengths, the active force–length relationship was derived (Fabiato and Fabiato, 1978).

Similar active force–pCa results generated in this laboratory are shown in Fig. 10 (Section IV) from chemically skinned WKY rat cells. The data in Fig. 10 show the typical sigmoid relationship between developed force and calcium characteristic of all striated muscle. At a sarcomere length of 2.11 μm, this cell's relationship has a pCa_{50} (the pCa at which 50% of maximal force is developed) of 5.58 and a Hill coefficient (n_H) of 2.62. Normalized active force–pCa data from cardiac trabeculae (Kentish et al., 1986) and myofibrils (Linke et al., 1994) are virtually identical to those from these myocytes (see Fig. 10) and those from other laboratories. Interestingly, oscillatory contractile behavior similar to Fabiato's was noted at low levels of calcium buffering in myofibrils (Linke et al., 1993). Thus, active force–pCa data generated from skinned multicellular, myocyte, and myofibrillar preparations are qualitatively the same. However, quantitatively there can be differences in the force–pCa relationship between skinned cells evaluated under different experimental conditions.

A number of studies, predominantly from the laboratory of R.L. Moss, have used force–pCa relations as a tool to examine the functional roles of regulatory proteins, receptors, temperature, and sarcomere length in skinned myocytes glued to transducers. As previously discussed (Section IV), cardiac muscle expresses a length-dependent calcium sensitivity in both skinned and unskinned multicellular preparations (Allen and Kentish, 1985, 1988; Fuchs, 1995; Hofmann and Fuchs, 1988; Kentish, 1986; Solaro et al., 1993). This length dependency also has been demonstrated in skinned isolated myocytes under various conditions (Fabiato and Fabiato, 1978; Hofmann et al., 1991; McDonald et al., 1995; McDonald and Moss, 1995; Vannier et al., 1996). Of particular note is the work of McDonald and Moss (1995), which clearly related this length sensitivity to changes in myofilament lattice spacing with length and with osmotic compression. These findings are consistent with their transgenic mouse data (McDonald et al., 1995), which show that length-dependent calcium sensitivity was not altered by expression of skeletal instead of the cardiac TnC. Thus, these data are consistent with the interpretation of Fuchs and Wang (1996) that the length-dependent calcium sensitivity of cardiac muscle (Section IV) is most likely an inherent feature of the myofibrillar structure and function at the cellular level.

In other work, alterations in the force–pCa relationship have been characterized following various interventions. Sweitzer and Moss (1990) observed that the force–pCa relationship in rat cardiac myocytes was more sensitive to temperature than in skeletal muscle cells at a given sarcomere length. As previously discussed in Section IV, Hofmann et al. (1991) evaluated the regulatory role of C-protein in myocytes. They reported a diminished calcium sensitivity (rightward shift in the curve) and cooperativity (flattening of the curve) in response to partial extraction of C-protein in skinned cells. It was fully restored by its reconstitution. Using similar methods, β-adrenergic stimulation by PKA or isoproterenol also resulted in a decreased calcium sensitivity (Hofmann and Lange, 1994; Strang *et al.*, 1994), whereas there was no change in sensitivity with α_1-adrenergic receptor stimulation with phenylephrine (Strang and Moss, 1995). These data are consistent with those from multicellular cardiac preparations (Endoh and Blinks, 1988) and indicate that the increased contractility characteristic of β-adrenergic stimulation (Opie, 1995) is more likely due to an increased cycling rate of cross-bridges rather than an increased number of cross-bridges. Furthermore, the reduction of calcium sensitivity during β-adrenergic stimulation could explain the increased rate of relaxation by accelerating thin filament inactivation at the end of a contraction (Strang *et al.*, 1994). Despite this evidence for decreased myofilament calcium sensitivity in skinned cells, it is well known that β-adrenergic stimulation is positively inotropic in the intact tissue. This must mean that increased calcium entry at the channel level, effects on phospholamban at the SR level, or other possible effects compensate for these changes in myofilament sensitivity (see Chapters 3 and 5).

Because force-generation capabilities are altered in response to hemodynamic changes associated with hypertrophied and failing heart, the myofilament's calcium sensitivity might also be expected to change. In cardiomyopathic hamster heart, which has a diminished force-generating capability, the force–pCa relationship indicates an increased myofilament calcium sensitivity (leftward shift in curve) in multicellular preparations that was attributed to changes in troponin isoforms (Heyder *et al.*, 1995; Veksler and Ventura-Clapier, 1994). In biopsies obtained from dilated cardiomyopathic canine heart exhibiting diminished function, Wolff *et al.* (1995b) also noted a leftward shift in the force–pCa relationship. Furthermore, they reported an even greater rightward shift (diminished sensitivity) following β-adrenergic stimulation with PKA in cells from these failing dog hearts as compared to those from their controls and from the rat cell data of Strang *et al.* (1994). In nonfailing but hypertrophied SHR rat heart where the developed force is greater than control, Perreault *et al.* (1990) found no difference in multicellular-based force–pCa curves. However, preliminary data from this laboratory indicates a rightward shift (decreased sensitivity) and a steeper (increased cooperativity) force–pCa relationship generated from myocytes isolated from nonfailing, hypertrophied SHR hearts as compared to control WKY hearts. Although more work is necessary for a complete evaluation, these data from hypertrophied and failing heart suggest an inverse relationship between the heart's changes in contractile function and the myofilament's sensitivity to calcium. Calcium sensitivity modulation is likely one of many compensatory responses to the increased demand on the heart in these models of pressure overload hypertrophy.

Quantitatively, the differences in calcium sensitivity and cooperativity seen in force–pCa relationships from skinned cells measured under different experimental conditions can be confounded by a number of factors. Gao *et al.* (1994) and Kentish *et al.* (1986) reported differences in calcium sensitivity characteristics between skinned and unskinned preparations under the same conditions. This may be due to unknown changes in myofilament lattice spacing or in sarcomere length. Indeed, transducer compliance could also alter sarcomere length to an unknown degree during the course of an experiment. Few cell force measurements systems have concurrent sarcomere length monitoring capabilities sufficient to evaluate this potential problem (Palmer *et al.*, 1996). Furthermore, no one has yet developed a feedback system to clamp sarcomere length in skinned myocytes during an activation, as has been achieved in multicellular preparations (Hancock *et al.*, 1993, 1996; ter Keurs *et al.*, 1980a).

In addition to force–pCa studies that utilize the maximal force developed by a cell at a fixed length, a few studies have attempted to generate force–velocity relationships to characterize dynamic cross-bridge function at various loads and shortening velocities. DeClerck *et al.* (1977) obtained records from dissected papillary muscle strips attached to a force transducer and length controller. They were able to generate a hyperbolic force–velocity relationship over a limited range of values. Unfortunately due to technical limitations, their velocities were significantly lower than that obtained from similar multicellular papillary muscles in other labs. More recently, Sweitzer and Moss (1993) obtained hyperbolic force–velocity relationships similar to those of multicellular preparations from activated single myocytes subjected to ramp changes in length. They also demonstrated that the velocity of shortening depended on the myosin isoform and the degree of activation. Interestingly, at low levels of activation, the velocity of shortening against a constant load declined with time in a curvilinear fashion. These data suggested a shortening inactivation mechanism that was tied to the length-dependent calcium sensitivity and/or to an exponentially increasing passive restoring force. Thus, the single myocyte demonstrates the classical force–velocity relationship characteristic of the myocardium and all striated muscle.

2. Transient Force Measurements from Skinned Cells

The studies of skinned cardiac myocytes described previously are restricted to steady-state measures of mechanical function following changes in length or in the external medium. To study faster kinetic processes of cardiac contraction and relaxation, rapid length transient and flash photolysis methods have been developed. Although these have been mostly applied to the study of skeletal muscle fiber kinetics, there are some studies that characterize cross-bridge kinetics in various cardiac preparations.

The maximal velocity of unloaded shortening (V_{max}) can be estimated from the time course of the redevelopment of tension following a rapid shortening of the muscle to zero load during activation—the "slack test." Because V_{max} is generally believed to be determined by the rate of cross-bridge detachment (A.F. Huxley, 1957, 1988, 1995), these data provide an index of the cross-bridge cycling rate under the specified experimental conditions. Hofmann and Moss (1992) evaluated the rate of unloaded shortening velocity using the slack test method in frog atrial and rat ventricular cells and reported a calcium concentration dependence in both preparations. In a follow-up study, Hofmann and Lange (1994) observed no change in unloaded shortening velocity of rat myocytes in response to β-adrenergic stimulation even though their calcium sensitivity was diminished (see previous discussion). Conversely, Strang et al. (1994) found a significant increase in the unloaded shortening velocity with β-adrenergic stimulation in the same preparation in similar experiments. Later, they observed a significant decrease in the unloaded shortening velocity with α_1-adrenergic stimulation without changes in calcium sensitivity (Strang and Moss, 1995). Although still unresolved, these later two studies suggest a correlation between adrenergic stimulation and cross-bridge cycling rates that can alter the inotropic state of the myocyte through thick filament regulatory mechanisms (Section IV).

Cross-bridge detachment and attachment kinetics can be examined by evaluating the redevelopment of active force resulting from a rapid shortening of an activated preparation to zero load followed by an immediate restretch back to the initial length. The time course of the force transient reflects the cross-bridge kinetics and is described by the exponential rate constant K_{tr} (Brenner et al., 1995). Considerable work has been done using this investigative tool in skeletal muscle where K_{tr} varies 10-fold from low to high activation levels in a nonlinear fashion. However, the situation in cardiac muscle remains controversial and mostly limited to data from multicellular preparations. Hancock et al. (1993, 1996) found no changes in K_{tr} with activation levels in ferret papillary and rat trabeculae. These data suggest that calcium does not modulate the kinetics of cross-bridge attachment or detachment in cardiac muscle as it does in skeletal muscle. Conversely, Wolff et al. (1995a) observed that K_{tr} varied linearly over a wide range of calcium concentrations in rat trabeculae. Also, the range of calcium sensitivity is much greater in these cardiac cells than that seen in fast skeletal muscle and is similar to that of slow skeletal muscle. These data are consistent with a model in which calcium is a graded regulator of both the extent and rate of cross-bridge binding to the thin filament, particularly at low levels of activating calcium. Finally, in isolated rat myocytes, Vannier et al. (1996) have reported a calcium dependence in K_{tr} similar to that of Wolff et al. (1995a). The differences in K_{tr} results between the various laboratories are likely due to significant differences in the methodology.

The development of a technique to flash photolyze "caged compounds" within skinned muscle (Ellis-Davies *et al.*, 1996; Homsher and Millar, 1990) has overcome the diffusion delay associated with solution exchange. The principle behind this method is to allow an inactive photolabile precursor to diffuse into a skinned cell. Upon illumination with a brief pulse of light at the appropriate wavelength, the precursor undergoes a rapid photochemical change to release the desired product, such as ATP or calcium, on a millisecond time scale in the immediate environment of the myofibrils. This photolysis method has enabled the determination of the rate constants in many steps in the ATP-driven contractile dynamics in skeletal muscle (Section III). In contrast, this same approach has not, as yet, been widely applied to reveal the cross-bridge kinetics in cardiac muscle.

Caged ATP has been successfully used in skinned trabeculae to reveal the complex multiexponential relaxation processes from rigor (step 1 in cross-bridge cycle, Section III) and subsequent development of tension at various calcium concentrations. In guinea pig heart, Martin and Barsotti (1994a, 1994b) found that the cross-bridge detachment rate from rigor was calcium dependent following a step increase in ATP concentration. In swine heart, Morano *et al.* (1995b) were able to correlate cross-bridge kinetics to myosin isoforms in both atrial and ventricular samples. Similarly, both caged calcium and the caged calcium chelator, diazo-2, have been used in guinea pig trabeculae (Simnett *et al.*, 1993) and in isolated rat myocytes (Araujo and Walker, 1994, 1996). These data suggest a difference in the calcium and magnesium dependency of tension development between skeletal and cardiac muscle that are not attributable solely to differences in troponin isoforms. Although still limited in number, these flash photolysis studies from cardiac muscle preparations demonstrate a clear difference from skeletal muscle's cross-bridge kinetics.

3. Force Measurements from Intact Cells

Although skinned cells provide considerable insight into the mechanisms of contraction (discussed previously), the mechanical evaluation of active force from intact (unskinned) cells potentially provides data less affected by artifact due to skinning and more comparable to whole organ. Despite considerable effort, little progress has been made in fully evaluating the force characteristics from unskinned mammalian ventricular myocytes activated through the normal membrane depolarization and E–C coupling mechanisms. Brady *et al.* (1979) pioneered the single-barreled micropipette method to measure paced twitch from field-stimulated rat myocytes. Using single-barreled pipettes with continuous vacuum suction and no glue, they were able to record stable twitch forces in the 4-μN range from unskinned rat myocytes in 1 mM external calcium. Shortly afterward, Fabiato (1981) adapted his microtool attachment method to record forces from intact rat and rabbit myocytes in the μN range at physiological (2.5 mM) levels of calcium. When normalized to the cross-sectional area of the cells, the data from both Fabiato's and Brady's experiments were comparable to that of papillary muscle studies (Brady *et al.*, 1981; Brady, 1984, 1991b).

More recently, several investigators have measured forces from un-skinned cells glued to glass needles, plates, or carbon fibers. Copelas *et al.* (1987) used fibrin to attach toad and ferret ventricular myocytes and mea-sure force at short sarcomere lengths. Shepherd *et al.* (1990) used poly-L-lysine to stick the sides of guinea pig cells between two rods whose deflection indicated force. They simultaneously measured auxotonic force in the mN range and ionic currents from voltage clamped cells. Bluhm *et al.* (1995) stuck the sides of rabbit myocytes with poly-L-lysine to a force transducer and measured the active force–interval and force–length relationships in calcium-tolerant rabbit myocytes. Their data were comparable to the me-chanical properties of rabbit papillary muscle under similar conditions. A more productive approach has been to stick the sides of intact guinea pig myocytes to carbon fibers and determine force from the degree of deflection during electrically stimulated contractions (Gannier *et al.*, 1993; Garnier, 1994; White *et al.*, 1995). These highly compliant fibers permit the assess-ment of auxotonic (both force and length vary with time) force over a limited range of sarcomere lengths (Gannier *et al.* 1993; Garnier, 1994). Unfortunately, there is also an unknown amount of nonuniform sarcomere strain resulting from side attachment of myocytes to the transducers that restrains the sarcomeres on the attached side more than those on the opposite side. The degree of attachment-induced strain may also vary with time during a contraction. Thus, the measurements are generally limited to auxotonic contractions over a limited range of force with potentially nonuniform changes in sarcomere length with time.

Despite these concerns, White *et al.* (1995) measured auxotonic contrac-tions simultaneously with calcium dynamics (fura-2) from voltage-clamped guinea pig myocytes. Using less compliant carbon fibers than previously available, they reported that mechanical loading of cells decreased the amount of shortening (and therefore force) and abbreviated the contraction duration. But there were no consistent changes in the action potential or the calcium transients. Furthermore, increases in diastolic cell length (increased preload) or in the stimulation rate increased the amount of auxotonic shortening (positive staircase) but again did not alter the calcium transients. Thus, these data suggest the mechanisms underlying mechanical loading and length change responses in cardiac muscle are the same in both multicellular and isolated myocyte preparations.

In summary, the isolated cardiac myocyte and its molecular constituents provide a simplified structural and functional model for the evaluation of myocardial mechanics that cannot be achieved with whole heart or multicellular preparations. However, the evaluation of mechanical function from cardiac myocytes in terms of force, length, and time has been limited by technical constraints arising from the small size of the preparation and its lack of natural attachment sites. Thus, numerous methods have been utilized by various investigators to determine cell length, sarcomere periodicity, force, and stiffness from skinned or unskinned cardiac myocytes. The diastolic (passive) properties expressed in isolated cardiac myocytes have been found to be a significant contributor to the heart's overall mechanical function. Furthermore, data from hypertrophied heart suggest that cardiomyopathic changes in the cytoskeleton can alter the diastolic function in cells and therefore the heart. Cytoskeletal structures such as titin and the intermediate filaments are the primary candidates for the source of the passive forces resisting stretch and the restoring forces resist shortening in cardiac myoctes. Resting sarcomere lengths and their uniformity in cells provide insight into the balance of internal forces from the cytoskeleton and myofilaments. Similarly, sarcomere length dynamics in unloaded contracting cells, free from the confounding effect of the extracellular matrix, provides a more direct measure of contractile function at the cross-bridge level. Although contractile dynamics of unloaded cells may still be confounded by a number of factors, qualitative changes resulting from inotropic interventions or cardiac hypertrophy assist in our understanding of myocardial function at multiple levels. Finally, the assessment of steady-state and transient forces from skinned or unskinned isolated cardiac myocytes has provided considerable new information regarding the systolic function of the heart under externally loaded conditions at the cellular and molecular levels. Although limited in number, transient studies reveal significant differences between cardiac and skeletal muscle function at the cross-bridge level. Combined, these force, length, and time evaluations from isolated cardiac myocytes have provided significant new insights into the systolic and diastolic function in the mammalian heart.

ACKNOWLEDGMENTS

The author sincerely appreciates Dr. Langer's invitation to contribute to this book, his excellent editorial guidance, and his extraordinary patience throughout this project. The author is also indebted to the many other investigators who provided micrographs and figures of their data for inclusion here. I am particularly grateful to Dr. Roy E. Palmer who provided original unpublished data. In addition, I appreciate the careful reading of this chapter and comments provided by Drs. Terrence W. Miller and Roy E. Palmer. Finally, I appreciate the intellectual contributions of Dr. Allan J. Brady whose work with me over the decades helped shape the content of this chapter. Original data for this chapter were funded by NIH HL-47065 (KPR).

REFERENCES

Allen, D.G., and Kentish, J.C. (1985). The cellular basis of the length–tension relation in cardiac muscle. *J. Mol. Cell Cardiol.* 17:821–840.

Allen, D.G., and Kentish, J.C. (1988). Calcium concentration in the myoplasm of skinned ferret ventricular muscle following changes in muscle length. *J. Physiol.* (London) 407:489–503.

Allen, T.S., Ling, N., Irving, M., and Goldman, Y.E. (1996). Orientation changes in myosin regulatory light chains following photorelease of ATP in skinned muscle fibers. *Biophys. J.* 70:1847–1862.

Amos, L. A. (1985). Structure of muscle filaments studied by electron microscopy. *Annu. Rev. Biophys. Chem.* 14:291–313.

Anderson, P.A.W., Greig, A., Mark, T.M., Malouf, N.N., Oakeley, A.E., Ungerleider, R.M., Allen, P.D., and Kay, B.K. (1995). Molecular basis of human cardiac troponin T isoforms expressed in the developing adult and failing heart. *Circ. Res.* 76:681–686.

Araujo, A., and Walker, J.W. (1994). Kinetics of tension development in skinned cardiac myocytes measured by photorelease of Ca^{2+}. *Am. J. Physiol.-Heart Circ. Phys.* 36:H1643–H1653.

Araujo, A., and Walker, J. W. (1996). Phosphate release and force generation in cardiac myocytes investigated with caged phosphate and caged calcium. *Biophys. J.* 70:2316–2326.

Ausma, J., Furst, D., Thone, F., Shivalkar, B., Flameng, W., Weber, K., Ramaekers, F., and Borgers, M. (1995). Molecular changes of titin in left ventricular dysfunction as a result of chronic hibernation. *J. Mol. Cell Cardiol* 27:1203–1212.

Babu, A., Rao, V.G., Su, H., and Gulati, J. (1993). Critical minimum length of the central helix in troponin-C for the Ca^{2+} switch in muscular contraction. *J. Biol. Chem.* 268:19232–19238.

Banos, F.G.D., Bordas, J., Lowy, J., and Svensson, A. (1996). Small segmental rearrangements in the myosin head can explain force generation in muscle. *Biophys. J.* 71:576–589.

Block, S. M. (1990). Optical tweezers: A new tool for biophysics. *In* "Non-Invasive Techniques in Cell Biology" (Foskett, J.K., and Ginstein, S. eds.), pp. 375–402. Wiley-Liss, New York.

Bluhm, W.F., McCulloch, A.D., and Lew W.Y.W. (1995). Active force in rabbit ventricular myocytes. *J. Biomech.* 28:1119–1122.

Blumberg, F., Hort, C., and Hort, W. (1995). Distribution of myofibrils and mitochondria in ventricular myocytes. A quantitative light and electron microscopic study in dog and chicken hearts. *Z. Kardiol.* 84:154–162. (In German).

Bobkov, A.A., Bobkova, E.A., Lin, S.H., and Reisler, E. (1996). The role of surface loops (residues 204–216 and 627–646) in the motor function of the myosin head. *Proc. Natl. Acad. Sci. USA* 93:2285–2289.

Boyett, M.R., Moore, M., Jewell, B.R., Montogmery, R.A.P., Kirby, M.S., and Orchard, C.H. (1988). An improved apparatus for the optical recording of contraction of single heart cells. *Pflugers Arch.* 413:197–205.

Brady, A. J. (1974). Mechanics of the myocardium. *In* "The Mammalian Myocardium" (Langer, G.A. and Brady, A.J., eds.), pp. 163–192. John Wiley & Sons, New York.

Brady, A.J. (1984). Contractile and mechanical properties of the myocardium. *In* "Physiology and Pathophysiology of the Heart" (Sperelakis, N., ed.), pp. 279–299. Martinus Nijhoff Publishing, Boston.

Brady, A.J. (1990). Passive elastic properties of cardiac myocytes relative to intact cardiac tissue. *Issues Biomed.* 13:37–52.

Brady, A.J. (1991a). Length dependence of passive stiffness in single cardiac myocyts. *Am. J. Physiol.* 260:H1062–H1071.

Brady, A.J. (1991b). Mechanical properties of isolated cardiac myocytes. *Physiol. Rev.* 71:413–428.

Brady, A.J., and Farnsworth, S.P. (1986). Cardiac myocyte stiffness following extraction with detergent and high salt solutions. *Am. J. Physiol.* 250:H932–H943.

Brady, A.J., Tan, S.T., and Ricchiuti, N.V. (1979). Contractile force measured in unskinned isolated adult rat heart fibres. *Nature* 282:728–729.

Brady, A.J., Tan, S.T., and Ricchiuti, N.V. (1981). Perturbation measurements of papillary muscle elasticity. *Am. J. Physiol.* 241:H155–H173.

Bremel, R.D., and Weber, A. (1972). Cooperation within actin filament in vertebrate skeletal muscle. *Nature New Biol.* 238:97–101.

Brenner, B., and Eisenberg, E. (1987). The mechanism of muscle contraction. Biochemical, mechanical, and structural approaches to elucidate cross-bridge action in muscle. *Basic Res. Cardiol.* 82:3–16.

Brenner, B., Chalovich, J.M., and Yu, L.C. (1995). Distinct molecular processes associated with isometric force generation and rapid tension recovery after quick release. *Biophys. J.* 68:S106–S111.

Brooksby, P., Levi, A.J., and Jones, J.V. (1992). Contractile properties of ventricular myocytes isolated from spontaneously hypertensive rat. *J. Hypertens.* 10:521–527.

Burton, K. (1992). Myosin step size–Estimates from motility assays and shortening muscle. *J. Muscle Res. Cell Motil.* 13:590–607.

Butters, C.A., Willadsen, K.A., and Tobacman, L.S. (1993). Cooperative interactions between adjacent troponin–tropomyosin complexes may be transmitted through the actin filament. *J. Biol. Chem.* 268:15565–15570.

Cecchi, G., Colomo, F., Poggesi, C., and Tesi, C. (1993). A force transducer and a length-ramp generator for mechanical investigations of frog-heart myocytes. *Pflugers Arch.-Eur. J. Physiol.* 423:113–120.

Censullo, R., and Cheung, H.C. (1994). Tropomyosin length and two-stranded F-actin flexibility in the thin filament. J. Mol. Biol. 243:520–529.

Chalovich, J. M. (1992). Actin mediated regulation of muscle contraction. *Pharmacol. Ther.* 55:95–148.

Cheney, R.E., Riley, M.A., and Mooseker, M.S. (1993). Phylogenetic analysis of the myosin superfamily. *Cell Motil. Cytoskeleton* 24:215–223.

Collins, J.F., Pawloski-Dahm, C., Davis, M.G., Ball, N., Dorn, G.W., and Walsh, R.A. (1996). The role of the cytoskeleton in left ventricular pressure overload hypertrophy and failure. *J. Mol. Cell. Cardiol.* 28:1435–1443.

Conrad, C.H., Brooks, W.W., Hayes, J.A., Sen, S., Robinson, K.G., and Bing, O.H.L. (1995). Myocardial fibrosis and stiffness with hypertrophy and heart failure in the spontaneously hypertensive rat. *Circulation* 91:161–170.

Cooke, R. (1995). The actomyosin engine. FASEB J. 9:636–642.

Cope, M.J.T.V., Whisstock, J., Rayment, I., and Kendrick-Jones, J. (1996). Conservation within the myosin motor domain: Implications for structure and function. *Structure* 4:969–987.

Copelas, L., Briggs, M., Grossman, W., and Morgan, J.P. (1987). A method for recording isometric tension development by isolated cardiac myocytes: Transducer attachment with fibrin glue. *Pflugers Arch.* 408:315–317.

Cuda, G., Fananapazir, L., Zhu, W.S., Sellers, J.R., and Epstein, N.D. (1993). Skeletal muscle expression and abnormal function of beta-myosin in hypertrophic cardiomyopathy. *J. Clin. Invest.* 91:2861–2865.

Dantzig, J.A., Goldman, Y.E., Millar, N.C., Lacktis, J., and Homsher, E. (1992). Reversal of the cross-bridge force-generating transition by photogeneration of phosphate in rabbit psoas muscle fibres. *J. Physiol.* (London) 451:247–278.

DeClerck, N. M., Claes, V.A., and Brutsaert, D.L. (1977). Force velocity relations of single cardiac muscle cells. *J. Gen. Physiol.* 69:221–241.

DeClerck, N.M., Claes, V.A., and Brutsaert, D.L. (1984). Uniform sarcomere behaviour during twitch of intact single cardiac cells. *J. Mol. Cell. Cardiol.* 16:735–745.

Delbridge, L.M.D., and Roos, K.P. (1997). Optical methods to evaluate the contractile function of unloaded isolated cardiac myocytes. *J. Mol. Cell. Cardiol.* 29:11–25.

Delbridge, L.M., Harris, P.J., and Morgan, T.O. (1989). Characterization of single heart cell contracility by rapid imaging. *Clin. Phys. Physiol. Meas.* 16:179–184.

Delbridge, L.M., Connell, P.J., Morgan, T.O., and Harris, P.J. (1996). Contractile function of cardiomyocytes from the spontaneously hypertensive rat. *J. Mol. Cell. Cardiol.* 28:723–733.

de Tombe, P.P., and ter Keurs, H.E.D.J. (1992). An internal viscous element limits unloaded velocity of sarcomere shortening in ray myocardium. *J. Physiol.* (London). 454:619–642.

Donald, T. C., Reeves, D.N.S., Reeves, R.C., Walker, A.A., and Hefner, L.L. (1980). Effects of damaged ends in papillary muscle preparations. *Am. J. Physiol.* 238:H14–H23.

Dong, W.J., and Cheung, H.C. (1996). Calcium-induced conformational change in cardiac troponin C studied by fluorescence probes attached to Cys-84. *Bba-Protein Struct. Mol. Enzym.* 1295:139–146.

Dool, J.S., Mak, A.S., Friberg, P., Wahlander, H., Hawrylechko, A., and Adams, M.A. (1995). Regional myosin heavy chain expression in volume and pressure overload induced cardiac hypertrophy. *Acta Physiol. Scand.* 155:397–404.

Duthinh, V., and Houser, S.R. (1988). Contractile properties of single isolated feline ventriculocytes. *Am. J. Physiol.* 254:H59–H66.

Eisenberg, B.R. (1983) *Skeletal muscle.* In "Handbook of Physiology" (Peachey, L.D., Adrian, R.H., and Geiger, S.R., eds.), pp. 73–112. American Physiological Society, Bethesda, MD.

Eisenberg, E., Hill, T. L., and Chen, Y. (1980). Cross-bridge model of muscle contraction. *Biophys. J.* 29:195–227.

Ellis-Davies, G.C.R., Kaplan, J.H., and Barsotti, R.J. (1996). Laser photolysis of caged calcium: Rates of calcium release by nitrophenyl-EGTA and DM-nitrophen. *Biophys. J.* 70:1006–1016.

Endoh, M., and Blinks, J.R. (1988). Actions of sympathomimetic amines on the Ca^{2+} transients and contractions of rabbit myocardium: Reciprocal changes in myofibrillar responsiveness to Ca^{2+} mediated through α- and β-adrenoceptors. *Circ. Res.* 62:247–265.

Fabiato, A. (1981). Myoplasmic free calcium concentration reached during the twitch of an intact isolated cardiac cell and during calcium-induced release of calcium from the sarcoplasmic reticulum of a skinned cardiac cell from the adult rat or rabbit ventricle. *J. Gen. Physiol.* 78:457–497.

Fabiato, A., and Fabiato, F. (1975). Contractions induced by a calcium-triggered release of calcium from the sarcoplasmic reticulum of single skinned cardiac cells. *J. Physiol.* (London) 249:469–495.

Fabiato, A., and Fabiato, F. (1976). Dependence of calcium release, tension generation and restoring forces on sarcomere length in skinned cardiac cells. *Eur. J. Cardiol.* 4/Suppl: 13–27.

Fabiato, A., and Fabiato, F. (1978). Myofilament-generated tension oscillations during partial calcium activation and activation dependence of the sarcomere length-tension relation of skinned cardiac cells. *J. Gen. Physiol.* 72:667–699.

Fananapazir, L., and Epstein, N.D. (1994). Genotype-phenotype correlations in hypertrophic cardiomyopathy—Insights provided by comparisons of kindreds with distinct and identical beta-myosin heavy chain gene mutations. *Circulation* 89:22–32.

Finer, J.T., Simmons, R.M., and Spudich, J.A. (1994). Single myosin molecule mechanics—Piconewton forces and nanometre steps. *Nature* 368:113–119.

Fish, D., Orenstein, J., and Bloom, S. (1984). Passive stiffness of isolated cardiac and skeletal myocytes in the hamster. *Circ. Res.* 54:267–276.

Ford, L.E., Huxley, A.F., and Simmons, R.M. (1977). Tension responses to sudden length change in stimulated frog muscle fibres near slack length. *J. Physiol.* (London) 269:441–515.

Frank, J.S. (1989). Ultrastructure of the sarcolemma of isolated cardiomyocytes. *In* "Isolated Adult Cardiomyocytes" (Piper, H.M., and Isenberg, G., eds.), pp. 125–143. CRC Press, Boca Raton.

Frank, J.S., Brady, A.J., Farnsworth, S.P., and Mottino, G. (1986). Ultrastructure and function of isolated myocytes after calcium depletion and repletion. *Am. J. Physiol.* 250:H265–H275.

Fraser, I.D.C., and Marston, S.B. (1995). *In vitro* motility analysis of actin-tropomyosin regulation by troponin and calcium—The thin filament is switched as a single cooperative unit. *J. Biol. Chem.* 270:7836–7841.

Freiburg, A., and Gautel, M. (1996). A molecular map of the interactions between titin and myosin-binding protein C—Implications for sarcomeric assembly in familial hypertrophic cardiomyopathy. *Eur. J. Biochem.* 235:317–323.

Fuchs, F. (1995). Mechanical modulation of the Ca^{2+} regulatory protein complex in cardiac muscle. *NIPS* 10:6–11.

Fuchs, F., and Wang, Y.P. (1996). Sarcomere length versus interfilament spacing as determinants of cardiac myofilament Ca^{2+} sensitivity and Ca^{2+} binding. *J. Mol. Cell Cardiol.* 28:1375–1383.

Furst, D.O., and Gautel, M. (1995). The anatomy of a molecular giant: How the sarcomere cytoskeleton is assembled from immunoglobulin superfamily molecules. *J. Mol. Cell Cardiol.* 27:951–959.

Gannier, F., Bernengo, J.C., Jacquemond, V., and Garnier, D. (1993). Measurements of sarcomere dynamics simultaneously with auxotonic force in isolated cardiac cells. *IEEE Trans. Biomed. Eng.* 40: 1226–1232.

Gao, W.D., Backx, P.H., Azan-Backx, M., and Marban, E. (1994). Myofilament Ca^{2+} sensitivity in intact versus skinned rat ventricular muscle. *Circ. Res.* 74:408–415.

Garnier, D. (1994). Attachment procedures for mechanical manipulation of isolated cardiac myocytes: A challenge. *Cardiovasc. Res.* 28:1758–1764.

Gautel, M., and Goulding, D. (1996). A molecular map of titin/connectin elasticity reveals two different mechanisms acting in series. *FEBS LETT.* 385:11–14.

Gautel, M., Zuffardi, O., Freiburg, A., and Labeit, S. (1995). Phosphorylation switches specific for the cardiac isoform of myosin binding protein-C: A modulator of cardiac contraction? *EMBO. J.* 14:1952–1960.

Geeves, M.A., and Conibear, P.B. (1995). The role of three-state docking of myosin S1 with actin in force generation. *Biophys. J.* 68:S196–S201.

Gerdes, A. M., Kellerman, S.E., Malec, K.B., and Schocken, D.D. (1994). Transverse shape characteristics of cardiac myocytes from rats and humans. *Cardioscience* 5:31–36.

Gibbs, C.L., and Barclay, C.J. (1995). Cardiac efficiency. *Cardiovasc. Res.* 30:627–634.

Goldstein, M.A., Michael, L.H., Schroeter, J.P., and Sass, R.L. (1989). Two structural states of Z-bands in cardiac muscle. *Am. J. Physiol.* 256:H552–H559.

Gordon, A.M., Huxley, A.F., and Julian, F.J. (1966). The variation in isometric tension with sarcomere length in vertebrate muscle fibres. *J. Physiol.* (London) 184:170–192.

Granzier, H.L.M., and Irving, T.C. (1995). Passive tension in cardiac muscle: Contribution of collagen, titin, microtubules, and intermediate filaments. *Biophys. J.* 68:1027–1044.

Granzier, H., Helmes, M., and Trombitas, K. (1996). Nonuniform elasticity of titan in cardiac myocytes: A study using immunoelectron microscopy and cellular mechanics. *Biophys. J.* 70:430–442.

Guccione, J.M., McCulloch, A.D., and Hunter, W.C. (1993). Three-dimensional finite element analysis of anterior-posterior variations in local sarcomere length and active fiber stress during left ventricular ejection. *Adv. Bioeng.* 26:571–574.

Gulati, J., Sonnenblick, E., and Babu, A. (1991). The role of troponin C in the length dependence of Ca-sensitive force of mammalian skeletal and cardiac muscles. *J. Physiol.* (London) 441:305–324.

Gulati, J., Babu, A., and Su, H. (1992). Functional delineation of the Ca^{2+} deficient EF-hand in cardiac muscle, with genetically engineered cardiac-skeletal chimeric troponin C. *J. Biol. Chem.* 267:25073–25077.

Gulati, J., Akella, A.B., Nikolic, S.D., Starc, V., and Siri, F. (1994). Shifts in contractile regulatory protein subunits troponin T and troponin I in cardiac hypertrophy. *Biochem. Biophys. Res. Commun.* 202:384–390.

Guo, X.D., Wattanapermpool, J., Palmiter, K.A., Murphy, A.M., and Solaro, R.J. (1994). Mutagenesis of cardiac troponin I–Role of the unique NH2-terminal peptide in myofilament activation. *J. Biol. Chem.* 269:15210–15216.

Hackney, D.D. (1996). The kinetic cycles of myosin, kinesin, and dynein. *Annu. Rev. Physiol.* 58:731–750.

Hancock, W.O., Martyn, D.A., and Huntsman, L.L. (1993). Ca^{2+} and segment length dependence of isometric force kinetics intact ferret cardiac muscle. *Circ. Res.* 73:603–611.

Hancock, W.O., Martyn, D.A., Huntsman, L.L., and Gordon, A.M. (1996). Influence of Ca^{2+} on force redevelopment kinetics in skinned rat myocadium. *Biophys. J.* 70:2819–2829.

Harding, S.E., Vescovo, G., Kirby, M.S., Jones, S.M., Gurden, J., and Poole-Wilson, P.A. (1988). Contractile responses of isolated adult rat and rabbit cardiac myocytes to isoproterenol and calcium. *J. Mol. Cell. Cardiol.* 20:635–647.

Harris, D.E., Work, S.S., Wright, R.K., Alpert, N.R., and Warshaw, D.M. (1994). Smooth cardiac and skeletal muscle myosin force and motion generation assessed by cross-bridge mechanical interactions *in vitro. J. Muscle Res. Cell Motil.* 15:11–19.

Haworth, R.A., Griffin, P., Saleh, B., Goknur, A.B., and Berkoff, H.A. (1987). Contractile function of isolated young and adult rat heart cells. *Am. J. Physiol.* 253:H1484–H1491.

Hein, S., Scholz, D., Fujitani, N., Rennollet, H., Brand, T., Friedl, A., and Schaper, J. (1994). Altered expression of titin and contractile proteins in failing human myocardium. *J. Mol. Cell Cardiol.* 26:1291–1306.

Helmes, M., Trombitas, K., and Granzier, H. (1996). Titin develops restoring force in rat cardiac myocytes. *Circ. Res.* 79:619–626.

Heyder, S., Malhotra, A., and Ruegg, J.C. (1995). Myofibrillar Ca^{2+} sensitivity of cardiomyopathic hamster hearts. *Pflugers Arch.-Eur. J. Physiol.* 429:539–545.

Hibberd, M.G., and Trentham, D.R. (1996). Relationships between chemical and mechanical events during muscular contraction. *Ann. Rev. Biophys. Biophys. Chem.* 15:119–161.

Higuchi, H., and Goldman, Y.E. (1995). Sliding distance per ATP molecule hydrolyzed by myosin heads during isotonic shortening of skinned muscle fibers. *Biophys J.* 69:1491–1507.

Higuchi, H., Yanagida, T., and Goldman, Y.E. (1995). Compliance of thin filaments in skinned fibers of rabbit skeletal muscle. *Biophys. J.* 69:1000–1010.

Hill, A.V. (1910). The possible effects of the aggregation of the molecules of haemoglobin on its dissociation curve. *J. Physiol* (London) 40:4–7.

Hitchcock-DeGregori, S.E., and Varnell, T.A. (1990). Tropomyosin has discrete actin-binding sites with sevenfold and fourteenfold periodicities. *J. Mol. Biol.* 214:885–896.

Hoenger, A., Sablin, E.P., Vale, R.D., Fletterick, R.J., and Milligan, R.A. (1995). Three-dimensional structure of a tubulin–motor–protein complex. *Nature* 376:271–274.

Hofmann, P.A., and Fuchs, F. (1988). Bound calcium and force development in skinned cardiac muscle bundles: Effects of sarcomere length. *J. Mol. Cell Cardiol.* 20:667–677.

Hofmann, P.A., and Moss, R.L. (1992). Effects of calcium on shortening velocity in frog chemically skinned atrial myocytes and in mechanically disrupted ventricular myocardium from rat. *Circ. Res.* 70:885–892.

Hofmann, P.A., and Lange, J.H. (1994). Effects of phosphorylation of troponin-I and C protein on isometric tension and velocity of unloaded shortening in skinned single cardiac myocytes from rats. *Circ. Res.* 74:718–726.

Hofmann, P.A., Hartzell, H.G., and Moss, R.L. (1991). Alterations in Ca sensitive tension due to partial extraction of C-protein from rat skinned cardiac myocytes and rabbit skeletal muscle fibers. *J. Gen. Physiol.* 97:1141–1163.

Hoh, J.F.Y., McGrath, P.A., and Hale, P.T. (1977). Electrophoretic analysis of multiple forms of rat cardiac myosin: Effects of hypophysectomy and thyroxine replacement. *J. Mol. Cell Cardiol.* 10:1053–1076.

Holmes, J., Takamaya, Y., LeGrice, I., and Covell, J.W. (1995). Depressed regional deformation near anterior papillary muscle. *Am. J. Physiol. Heart Circ. Phys.* 38:H262–H270.

Holmes, K.C. (1995). The actomyosin interaction and its control by tropomyosin. *Biophys. J.* 68:S2–S7.

Holms, K.C., Popp, D., Gebhardt, W., and Kabsch, W. (1990). Atomic model of the actin filament. *Nature* 365:810–816.

Homsher, E. (1987). Muscle enthalpy production and its relationship to actomyosin ATPase. *Annu. Rev. Physiol.* 49:673–690.

Homsher, E., and Millar, N.C. (1990). Caged compounds and striated muscle contraction. *Annu. Rev. Physiol.* 52:875–896.

Homsher, E., Wang, F., and Sellers, J.R. (1992). Factors affecting movement of F-actin filaments propelled by skeletal muscle heavy meromyosin. *Am. J. Physiol.* 262:C714–C723.

Homsher, E., Kim, B., Bobkova, A., and Tobacman, L.S. (1996). Calcium regulation of thin filament movement in an *in vitro* motility assay. *Biophys. J.* 70:1881–1892.

Horowits, R., and Podolsky, R.J. (1987). The positional stability of thick filaments in activated skeletal muscle depends on sarcomere length: Evidence for the role of titin filaments. *J. Cell. Biol.* 105:2217–2223.

Houmeida, A., Holt, J., Tskhovrebova, L., and Trinick, J. (1995). Studies of the interaction between titin and myosin. *J. Cell Biol.* 131:1471–1481.

Howard, J. (1995). The mechanics of force generation by kinesin. *Biophys. J.* 68:S245–S255.

Hunt, A.J., Gittes, F., and Howard, J. (1994). The force exerted by a single kinesin molecule against a viscous load. *Biophys. J.* 67:766–781.

Huxley, A.F. (1957). Muscle structure and theories of contraction. *Prog. Biophys. Mol. Biol.* 7:255–318.

Huxley, A.F. (1988). Prefatory chapter: Muscular contraction. *Annu. Rev. Physiol.* 50:1–16.

Huxley, A.F. (1995). Muscle contraction mechanism. *Cardiovasc. Res.* 29:747–748.

Huxley, A.F., and Niedergerke, R. (1954). Structural changes in muscle during contraction. *Nature* 173:971–973.

Huxley, A.F., and Simmons, R.M. (1971). Proposed mechanism of force generation in striated muscle. *Nature* 233:533–538.

Huxley, H.E. (1957). The double array of filaments in cross-striated muscle. *J. Biophys. Biochem. Cyt.* 3:631–647.

Huxley, H.E. (1996). A personal view of muscle and motility mechanisms. *Annu. Rev. Physiol.* 58: 1–19.

Huxley, H.E., and Hanson, J. (1954). Changes in the cross-striations of muscle during contraction and stretch and their structural interpretation. *Nature* 173:973–976.

Improta, S., Politou, A.S., and Pastore, A. (1996). Immunoglobulin-like modules from titin I-band: Extensible components of muscle elasticity. *Structure* 4:323–337.

Irving, M., Allen, T.S., Sabido-David, C., Craik, J.S., Brandmeier, B., Kendrick-Jones, J., Corrie, J.E.T., Trentham, D.R., and Goldman, Y.E. (1995). Tilting of the light-chain region of myosin during step length changes and active force generation in skeletal muscle. *Nature* 375:688–691.

Ishijima, A., Kojima, H., Higuchi, H., Harada, Y., Funatsu, T., and Yanagida, T. (1996). Multiple-and single-molecule analysis of the actomyosin motor by nanometer piconewton manipulation with a microneedle: Unitary steps and forces. *Biophys. J.* 70:383–400.

Isobe, Y., Warner, F.D., and Lemanski, L.F. (1988). Three-dimensional immunogold localization of α-actinin within the cytoskeleton networks of cultured cardiac muscle and nonmuscle cells. *Proc. Nat. Acad. Sci. USA* 85:6758–6762.

Jin, J.P. (1995). Cloned rat cardiac titin class I and class II motifs–Expression, purification, characterization, and interaction with F-actin. *J. Biol. Chem.* 270:6908–6916.

Johnson, K.A. (1985). Pathway of the microtubule-dynein ATPase and the structure of dynein: A comparison with actomyosin. *Annu. Rev. Biophys. Chem.* 14:161–188.

Kawai, M., Saeki, Y., and Zhao, Y. (1993). Crossbridge scheme and the kinetic constants of elementary steps deduced from chemically skinned papillary and trabecular muscles of the ferret. *Circ. Res.* 73:35–50.

Keller, C.S. (1995). Structure and function of titin and nebulin. *Curr. Opin. Cell Biol.* 7:32–38.

Kent, R.L., Mann, D.L., Urabe, Y., Hisano, R., Hewett, K.W., Loughnane, M., and Cooper, G. (1989). Contractile function of isolated feline cardiocytes in response to viscous loading. *Am. J. Physiol.* 257:H1717–1727.

Kentish, J.C., ter Keurs, H.E.D.J., Ricciardi, L., Bucx, J.J.J., and Noble, M.I.M. (1986). Comparison between the sarcomere length–force relations of intact and skinned trabeculae from rat right ventricle. Influence of calcium concentrations of these relations. *Circ. Res.* 58:755–768.

Kikkawa, M., Ishikawa, T., Wakabayashi, T., and Hirokawa, N. (1995). Three-dimensional structure of the Kinesin head-microtubule complex. *Nature* 376:274–277.

Kinose, F., Wang, S.X., Kidambi, U.S., Moncman, C.L., and Winkelmann, D.A. (1996). Glycine 699 is pivotal for the motor activity of skeletal muscle myosin. *J. Cell Biol.* 134:895–909.

Kishino, A., and Yanagida, T. (1988). Force measurements by micromanipulation of a single actin filament by glass needles. *Nature* 334:74–76.

Kobayashi, T., Hamada, M., Okayama, H., Shigematsu, Y., Sumimoto, T., and Hiwada, K. (1995). Contractile properties of left ventricular myocytes isolated from spontaneously hypertensive rats: Effect of angiotension II. *J. Hypertens.* 13:1803–1807.

Kron, S.J., and Spudich, J.A. (1986). Fluorescent actin filaments move on myosin fixed to a glass surface. *Proc. Natl. Acad. Sci. USA* 83:6272–6276.

Krueger, J.W. (1988). Measurement and interpretation of contraction in isolated cardiac myocytes. *In* "Biology of Isolated Adult Cardiac Myocytes" (Clark, W.A., Decker, R.S., and Borg, T.K., eds.), pp. 172–186. Elsevier Science, New York.

Krueger, J.W., and Denton, A. (1992). High resolution measurement of striation patterns and sarcomere motions in cardiac muscle cells. *Biophys. J.* 61:129–144.

Krueger, J.W., and Pollack, G.H. (1975). Myocardial sarcomere dynamics during isometric contraction. *J. Physiol.* (London) 251:627–643.

Krueger, J.W., Forletti, D., and Wittenberg, B.A. (1980). Uniform sarcomere shortening behavior in isolated cardiac muscle cells. *J. Gen. Physiol.* 76:587–607.

Krueger, J.W., Denton, A., and Siciliano, G. (1992). Nature of motions between sarcomeres in asynchronously contracting cardiac muscle cells. *Biophys.J.* 61:145–160.

Kull, F.J., Sablin, E.P., Lau, R., Fletterick, R.J., and Vale, R.D. (1996). Crystal structure of the Kinesin motor domain reveals a structural similarity to myosin. *Nature* 380:550–555.

Labeit, S., and Kolmerer, B. (1995). Titins: Giant proteins in charge of muscle ultrastructure and elasticity. *Science* 270:293–296.

Lankford, E.B., Epstein, N.D., Fananapazir, L., and Sweeny, H.L. (1995). Abnormal contractile properties of muscle fibers expressing beta-myosin heavy chain gene mutations in patients with hypertrophic cardiomyopathy—Rapid publication. *J. Clin. Invest.* 95:1409–1414.

Lehrer, S.S. (1994). The regulatory switch of the muscle thin filament Ca^{2+} or myosin heads? *J. Muscle Res. Cell Motil.* 15:232–236.

Leung, A.F. (1983a). Light diffractory for determining the sarcomere length of striated muscle: An evaluation. *J. Muscle Res. Cell Motil* 4:473–484.

Leung, A.F. (1983b). Sarcomere dynamics in single myocardial cells as revealed by high-resolution light diffractometry *J. Muscle Res. Cell Motil.* 4:485–502.

Levine, R.J.C., Kensler, R.W. Yang, Z.H. Stull, J.T., and Sweeny, H.L. (1996). Myosin light chain phosphorylation affects the structure of rabbit skeletal muscle thick filaments. *Biophys. J.* 71:898–907.

Lieber, R.L. (1992). "Skeletal Muscle Structure and Function" Williams & Wilkins, Baltimore.

Lieberman, M., Hauschka, S.D., Hall, Z.W., Eisenberg, B.R., Horn, R., Walsh, J.V., Tsien, R.W., Jones, A.W., Walker, J.L., Poenie, M., Fay, F.S., Fabiato, F., and Ashley, C.C. (1987). Isolated muscle cells as a physiological model. *Am. J. Physiol.* 253:C349–C363.

Lin, D., Bobkova, A., Homsher E., and Tobacman, L.S. (1996). Altered cardiac troponin T *in vitro* function in the presence of a mutation implicated in familial hypertrophic cardiomyopathy. *J. Clin. Invest.* 97:1–8.

Lin G., Pister, K.S.J., and Roos, K.P. (1996). Standard CMOS piezoresistive sensor to quantify heart cell contractile forces. *Proc. IEEE.* MEMS-96:150–155.

Ling, N., Shrimpton, C., Sleep, J., Kendrick-Jones, J., and Irvin, M. (1996). Fluorescent probes of the orientation of myosin regulatory tight chains in relaxed, rigor, and contracting muscle. *Biophys. J.* 70:1836–1846.

Linke, W.A., Bartoo, M.L., and Pollack, G.H. (1993). Spontaneous sarcomeric oscillations at intermediate activation levels in single isolated cardiac myofibrils. *Circ. Res.* 73:724–734.

Linke, W.A., Popov, V.I., and Pollack, G.H. (1994). Passive and active tension in single cardiac myofibrils. *Biophys. J.* 67:782–792.

Linke, W.A., Ivemeyer, M., Olivieri, N., Kolmerer, B., Reugg, J.C., and Labeit, S. (1996). Towards a molecular understanding of the elasticity of titin. *J. Mol. Biol.* 261:62–71.

Little, W.C., Ohno, M., Kitzman, D.W., Thomas, J.D., and Cheng, C.P. (1995). Determination of left ventricular chamber stiffness from the time for deceleration of early left ventricular filling. *Circulation* 92:1933–1939.

Lowey, S., and Trybus, K. (1995). Role of skeletal and smooth muscle myosin light chains. *Biophys. J.* 68:120s–127s.

Lowey, S., Waller, G. S., and Trybus, K. M. (1993a). Function of skeletal muscle myosin heavy and light chain isoforms by an *in vitro* motility assay. *J. Biol. Chem.* 268:20414–20418.

Lowey, S., Waller, G.S., and Trybus, K.M. (1993b). Skeletal muscle myosin light chains are essential for physiological speeds of shortening. *Nature* 365:454–456.

Lundblad, A., Gonzalez-Serratos, H., Inesi, G., Swanson, J., and Paolini, P. (1986). Patterns of sarcomere activation, temperature dependence, and effect of ryanodine in chemically skinned cardiac fibers. *J. Gen. Physiol.* 87:885–905.

Lymn, R. W., and Taylor, E. W. (1971). Mechanism of adenosine triphosphate hydrolysis by actomyosin. *Biochemistry* 10:4617–4623.

MacKenna, D.A., Omens, J.H., McCulloch, A.D., and Covell, J.W. (1994). Contribution of collagen matrix to passive left ventricular mechanics in isolated rat hearts. *Am. J. Physiol.* 266:H1007–H1018.

MacKenna, D.A., Omens, J.H., and Covell, J.W. (1996). Left ventricular perimysial collagen fibers uncoil rather than stretch during diastolic filling. *Basic Res. Cardiol.* 91:111–122.

Malhotra, A. (1994). Role of regulatory proteins (troponin–tropomyosin) in pathologic states. *Mol. Cell Biochem.* 135:43–50.

Mann, D.L., Urabe, Y., Kent, R.L., Vinciguerra, S., and Cooper G. (1991). Cellular versus myocardial basis for the contractile dysfunction of hypertrophied myocardium. *Circ. Res.* 68:402–415.

Margossian, S.S., Krueger, J.W., Sellers, J.R., Cuda, G., Caulfield, J.B., Norton, P., and Slayter, H.S. (1991). Influence of the cardiac myosin hinge region on contractile activity. *Proc. Nat. Acad. Sci. USA* 88:4941–4945.

Margossian, S.S., White, H.D., Caufield, J.B., Norton, P., Taylor, S., and Slayter, H.S. (1992). Light chain-2 profile and activity of human ventricular myosin during dilated cardiomyopathy—Identification of a causal agent for impaired myocardial function. *Circulation* 85:1720–1733.

Marston, S.B., and Taylor, E.W. (1980). Comparison of the myosin and actomyosin ATPase mechanisms of the four types of vertebrate muscles. *J. Mol. Biol.* 139:573–600.

Martin, H., and Barsotti, R.J. (1994a). Relaxation from rigor of skinned trabeculae of the guinea pig induced by laser photolysis of caged ATP. *Biophys. J.* 66:1115–1128.

Martin, H., and Barsotti, R.J. (1994b). Activation of skinned trabeculae of the guinea pig induced by laser photolysis of caged ATP. *Biophys. J.* 67:1933–1941.

Maruyama, K., Natori, R., and Nonomura, Y. (1976). New elastic protein from muscle. *Nature* 262:58–60.

Maruyama, K. (1994). Connectin, an elastic protein of striated muscle. *Biophys. Chem.* 50:73–85.

Matsubara, S., and Maruyama, K. (1977). Role of connectin in the length–tension relation of skeletal and cardiac muscles. *Jpn. J. Physiol.* 27:589–600.

McDonald, K.S., and Moss, R.L. (1995). Osmotic compression of single cardiac myocytes eliminates the reduction in Ca^{2+} sensitivity of tension at short sarcomere length. *Circ. Res.* 77:199–205.

McDonald, K.S., Field, L.J., Parmacek, M.S., Soonpaa, M., Leiden, J.M., and Moss, R.L. (1995). Length dependence of Ca^{2+} sensitivity of tension in mouse cardiac myocytes expressing skeletal troponin C. *J. Physiol.* (London) 485:131–139.

McNally, E.M., Kraft, R., Bravo-Zehnder, M., Taylor, D.A., and Leinwand, L.A. (1989). Full-length rat alpha and beta cardiac myosin heavy chain sequences. *J. Mol. Biol.* 210:665–671.

Metzger, J.M. (1995). Myosin binding-induced cooperative activation of the thin filament in cardiac myocytes and skeletal muscle fibers. *Biophys. J.* 68:1430–1442.

Mijailovich, S.M., Fredberg, J.J., and Butler, J.P. (1996). On the theory of muscle contraction: Filament extensibility and the development of isometric force and stiffness. *Biophys. J.* 71:1475–1484.

Millar, N.C., and Homsher, E. (1992). Kinetics of force generation and phosphate release in skinned rabbit soleus muscle fibers. *Am. J. Physiol.* 262:C1239–C1245.

Milner, D.J., Weitzer, G., Tran, D., Bradley, A., and Capetanaki, Y. (1996). Disruption of muscle architecture and myocardial degeneration in mice lacking desmin. *J. Cell Biol.* 134:1255–1270.

Miyata, H., Hakozaki, H., Yoshikawa, H., Suzuki, N., Kinosita, K., Nishizaka, T., and Ishiwata, S. (1994). Stepwise motion of an actin filament over a small number of heavy meromyosin molecules is revealed in an *in vitro* motility assay. *J. Biochem.* (Tokyo) 115:644–647.

Molloy, J.E., Burns, J.E., Kendrick-Jones, J., Tregear, R.T., and White, D.C.S. (1995). Movement and force produced by a single myosin head. *Nature* 378:209–212.

Moncman, C.L., and Wang, K. (1995). Nebulette: A 107 kD nebulin-like protein in cardiac muscle. *Cell Motil. Cytoskeleton* 32:205–225.

Morano, I., and Ruegg, J.C. (1986). Calcium sensitivity of myofilaments in cardiac muscle: Effect of myosin phosphorylation. *In* "Controversial Issues in Cardiac Pathophysiology" (Jacob, R., ed.), pp. 17–23. Steinkopff Verlag, Darmstadt, Germany.

Morano, I., Hadicke, K., Grom, S., Koch, A., Schwinger, R.H.G., Bohm, M., Bartel, S., Erdman, E., and Krause, E.G. (1994). Titin, myosin light chains and C-protein in the developing and failing human heart. *J. Mol. Cell Cardiol.* 26:361–368.

Morano, I., Ritter, O., Bonz, A., Timek, T., Vahl, C.F., and Michel, G. (1995a). Myosin light chain actin interaction regulates cardiac contractility. *Circ. Res.* 76:720–725.

Morano, I., Osterman, A., and Arner, A. (1995b). Rate of active tension development from rigor in skinned atrial and ventricular cardiac fibres from swine following photolytic release of ATP from caged ATP. *Acta Physiol. Scand.* 154:343–353.

Morkin, E. (1993). Regulation of myosin heavy chain genes in the heart. *Circulation* 87:1451–1460.

Moss, R.L. (1992). Ca^{2+} regulation of mechanical properties of striated muscle: Mechanistic studies using extraction and replacement of regulatory proteins. *Circ. Res.* 70:865–884.

Moss, R.L., Lauer, M.R., Giulian, G.G., and Greaser, M.L. (1986). Altered Ca^{2+} dependence of tension development in skinned skeletal muscle fibers following modification of troponin by partial substitution with cardiac troponin C. *J. Biol. Chem.* 261:6096–6099.

Moss, R.L., Nwoye, L.O., and Greaser, M.L. (1991). Substitution of cardiac troponin C into rabbit muscle does not alter the length dependence of Ca sensitivity of tension. *J. Physiol.* (London) 440:273–289.

Mukherjee, R., Crawford, F.A., Hewett, K.W., and Spinale, F.G. (1993). Cell and sarcomere contractile performance from the same cardiocyte using video microscopy. *J. Appl. Physiol.* 74:2023–2033.

Nassar, R., Reedy, M.C., and Anderson, A.W. (1987). Development changes in the ultrastructure and sarcomere shortening of the isolated rabbit ventricular myocyte. *Circ. Res.* 61:465–483.

Nassar, R., Malouf, N.N., Kelly, M.B., Oakeley, A.E., and Anderson, P.A.W. (1991). Force–pCa relation and troponin T isoforms of rabbit myocardium. *Circ. Res.* 69:1470–1475.

Niggli, E., and Lederer, W.J. (1991). Restoring forces in cardiac myocytes insight from relaxations induced by photolysis of caged ATP. *Biophys. J.* 59:1123–1135.

Obermann, W.M.J., Plessmann, U., Weber, K., and Furst, D.O. (1995). Purification and biochemical characterization of myomesin, a myosin-binding and titin-binding protein, from bovine skeletal muscle. *Eur. J. Biochem.* 233:110–115.

Opie, L.H. (1995). Regulation of myocardial contractility. *J. Cardiovasc. Pharmacol.* 26:S1–S9.

Omens, J.H., Rodriquez, E.K., and McCulloch, A.D. (1996). Transmural changes in stress-free myocyte morphology during pressure overload hypertrophy in the rat. *J. Mol. Cell Cardiol.* 28:1975–1983.

Ostap, E.M., Barnett, V.A., and Thomas, D.D. (1995). Resolution of three structural states of spin-labeled myosin in contracting muscle. *Biophys. J.* 69:177–188.

Pagani, E.D., and Julian, F.J. (1984). Rabbit papillary muscle myosin isozymes and the velocity of muscle shortening. *Circ. Res.* 54:586–594.

Page, E. (1978). Quantitative ultrastructural analysis in cardiac membrane physiology. *Am. J. Physiol.* 235:C147–C158.

Palmer, R.E., Brady, A.J., and Roos, K.P. (1996). Mechanical measurements from isolated cardiac myocytes using a pipette attachment system. *Am. J. Physiol-Cell Physiol.* 270:C697–C704.

Palmer, S., and Kentish, J.C. (1994). The role of troponin C in modulating the Ca^{2+} sensitivity of mammalian skinned cardiac and skeletal muscle fibres. *J. Physiol.* (London) 480:45–60.

Park, S., Ajtai, K., and Burghardt, T.P. (1996). Cleft containing reactive thiol of myosin closes during ATP hydrolysis. Bba-Protein. *Struct. Mol. Enzym.* 1296:1–4.

Patel, J.R., Diffee, G.M., and Moss, R.L. (1996). Myosin regulatory light chain modulates the Ca^{2+} dependence of the kinetics of tension development in skeletal muscle fibers. *Biophys. J.* 70:2333–2340.

Patterson, S.W., and Starling, E.H. (1914). On the mechanical factors which determine the output of the ventricles. *J. Physiol.* (London) 48:357–379.

Perreault, C.L., Bing, O.H.L., Brooks, W.W., Ransil, B.J., and Morgan, J.P. (1990). Differential effects of cardiac hypertrophy and failure on right versus left ventricular calcium activation. *Circ. Res.* 67:707–712.

Politou, A.S., Thomas, D.J., and Pastore, A. (1995). The folding and stability of titin immuno-globulin-like modules, with implications for the mechanism of elasticity. *Biophys. J.* 69:2601–2610.

Pollack, G.H., and Krueger, J.W. (1976). Sarcomere dynamics in intact cardiac muscle. *Eur. J. Cardiol.* 4/suppl:53–65.

Price, M.G. (1984). Molecular analysis of intermediate filament cytoskeleton: A putative load-bearing structure. *Am. J. Physiol.* 246:H566–H572.

Price, M.G. (1991). Striated muscle endosarcomeric and exosarcomeric lattices. *Adv. Struct. Biol.* 1:175–207.

Rao, V.G., Akella, A., Su, H., and Gulati, J. (1995). Molecular mobility of the Ca^{2+} deficient EF-hand of cardiac troponin C as revealed by fluorescence polarization of genetically inserted tryptophan. *Biochem.* 34:562–568.

Rayment, I. (1996a). Kinesin and myosin: Molecular motors with similar engines. *Structure* 4:501–504.

Rayment, I. (1996b). The structural basis of the myosin ATPase activity. *J. Biol. Chem.* 271:15850–15853.

Rayment, I., and Holden, H.M. (1994). The three-dimensional structure of a molecular motor. *Trends. Biochem. Sci.* 19:129–134.

Rayment, I., Rypneiwski, W.R., Schmidt-Base, K., Smith, R., Tomchick, D.R., Benning, M., Winkelmann, D.A., Wesenberg, G., and Holden, H.M. (1993a). Three-dimensional structure of myosin subfragment-1: A molecular motor. *Science* 261:50–58.

Rayment, I., Holden, H.M., Whittaker, M., Yohn, C.B., Lorenz, M., Holmes, K.C., and Milligan, R.A. (1993b). Structure of the actin-myosin complex and its implications for muscle contraction. *Science* 261:58–65.

Rayment, I., Holden, H.M., Sellers, J.R., Fananapazir, L., and Epstein, N.D. (1995). Structural interpretation of the mutations in the beta-cardiac myosin that have been implicated in familial hypertrophic cardiomyopathy. *Proc. Natl. Acad. Sci. USA* 92:3864–3868.

Rayment, I., Smith, C., and Yount, R.G. (1996). The active site of myosin. *Annu. Rev. Physiol.* 58:671–702.

Regnier, M., Morris, C., and Homsher, E. (1995). Regulation of the cross-bridge transition from a weakly to strongly bound state in skinned rabbit muscle fibers. *Am. J. Physiol. Cell Physiol.* 38:C1532–C1539.

Rich, T.L., Langer, G.A., and Klassen, M.G. (1988). Two components of coupling calcium in single ventricular cell of rabbits and rats. *Am. J. Physiol.* 254:H937–H946.

Rodriquez, E.K., Hunter, W.C., Royce, M.J., Leppo, M.K., Douglas, A.S., and Weisman, H.F. (1992). A method to reconstruct myocardial sarcomere lengths and orientations at transmural sites in beating canine hearts. *Am. J. Physiol.* 263:H293–H306.

Rodriquez, E.K., Omen, J.H., Waldman, L.K., and McCulloch, A.D. (1993). Effect of residual stress on transmural sarcomere length distributions in rat left ventricle. *Am. J. Physiol.* 264:H1048–H1056.

Romberg, L., and Vale, R.D. (1993). Chemomechanical cycle of kinesin differs from that of myosin. *Nature* 361:168–170.

Roos, K.P. (1986a). Length, width, and volume changes in osmotically stressed myocytes. *Am. J. Physiol.* 251:H1373–H1378.

Roos, K.P. (1986b). Three-dimensional reconstructions of optically imagined single heart cell striation patterns. *Biophys. J.* 49:44–46.

Roos, K.P. (1987). Sarcomere length uniformity determined from three-dimensional reconstructions of resting isolated heart cell striation patterns. *Biophys. J.* 52:317–327.

Roos, K.P., and Brady, A.J. (1982). Individual sarcomere length determination from isolated cardiac cells using high-resolution optical microscopy and digital image processing. *Biophys. J.* 40:233–244.

Roos, K.P., Brady, A.J., and Tan, S.T. (1982). Direct measurement of sarcomere length from isolated cardiac cells. *Am. J. Physiol.* 242:H68–H78.

Roos, K.P., and Leung, A.F. (1987). Theoretical fraunhofer light diffraction patterns calculated from three-dimensional sarcomere arrays imaged from isolated cardiac cells at rest. *Biophys. J.* 52:329–341.

Roos, K.P., and Brady, A.J. (1989). Stiffness and shortening changes in myofilament-extracted rat cardiac myocytes. *Am. J. Physiol.* 256:H539–H551.

Roos, K.P., and Taylor, S.R. (1989). Striation dynamics in isolated rat cardiac myocytes revealed by computer vision. *J. Physiol.* (London) 418:50P (abstract).

Roos, K.P., and Brady, A.J. (1990). Osmotic compression and stiffness changes in relaxed skinned cardiac myocytes in PVP-40 and Dextran T-500. *Biophys. J.* 58:1273–1283.

Roos, K.P., and Taylor, S.R. (1993). High speed video imaging and digital analysis of microscopic features in contracting striated muscle cells. *Opt. Eng.* 32:306–313.

Ruppel, K.M., and Spudich, J.A. (1995). Myosin motor function: Structural and mutagenic approaches. *Curr. Opin. Cell Biol.* 7:89–93.

Ruppel, K.M., Uyeda, T.Q.P., and Spudich, J.A. (1994). Role of highly conserved lysine 130 of myosin motor domain. *In vivo* and *in vitro* characterization of site specifically mutated myosin. *J. Biol. Chem.* 269:18773–18780.

Saba, Z., Nassar, R., Ungerleider, R.M., Oakeley, A.E., and Anderson, P.A.W. (1996). Cardiac troponin T isoform expression II correlates with pathophysiological descriptors in patients who underwent corrective surgery for congenital heart disease. *Circulation* 94:472–476.

Sata, M., Sugiura, S., Yamashita, H., Momomura, S., and Serizawa, T. (1993). Dynamic interaction between cardiac myosin isoforms modifies velocity of actomyosin sliding *in vitro*. *Circ. Res.* 73:696–704.

Sata, M., Yamashita, H., Sugiura, S., Fujita, H., Momomura, S., and Serizawa, T. (1995). A new *in vitro* motility assay technique to evaluate calcium sensitivity of the cardiac contractile proteins. *Pflugers Arch.-Eur. J. Physiol.* 429:443–445.

Schiaffino, S., Gorza, L., and Ausoni, S. (1993). Troponin isoform switching in the developing heart and its functional consequences. *Trend. Cardiovasc. Med.* 3:12–17.

Sellers, J.R. (1996). Kinesin and NCD, two structural cousins of myosin. *J. Muscle Res. Cell Motil.* 17:173–175.

Sellers, J.R., and Kachar, B. (1990). Polarity and velocity and sliding filaments: Control of direction by actin and of speed by myosin. *Science* 249:406–408.

Sellers, J.R., Cuda, G., Wang, F., and Homsher, E. (1993). Myosin-specific adaptations of the motility assay. *Methods Cell Biol.* 39:23–49.

Sheetz, M.P., and Spudich, J.A. (1983). Movement of myosin-coated fluorescent beads on actin cables *in vitro*. *Nature* (London) 303:31–35.

Shepherd, N., Vornanen, M., and Isenberg, G. (1990). Force measurements from voltage-clamped guinea pig ventricular myocytes. *Am. J. Physiol.* 258:H452–H459.

Simmons, R.M., Finer, J.T., Chu, S., and Spudich, J.A. (1996). Quantitative measurements of force and displacement using an optical trap. *Biophys. J.* 70:1813–1822.

Simnett, S.J., Lipscomb, S., Ashley, C.C., and Mulligan, I.P. (1993). The effect of EMD 57033, a novel cardiotonic agent, on the relaxation of skinned cardiac and skeletal muscle produced by photolysis of diazo-2, a caged calcium chelator. *Pflugers Arch.-Eur. J. Physiol.* 425: 175–177.

Siri, F.M., Krueger, J., Nordin, C., Ming, Z., and Aronson, R.S. (1991). Depressed intracellular calcium transients and contraction in myocytes from hypertrophied and failing guinea pig hearts. *Am. J. Physiol.* 261:H514–H530.

Smith, D.A., and Geeves, M.A. (1995). Strain-dependent cross-bridge cycle for muscle. *Biophys. J.* 69:524–537.

Solaro, R.J., and van Eyk, J. (1996). Altered interactions among thin filament proteins modulate cardiac function. *J. Mol. Cell Cardiol.* 28:217–230.

Solaro, R.J., Powers, F.M., Gao, L.Z., and Gwathmey, J.K. (1993). Control of myofilament activation in heart failure. *Circulation* 87:38–43.

Sollott, S.J., and Lakatta, E.G. (1994). Novel method to alter length and load in isolated mammalian cardiac myocytes. Special Communication. *Am. J. Physiol.-Heart Circ. Phys.* 36:H1619–H1629.

Sollott, S.J., Ziman, B.D., Warshaw, D.M., Spurgeon, H.A., and Lakatta, E.G. (1996). Actomyosin interaction modulates resting length of unstimulated cardiac ventricular cells. *Am. J. Physiol.-Heart Circ. Phys.* 40:H896–H905.

Sommer, J.R., and Johnson, E.A. (1979). Ultrastructure of cardiac muscle. *In* "Handbook of Physiology, Section 2: The Cardiovascular System, Volume 1: The Heart" (Berne, R.M., Sperelakis, N., and Geiger, S.R., eds.), pp. 113–186. American Physiological Society, Bethesda.

Sorenson, A.L., Tepper, D., Sonnenblick, E.H., Robinson, T.F., and Capasso, J.M. (1985). Size and shape of enzymatically isolated ventricular myocytes from rats and cardiomyopathic hamsters. *Cardiovasc. Res.* 19:793–799.

Sosa, H., Popp, D., Ouyang, G., and Huxley, H.E. (1994). Ultrastructure of skeletal muscle fibers studied by a plunge quick freezing method: Myofilament lengths. *Biophys. J.* 67: 283–292.

Spinale, F.G., Fulbright, B.M., Mukherjee, R., Tanaka, R., Hu, J., Crawford, F.A., and Zile, M.R. (1992). Relation between ventricular and myocyte function with tachycardia-induced cardiomyopathy. *Circ. Res.* 71:174–187.

Spudich, J.A. (1994). How molecular motors work. *Nature* 372:515–518.

Spurgeon, H.A., Stern, M.D., Baartz, G., Raffaeli, S., Hansford, R.G., Talo, A., Lakatta, E.G., and Capogrossi, M.C. (1990) Simultaneous measurement of Ca^{++}, contraction, potential in cardiac myocytes. *Am. J. Physiol.* 258:H574–H586.

Squire, J.M. (1981). "The Structural Basis of Muscular Contraction." Plenum Press, New York.

Squire, J.M., Harford, J.J., and Alkhayat, H.A. (1994). Molecular movements in contracting muscle: Towards "muscle—the movie". *Biophys. Chem.* 50:87–96.

Steadman, B.W., Moore, K.B., Spitzer, K.W., and Bridge, J.H.B. (1988). A video system for measuring motion in contracting heart cells. *IEEE Trans. Biomed. Eng.* 35:264–272.

Strang, K.T., and Moss, R.L. (1995). Alpha(1)-Adrenergic receptor stimulation decreases maximum shortening velocity of skinned single ventricular myocytes from rats. *Circ. Res.* 77:114–120.

Strang, K.T., Sweitzer, N.K., Greaser, M.L., and Moss, R.L. (1994). β-adrenergic receptor stimulation increases unloaded shortening velocity of skinned single ventricular myocytes from rats. *Circ. Res.* 74:542–549.

Streeter, D.D., Spotnitz, H.M., Patel, D.J., Ross Jr., J., and Sonnenblick, E.H. (1969). Fiber orientation in the canine left ventricle during diastole ad systole. *Circ. Res.* 24:339–347.

Sugiura, S., Kobayakawa, N., Momomura, S., Chaen, S., Omata, M., and Sugi, H. (1996). Different cardiac myosin isoforms exhibit equal force-generating ability *in vitro*. *Bba-Bioenergetics* 1273:73–76.

Svoboda, K., Schmidt, C.F., Schnapp, B.J., and Block, S.M. (1993). Direct observation of kinesin stepping by optical trapping interferometry. *Nature* 365:721–727.

Swartz, D.R., Moss, R.L., and Greaser, M.L. (1996). Calcium alone does not fully activate the thin filament for S1 binding to rigor myofibrils. *Biophys. J.* 71:1891–1904.

Sweeney, H.L., and Holzbaur, E.L.F. (1996). Mutational analysis of motor proteins. *Annu. Rev. Physiol.* 58:751–792.

Sweeney, H.L., Brito, R.M., Rosevear, P.R., and Putkey, J.A. (1990). The low-affinity Ca^{2+}-binding sites in cardiac/slow skeletal muscle troponin C perform distinct functions: Site I alone cannot trigger contraction. *Proc. Natl. Acad. Sci. USA* 87:9538–9542.

Sweeney, H.L., Bowman, B.F., and Stull, J.T. (1993). Myosin light chain phosphorylation in vertebrate striated muscle. Regulation and function. *Am. J. Physiol.* 264:C1085–C1095.

Sweitzer, N.K., and Moss, R.L. (1990). The effect of altered temperature on Ca^{2+} sensitive force in permeabilized myocardium and skeletal muscle. Evidence for force dependence of thin filament activation. *J. Gen. Physiol.* 96:1221–1245.

Sweitzer, N.K., and Moss, R.L. (1993). Determinants of loaded shortening velocity in single cardiac myocytes permeabilized with alpha-hemolysin. *Circ. Res.* 73:1150–1162.

Tagawa, H., Rozich, J.D., Tsutsui, H., Narishige, T., Kuppuswamy, D., Sato, H., McDermott, P.J., Koide, M., and Cooper, G. (1996). Basis for increased microtubules in pressure-hypertrophied cardiocytes. *Circ.* 93:1230–1243.

Tanamura, A., Takeda, N., Iwai, T., Tuchiya, M., Arino, T., and Nagano, M. (1993). Myocardial contractility and ventricular myosin isoenzymes as influenced by cardiac hypertrophy and its regression. *Basic. Res. Cardio.* 88:72–79.

Tarr, M., Trank, J.W., Leiffer, P., and Shepherd, N. (1979). Sarcomere length-resting tension relation in single frog atrial cardiac cells. *Circ. Res.* 45:554–559.

ter Keurs, H.E.D.J., Rijnsburger, W.H., van Heuningen, R., and Nagelsmit, M.J. (1980a). Tension development and sarcomere length in rat cardiac trabeculae. *Circ. Res.* 46:703–714.

ter Keurs, H.E.D.J., Rijnsburger, W.H., and van Heuningen, R. (1980b). Restoring forces and relation of rat cardiac muscle. *Eur. Heart J.* 1:67–80.

Tobacman, L.S. (1996). Thin filament-mediated regulation of cardiac contraction. *Annu. Rev. Physiol.* 58:447–481.

Toyoshima, Y.Y., Kron, S.J., McNally, E.M., Niebling, K.R., Toyoshima, C., and Spudich, J.A. (1987). Myosin subfragment-1 is sufficient to move actin filaments *in vitro*. *Nature* 328:536–539.

Trinick, J. (1994). Titin and nebulin: Protein rulers in muscle? *Trends Biochem. Sci.* 19:405–409.

Trinick, J. (1996). Cytoskeleton. Titin as a scaffold and spring. *Curr. Biol.* 6:258–260.

Trybus, K.M. (1994). Role of myosin light chains. *J. Muscle Res. Cell Motil.* 15:587–594.

Tung, L. (1986). An ultrasensitive transducer for measurement of isometric contractile force from single heart cells. *Pflugers. Arch.* 407:109–115.

Urabe, Y., Hamada, Y., Spinale, F.G., Carabello, B.A., Kent, R.L., Cooper, G., and Mann, D.L. (1993). Cardiocyte contractile performance in experimental biventricular volume-overload hypertrophy. *Am. J. Physiol.* 264:H1615–H1623.

Uyeda, T.Q.P., Warrick, H.M., Kron, S.J., and Spudich, J.A. (1991). Quantized velocities at low myosin densities in an *in vitro* motility assay. *Nature* 352:307–311.

Vallee, R.B., and Sheetz, M.P. (1996). Targeting of motor proteins. *Science* 271:1539–1544.

VanBuren, P., Waller, G.S., Harris, D.E., Trybus, K.M., Warshaw, D.M., and Lowey, S. (1994). The essential light chain is required for full force production by skeletal muscle myosin. *Proc. Natl. Acad. Sci. USA* 91:12403–12407.

VanBuren, P., Harris, D.E., Alpert, N.R., and Warshaw, D.M. (1995). Cardiac V-1 and V-3 myosins differ in their hydrolytic and mechanical activities *in vitro*. *Circ. Res.* 77:439–444.

Van Leuven, S.L., Waldman, L.K., McCulloch A.D., and Covell, J.W. (1994). Gradients of epicardial strain across the perfusion boundary during acute myocardial ischemia. *Am. J. Physiol.* 267:H2348–H2362.

Vannier, C., Chevassus, H., and Vassort, G. (1996). Ca-dependence of isometric force kinetics in single skinned ventricular cardiomyocytes from rats. *Cardiovasc. Res.* 32:580–586.

Veksler, V., and Ventura-Clapier, R. (1994). *In situ* study of myofibrils, mitochondria and bound creatine kinases in experimental cardiomyopathies. *Mol. Cell. Biochem.* 133/134:287–298.

Vescovo, G., Harding, S.E., Jones, M., Libera, L.D., Pessina, A.C., and Poole-Wilson, P.A. (1989). Contractile abnormalities of single right ventricular myocytes isolated from rats with right ventricular hypertrophy. *J. Mol. Cell. Cardiol.* 21:103–111.

Vinkemeier, U., Obermann, W., Weber, K., and Furst, D.O. (1993). The globular head domain of titin extends into the center of the sarcomeric M-band—cDNA cloning, epitope mapping and immunoelectron microscopy of two titin-associated proteins. *J. Cell Sci.* 106:319–330.

Vulpis, V., Seccia, T.M., Nico, B., Ricci, S., Roncali, L., and Pirrelli, A. (1995). Left ventricular hypertrophy in spontaneously hypertensive rat: Effects of ACE-inhibition on myocardio-cyte ultrastructure. *Pharmacol. Res.* 31:375–381.

Wang, K. (1985). Sarcomere-associated cyoskeletal lattices in striated muscle. *In* "Cell and Muscle Motility" (Shay, J.W., ed.), Vol. 6, pp. 315–369. Plenum Press, New York.

Wang, K., and Ramirez-Mitchell, R. (1983). A network of transverse and longitudinal intermediate filaments is associated with sarcomeres of adult vertebrate skeletal muscle. *J. Cell Biol.* 96: 562–570.

Wang, K., Ramirez-Mitchell, R., and Palter, D. (1984). Titin is an extraordinarily long, flexible, and slender myofibrillary protein. *Proc. Nat. Acad. Sci. USA* 81:3685–3689.

Wang, K., McCarter, R., Wright, J., Beverly, J., and Ramirez-Mitchell, R. (1991). Regulation of skeletal muscle stiffness and elasticity by titin isoforms. A test of the segmental extension model of resting tension. *Proc. Nat. Acad. Sci. USA* 88:7101–7105.

Wang, K., McCarter, R., Wright, J., Beverly, J., and Ramirez-Mitchell, R. (1993). Viscoelasticity of the sarcomere matrix of skeletal muscles. The titin-myosin composite filament is a dual-stage molecular spring. *Biophys. J.* 64:1161–1177.

Wang, S-M., and Greaser, M.L. (1985). Immunocytochemical studies using a monoclonal antibody to bovine cardiac titin on intact and extracted myofibrils. *J. Muscle Res. Cell Motil.* 6:293–312.

Wang, Z., Kahn, S., and Sheetz, M.P. (1995). Single cytoplasmic dynein molecule movements: Characterization and comparison with kinesin. *Biophys. J.* 69:2011–2023.

Wang, Z.L., Mukherjee, R., Lam, C.F, and Spinale, F.G. (1996). Spatial characterization of contracting cardiac myocytes by computer-assisted, video-based image processing. Special communication. *Am. J. Physiol.-Heart Circ. Phy.* 39:H769–H779.

Warrick, H.M., and Spudich, J.A. (1987). Myosin structure and function in cell motility. *Annu. Rev. Cell Biol.* 3:379–421.

Warrick, H.M., Simmons, R.M., Finer, J.T., Uyeda, T.Q.P., Chu, S., and Spudich, J.A. (1993). *In vitro* methods for measuring force and velocity of the actinl–myosin interaction using purified proteins. *Methods Cell Biol.* 39:2–21.

Warshaw, D.M. (1996). The *in vitro* motility assay: A window into the myosin molecular motor. *News Physiol. Sci.* 11:1–7.

Watkins, H., Anan, R., Coviello, D.A., Spirito, P., Seidman, J.G., and Seidman, C.E. (1995). A *de novo* mutation in alpha-tropomyosin that causes hypertrophic cardiomyopathy. *Circ.* 91:2303–2305.

Watkins, S.C., Samuel, J.L., Marotte, F., Bertier-Savalle, B., and Rappaport, L. (1987). Microtubules and desmin filaments during onset of heart hypertrophy in rat: A double immunoelectron microscope study. *Circ. Res.* 60:327–336.

Watson, P.A., Hannan, R., Carl, L.L., and Giger, K.E. (1996). Desmin gene expression in cardiac myocytes is responsive to contractile activity and stretch. *Am. J. Physiol-Cell Physiol.* 39:C1228–C1235.

Weber, K.T., Sun, Y., Tyagi, S.C., and Cleutjens, J.P.M. (1994). Review—Collagen network of the myocardium—Function, structural remodeling and regulatory mechanisms. *J. Mol. Cell Cardiol.* 26:279–292.

Weisberg, A., and Winegard, S. (1996). Alteration of myosin cross bridges by phosphorylation of myosin-binding protein C in cardiac muscle. *Proc. Natl. Acad. Sci. USA* 93:8999–9003.

White, E., Boyett, M.R., and Orchard, C.H. (1995). The effects of mechanical loading and changes of length on single guinea-pig ventricular myocytes. *J. Physiol.-London.* 482:93–107.

Wolff, M.R., McDonald, K.S., and Moss, R.L. (1995a). Rate of tension development in cardiac muscle varies with level of activator calcium. *Circ. Res.* 76:154–160.

Wolff, M.R., Whitesell, L.F., and Moss, R.L. (1995b). Calcium sensitivity isometric tension is increased in canine experimental heart failure. *Circ. Res.* 76:781–789.

Wussling, M., Schenk, W., and Nilius, B. (1987). A study of dynamic properties in isolated myocardiacl cells by the laser diffraction method. *J. Mol. Cell. Cardiol.* 19:897–907.

Yanagida, T., and Ishijima, A. (1995). Forces and steps generated by single myosin molecules. *Biophys. J.* 68:S312–S320.

Yanagida, T., Nakase, M., Nishiyama, K., and Oosawa, F. (1984). Direct observation of motion of single F-actin filaments in the presence of myosin. *Nature* (London) 307:58–60.

Yamashita, H., Sugiura, S., Serizawa, T., Sugimoto, T., Izuka, M., Katayama, E., and Shimmen, T. (1992). Sliding velocity of isolated rabbit cardiac myosin correlates with isozyme distribution. *Am. J. Physiol.* 263:H464–H472.

Yates, L.D., and Greaser, M.L. (1983). Quantitative determination of myosin and actin in rabbit skeletal muscle. *J. Mol. Biol.* 168:123–141.

Zhang, R., Zhao, J.J., Mandveno, A., and Potter, J.D. (1995). Cardiac troponin I phosphorylation increases the rate of cardiac muscle relaxation. *Circ. Res.* 76:1028–1035.

Zhao, L., Pate, E., Baker, A.J., and Cooke, R. (1995). The myosin catalytic domain does not rotate during the working power stroke. *Biophys. J.* 69:994–999.

Zile, M.R., Buckely, J.M., Richardson, K.E., and Cooper, G. (1994). Passive stiffness and viscous damping in the hypertrophied cardiocyte. *Circ.* 90:I-432 (Abstract).

Zile, M.R., Mukherjee, R., Clayton, C., Kato, S., and Spinale, F.G. (1995). Effects of chronic supraventricular pacing tachycardia on relaxation rate in isolated cardiac muscle cells. *Amer. J. Physiol.-Heart Circ. Phy.* 37:H2104–H2113.

Zot, A.S., and Potter, J.D. (1987). Structural aspects of troponin–tropomyosin regulation of skeletal muscle contraction. *Annu. Rev. Biophys. Biophys. Chem.* 16:535–539.

7

METABOLISM IN NORMAL AND ISCHEMIC MYOCARDIUM

JOSHUA I. GOLDHABER

I. INTRODUCTION

The heart performs mechanical work 24 hours a day for a lifetime. Normal cardiac output, measured in liters of blood pumped per minute, ranges from 4.5 at rest to 35 during exercise in a trained athlete. With each beat, depolarization of the membrane via Na^+ channels leads to the influx of Ca^{2+} predominately via voltage-gated sarcolemmal Ca^{2+} channels. This Ca^{2+} serves as the trigger that releases a far greater amount of Ca^{2+} [stored in the sarcoplasmic reticulum (SR)] into the cytoplasm where it activates myofilaments to cycle cross-bridges that develop force sufficient to eject blood into the aorta (see Chapter 6). This extraordinary amount of continuous work demands an uninterrupted supply of energy, in the form of ATP, which is normally consumed at a rate of about 60 mmol/min (Jacobus, 1985). The actual process of contraction accounts for roughly 60% of the heart's energy expenditure (Fig. 1). Some of this energy is in the form of ionic concentration gradients (extracellular to intracellular Na^+ and Ca^{2+}, SR to cytosolic Ca^{2+}), but a far greater proportion is direct consumption of ATP for cycling (attachment followed by release) of myofilament cross-bridges. Although relaxation is initiated by activation of a specific K^+ conductance (see Chapter 3), removal of Ca^{2+} from the cytoplasm is required for mechanical relaxation to occur. The Ca^{2+} must be pumped against a concentration gradient into the SR by its Ca^{2+} ATPase, a process that consumes approximately 15% of the metabolic energy used by the cell. Restoration of Na^+ and K^+ gradients by the Na^+–K^+ pump accounts for about 5% of the heart's energy consumption. The last 20% (non-E–C) is

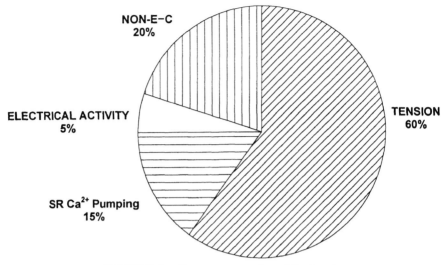

FIGURE 1   Energy consumption of the heart.

used to maintain basic cellular functions, such as protein synthesis (Gibbs and Chapman, 1979).

II. NORMAL MYOCARDIAL METABOLISM

A. METABOLIC SUBSTRATES

Normal myocardium generates more than 90% of its ATP by oxidative metabolism and less than 10% by anaerobic glycolysis (Kobayashi and Neely, 1979). Fatty acids, particularly oleic and palmitic acid, in the form of fatty (acyl) esters containing acyl-CoA are the preferred metabolic substrate (>60%), with glucose as the remainder accounting for about 30% (Neely and Morgan, 1974; Saddik and Lopaschuk, 1991). Under aerobic conditions, the oxidation of fatty acids with increased production of acetyl-CoA inhibits pyruvate dehydrogenase (Fig. 2). This in turn leads to accumulation of glucose-6-phosphate, which inhibits hexokinase and thereby decreases phosphorylation of glucose. Fatty acid oxidation also inhibits membrane transport of glucose into myocytes, reduces the contribution of glycolytic ATP production to overall breakdown of glycogen and glucose, and reduces glucose oxidation (Finegan *et al.*, 1992).

Glycogen is a storage form of glucose that can be metabolized by the heart as an alternative to exogenous glucose. During glycogenolysis, the heart breaks down stored glycogen into a phosphorylated form of glucose (glucose-1-P) that can then be converted to glucose-6-P for subsequent metabolism to pyruvate. The extent of glycogenolysis during normal metabolism was previously believed to be insignificant (Neely and Morgan, 1974), rising in the setting of increased workloads (Neely *et al.*, 1970) or under experimental conditions where insulin or fatty acids were omitted from perfusate (Goodwin *et al.*, 1995). However, recent evidence suggests that substantial glycogen turnover occurs in the aerobic heart in the presence of net glycogenolysis, which accounts for 41% of the ATP produced from myocardial glucose (Henning *et al.*, 1996) in an isolated heart preparation. Furthermore, whereas less than 20% of exogenous glucose passing through glycolysis is oxidized, more than 50% of glucose from glycogen passing through glycolysis is oxidized (Henning *et al.*, 1996).

B. OXIDATIVE METABOLISM

Oxidative metabolism takes place in the mitochondria (see Fig. 2); therefore, all of the substrates, metabolites, and cofactors must cross the membrane that separates the mitochondrial space from the cytosolic space. The tricarboxylic acid cycle, which takes place in the mitochondrial matrix, yields only one high-energy phosphate bond per 2-carbon fragment by substrate-level phosphorylation. Most of the ATP generated in the heart is generated by the next step, which is oxidative or respiratory chain-linked phosphorylation in the inner mitochondrial membrane. These oxidation reactions occur when electrons carried by the reduced coenzymes nicotinamide adenine dinucleotide (NADH) and flavin adenine dinucleotide (FADH$_2$) are transferred to molecular oxygen in a tightly regulated fashion.

$$NADH + H^+ + \tfrac{1}{2}O_2 \rightarrow NAD^+ + H_2O \qquad (1)$$

$$FADH_2 + \tfrac{1}{2}O_2 \rightarrow FAD + H_2O \qquad (2)$$

This tight regulation is critical because incomplete reduction of oxygen by NADH can result in the generation of oxygen-free radicals (see following discussion) that can have serious deleterious effects on cardiac function.

ATP generated within the mitochondrial matrix is impermeable to the inner mitochondrial membrane. For the energy within the phosphate bond of ATP to be exported to the cytoplasm, ATP binds to the enzyme adenine nucleotide translocase (also known as ATP-ADP transferase) located on the inner mitochondrial membrane and is transported across the membrane in exchange for ADP. Once across the inner mitochondrial membrane, ATP phosphorylates creatine to creatine phosphate in a reaction catalyzed by mitochondrial creatine kinase.

$$ATP + creatine \rightarrow ADP + creatine\ phosphate \qquad (3)$$

ADP is then returned to the mitochondrial matrix by adenine nucleotide translocase where it can be rephosphorylated by oxidative metabolism.

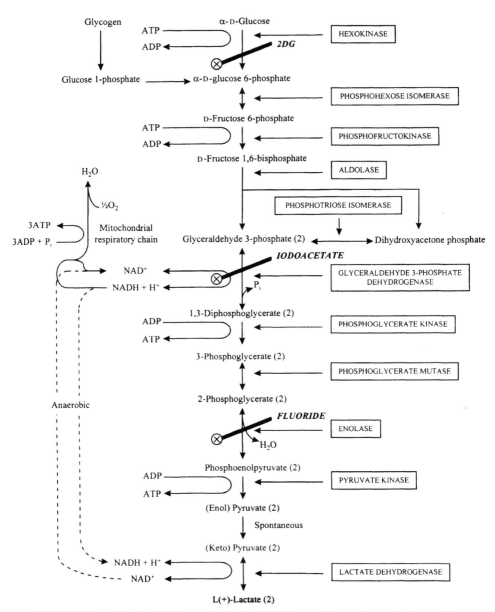

FIGURE 2 Glycolytic (left) and oxidative (right) metabolic pathways in the cardiac myocyte. Sites of inhibition are denoted by ⊗, and inhibitors are labeled in italic.

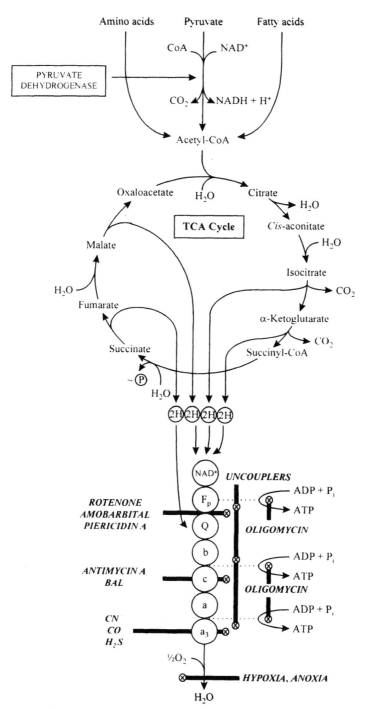

FIGURE 2 (*Continued*)

Creatine phosphate is free to diffuse into the cytoplasm where it phosphorylates ADP to give ATP in the reverse reaction by cytoplasmic creatine kinases near sites of ATP hydrolysis.

$$\text{creatine phosphate} + \text{ADP} \rightarrow \text{creatine} + \text{ATP} \qquad (4)$$

In this fashion, the creatine kinase reaction shuttles high-energy phosphate groups from the mitochondria to the cytoplasm where it can be used to phosphorylate ADP to ATP. This mechanism also serves to buffer intracellular ATP during transitions to new workload states (Bittl et al., 1987). The high levels of creatine phosphate afforded by the creatine kinase shuttle (about five times that of ATP) ensure that any free ADP will be rapidly phosphorylated to ATP. Consequently, free ADP levels can be kept low in myocardium. Because the free energy of ATP hydrolysis is proportional to log $[(\text{ATP})/(\text{ADP}) \times (P_i)]$, low levels of cytosolic ADP and inorganic phosphate favor a high free energy state in the muscle.

C. ANAEROBIC METABOLISM

Anaerobic metabolism is considerably less efficient than oxidative metabolism. A single glucose molecule generates only 2 ATP molecules while being metabolized to 2 pyruvate molecules via anaerobic glycolysis, whereas subsequent oxidative metabolism of the pyruvates via the tricarboxylic acid cycle yields 34 ATP. Furthermore, the glycolytic pathway contains an oxidative step that reduces the oxidized coenzyme NAD to NADH. To restore NAD for continued glycolysis, NADH must subsequently be reoxidized. Under aerobic conditions, this process usually occurs in the mitochondria (see equation 1) and generates ATP; however, in the absence of mitochondrial metabolism, oxidation of NADH to NAD must be coupled to the reduction of pyruvate to lactate. As this process usually accounts for less than 10% of normal myocardial metabolism, the amount of lactate formed is small. However, when oxidative metabolism is limited, such as during ischemia or hypoxia, the production of lactate may contribute to intracellular acidosis and possibly cellular K^+ loss, although this is controversial (see following discussion, and Shieh et al., 1994).

D. REGULATION OF OXYGEN DELIVERY

Under resting conditions, the heart extracts close to 75% of the oxygen delivered by the coronary arteries, compared to 20–25% by most other tissues. To increase oxygen consumption during exercise from its baseline of 9 ml/100 g/min (Gibbs and Chapman, 1979), the heart must increase oxygen delivery. Oxygen delivery by the coronary circulation is regulated by metabolic needs (Berne et al., 1983; Sparks and Bardenheuer, 1986). As cardiac work increases, consumption of ATP yields ADP and AMP. AMP is subsequently dephosphorylated by 5'-nucleotidase to yield adenosine, a potent vasodilator that increases coronary blood flow in proportion to oxygen requirements. It has been estimated that oxygen consumption can be increased significantly via this mechanism, with adenosine acting as a key homeostatic metabolite (for review, see Schrader, 1990). Because adenosine formation is increased during periods of energy deficit, this metabolite provides a sensitive link between cardiac metabolic activity, O_2 supply, and perhaps substrate utilization as well (Wyatt et al., 1989; Dale et al., 1991; Law and McLane, 1991; Finegan et al., 1992). However, adenosine-independent mediators of coronary flow, such as nitric oxide and oxygen tension, may also play an important role in matching coronary flow to energy consumption under pathophysiological conditions. In the case where blood flow is gradually reduced, such as during chronic ischemia or in the setting of myocardial hibernation (see following discussion), the myocardium can "down-regulate" its energy consumption to match the reduction in coronary flow (Bristow et al., 1991; Ito, 1995).

In summary, the mammalian myocardium generates most of its energy oxidatively, consuming fatty acids in preference to carbohydrates. Carbohydrate stores, in the form of glycogen, are continuously "turned over" and preferentially oxidized. Oxidative reactions take place in mitochondria and are tightly regulated to limit exposure of extramitochondrial organelles to free radicals. High-energy phosphate bonds are shuttled from the mitochondria to the cytoplasm by the creatine kinase shuttle, which also serves as a buffer for intracellular ATP. Anaerobic metabolism is far less efficient than oxidative metabolism and must produce lactate from pyruvate to replenish NAD. The heart extracts a high percentage of oxygen from blood delivered to it by the coronary arteries. Myocardial oxygen consumption and metabolism are tightly coupled to oxygen delivery, enabling cardiac output to respond to increased needs. Adenosine is believed to play a prominent role in the regulation of oxygen delivery.

III. METABOLIC REGULATION OF CELLULAR PROCESSES BY GLYCOLYTIC AND OXIDATIVE METABOLISM

The metabolic regulation of the cellular processes of the myocyte has been an intense area of scientific investigation. This is due in large part to the direct relevance of this subject to clinical ischemia in humans. During ischemia, oxidative metabolism is inhibited primarily and glycolytic metabolism is inhibited secondarily by accumulation of metabolic byproducts. Therefore, studies of both metabolic pathways are highly relevant to ischemia. The general approach to study a metabolic pathway is to inhibit it and then observe the consequences. This section summarizes the regulation of $[Ca^{2+}]_i$ and contractility, K^+, and Na^+ by glycolytic and oxidative metabolism.

A. STRATEGIES TO INHIBIT
MYOCARDIAL METABOLISM

1. Glycolysis

Methods of glycolytic inhibition include the use of iodoacetate (IAA), 2-deoxyglucose (2-DG), or glycogen depletion (Pirolo and Allen, 1986). 2-DG is a competitive substrate antagonist (Fig. 2), and as such its inhibition is incomplete and slow compared to IAA, a potent and irreversible inhibitor of glyceraldehyde-3-phosphate dehydrogenase (Webb, 1966). At high doses IAA may interfere with the function of a variety of sulfhydryl containing enzymes, although at low doses ($\leq 200 \, \mu M$) the effects are believed to be relatively specific (Chatham *et al.*, 1988). Both 2-DG and IAA cause accumulation of sugar phosphates (Kusuoka and Marban, 1994). Glycogen depletion is a slow process that can be achieved simply by omitting glucose from the perfusate, or it can be hastened by a brief (3 min) application of glucagon. To attain selective inhibition of glycolysis without affecting oxidative metabolism, substrates that directly enter the tricarboxylic acid cycle, such as pyruvate, acetate, or palmitate, must be provided. This enables oxidative metabolism to proceed in the absence of glycolysis.

2. Oxidative Metabolism

Block of ATP production by oxidative metabolism can be achieved in several different ways. *Removal of exogenous substrate* (e.g., glucose, fatty acids) will deplete endogenous stores, thereby inhibiting first glycolytic and then eventually oxidative metabolism. This method is slow and not very practical. A more selective and rapid inhibition of oxidative metabolism can be achieved by providing exogenous glucose during *anoxia* or under normoxic conditions while exposing the preparation to a specific blocker of oxidatively generated ATP (see Fig. 2). These include *inhibitors of electron transport* in the respiratory chain, including rotenone, which is an extremely toxic plant substance used as a fish poison that blocks electron transport between NADH and ubiquinone. Other agents in this class include the barbiturate amytal and the antibiotic piericidin A. Antimycin A, a bacterial product, blocks transport from cytochrome b to c. Cyanide (CN) blocks transport from cytochrome aa_3 to oxygen. *Uncouplers of oxidative phosphorylation* allow electron transport and respiration to continue but prevent the phosphorylation of ADP to ATP. These agents, which typically stimulate the rate of oxygen uptake by mitochondria, even in the absence of ADP, include 2,4-dinitrophenol (DNP), dicumarol, carbonylcyanide

p-trifluoro-methoxyphenylhydrazone (FCCP), m-chlorocarbonyl cyanide phenylhydrazone (CCCP) (Heytler and Prichard, 1962), 5-chloro-3-t-butyl-2'-chloro-4'-nitrosalicylanilide, and arsenate. An important property of some of the uncouplers is that they often promote the passage of hydrogen ions through the mitochondrial membrane, which is normally impermeable to them. This hydrogen ionophore activity may complicate the interpretation of experiments using these agents. *Inhibitors of oxidative phosphorylation* prevent both phosphorylation of ADP to ATP and the stimulation of oxygen consumption by ADP. Electron transport is indirectly inhibited by the build-up of high-energy intermediates. Examples include oligomycin, rutamycin, aurovertin, and triethyltin (Lehninger *et al.*, 1993). *Inhibitors of adenine–nucleotide translocase* prevent transport of ATP and ADP across the inner mitochondrial membrane. Examples include atractyloside, a toxic plant compound, and the antibiotic bongkrekic acid.

B. METABOLIC REGULATION OF INTRACELLULAR Ca ($[Ca^{2+}]_i$) AND E–C COUPLING

The importance of $[Ca^{2+}]_i$ regulation in the etiology of myocardial dysfunction, particularly during cardiac ischemia, cannot be overemphasized. The Ca^{2+} ion plays a central role in the control of multiple cellular functions, including contraction, enzyme activity, membrane stability, and metabolism. Abnormal $[Ca^{2+}]_i$ homeostasis, particularly *elevations* in $[Ca^{2+}]_i$, has been linked experimentally to disorders of E–C coupling, contracture development, activation of arrhythmogenic transient inward currents via Ca^{2+}-activated nonselective cation channels or Na^+-Ca^{2+} exchange, interference with energy production by mitochondria, and activation of intracellular proteases and phospholipases that damage cellular membranes and organelles. Selective inhibition of glycolytic and/or oxidative metabolism has been used experimentally to explore the metabolic regulation of $[Ca^{2+}]_i$. These studies have revealed a critical dependence of $[Ca^{2+}]_i$ regulation on metabolism.

1. Selective Inhibition of Oxidative Metabolism

Selective inhibition of oxidative metabolism has produced inconsistent effects on $[Ca^{2+}]_i$ and contraction, depending on the preparation, type of metabolic inhibitor, and technique for $[Ca^{2+}]_i$ measurement. In papillary muscles from rats, cats, and ferrets loaded with the $[Ca^{2+}]_i$ indicator aequorin, selective inhibition of oxidative metabolism by anoxia or exposure to CN caused a decrease in tension development but no reduction in the amplitude of the $[Ca^{2+}]_i$ transient (Allen and Orchard, 1983). In these experiments, there was no elevation of diastolic $[Ca^{2+}]_i$, even during substantial contracture. The presence of glucose in the perfusate prevented contracture in most of their experiments. Eisner *et al.* (1989) reported variable effects (both increases and decreases) on the $[Ca^{2+}]_i$ transient and contraction amplitude in isolated rat myocytes loaded with the $[Ca^{2+}]_i$ indicator fura-2 and exposed to CN with glucose present. There was no diastolic cell shortening or increase in diastolic $[Ca^{2+}]_i$ noted during the course of their experiments. In isolated ferret hearts loaded with aequorin, perfusion with hypoxic solution caused a decrease in peak $[Ca^{2+}]_i$ and an increase in diastolic $[Ca^{2+}]_i$, which paralleled changes in developed and resting tension, respectively (Kihara et al., 1989). Seki and MacLeod (1995) have reported that substrate-free chemical anoxia using $Na_2S_2O_4$ caused a significant decrease of both twitch and $[Ca^{2+}]_i$ transient amplitude within 3 minutes in isolated guinea pig ventricular myocytes loaded with the Ca^{2+} sensitive fluorescent indicator indo-1. Diastolic $[Ca^{2+}]_i$ increased after 10 minutes of anoxia, but contracture was not assessed. Silverman et al. (1991) found no change in diastolic $[Ca^{2+}]_i$, $[Ca^{2+}]_i$ transient amplitude, or relaxation after 10 minutes of substrate-free hypoxia in isolated rat ventricular myocytes loaded with indo-1.

Thus, selective inhibition of oxidative metabolism usually reduces tension development or twitch amplitude, but no consistent relationship between these changes and $[Ca^{2+}]_i$ is apparent. Furthermore, contracture is not universally observed during inhibition of oxidative metabolism, nor is it correlated with elevations in diastolic $[Ca^{2+}]_i$ when it is observed. These inconsistencies may be partly related to different capacities for anaerobic glycolysis between various preparations when oxidative metabolism is inhibited. In particular, the absence of glucose was more likely to be associated with the preparation developing contracture.

2. Combined Inhibition of Oxidative and Glycolytic Metabolism

Combined inhibition of oxidative and glycolytic metabolism has more consistently been found to cause both a decrease in the systolic $[Ca^{2+}]_i$ transient and an increase in diastolic $[Ca^{2+}]_i$. For example, in the study by Allen and Orchard (1983) discussed previously, when both glycolysis and oxidative metabolism were inhibited, $[Ca^{2+}]_i$ transients and developed tension fell consistently. Similarly, Eisner et al. (1989) found more consistent decreases in $[Ca^{2+}]_i$ transients and contraction amplitude with late increases in diastolic $[Ca^{2+}]_i$ when both oxidative metabolism and glycolysis were inhibited. The degree of glycolytic inhibition may critically influence the results of experiments examining the effects of combined metabolic inhibition. For example, Stern et al. (1988), using substrate-free hypoxia in isolated rat ventricular myocytes loaded with indo-AM, found an initial increase in contraction amplitude associated with a slight decrease in diastolic cell length, followed somewhat later by a fall in contraction amplitude and a decrease in diastolic cell length. The decrease in contraction amplitude was paralleled by a decrease in the amplitude of the $[Ca^{2+}]_i$ transient. When the cells were pretreated with 2-DG for 1 hour prior to the onset of anoxia, a uniformly rapid and complete fall in contractile and $[Ca^{2+}]_i$ transient amplitude was observed, which suggests that the degree of glycolytic inhibition was an important determinant of $[Ca^{2+}]_i$ regulation.

The mechanism for the fall in the $[Ca^{2+}]_i$ transient has been investigated in patch-clamped guinea pig ventricular myocytes loaded with fura-2. When these cells were exposed to the mitochondrial uncoupler FCCP in the presence of 2-DG, there was a marked reduction in contraction and $[Ca^{2+}]_i$ transient amplitudes, which corresponded to the reduction in the amplitude of L-type Ca^{2+} current (Goldhaber et al., 1991). Because there was no reduction in the caffeine-releasable store of $[Ca^{2+}]_i$ following metabolic inhibition, it was concluded that the reduction in L-type Ca^{2+} current, which triggers the release of Ca^{2+} from the SR (see Chapter 4), was the primary cause of the depressed $[Ca^{2+}]_i$ transient and contractile amplitude observed under these conditions. An important finding in that study was that contraction could be restored by augmenting the Ca^{2+} current using the Ca^{2+} channel agonist BayK 8644, indicating that fundamental components of E–C coupling remain intact during severe combined metabolic inhibition (Fig. 3). Similar effects on Ca^{2+} current and contraction have been reported in rat ventricular myocytes exposed to CN and 2-DG (Lederer et al., 1989).

Diastolic $[Ca^{2+}]_i$ rises in most preparations following inhibition of both glycolytic and oxidative metabolism. The timing of the increase in diastolic $[Ca^{2+}]_i$ in relation to the development of contracture remains controversial, however (Dahl and Isenberg, 1980; Bers and Ellis, 1982; Allen and Orchard, 1983; Cobbold and Bourne, 1984; Snowdowne et al., 1985; Barry et al., 1987; Guarnieri, 1987; Haworth et al., 1987; Kim and Smith, 1988; Murphy et al., 1988, Allen et al., 1989; Eisner et al., 1989; Lee and Allen, 1992; Goldhaber and Liu, 1994). Differences between various studies may be related to how completely glycolytic metabolism was inhibited, which ranged from simply removing exogenous glucose (incomplete inhibition because glycogenolysis may still proceed) to adding iodoacetate (severe inhibition and possibly nonspecific effects depending on the dose used). Because complete metabolic inhibition reduces Ca^{2+} entry via the L-type Ca^{2+} current, increases in diastolic $[Ca^{2+}]_i$ must be caused by sarcolemmal "leak" (Wang et al., 1995) or reductions in the effectiveness of cellular mechanisms that remove Ca^{2+} from the cytosol. These include the ATP-dependent Ca^{2+} pump of the SR; the Na–Ca^{2+} exchanger; and, to a lesser extent, the ATP-dependent Ca^{2+} pump of the sarcolemma. The physiological behaviors of these Ca^{2+} removal mechanisms are discussed in detail in Chapter 4. Goldhaber et al. (1991) have shown that combined metabolic inhibition with FCCP and 2-DG slows the rate of decline of the $[Ca^{2+}]_i$ transient following a voltage clamp pulse by 64%, which is consistent with a defect in Ca^{2+} reuptake or extrusion mechanisms. However, they found no evidence that Na^+–Ca^{2+} exchange was inhibited under the same conditions, even though ATP has been shown to modulate Na^+–Ca^{2+} exchange activity in giant patches (Hil-

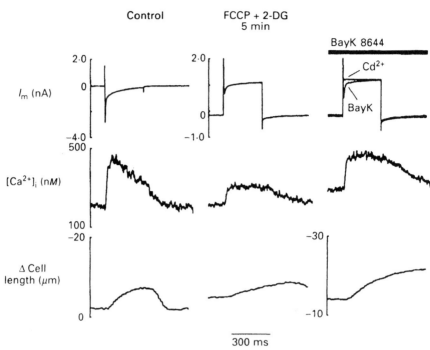

FIGURE 3   Effects of metabolic inhibition using FCCP (1 μM) and 2-DG (10 mM) on membrane current (I_m), $[Ca^{2+}]_i$, and cell length change during 300 ms voltage clamps from a holding potential of -40 mV to 0 mV in a whole cell patch-clamped guinea pig ventricular myocyte. The pipette solution contained 0.1 mM pentapotassium fura-2 for measurement of $[Ca^{2+}]_i$. After 5 minutes of metabolic inhibition (middle panels), there was a marked decrease in I_{Ca}, the $[Ca^{2+}]_i$ transient, and cell shortening amplitude. Subsequent exposure to the L-type Ca^{2+} channel agonist BayK 8644 (5 μM) + elevated extracellular Ca^{2+} (10 mM) increased I_{Ca} as well as the $[Ca^{2+}]_i$ transient and cell shortening amplitude, suggesting that the fundamental components of E–C coupling remained intact despite metabolic inhibition severe enough to cause rigor. (Reprinted with permission from Goldhaber *et al.*, 1991.)

gemann *et al.*, 1992). A similar lack of inhibition of Na^+–Ca^{2+} exchange by glycolytic inhibition has been reported in neonatal rat cardiomyocytes (Barrigon *et al.*, 1996), although this has not been a universal finding (Satoh *et al.*, 1995). Several investigators have demonstrated that SR Ca^{2+} stores, as assessed by caffeine release, are not depleted by complete metabolic inhibition (Lederer *et al.*, 1989; Goldhaber *et al.*, 1991; Seki and MacLeod, 1995), and in fact they may sequester excess Ca^{2+}. Nevertheless, the slowed decline of the $[Ca^{2+}]_i$ transient under these conditions suggests that the rate of SR Ca^{2+} uptake is reduced. It is well known that the SR Ca^{2+} pump is very sensitive to ATP levels (Shigekawa *et al.*, 1978; Schuurmans Stekhoven and Bonting, 1981) and phosphorylation potential (Kammermeier *et al.*, 1982). The continued uptake of Ca^{2+} by the SR despite the onset of rigor during metabolic inhibition implies that the energy requirements of the SR Ca^{2+} pump remain at least partially satisfied under those conditions. One potential explanation for continued SR Ca^{2+} pump activity is that glycolysis may not have been adequately inhibited by the experimental strategies used. This could be important if ATP derived from glycolysis is the preferred energy source of the SR Ca^{2+} pump (see the following discussion). Mitochondria increase their Ca^{2+} content during anoxia. High levels of mitochondrial Ca^{2+} during anoxia are correlated with decreased cell survival following reoxygenation (Miyata *et al.*, 1992).

Studies in patch-clamped single ventricular myocytes have consistently demonstrated that metabolic inhibition reduces the amplitude of L-type Ca^{2+} current (Lederer *et al.*, 1989; Goldhaber *et al.*, 1991). This is not a surprising result considering the dependence of the Ca^{2+} channel on phosphorylation (Irisawa and Kokubun, 1983; Taniguchi *et al.*, 1983; Noma and Shibasaki, 1985), and it is consistent with studies showing a decrease in $^{47}Ca^{2+}$ uptake during glucose-free hypoxia (Nayler *et al.*, 1979). Because cardiac myocytes already decrease their influx of Ca^{2+} via voltage-sensitive Ca^{2+} channels, the additional benefit of lowering extracellular Ca^{2+} or pretreating cells with Ca^{2+} channel blockers during hypoxia and metabolic inhibition (Nayler *et al.*, 1979; Kohmoto and Barry, 1989) is most likely due to a *further* reduction in Ca^{2+} entry. This results in decreased Ca^{2+} delivery to the SR and mitochondria, delayed Ca^{2+} overload of these organelles, and finally a delayed increase in diastolic $[Ca^{2+}]_i$. A further significant increase in sarcolemmal "leak" of Ca^{2+} may be a late phenomenon that occurs after rigor and increases in diastolic $[Ca^{2+}]_i$ have already been established.

These results indicate that combined inhibition of oxidative and glycolytic metabolism in myocardial cells has a more predictable outcome than selective inhibition of oxidative metabolism. The earliest effect is a decrease in I_{Ca}, the $[Ca^{2+}]_i$ transient, and twitch amplitude. These early effects are followed by the onset of rigor, which is associated with an increase in SR and mitochondrial Ca^{2+} and a decrease in the rate of return of the $[Ca^{2+}]_i$ transient to diastolic levels. $Na^+–Ca^{2+}$ exchange activity and fundamental E–C coupling relationships are maintained. The extent of rise in diastolic $[Ca^{2+}]_i$ is variable at this point depending on the strategy for metabolic inhibition and the model used, with increases in diastolic $[Ca^{2+}]_i$ more likely to occur earlier in models with increased workloads. Significant increases in diastolic $[Ca^{2+}]_i$ are more consistently observed at later stages after rigor is fully developed. Sarcolemmal Ca^{2+} "leak" may play a more active role at this stage.

3. Selective Inhibition of Glycolysis

The studies described previously suggest that diastolic $[Ca^{2+}]_i$ and the $[Ca^{2+}]_i$ transient are reasonably maintained during selective inhibition of oxidative phosphorylation when glycolysis is allowed to proceed and that they deteriorate when glycolysis is inhibited. In contrast, selective inhibition of glycolysis with IAA causes severe contracture in cardiac muscle, even when normal high-energy phosphate concentrations are maintained by supplying the heart with pyruvate as a substrate for oxidative phosphorylation (Weiss and Hiltbrand, 1985). The degree of contracture and $[Ca^{2+}]_i$ overload may depend somewhat on the experimental conditions, especially the workload and extent of Ca^{2+} influx, because a more recent study using a similar model showed no depletion of high-energy phosphates, functional abnormalities, or elevations in diastolic $[Ca^{2+}]_i$ until the heart was stimulated with the β-agonist isoproterenol (Nakamura et al., 1993). In both neonatal- and adult-isolated myocyte models, IAA has been shown to increase cellular Ca^{2+} content (Buja et al., 1985; Barrigon et al., 1996), depress the $[Ca^{2+}]_i$ transient, elevate diastolic $[Ca^{2+}]_i$, and induce contracture (Eisner et al., 1989). Similar results have been observed in cultured chick embryonic cells and paced adult rabbit ventricular myocytes exposed to 2-DG in nonquiescent myocytes (Ikenouchi et al., 1991). Wang et al. (1995) have shown that IAA and 2-DG together activate a Ca^{2+} leak channel on the sarcolemmal membrane of patch-clamped adult rat ventricular myocytes. The same channel could also be activated by hydrogen peroxide, an agent that may selectively inhibit glycolysis (Goldhaber et al., 1989; Corretti et al., 1991). The sensitivity of $[Ca^{2+}]_i$ and diastolic contractile function to glycolytic inhibition observed in these studies raises the possibility that glycolysis plays a special role in maintaining certain cellular functions in cardiac myocytes, for example, the Ca^{2+} pump of the SR. Another possibility is that excess H^+ generated during selective glycolytic inhibition is mainly extruded via Na^+-H^+ exchange, resulting in increased Na^+ entry. The increased $[Na^{2+}]_i$ would in turn promote accelerated Ca^{2+} entry via reverse Na^+-Ca^{2+} exchange. However, excess H^+ generated during anaerobic metabolism can be extruded along with lactate via its transporter, thereby avoiding the Na^+-H^+ exchange route and its indirect effects on $[Ca^{2+}]_i$. Preliminary evidence in support of this hypothesis has been obtained (G. Langer, personal communication, August, 1996). Alternatively, inhibition of glycolytic metabolism may generate metabolites that are particularly deleterious to E–C coupling and $[Ca^{2+}]_i$. For example, Kusuoka and Marban (1994) have suggested that diastolic $[Ca^{2+}]_i$ overload and increased diastolic tone are more closely associated with accumulation of sugar phosphates than reductions in global high-energy phosphate levels or source of ATP production (glycolytic vs oxidative).

C. METABOLIC REGULATION OF CELLULAR K$^+$

Cellular potassium is regulated by the Na^+-K^+ pump (Chapter 4), as well as a large number of potassium-selective ion channels (Chapter 3). The potassium gradient and large diastolic potassium conductance are critical determinates of resting membrane potential and repolarization of the action potential. Abnormal regulation of cellular K^+, particularly cellular K^+ loss during ischemia, leads to conditions that favor the development of lethal reentrant ventricular arrhythmias (see Chapter 3). Thus, the metabolic regulation of K^+ has been an area of intense investigation for many years.

It is well known that a marked increase in cellular K^+ efflux occurs very early during anoxia, combined inhibition of oxidative and glycolytic metabolism, or selective glycolytic inhibition. The increase in K^+ efflux has been proposed to be the dominant cause of net cellular K^+ loss (Rau et al., 1977; Weiss and Hiltbrand, 1985), although reduced K^+ influx due to partial inhibition of the Na^+-K^+ ATPase is also possible (Wilde and Kleber, 1986). Although there is agreement that unidirectional K^+ efflux increases under conditions of metabolic stress due to activation of metabolically sensitive K^+ channels, their role in causing net K^+ loss is controversial (Wilde and Aksnes, 1995). Other mechanisms that may alter K^+ balance in ischemia include Na^+-activated K^+ channels, arachadonic acid activated K^+ channels, lactate-K^+ cotransport, P_i-K^+ cotransport, and Na^+-K^+-$2Cl^-$ transport.

1. ATP-Sensitive K^+ Channels

In the early 1980s, voltage clamp studies in papillary muscles (Vleugels et al., 1980) and single ventricular myocytes (Isenberg et al., 1983) identified an increase in a time-dependent membrane K^+ conductance as the cause of action potential duration (APD) shortening during hypoxia and metabolic inhibition. Subsequently, the discovery of cardiac ATP-sensitive K^+ channels (Noma, 1983) provided an attractive explanation for increased K^+ efflux and APD shortening under these conditions. These channels are blocked by ATP at the cytoplasmic surface and are normally closed, but they are open when the cytosolic ATP concentration falls below a critical threshold. Block by ATP does not involve phosphorylation because nonmetabolizable ATP derivatives block ATP-sensitive K^+ channels with equivalent potency (Findlay, 1988a; Lederer and Nichols, 1989; Ashcroft and Ashcroft, 1990, but see Ribalet et al., 1989). In intact cells, the ATP levels at which ATP-sensitive K^+ channels are activated during hypoxia or metabolic inhibition are unknown, yet in excised membrane patches the ATP concentration producing half-maximal suppression of ATP-sensitive K^+ channels (K_d) is 25–100 μM (Noma, 1983; Lederer and Nichols, 1989; Deutsch and Weiss, 1993). In contrast, in intact cardiac muscle the cytosolic ATP concentration is estimated at 5–10 mM, and during the first 10 minutes of hypoxia or ischemia it only falls by about 30% (Rovetto et al., 1973) (Fig. 4).

a. Regulation of I_{KATP} by Other Metabolic Products

It has been difficult to reconcile the large discrepancy between the millimolar levels of cytosolic ATP during early hypoxia in intact heart and the micromolar concentrations of ATP needed to suppress ATP-sensitive K^+ channels in excised membrane patches. However, only a slight increase in the open probability of ATP-sensitive K^+ channels is needed to account for the degree of APD shortening and increased K^+ efflux observed during hypoxia. Shifts in the sensitivity of ATP-sensitive K^+ channels to ATP can be caused by a number of metabolic products, including H^+, P_i, Ca^{+2}, NAD and NADH, lysophosphoglycerides, lactate, exogenous free radicals, and various nucleotides and their precursors (see Terzic *et al.*, 1995, for review). Of these factors, ADP has been shown to cause the most prominent increase in the K_d, from 25 to about 100 μM, whereas GDP and acidosis had more modest effects (Findlay, 1988b; Lederer and Nichols, 1989; Weiss *et al.*, 1992). Because of the high density of ATP-sensitive K^+ channels in heart, it has been shown that activation of only a small percentage of channels ($<0.5\%$) is sufficient to account for the observed APD shortening and increased K^+ efflux during hypoxia and ischemia (Weiss *et al.*, 1992). The increase in the K_d of ATP-sensitive K^+ channels produced by increases in free ADP concentrations during hypoxia and other forms of metabolic inhibition (as well as ischemia) is quantitatively sufficient to account for the magnitude and time course of the increase in cellular K^+ loss and APD shortening (Weiss *et al.*, 1992). It has also been shown that the ATP sensitivity of these channels is markedly reduced during metabolic inhibition, with the K_d increasing from ~50 to ~300 μM, presumably because of irreversible modification of the channel protein (Deutsch and Weiss, 1993). Furthermore, the desensitization of ATP-sensitive K^+ channels to ATP could be prevented by eliminating $[Ca^{2+}]_i$ overload during metabolic inhibition, which suggests that activation of a Ca^{2+}-dependent protease or phospholipase may be responsible. Consistent with this idea, treatment of the cytoplasmic surface of inside-out membrane patches with trypsin and, to a lesser extent, with chymotrypsin or with phospholipases A2, C, or D was shown to cause a similar desensitization to ATP (Deutsch and Weiss, 1994). It is conceivable that proteases, phospholipases, or other intracellular enzymes activated during metabolic inhibition might alter the structure or environment of ATP-sensitive K^+ channels in a way that reduces their sensitivity to ATP and thus causes them to open earlier in the course of metabolic inhibition. This mechanism may be an important step in "ischemic preconditioning" (as discussed in Section V).

b. Preferential Regulation of ATP-Sensitive K^+ Channels by Glycolysis

Several lines of evidence suggest that glycolysis is a preferential metabolic pathway regulating cardiac ATP-sensitive K^+ channels. First, inhibition of anaerobic glycolysis in the isolated rabbit interventricular septum caused marked cellular K^+ loss but did not deplete tissue ATP content (Weiss and Hiltbrand, 1985). Furthermore, cellular K^+ loss during selective inhibition of glycolysis in the intact rabbit ventricular septum was blocked by glibenclamide, a sulfonylurea that blocks ATP-sensitive K^+ channels (Weiss and Lamp, 1989). In patch-clamped guinea pig ventricular myocytes permeabilized at one end with saponin, ATP-sensitive K^+ channels recorded from a cell-attached patch could be suppressed by substrates for ATP production from anaerobic glycolysis or oxidative metabolism or by creatine phosphate via the creatine kinase reaction, which indicates that the metabolic machinery of the permeabilized cells remained intact. However, when cytoplasmic ATP utilization rates were stimulated, glycolytic substrates, even in the presence of FCCP, were superior to substrates for either oxidative phosphorylation or creatine phosphate at suppressing ATP-sensitive K^+ channel openings. In contrast, the latter two groups of substrates were slightly (although not significantly) more effective at preventing cell shortening due to rigor. The preferential regulation of ATP-sensitive K^+ channels by substrates for glycolytic metabolism, even in a significant proportion of excised membrane patches, has been interpreted to suggest that glycolytic enzymes might be associated with ATP-sensitive K^+ channels in the sarcolemma, thereby preferentially supplying glycolytically derived ATP to the channels (Weiss and Lamp, 1989). Alternatively, glycolytic enzymes may preferentially suppress ATP-sensitive K^+ channels by minimizing free $[ADP]_i$ in the vicinity of the channels rather than by supplying a critical pool of subsarcolemmal ATP to the channels (Weiss and Venkatesh, 1993). Although these findings may account for the glibenclamide-sensitive increase in K^+ efflux during selective inhibition of glycolysis, its relevance to K^+ loss during early ischemia and hypoxia is unclear because glycolytic flux increases markedly under these conditions (see following discussion, and Rovetto et al., 1973).

An interesting consequence of the special relationship between glycolytic metabolism and regulation of I_{KATP} is the cyclical activation of I_{KATP} observed during partial inhibition of glycolysis in patch-clamped guinea pig ventricular myocytes (O'Rourke et al., 1994). The oscillations of I_{KATP} were always preceded by transient decreases in NADH, which suggests that a change in the rate of metabolism initiated the phenomenon. Voltage records from oscillating cells revealed cyclical changes in the action potential duration, including periods of inexcitability. This kind of behavior may contribute to the pathogenesis of lethal ventricular arrhythmias during myocardial infarction.

D. METABOLIC REGULATION OF $[Na^+]_i$

$[Na^+]_i$ plays a critical role in the regulation of normal cardiac function. Through Na^+–Ca^{2+} (Chapters 4 and 5) and Na^+–H^+ exchange (Chapter 4), $[Na^+]_i$ is involved in the regulation of both $[Ca^{2+}]_i$ and pH. Increases in

[Na$^+$]$_i$ during metabolic inhibition have been observed in most, but not all, cardiac preparations (Guarnieri, 1987; Murphy *et al.*, 1988; Anderson *et al.*, 1990; Donoso *et al.*, 1992; Haigney *et al.*, 1992). Increases in [Na$^+$]$_i$ precede contracture, which is similar to what has been recorded using ion selective microelectrodes in Purkinje fibers exposed to hypoxic superfusate (Mac-Leod, 1989). In the latter study, the presence of glucose in the perfusate inhibited the rise in [Na$^+$]$_i$ similar to the results of Guarnieri in ferret papillary muscle (Guarnieri, 1987), but some variability has been reported (Wilde and Kleber, 1986; Ellis and Noireaud, 1987; Stewart *et al.*, 1994). In one study of [Na$^+$]$_i$ in hypoxic ferret hearts (Neubauer *et al.*, 1992), ventricular fibrillation was observed only when [Na$^+$]$_i$ increased (about half the hearts). During metabolic inhibition, interventions designed to inhibit glycolysis caused a more rapid increase in [Na$^+$]$_i$ than CN alone (MacLeod, 1989). Interestingly, inhibition of the Na$^+$ pump with strophanthidin increased [Na$^+$]$_i$, but it did not cause diastolic shortening even after CN was added to the perfusate. The causes of [Na$^+$]$_i$ overload are uncertain, but similar to [Ca^{2+}]$_i$ overload they could either be due to an increase in Na$^+$ influx, a decrease in Na$^+$ efflux, or both. In heart, the major Na$^+$ influx mechanism is the voltage-dependent Na$^+$ current, and the major efflux mechanism is the Na$^+$ pump; other Na$^+$ regulatory mechanisms include Na$^+$–H$^+$ exchange, which is also involved in the regulation of intracellular pH and Na$^+$–K$^+$-Cl$^-$ cotransport. The effect of metabolic inhibition on each of these mechanisms is discussed in turn.

1. Voltage-Sensitive Na$^+$ Channels

The role of Na$^+$ channels in increasing [Na$^+$]$_i$ during metabolic inhibition is uncertain. A single Na$^+$ channel can admit more than 10 million Na$^+$ ions per second into the cytoplasm. Furthermore, the density of Na$^+$ channels is so high that only a slight increase in Na$^+$ channel open probability would significantly increase the amount of Na$^+$ admitted into the cell. Na$^+$ channel blockers have been shown to attenuate increases in [Na$^+$]$_i$ (and [Ca^{2+}]$_i$) in electrically stimulated hypoxic (no glucose) isolated rat ventricular myocytes (Haigney *et al.*, 1994) but not in myocytes at rest (Satoh *et al.*, 1995). Furthermore, direct measurements in patch-clamped guinea pig ventricular myocytes exposed to metabolic inhibitors (0.2 mM 2,4-dinitrophenol, or 1 mM IAA) failed to show any alteration in Na$^+$ currents (Mejia-Alvarez and Marban, 1992), arguing against a close coupling of metabolism and Na$^+$ channels. Na$^+$ channel blockers may be beneficial only inasmuch as they reduce the overall burden of [Na$^+$]$_i$ to the cell. The situation may be more complicated in ischemia when other metabolic products accumulate, as lysophospholipids have been shown to induce Na$^+$ channel openings at diastolic membrane potentials (see the subsection "Amphiphiles" in Section V).

2. Na⁺–H⁺ Exchange

Metabolic inhibition causes an intracellular acidosis (Eisner *et al.*, 1989; Smith *et al.*, 1993) that affects multiple aspects of E–C coupling (for review, see Orchard and Kentish, 1990). Although lactate and H^+ production are a consequence of stimulated anaerobic glycolysis in the setting of inhibited oxidative metabolism (Allen *et al.*, 1985), hydrolysis of high-energy phosphates can also cause significant acidosis during inhibition of glycolytic metabolism when lactate levels are low (Smith *et al.*, 1993). This may not be the case during ischemia (Garlick *et al.*, 1979) or in the setting of hypoxia where extracellular pH declined less when glycolysis and lactate production were reduced by glycogen depletion (Weiss *et al.*, 1984).

Decreases in pH during metabolic inhibition could increase $[Na^+]_i$ by driving Na^+–H^+ exchange. Combined inhibition of oxidative and glycolytic metabolism using 3.3 mM Amytal and 5 μM CCCP in glucose-free solution (Satoh *et al.*, 1995) caused an increase in $[Na^+]_i$ (from 6.2 to 18.6 mM in the initial 20 min), which preceded diastolic cell shortening and increases in $[Ca^{2+}]_i$. These changes could be prevented by hexamethylamiloride, a blocker of Na^+–H^+ exchange. However, the specificity of amiloride derivatives for Na^+–H^+ exchange has been questioned by Haigney *et al.* (1994), who found that ethylisopropyl amiloride could block $[Na^+]_i$ accumulation caused by exposure to a Na^+ channel opener under normoxic conditions when acidosis was presumably not present. Further studies are needed to clarify the role of the Na^+–H^+ exchanger in $[Na^+]_i$ increases during metabolic inhibition.

3. Na⁺–K⁺ Pump

The Na^+–K^+ pump plays a critical role in the maintenance of Na^+ and K^+ gradients; however, its contribution to $[Na^+]_i$ accumulation during metabolic inhibition has been difficult to quantify. Donoso *et al.* (1992) have calculated that the Na^+ pump would have to be inhibited by 63%, corresponding to an ATP concentration of 170 μM, to account for the increase in $[Na^+]_i$ they observed in rat ventricular myocytes exposed to CN in the absence of glucose (although this may be an underestimate if P_i increases under these conditions). This concentration of ATP is far less than what has been reported during hypoxia (or ischemia). Similarly, Na^+ pump inhibition cannot be predicted on the basis of changes in the free energy of ATP hydrolysis during hypoxia (Allen *et al.*, 1985). Under normoxic conditions, assuming $[Na^+]_i = 9$ mM, $[K^+]_i = 140$ (at 35°), $V_m = -70$, and a 3:2 Na^+:K^+ exchange stoichiometry per ATP hydrolyzed, then the pump requires about 44 kJ mol⁻¹. Normal delta G is 61.5 kJ mol⁻¹. During hypoxia with glycolysis active, delta G decreases to 48.6 kJ; with glycolysis inhibited as well, delta G falls to 46.9 kJ.

It is somewhat surprising then that Na^+ pump activity is reduced during metabolic inhibition. When measured as Rb^+ uptake, Na^+ pump activity is reduced in perfused rat hearts exposed to 30 minutes of substrate-free hypoxia (Grinwald, 1992) and fails to recover at reoxygenation. Similar results have been obtained in sheep Purkinje fiber cardioballs studied with the patch-clamp technique (Glitsch and Tappe, 1993). In that study, inhibition of oxidative metabolism with the uncoupler CCCP in substrate-free perfusate caused a marked decrease in Na^+ pump current. This inhibition could be prevented by including glucose and ATP in the perfusate or accelerated by adding a high dose of either iodoacetate (2 mM) or 2-DG to inhibit glycolysis. IAA with supplemental pyruvate markedly accelerated inhibition of the pump compared with CCCP alone. Inhibition of the pump could be reversed by internal perfusion with ATP.

Although direct measures of Rb^+ uptake and pump current suggest that Na^+ pump activity is reduced during metabolic inhibition, it is unlikely that reduced ATP levels of free energy changes are solely responsible. Consequently, one must postulate that other metabolic products contribute to pump inhibition. Increased H^+ and P_i (Deitmer and Ellis, 1980; Eisner and Richards, 1982; Huang and Askari, 1984) are likely candidates because both have been shown to have inhibitory effects on the Na^+ pump.

4. Na^+ Entry via a Nonselective Current

In the Cardiovascular Research Laboratory and Division of Cardiology at the UCLA School of Medicine (hereafter referred to as "this laboratory"), we have recently identified a nonselective current that is activated by CCCP or rotenone (in the absence of glucose) in patch-clamped rabbit ventricular myocytes (Goldhaber et al., 1996a). The current does not require elevated $[Ca^{2+}]_i$ for activation and is reversible at washout of the metabolic inhibitor. Activation of this channel could therefore precede increases in $[Ca^{2+}]_i$ caused by metabolic inhibition and contribute to $[Na^+]_i$ loading.

E. IS METABOLISM COMPARTMENTALIZED?

The observation that selective inhibition of glycolysis has more pronounced and reliable effects on $[Ca^{2+}]_i$, diastolic function, K^+ loss, and $[Na^+]_i$ accumulation than selective inhibition of oxidative metabolism alone is consistent with a growing body of evidence suggesting that glycolysis may be a preferential source of ATP for certain aspects of cardiac function in the heart (McDonald and MacLeod, 1973; Bricknell et al., 1981; Higgins et al., 1981; Hasin and Barry, 1984; Weiss and Hiltbrand, 1985; Weiss and Lamp, 1987; Weiss and Lamp, 1989) as well as other tissues, such as vascular smooth muscle (Lynch and Paul, 1983); erythrocytes (Parker and Hoffman, 1967; Mercer and Dunham, 1981); hepatocytes (Jones, 1986); brain (Lipton and Robacker, 1983); and cultured cells (Balaban and Bader, 1984; Lynch and Balaban, 1986). Generally these studies have concluded that glycolytically derived ATP is preferentially utilized to support membrane-associated

ATPases such as the Na^+-K^+ pump, whereas oxidatively derived ATP is preferentially used by the cytosolic organelles such as the contractile apparatus. Glycogenolytic and glycolytic enzyme complexes have been shown to be specifically associated with the SR in the heart (Entman et al., 1976; Pierce and Philipson, 1985), and glycolytic enzymes are functionally coupled to the SR Ca^{2+} pump (Xu et al., 1995). Glucose from glycogen is preferentially oxidized compared with exogenous glucose (Henning et al., 1996). Glycolysis may support the maintenance of the mitochondrial membrane potential during anoxia and metabolic inhibition in adult rat cardiac myocytes (Di Lisa et al., 1995). In skeletal muscle triads, compartmentalized glycolytic ATP production has been demonstrated (Han et al., 1992), and in vascular smooth muscle it has been shown that high-energy phosphates derived from exogenous glucose are utilized differently than high-energy phosphates derived from glycogenolysis (Lynch and Paul, 1983). Similar results have been obtained in the isolated arterially perfused rabbit interventricular septum. In this model, metabolism of exogenous glucose is more effective than glycogenolysis at supporting electromechanical function during hypoxia (Runnman et al., 1990). These findings would be consistent with the localization of key glycolytic enzymes at specific sites in the cell responsible for maintenance of critical functions. This would enable glycolysis to maintain adenine nucleotides locally at concentrations not necessarily reflected by total tissue levels of high-energy phosphates. In this regard, Opie (1983) has shown a 240-fold greater ATP concentration remaining in the intact cell, which is in rigor contraction, as compared to the concentration required to place contractile proteins in a rigor complex in vitro. It is possible that if phosphorylase and phosphoglucomutase, the enzymes necessary to convert glycogen to glucose-6-phosphate, were not present along with the other key glycolytic enzymes at these same sites, then glycogen might not be able to substitute effectively for exogenous glucose as a metabolic substrate. Alternatively, it is possible that the progressive depletion of glycogen during hypoxia with a physiologic glucose concentration present may reduce high-energy phosphate supply to a structure preferentially dependent on glycogenolysis, such as the SR.

Nevertheless, the compartmentation hypothesis remains controversial because measured intracellular ATP diffusion coefficients seem inconsistent with the development of large local gradients within the cytosol. There are no obvious anatomic barriers to impair cytoplasmic ATP diffusion, and it is difficult to exclude secondary effects of metabolic inhibition as potential causes of phenomena interpreted as evidence of metabolic compartmentation. For example, Kusuoka and Marban (1994) have suggested an alternative mechanism to explain why glycolytic inhibition leads to elevation of diastolic $[Ca^{2+}]_i$ without invoking compartmentation. In isolated perfused ferret hearts, diastolic $[Ca^{2+}]_i$ overload occurred during selective inhibition of glycolysis with 2-DG or IAA, only in association with accumulation of sugar phosphates. If the accumulation of these glycolytic intermediates was prevented, $[Ca^{2+}]_i$ overload did not occur. These findings led them to speculate that sugar phosphates might have a direct toxic effect on Ca^{2+} handling mechanisms.

In summary, the metabolic regulation of cellular processes is directly relevant to clinical ischemia in humans. Specific inhibitors of glycolytic and oxidative metabolism are available and can be used selectively or in combination to inhibit both oxidative and glycolytic metabolism.

Abnormal $[Ca^{2+}]_i$ homeostasis is detrimental to cellular function. Metabolic inhibition, particularly inhibition of glycolysis, results in depressed contractility, elevations in diastolic $[Ca^{2+}]_i$, and increased diastolic pressure or shortening. Even at the point of rigor, fundamental components of E–C coupling remain functional, including the Ca^{2+}-induced Ca^{2+} release mechanism, $Na^+–Ca^{2+}$ exchange, and SR Ca^{2+} pumping, although the latter may be moderately compromised. Because diastolic cell shortening or increases in diastolic pressure precede diastolic $[Ca^{2+}]_i$ overload in several models of metabolic inhibition, the exact relationship of cytosolic Ca^{2+} to early diastolic contractile abnormalities is not certain. Energy depletion and its effects on actomyosin cross-bridge function is therefore a more attractive mechanism to explain the onset of diastolic cell shortening and elevated diastolic pressure during the early stages of metabolic inhibition. Elevated $[Ca^{2+}]_i$ certainly contributes to diastolic abnormalities later in metabolic inhibition.

Cellular K^+ regulation is a critical determinate of membrane potential and action potential duration. A marked increase in cellular K^+ conductance and efflux occurs early during metabolic inhibition and contributes to tissue inexcitability. ATP-sensitive K^+ channels are activated during metabolic inhibition, in part because ATP levels fall but perhaps, more important, because their sensitivity to ATP is decreased by the accumulation of metabolic products such as ADP. Glycolysis may preferentially supply ATP to these channels. Failure of the Na^+ pump may contribute to K^+ loss during the late stages of metabolic inhibition.

Metabolic inhibition is associated with $[Na^+]_i$ accumulation and the alteration of a number of $[Na^+]_i$ handling mechanisms, however, the relative contribution of each of these mechanisms to $[Na^+]_i$ accumulation is uncertain. Voltage-dependent Na^+ channels are not inhibited by metabolic inhibition and can therefore provide continued influx of Na^+. Acidosis can also promote Na^+ influx via $Na^+–H^+$ exchange, but its role during metabolic inhibition is controversial because $Na^+–H^+$ blockers have nonspecific effects on other Na^+ influx mechanisms. Direct measurements of Na^+ pump current indicate that the pump is impaired during metabolic inhibition, although this may be a late effect. Finally, a nonselective current activated by metabolic inhibition may contribute to pathological Na^+ entry. Most likely a combination of these mechanisms is responsible for the increased $[Na^+]_i$ observed in metabolic inhibition.

Several lines of evidence suggest that metabolism may be compartmentalized in heart. Glycogenolytic and glycolytic enzyme complexes have been shown to be specifically associated with intracellular organelles. Glucose produced by glycogenolysis is metabolized differently than exogenous glucose. Selective inhibition of glycolysis causes K^+ loss without decreasing ATP levels, and substrates for ATP production from anaerobic glycolysis are superior to substrates for oxidative phosphorylation or creatine phosphate at suppressing ATP-sensitive K^+ channels.

IV. MYOCARDIAL ISCHEMIA

A. ACUTE ISCHEMIA

1. Metabolic Changes

Interruption of blood flow to myocardium has numerous profound effects on metabolism and function. In less than 10 seconds the supply of oxygen available to the myocardium is depleted (Jennings et al., 1986). Because oxygen is no longer provided, oxidative metabolism ceases and activated long-chain nonesterified fatty acid intermediates, such as acyl CoA and acyl carnitine CoA, accumulate via inhibition of β-oxidation. Within 1–3 minutes, the rate of anaerobic glycolysis increases dramatically (Neely and Morgan, 1974; Oram et al., 1975; Myears et al., 1987) in an effort to compensate for the loss of high-energy phosphate production. The source of glucose is entirely from glycogenolysis (Owen et al., 1990), whereas exogenous glucose is used preferentially for glycolysis associated with less severe ischemia (Bricknell and Opie, 1978). Because even maximum glycolytic rates cannot adequately compensate for the loss of oxidative ATP generation [estimated by Neely et al. (1975) to produce only ~20% of the ATP required by aerobic tissue], creatine phosphate is rapidly consumed in an effort to preserve ATP levels (Fig. 4) (Hearse, 1979; Gard et al., 1985; Clarke et al., 1987). Covell et al. (1967) noted a 40% decrease in CP but no change in ATP an average of 47 seconds into ischemia. Others noted 10–20% changes in ATP content during the first 60 seconds of ischemia, but these changes were not statistically significant (Vial et al., 1978; Prinzen et al., 1986). Thanks in part to the "buffering" effect of creatine phosphate, after 10 minutes of ischemia, total ATP has fallen only by one-third (Hearse, 1979). Once CP stores are depleted, ATP levels inexorably decline, and as high-energy phosphate bonds are hydrolyzed, inorganic phosphate and protons accumulate (Clarke et al., 1987). However, some caution should be exercised when interpreting early transmural changes in high-energy phosphates that may underestimate subendocardial changes. For example, Arai et al. (1992)

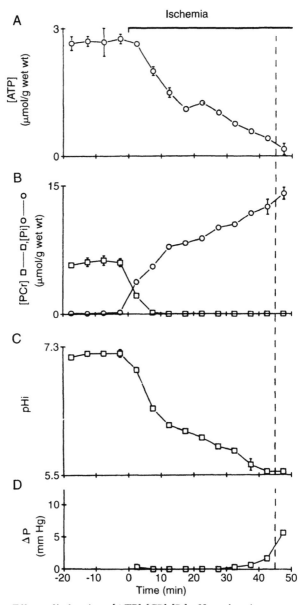

FIGURE 4   Effects of ischemia on [ATP], [CP], [P_i], pH_i, and resting pressure development shown by the extent of increase from minimal resting pressure (ΔP) in three ferret hearts perfused in Langendorff fashion and studied with the nuclear magnetic resonance technique. Note the preservation of [ATP] at 2 minutes of ischemia despite the rapid reduction in [CP] by greater than 50%. By 8 minutes of ischemia, [CP] levels are undetectable, but ATP levels have only decreased slightly. (Adapted by permission from Koretsune and Marban, 1990.)

found that *subendocardial* levels of CP and ATP fall by 23 and 16%, respectively, within 15 beats of the onset of a 50% reduction in coronary flow, compared to an 8% decrease in *transmural* ATP. In that study, on a molar basis, the amount of subendocardial ATP lost (0.73 μmol/g wet wt) was about half the amount of CP consumed (1.58 μmol/g wet wt) during the first 15 beats of ischemia).

Elevation of $[Ca^{2+}]_i$ is buffered partially by the mitochondria, which continue to accumulate Ca^{2+} (Shen and Jennings, 1972; Pesaturo and Gwathmey, 1990; Miyata *et al.*, 1992; Barrigon *et al.*, 1996). As a result, there is dissipation of the H^+ gradient (Parr *et al.*, 1975) generated by electron transport (Pesaturo and Gwanthmey, 1990). This may lead to uncoupling of oxidative phosphorylation and reduction of ATP synthesis consistent with other studies in heart (Kitakaze *et al.*, 1988b).

In the continued absence of oxygen, anaerobic glycolysis produces lactate rather than acetyl-Co-A. The build-up of lactate, H, and NADH (Rovetto *et al.*, 1973; Kingsley-Hickman *et al.*, 1990) progressively inhibits glycolysis. $[Ca^{2+}]_i$ accumulation also inhibits glycolysis by facilitating binding of calmodulin to phosphofructokinase, considered the key rate-limiting enzyme in the glycolytic pathway (Mayr, 1984; Buschmeier *et al.*, 1987; Auffermann *et al.*, 1990). The end result is a complete shut-down of myocardial metabolism.

2. Consequences of Ischemia on Cardiac Function

a. Overview

Although the sequence of events during ischemia that results in ATP depletion has been well characterized, the functional consequences of ischemia do not naturally follow the levels of ATP in the cell (see, however, Arai *et al.*, 1992). Whereas it takes several minutes for ATP levels to fall significantly during ischemia, cardiac function deteriorates within seconds; it is initially manifest as an increase in the rate of cellular K loss (Shine *et al.*, 1976; Hill and Gettes, 1980; Hirche *et al.*, 1980; Kleber, 1983) and a decrease in the rate of twitch relaxation (Shine *et al.*, 1976). Contractile amplitude decreases within 60 seconds in isolated hearts (Shine *et al.*, 1976), but more rapidly *in vivo* during coronary occlusion (Kubler and Katz, 1977), and it is followed by inexcitability and ultimately contracture.

b. Contraction and $[Ca^{2+}]_i$

i. Tension Development   The initial decrement in contractile amplitude during very early ischemia does not appear to be directly related to metabolic failure. Instead, vascular collapse, also known as the "garden hose" effect (Koretsune *et al.*, 1991), and alterations in the response of myofilaments to activator Ca^{2+} have been implicated. According to the "garden hose" effect hypothesis, reductions in coronary pressure during ischemia lead to reduced tension development due to a Frank-Starling mechanism (i.e., reduced sarcomere length of the myocardial cells that surround the vessels, which results in a fall in tension development (Fig. 5). Alterations in the response of myofilaments are caused by the accumulation of inorganic phosphate, ADP, and hydrogen ion (Fabiato and Fabiato, 1978; Kentish, 1986), resulting from hydrolysis of high-energy phosphate compounds during ischemia. H^+ and P_i are known to decrease the sensitivity of myofilaments to $[Ca^{2+}]_i$, and elevated ADP slows relaxation. Changes in phosphorylation potential or creatine phosphate to inorganic phosphate ratios may also contribute to early contractile failure (Clarke *et al.*, 1987; Schaefer *et al.*, 1990; Schwartz *et al.*, 1990; Arai *et al.*, 1992). As ischemia progresses, shortening of the APD further abbreviates contraction.

ii. [Ca²⁺]ᵢ *and Contraction during Acute Ischemia* $[Ca^{2+}]_i$ overload has been viewed as a final common pathway leading to irreversible tissue injury during ischemia, as well as during reperfusion of ischemic myocardium and other conditions including the oxygen and calcium paradoxes. Unfortunately, genuinely ischemic heart is too complex to determine in detail which of the cellular processes normally regulating $[Ca^{2+}]_i$ levels becomes disabled. In simpler cardiac preparations in which such detailed information is obtainable, metabolic inhibition has been relied on to simulate certain important aspects of ischemia, but accumulation of metabolic by-products from the lack of vascular washout during ischemia cannot be accurately re-created. During ischemia, additional factors may affect $[Ca^{2+}]_i$ regulating mechanisms, including acidosis, lysophosphoglyceride and fatty acid ester accumulation, free radicals, catecholamines, cytokine production, and ATP release from nerve endings.

Early radioisotopic flux studies in ischemic and hypoxic heart suggested that elevations in $[Ca^{2+}]_i$ occurred primarily during reperfusion or reoxygenation (Nayler *et al.*, 1979), but more recent studies using newer techniques indicate that $[Ca^{2+}]_i$ increases soon after the onset of ischemia 10-15' Steenbergen *et al.*, 1987); 10-15' (Marban *et al.*, 1988); 90s-15m (Mohabir *et al.*, 1991; Fig. 6); and 2' (Kihara *et al.*, 1989). In isolated perfused rabbit hearts loaded with the $[Ca^{2+}]_i$ indicator indo-1, $[Ca^{2+}]_i$ transient amplitude remains normal after 1.5 minutes of ischemia, suggesting that Ca^{2+} currents and action potential duration (APD) are also close to normal, despite almost complete cessation of contractile activity. Further elevations in diastolic $[Ca^{2+}]_i$ do not clearly lead to increases in diastolic pressure, which seem to be more closely correlated with declines in tissue ATP (Koretsune and Marban, 1990; Steenbergen *et al.*, 1990). The ATP depletion can be slowed by Ca^{2+} channel blockade prior to ischemia. Even after 15 minutes of ischemia, $[Ca^{2+}]_i$ can return to normal at reflow, but ATP and developed pressure remain low (Marban *et al.*, 1990).

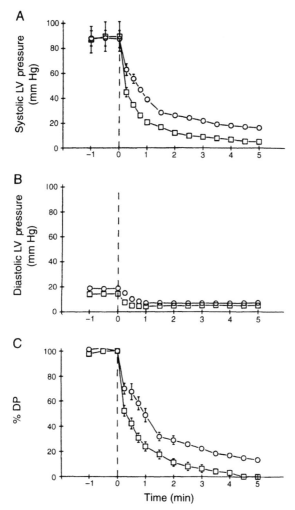

FIGURE 5 Systolic left ventricular (LV) pressure (A), diastolic LV pressure (B), and percent of control developed pressure (DP) in perfused ferret hearts during global ischemia (□) or during microembolization (○) by bolus injection of carbon microspheres (13.7 ± 0.8 μm in diameter). In this model, microembolization caused reductions in coronary flow and tissue metabolites, including high energy phosphates, comparable to ischemia. However, coronary perfusion pressure was preserved. Although function declines in both groups, the changes in systolic (panel A) and developed pressure (panel C) were slower with microembolization as compared with total global ischemia ($p < 0.01$ by multivariate analysis of variance). Diastolic pressure changes were similar in the two groups. The observation of much slower functional decline in the microembolization group is consistent with the hypothesis that intravascular collapse makes a major contribution to early contractile failure. (Reproduced with permission from Koretsune et al., 1991.)

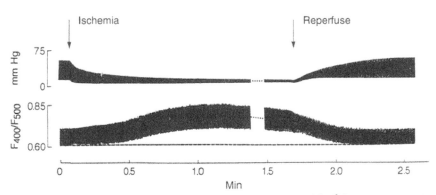

FIGURE 6   Effect of ischemia on left ventricular pressure and $[Ca^{2+}]_i$ transients and in a Langendorff-perfused rabbit heart loaded with the $[Ca^{2+}]_i$ indicator indo-1. Ischemia caused a rapid decline in systolic and diastolic pressure (top tracing), along with early elevation of both systolic and diastolic indo fluorescence ratios (F_{400}/F_{500}) (bottom tracing). The amplitude of the $[Ca^{2+}]_i$ transient also increased during the ischemic period. All parameters returned to baseline within 1 minute of reperfusion. (Reproduced with permission from Mohabir *et al.*, 1991.)

The cumulative evidence from experiments using metabolic inhibitors presented earlier suggested that increases in $[Ca^{2+}]_i$ during energy deprivation, even at the onset of rigor, are not likely to be the result of a massive influx of extracellular Ca^{2+} due to loss of sarcolemmal integrity, but from a failure of the processes regulating $[Ca^{2+}]_i$ removal from the cytosol, with perhaps a minor role for additional Ca^{2+} entry via Ca^{2+} leak channels.

Prolonged ischemia (>40 min) with its associated depletion of high-energy phosphates (Koretsune and Marban, 1990; Steenbergen *et al.*, 1990) allows for the formation of rigor cross-bridges, elevations in $[Ca^{2+}]_i$, and increased diastolic pressure or tension (Allen and Orchard, 1987). The exact cause of increased diastolic pressure during ischemia has been a matter of controversy because of the difficulty in separation of the effects of high-energy phosphate depletion (and development of rigor cross-bridges) versus elevated $[Ca^{2+}]_i$ (contracture). The evidence from studies using metabolic inhibitors described earlier supports the hypothesis that the earliest increases in diastolic tension or pressure during ischemia are more closely related to energy deprivation rather than $[Ca^{2+}]_i$ overload per se. Also consistent with the data from studies evaluating the role of metabolism in contracture is the observation that the development of contracture during ischemia is accelerated when glycolysis, as opposed to oxidative metabolism, is inhibited (Bricknell *et al.*, 1981; Lipasti *et al.*, 1984). Providing substrate for glycolysis was more effective than substrates for oxidative metabolism (e.g., pyruvate) at maintaining $[Ca^{2+}]_i$ homeostasis and promoting functional recovery during reperfusion of ischemic rabbit hearts studied with ^{19}F-NMR spectroscopy (Jeremy *et al.*, 1992). Increases in diastolic pressure in ischemic Langendorff-perfused rat hearts have been shown to coincide with the plateauing of the decrease in pH (Kingsley *et al.*, 1991), which is interpreted as being indicative of depletion of utilizable glycogen (Bailey *et al.*, 1982) and cessation of significant glycolytic production of ATP. In this model, global ATP measured by ^{31}P-NMR fell monotonically, did not correlate with the onset of increased diastolic pressure, and was in the mM range. These data suggest that cessation of glycolysis is a more important predictor of diastolic contractile abnormalities than reductions in global ATP.

iii. K^+ Balance during Ischemia   Net cellular K^+ loss occurs within 15–30 seconds after the onset of myocardial ischemia, hypoxia, and exposure to metabolic inhibitors (Rau and Langer, 1978; Hill and Gettes, 1980; Kleber, 1984; Weiss and Hiltbrand, 1985). During ischemia, the resulting accumulation of extracellular K^+ due to limited washout by reduced or absent blood flow contributes importantly to the marked electrophysiological abnormalities that predispose the heart to the development of reentrant arrhythmias (for review, see Janse and Wit, 1989). Action potential shortening occurs prior to any significant fall in ATP (Elliot *et al.*, 1989). As discussed earlier, net K^+ loss during the earlier stages of metabolic inhibition is predominantly due to an increase in K^+ efflux rather than a decrease in K^+ influx due to decreased Na^+ pump activity. Although the cause of increased K^+ efflux is not clearly resolved, activation of metabolically sensitive K^+ channels and anion-coupled K^+ efflux have both been proposed. Shrinkage of the extracellular space may also effectively concentrate extracellular potassium (Yan *et al.*, 1996). Longer durations of ischemia are associated with loss of membrane excitability due to further activation of metabolically sensitive K^+ currents and consequent shortening of the action potential (Stern, 1992). Because $[K^+]_o$ accumulation plays such a major role in arrhythmogenesis during acute myocardial ischemia (which remains the leading cause of death from coronary artery disease), resolution of the precise mechanisms involved in cellular K^+ loss during ischemia and other forms of metabolic inhibition is essential for developing improved approaches to this critical health problem.

a. ACTIVATION OF I_{KATP} DURING ISCHEMIA   Although decreases in cellular ATP concentration have been hypothesized to result in activation of ATP-sensitive K^+ channels during ischemia, cellular K^+ loss begins well before significant reductions in global ATP levels have occurred. A reduced sensitivity of the channels to ATP during early ischemia could explain how these channels activate in this setting despite preserved ATP levels. As described earlier, various components of the ischemic environment have been shown to affect the ATP sensitivity of ATP-sensitive K^+ channels in excised patches. These include H^+, P_i, Ca^{2+}, NAD and NADH, lysophosphoglycerides, lactate, exogenous free radicals, adenosine, and various nucleotides. Under physiological conditions relevant to severe ischemia, cardiac ATP-sensitive K^+ channels may also be modified irreversibly by proteases in a manner that contributes to persistent electrophysiological abnormalities and cellular K^+ loss after reperfusion. Blockers of I_{KATP} have been shown to partially reduce net cellular K^+ loss during ischemia (Bekheit *et al.*, 1990; Kantor *et al.*, 1990; Wilde *et al.*, 1990; Venkatesh *et al.*, 1991). However, pharmacological activation of I_{KATP} with cromakalim under normoxic conditions does not lead to net cellular K^+ loss despite increases in K^+ efflux rate and APD shortening approximating that seen with hypoxia (Deutsch *et al.*, 1994). These results imply that activation of I_{KATP} alone is not sufficient to explain K^+ loss during ischemia. One possibility is that an as yet unidentified inward current is activated during ischemia that opposes the hyperpolarization caused by activation of I_{KATP}. Such a current would promote K^+ loss through open ATP-sensitive K^+ channels during ischemia.

b. ANION-COUPLED K^+ EFFLUX DURING ISCHEMIA   Prior to the discovery of ATP-sensitive K^+ channels, and even since then in light of the difficulty explaining net cellular K^+ loss on the basis of activation of I_{KATP}, other mechanisms have been explored, including coupling of K^+ to efflux of anions (Mathur and Case, 1973; Kleber, 1984), particularly lactate and inorganic phosphate. Na^+-K^+-Cl^- cotransport (Mitani and Shattock, 1992) has also been proposed as an additional minor route of K^+ efflux.

Lactate accumulation can be detected within 15 beats of the onset of a 50% reducton in coronary flow (Arai *et al.*, 1992). According to the anion-coupled K^+ efflux hypothesis, as lactate and phosphate (P_i) anions generated intracellularly during ischemia or hypoxia diffuse into the extracellular space, they must be accompanied by an equal and opposite charge movement to maintain transsarcolemmal electroneutrality. It has been proposed that K^+, the most abundant intracellular cation, might serve as the charge balancing cation, resulting in increased K^+ efflux independent of any change in membrane K^+ permeability. In support of this hypothesis, the efflux of lactate and P_i during ischemia and hypoxia exceeds several fold the efflux of K^+ on a mole-to-mole basis, the time course of anion efflux and K^+ efflux are similar (Mathur and Case, 1973; Weiss *et al.*, 1989), and K^+ efflux associated with efflux of anions such as dimethyloxazolidinedione has been documented in the heart (Gaspardone *et al.*, 1986). Prominent effects of acidosis on ischemic $[K^+]_o$ accumulation (Kleber, 1984) could be more easily explained by an anion-coupled mechanism than by activation of I_{KATP}. In the heart, similar to other tissues, lactate is transported via the H^+-lactate transporter, which is nonelectrogenic and can be blocked by α-cyano-4-hydroxycinnamate (α-HC). However, K^+ is not efficiently co-transported with L-lactate in heart compared with H^+ (Shieh *et al.*, 1994), making lactate-coupled K^+ efflux very unlikely. P_i transport is coupled to

Na$^+$ and is probably electroneutral (Medina and Illingworth, 1980; Medina and Illingworth, 1984). Whether K$^+$ can substitute for Na$^+$ has not yet been studied. In summary, there is little evidence to support a major role for anion-coupled K$^+$ efflux during ischemia.

c. ACTIVATION OF OTHER K$^+$ CHANNELS DURING ISCHEMIA   In addition to ATP-sensitive K$^+$ channels, another class of K$^+$ channels activated by arachidonic acid (K$_{AA}$) has been described (Kim and Clapham, 1989). Because arachidonic acid and nonesterified fatty acids (NEFA) are known to accumulate during ischemia it has been proposed that K$_{AA}$ channels might also be involved in ischemic K$^+$ loss. However, significant increases in NEFA are not observed until 30–60 minutes of ischemia (Chien *et al.*, 1984; van Bilsen *et al.*, 1989), making it unlikely that K$_{AA}$ channels are involved in the early increase in cellular K$^+$ efflux. Na$^+$-activated K$^+$ channels (Kameyama *et al.*, 1984) are also unlikely to be activated during very early ischemia.

In conclusion, the mechanism of K$^+$ loss during ischemia is unknown. ATP-sensitive K$^+$ channels likely play a critical role, but their activation alone is insufficient to explain the K$^+$ loss observed.

iv. pH during Ischemia   Intracellular pH has been shown to fall rapidly after the onset of ischemia (Mohabir *et al.*, 1991), declining from a control level of 7.03 to 6.83 after 2 minutes, 6.32 after 10 minutes, and 6.11 after 15 minutes (Fig. 7; see also Fig. 4), although this degree of intracellular acidosis is lower than observed in the presence of erythrocytes that buffer H$^+$ (Yan and Kleber, 1992). The fall in pH has been attributed to persistent ATP hydrolysis and increased anaerobic glycogenolysis and glycolysis (Garlick *et al.*, 1979; Bailey *et al.*, 1981; Bailey *et al.*, 1982; Wolfe *et al.*, 1988). Anaerobic glycolysis is associated with lactate acid production and decreased proton consumption, as compared to mitochondrial ATP resynthe-

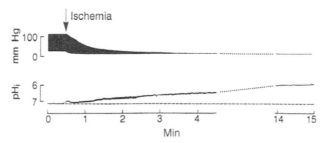

FIGURE 7   Effect of ischemia on left ventricular pressure (top tracing) and pH$_i$ in a Langendorff-perfused rabbit heart loaded with the fluorescent pH$_i$ indicator BCECF. Ischemia caused a rapid and steady decline in pH during the course of the 15-minute ischemic period. (Reproduced with permission from Mohabir *et al.*, 1991.)

sis and CP hydrolysis. These latter two processes consume protons (Dennis *et al.*, 1991), which explains why combined inhibtion of aerobic and anaerobic metabolism also produces an intracellular acidosis (Smith *et al.*, 1993). Intracellular acidosis is believed to be a major cause of reduced tension development in ischemic cardiac muscle (for review, see Orchard and Kentish, 1990), presumably by competing for Ca^{2+} binding sites on troponin C. A corollary is that protons will tend to displace Ca^{2+} from troponin and increase $[Ca^{2+}]_i$, as observed by Mohabir *et al.*, (1991) and others (see Fig. 6). Acidosis has also been shown to decrease maximal Ca^{2+}-activated force and to inhibit energy production in myocardium (Neely and Morgan, 1974).

v. $[Na^+]_i$ during Ischemia $[Na^+]_i$ accumulation during ischemia may be a critical factor underlying major metabolic and functional abnormalities of ischemic myocardium. Increases in $[Na^+]_i$ occurring during ischemia have been observed in many different models using a variety of measurement techniques, including Na^+-sensitive microelectrodes, isotopes, and NMR. (Guarnieri, 1987; Murphy *et al.*, 1987; Tani and Neely, 1989; Pike *et al.*, 1990). ^{23}Na NMR reveals that 30 minutes of ischemia is associated with an increase in $[Na^+]_i$, which varies in extent by species from 17 to 51 mM (in rat) and from 10 to 15 mM in guinea pig (Ingwall, 1995). Figure 8 shows serial changes in $[Na^+]_i$ in ferret hearts exposed to 20 minutes of global

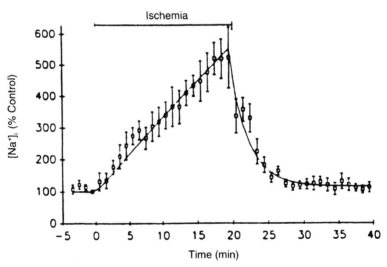

FIGURE 8   Effect of ischemia on $[Na^+]_i$ in eight ferret hearts. The hearts were perfused in Langendorff fashion, and $[Na^+]_i$ was measured using nuclear magnetic resonance spectroscopy. During the 20-minute period of global ischemia, there was a steady and significant rise in $[Na^+]_i$ and a rapid return to control levels at reperfusion. (Reproduced with permission from Pike *et al.*, 1990.)

ischemia. Note the steady rise in $[Na^+]_i$, which is about five times control at the end of the 20-minute period, returning to normal within 10 minutes of reperfusion (Pike *et al.*, 1990).

The mechanism of the $[Na^+]_i$ increase during ischemia is uncertain. Early increases in $[Na^+]_i$ and decreases in ATP and pH during ischemia can be delayed by lidocaine, thus implicating Na^+ entry via Na^+ channels (Butwell *et al.*, 1993). This is consistent with the aforementioned patch-clamp studies in isolated myocytes that indicate that Na^+ channels permit continued Na^+ influx during metabolic inhibition. Such continued Na^+ entry via voltage-gated channels could contribute to the overall Na^+ load of the myocyte during ischemia. Alternatively, other agents produced during ischemia—for example, free radicals (Bhatnagar *et al.*, 1990) or lysophospholipids (Burnasheve *et al.*, 1989; Undrovinas *et al.*, 1992)—may alter the behavor of Na^+ channels. For example, in frog myocytes exposed to oxidant stress, an increase in the amplitude, as well as a shift in the peak of the Na^+ "window" current toward the resting membrane potential, has been observed (Bhatnagar *et al.*, 1990).

Another potential mechanism receiving a great deal of attention has been acidosis-induced Na^+ influx via Na^+-H^+ exchange. In support of this hypothesis, numerous studies have reported protection against elevations in $[Na^+]_i$ (as well as $[Ca^{2+}]_i$) when cardiac preparations are treated with amiloride and its derivatives (Fig. 9) (Karmazyn, 1988; Tani and Neely, 1989; Murphy *et al.*, 1991). As discussed earlier, the poor specificity of amiloride derivatives weakens somewhat the conclusions of these studies (Haigney *et al.*, 1994). Nevertheless, the continued interest of the pharmaceutical industry in Na^+-H^+ blockers raises the possibility that newer and more specific agents will be developed. One of these, 3-methylsulfonyl-4-piperidinobenzoyl-guanidine methanesulfonate (HOE 694), has recently been shown to have a beneficial effect on cardiac function in ischemia/reperfusion (Scholz *et al.*, 1993). Recent observations in this laboratory (G. Langer, personal communication, August, 1996) indicate that Na^+-H^+ exchange is important in $[Na^+]_i$ accumulation and subsequent Ca^{2+} uptake.

Inhibition of the Na^+ pump is another candidate mechanism to explain $[Na^+]_i$ increases during ischemia. Na^+ pump inhibition would most likely occur late in the course of ischemia when ATP levels and phosphorylation potential have been reduced and metabolic products have accumulated. Although direct measurements of Na^+ pump activity during ischemia are not practical, indirect measurements in sarcolemmal vesicles obtained after 1 hour of ischemia in rabbit hearts suggest reduced pumping of Na^+ (Bersohn, 1995). These results are consistent with studies demonstrating inhibition of the Na^+ pump in preparations exposed to metabolic inhibitors (see previous discusson), perhaps in part due to accumulation of H^+ and P_i. Other toxic metabolites, such as oxygen free radicals or long chain acyl carnitines, also accumulate during ischemia and have been linked to Na^+

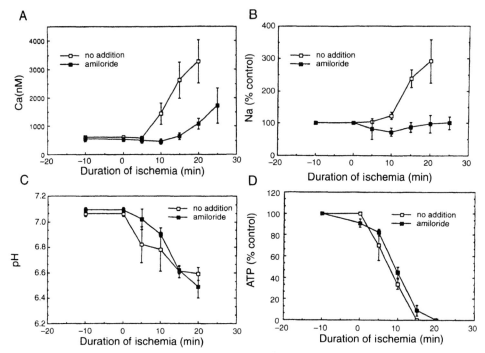

FIGURE 9 Effects of ischemia, with and without the addition of the Na^+–H^+ exchange blocker amiloride (1 mM, closed squares), on $[Ca^{2+}]_i$ (A), $[Na^+]_i$ (B), pH_i (C), and [ATP] (D) in Langendorff-perfused rat hearts. $[Ca^{2+}]_i$ and $[Na^+]_i$ were measured using ^{19}F NMR and ^{23}Na NMR spectra. ATP and pH_i were measured using ^{31}P NMR. Amiloride prevented the increase in $[Na^+]_i$, while delaying the increase in $[Ca^{2+}]_i$. There was no significant effect of amiloride on pH_i or ATP. (Reproduced with permission from Murphy *et al.*, 1991.)

pump inhibition (see following discussion, and Kakar *et al.*, 1987; Tanaka *et al.*, 1992; Shattock and Matsuura, 1993).

Regardless of the mechanism, which may be multifactorial, $[Na^+]_i$ accumulation during ischemia aggravates ATP depletion and acidosis by stimulating energy-dependent $[Na^+]_i$ handling mechanisms (e.g., the Na^+ pump prior to its inhibition). Elevated Na^+ also promotes reverse Na^+–Ca^{2+} exchange (Na^+ efflux and Ca^{2+} influx), thereby contributing to $[Ca^{2+}]_i$ overload and its attendant consequences. However, it must be noted that elevations in $[Ca^{2+}]_i$ lag behind $[Na^+]_i$ when measured by the same group in the same species (Marban *et al.*, 1990). This may be due to the continued capacity of Ca^{2+} extrusion or sequestration mechanisms to buffer increases in $[Ca^{2+}]_i$ or to decreased activity of the Na^+–Ca^{2+} exchanger during ischemia as a result of acidosis. Finally, the requirement for electroneutrality requires that elevated $[Na^+]_i$ be balanced by the efflux of another charged ion. A likely candidate would be K^+ because the K^+ conductance is elevated by activation of ATP-sensitive K^+ channels. Increased $[Na^+]_i$ has been proposed to be an important mechanism leading to K^+ loss during ischemia (Goldhaber *et al.*, 1996a).

B. CHRONIC ISCHEMIA—"HIBERNATING MYOCARDIUM"

A persistent reduction in blood flow to a region of myocardium can lead to depressed contractile function. This phenomenon has been termed *hibernating myocardium* (Braunwald and Rutherford, 1986) because tissue is still viable and recovers quickly once normal perfusion is restored. A hallmark of hibernating myocardium is *normal* metabolic activity, demonstrated clinically by assessment of glucose metabolism by positron emission tomography and evaluaton of 18-fluorodeoxyglucose (FDG) uptake. Normal FDG uptake is predictive of recovery once flow is restored (Tillisch *et al.*, 1986). The mechanism of contractile dysfunction is uncertain. One model of hibernating myocardium has been the isolated perfused ferret heart in which coronary perfusion pressure is decreased to the point where developed pressure is reduced without evidence of ischemia (to about 60 mM Hg, a 25% reduction from control) as assessed by measurements of lactate and inorganic phosphate production (Kitakaze and Marban, 1989). Under these conditions, ^{31}P NMR spectra showed little change in CP and ATP and only a slight increase in inorganic phosphate. $[Ca^{2+}]_i$ transients, measured by ^{19}F NMR in these 5F-BAPTA loaded hearts, were markedly reduced. Changes in $[Ca^{2+}]_i$ and force development reversed rapidly once normal coronary flow was restored. The mechanism of reduced $[Ca^{2+}]_i$ and force development in hibernating myocardium remains unknown. However, in another preparation, Langendorff-perfused rat hearts, coronary flow reductions of 25% (from 71 to 53 mM Hg) had no effect on $[Ca^{2+}]_i$ transients, pH, ATP, or CP, but did lead to significant increases in inorganic phosphate (Figueredo *et al.*, 1992). Thus, in contrast to the observations of Kitakaze and Marban, Figueredo observed an apparent decrease in myofilament calcium responsiveness. Further study is needed to distinguish between these two mechanisms.

In summary, myocardial ischemia results in an abrupt cessation of oxidative metabolism and a rapid increase in the rate of anaerobic glycolysis. Creatine phosphate levels fall rapidly to buffer ATP, which is depleted more slowly. Glycolytic end products, including lactate and NADH, accumulate and inhibit glycolysis. pH falls rapidly as well and contributes to further metabolic inhibition. Ca^{2+} accumulation also inhibits both glycolytic and any residual mitochondrial metabolism.

Although it takes several minutes for ATP levels to fall during ischemia, contractile function is reduced within seconds, probably due to vascular collapse and accumulation of H^+ and P_i, which decrease myofilament Ca^{2+}

responsiveness, and ADP, which slows relaxation. $[Ca^{2+}]_i$ transients do not decrease early in ischemia, but paradoxically increase, probably because H^+ competes for Ca^{2+} on troponin binding sites. Prolonged ischemia results in markedly increased diastolic pressure and $[Ca^{2+}]_i$. The earliest increases in diastolic pressure are most likely related to energy deprivation rather than $[Ca^{2+}]_i$ overload. Studies with metabolic inhibitors suggest that $[Ca^{2+}]_i$ overload is not caused primarily by energy deprivation, but may instead be caused by accumulation of metabolic by-products and other agents, such as free radicals, that are present during ischemia. $[Na^+]_i$ overload may also contribute to increased $[Ca^{2+}]_i$ (see following discussion).

Net cellular K^+ loss occurs within 15–30 seconds after the onset of ischemia and is caused by an increase in K^+ efflux, due in large part to activation of ATP-sensitive K^+ channels by decreases in ATP and accumulation of metabolic by-products. K^+ efflux coupled to anions and due to activation of K_{AA} and K_{Na} channels plays only a minor role.

$[Na^+]_i$ accumulation during ischemia is rapid. The mechanism is likely to be the same as during metabolic inhibition, for example, continued influx of Na^+ via voltage-dependent Na^+ channels, acidosis promoting Na^+–H^+ exchange, inhibition of the Na^+ pump, and perhaps Na^+ influx through nonselective channels. Accumulation of free radicals and fatty acid derivatives may also alter the properties of Na^+ channels and allow for increased Na^+ influx. It is likely that $[Na^+]_i$ increases are a major cause of $[Ca^{2+}]_i$ overload because the altered Na^+ gradient will reduce Ca^{2+} efflux by the Na^+–Ca^{2+} exchanger or even result in influx of Ca^{2+} by this mechanism. $[Na^+]_i$ increases may also promote K^+ loss during ischemia.

Persistent moderate reductions in coronary blood flow can lead to a chronic depression of contractile function, which is termed *hibernating myocardium*. This tissue remains viable and recovers quickly once normal perfusion is restored. Metabolic activity and high-energy phosphate levels are normal. Both decreases in $[Ca^{2+}]_i$ transients and reductions in myofilament Ca^{2+} responsiveness have been observed in experimental preparations that simulate hibernation.

V. REPERFUSION OF ISCHEMIC TISSUE: METABOLIC AND FUNCTIONAL ISSUES

At the cellular level, restoring the supply of oxygenated blood (reperfusion) does not necessarily result in improved myocyte survival or function, although the clinical observation in humans is that reperfusion within 24 hours of occlusion may provide some long-term survival benefit (Grines *et al.*, 1993), perhaps through mechanisms not specifically related to improved recovery of myocytes in the infarct zone (Lamas *et al.*, 1995). Two kinds of contractile dysfunction may occur with reperfusion. Reperfusion after brief periods of ischemia is associated with myocardial stunning, which is defined as impaired systolic function that gradually resolves during a period of hours to days. In marked contrast, reperfusion after prolonged ischemia, but before there is histological evidence of damage to cell structures, results in accelerated contracture and irreversible contractile failure. Disordered $[Ca^{2+}]_i$ homeostasis is believed to be of paramount importance in the pathogenesis of both types of contractile abnormalities associated with reperfusion. Repetitive but brief ischemic episodes, known as "preconditioning," may shorten the duration of contractile dysfunction associated with stunning and may also protect against irreversible injury during subsequent episodes of prolonged ischemia.

$[Ca^{2+}]_i$ overload of intracellular organelles, particularly mitochondria (Shen and Jennings, 1972), is a prominent histological feature of irreversibly injured reperfused myocardium. Recovery of myocyte function following a period of anoxia is more likely when mitochondrial calcium content is low (Miyata et al., 1992), which is consistent with the observation that reperfusion of ischemic preparations with low calcium solutions (Kusuoka et al., 1987) or high sodium solutions [which reduces $[Ca^{2+}]_i$ by enhancing forward Na^+–Ca^{2+} exchange (Kusuoka et al., 1993)] limits the extent of reperfusion injury. Another compelling piece of evidence is that $[Ca^{2+}]_i$ overload in the *absence* of ischemia can lead to functional abnormalities similar to those observed in the setting of reperfusion (Kitakaze et al., 1988b). Following is a discussion of the role of metabolism in the functional abnormalities of reperfusion.

A. MYOCARDIAL STUNNING

Brief periods of ischemia that are too short to cause myocardial necrosis may nonetheless result in contractile abnormalities that persist for hours, days, or weeks following restoration of normal flow. This phenomenon, termed *myocardial stunning* (Braunwald and Kloner, 1982; Kusuoka and Marban, 1992), has been observed both experimentally and clinically. The role of metabolic abnormalities in this phenomenon is not clear. Measurements of MVO_2 in reperfused stunned myocardium are variable, with some studies reporting an increase (Laxson et al., 1989) and others reporting a decrease (Myears et al., 1987; Liedtke et al., 1988; Demaison and Grynberg, 1994). Oxidative metabolism rapidly returns to normal at restoration of blood flow to stunned (as opposed to irreversibly injured) myocardium (Greenfield and Swain, 1987; Gorge et al., 1991), and CP levels quickly rise (Brooks and Willis, 1983), in some studies to levels even higher than their baseline values (Ichihara and Abiko, 1984). However, depletion of ATP and ADP stores is a consistent finding (DeBoer et al., 1980; Kloner et al., 1981; Jennings et al., 1985; Van Bilsen et al., 1989), and the timing of functional recovery is correlated with the timing of ATP recovery (Ellis et *al.*, 1983). The persistent depression of ATP during stunning has been attributed to washout of nucleotide precursors during the ischemic period (Reimer *et al.*, 1981). Nevertheless, while repletion of ATP precursors can restore ATP and ADP levels in stunned myocardium, contractile abnormalities do not recover in parallel (Ambrosio *et al.*, 1989).

There is good agreement that reperfused viable myocardium once again utilizes fatty acids preferentially as the substrate for energy production (Mickle *et al.*, 1986; Myears *et al.*, 1987; Liedtke *et al.*, 1988; Lopaschuk *et al.*, 1990). The source of these fatty acids is initially the metabolites accumulated in the myocardium during ischemia (Lopaschuk *et al.*, 1990). Although most studies show that the rate of fatty acid oxidation is similar to nonischemic controls after a 30-minute to 3-hour period of reperfusion (Schwaiger *et al.*, 1985b; Lopaschuk *et al.*, 1990; Gorge *et al.*, 1991; Demaison and Grynberg, 1994) (see Myears *et al.*, 1987, for dissent), decreased oxygen consumption (which raises the fatty acid oxidation : oxygen consumption ratio) results in some oxygen wasting that might decrease the energy available for contraction (Liedtke *et al.*, 1988; Demaison and Grynberg, 1994). Whether this mechanism contributes to disorders of postischemic mechanical function is unclear (Liedtke *et al.*, 1988; Lopaschuk *et al.*, 1990). Fatty acid intermediates may have nonmetabolic effects that may also contribute to ischemic injury (see following discussion).

Glucose uptake and metabolism following reperfusion are increased (Schwaiger *et al.*, 1985a, 1989; Myears *et al.*, 1987), although high concentrations of circulating fatty acids, similar to what is observed during ischemia, have been shown to reduce glucose oxidation (Lopaschuk *et al.*, 1990). Suppression of glucose oxidation by fatty acids during reperfusion could contribute to the mechanical effects of stunning. The beneficial effect of a carnitine palmitoyltransferase I inhibitor has been correlated with its ability to stimulate glucose oxidation in fatty acid perfused hearts (Lopaschuk *et al.*, 1988, 1990) although others have suggested a reduction in long chain acylcarnitine levels (Molaparast-Saless *et al.*, 1987). These results are consistent with more recent studies suggesting a requirement for glycolysis to ensure adequate functional recovery from episodes of ischemia in isolated perfused rabbit (Jeremy *et al.*, 1993) and rat hearts (Cross *et al.*, 1996). Interestingly, functional recovery from hypoxia may be enhanced when hearts utilize exogenous glucose rather than endogenous glycogen as the substrate for glycolysis (Runnman *et al.*, 1990). This beneficial effect was independent of tissue glycogen levels and did not depend on total glycolytic flux, suggesting that ATP produced by glycolysis of exogenous glucose is used to support key cellular functions that determine the severity of cardiac dysfunction during hypoxia and after reoxygenation. (Bricknell *et al.*, 1981; Taniguchi *et al.*, 1983; Cross *et al.*, 1996). Glucose and insulin have also been show to preserve diastolic function during underperfusion and subsequent reperfusion (Eberli *et al.*, 1991).

Although the metabolic characteristics of stunning have been well described, the observation that inotropic interventions (Ito et al., 1987; Bolli, 1992) improve contractility without restoring high-energy phosphate levels to normal argues against a specific metabolic explanation for stunning. Several lines of evidence suggest that the contractile abnormality of stunning is due to a decrease in myofilament Ca^{2+} responsiveness (Kusuoka et al., 1987; Soei et al., 1994) caused by a Ca^{2+}-dependent proteolysis of the contractile proteins (Gao et al., 1996). Stunning can be mimicked by perfusion of a normoxic preparation in high Ca^{2+} and can be prevented by protease inhibitors. This suggests that the increase in $[Ca^{2+}]_i$ during ischemia is the critical step underlying stunning and is consistent with the observation that low Ca^{2+} perfusate and extracellular acidosis (Kitakaze et al., 1988a) can protect against stunning, presumably by decreasing $[Ca^{2+}]_i$ during the ischemic period. Furthermore, the duration of myocardial stunning is usually days, matching the time course of myofilament turnover.

B. IRREVERSIBLE INJURY CAUSED BY REPERFUSION

Delayed reperfusion (after $\sim$ 40 min) of cardiac tissue with morphologic features suggesting viability leads to a rapid and massive increase in $[Ca^{2+}]_i$, accelerated contracture, cellular necrosis, and irreversible injury (Jennings and Ganote, 1974). The importance of $[Ca^{2+}]_i$ overload in this process is supported by the observation that low Ca^{2+} reperfusate is protective. It is notable that during ischemia the supply of Ca^{2+} is limited to the extracellular space (about 500 μmol/kg wet wt) and is relatively easily buffered. Upon reperfusion, however, the extracellular Ca^{2+} supply becomes infinite, which is quite a challenge to cellular Ca^{2+} handling mechanisms. Loss of sarcolemmal integrity does not appear to underlie the immediate increase in $[Ca^{2+}]_i$ that has been observed at reperfusion (Poole-Wilson et al., 1984), although a selective increase in Ca^{2+} permeability cannot be excluded (Clague and Langer, 1994). Depressed metabolism cannot fully explain the profound abnormalities of $[Ca^{2+}]_i$ regulation that complicate delayed reperfusion because, with the exception of run-down of I_{Ca}, the fundamental components of E–C coupling remain intact in experimental preparations exposed to direct inhibitors of oxidative and glycolytic metabolism see (Fig. 3) (Goldhaber et al., 1991). This suggests that other aspects of the ischemic milieu not well simulated by metabolic inhibition, or agents that are produced at the time of reperfusion, may make important contributions to abnormalities of $[Ca^{2+}]_i$ regulation and E–C coupling during reperfusion. In this regard it has been observed that potentially injurious compounds, ordinarily absent or at extremely low levels in the cellular environment under physiological conditions, increase to toxic concentrations during ischemia and reperfusion. Three classes of these components are of particular interest: oxygen free radicals (OFR), inflammatory cytokines, and amphiphiles. The evidence that suggests the involvement of these agents in reperfusion abnormalities is summarized in the next section.

1. Oxygen Free Radicals

Several lines of evidence suggest that OFR produced during ischemia and reperfusion play a major role in the pathogenesis of reperfusion injury. These highly reactive molecular compounds are by-products of oxidative metabolism, which accumulate to toxic levels in ischemic cardiac cells, in part due to increased production and in part because levels of endogenous OFR scavengers such as superoxide dismutase (SOD) and glutathione peroxidase decrease (Ferrari et al., 1985; Shlafer et al., 1987). Activated neutrophils (McCord, 1974) attracted to an ischemic or reperfused region of myocardium by chemotactic factors release additional OFR, such as superoxide (O_2^-), hydrogen peroxide (H_2O_2), or hypochlorous anion (OCL^-) into the infarct zone (Rowe et al., 1984). Neutrophils also release inflammatory cytokines (e.g., TNF-α, interleukins) that may indirectly elevate OFR levels by producing nitric oxide (NO) (Finkel et al., 1992), a free radical in its own right, and consequently another OFR species— peroxynitrite anion ($ONOO^-$)—that contributes to OH^- production (Beckman et al., 1990). Tissue OFR concentrations rise by 30% to 6 μM during ischemia and as high as 11 μM during reperfusion in arterially perfused rabbit hearts (Zweier et al., 1987). Many, but certainly not all, studies suggest that application of agents that scavenge OFR (e.g., SOD and catalase) prevent OFR formation (e.g., allopurinol, oxypurinol, and N-tert-butyl-a-phenylnitrone) or antagonize TNF-α (e.g., transforming growth factor-β), can reduce reperfusion injury in animal models; some have also been tried with limited success in humans (Werns et al., 1985; Ambrosio et al., 1986; Hearse and Tosaki, 1987; Puett et al., 1987; Mak and Weglicki, 1988; Przyklenk and Kloner, 1989; Lefer et al., 1990b; Flaherty et al., 1994). Application of antibodies against neutrophil adhesion proteins or neutrophil depletion in reperfusate also decreases OFR production during experimental ischemia and reperfusion and preserves myocardial integrity and function (Lucchesi, 1990).

Several studies have documented that OFR are capable of disrupting $[Ca^{2+}]_i$ homeostasis in both isolated myocytes and intact heart. Corretti et al. (1991) reported that diastolic $[Ca^{2+}]_i$ increased two- to three-fold during a 4-minute infusion of 0.75 mM H_2O_2 and iron chelate in isolated rabbit hearts studied with ^{19}F NMR and the Ca^{2+}-indicator 5F-BAPTA. An important element of this study was that the investigators used EPR spectroscopy in parallel experiments to demonstrate that the free radical generating systems (FRGS) used in their preparation produced tissue levels of OFR comparable to those observed in genuine ischemia/reperfusion. Barrington et al. (1988) provided indirect evidence that OFR can alter $[Ca^{2+}]_i$ regulation by demonstrating in isolated canine myocytes that FRGS cause electrophys-iologic changes characteristic of $[Ca^{2+}]_i$ overload, such as cellular depolar-ization, and triggered activity including delayed after-depolarizations. Simi-larly, Matsuura and Shattock (1991) demonstrated that an oxygen radical generating system activated the transient inward current (I_{TI}) and induced triggered repetitive action potential discharges. Josephson *et al.* (1991) found that H_2O_2 elevated diastolic $[Ca^{2+}]_i$ in adult rat ventricular myocytes loaded with the $[Ca^{2+}]_i$ sensitive dye indo-1. These authors also demon-strated that H_2O_2-induced contracture could be delayed by pefusing the cells in a low Ca^{2+} (0.5 mM) buffer or by including the Ca^{2+} channel blocker nitrendipine in the perfusate. Similarly, dialysis of the cytosol with EGTA delayed the electrophysiologic effects of FRGS in a patch-clamped feline ventricular myocyte model (Barrington, 1994), implying that $[Ca^{2+}]_i$ dynam-ics influence the cellular response to OFR. Other studies have suggested that OFR increase Ca^{2+} influx across the sarcolemma (Kaminishi and Kako, 1988) and that agents that scavenge OFR prevent the uptake of extracellular $^{45}Ca^{2+}$ during posthypoxic reoxygenation (Murphy *et al.*, 1988). None of these studies definitively pinpointed which $[Ca^{2+}]_i$ handling processes were responsible for the increase in $[Ca^{2+}]_i$. Despite the finding that Ca^{2+} channel blockers prevent $[Ca^{2+}]_i$ overload during exposure to FRGS, enhanced Ca^{2+} entry via I_{Ca} seems an unlikely mechanism of $[Ca^{2+}]_i$ overload for several reasons. Direct assessments of I_{Ca} using voltage clamp techniques (Fig. 10). suggest that I_{Ca} runs down at an accelerated rate during exposure to FRGS (Goldhaber *et al.*, 1989). Ca^{2+} channel blockers may simply reduce the overall burden of Ca^{2+} to the cell, thereby permitting compromised extru-sion or sequestration mechanisms to keep up with the influx of $[Ca^{2+}]_i$. Loss of sarcolemmal integrity is also an unlikely root cause of $[Ca^{2+}]_i$ overload by OFR because the sarcolemma remains impermeable to macro-molecules at a point when early increases in diastolic $[Ca^{2+}]_i$ occur (Joseph-son *et al.*, 1991). FRGS can activate Ca^{2+} leak channels in the sarcolemma (Wang *et al.*, 1995), but the extent of Ca^{2+} entry due to this mechanism during early reperfusion is probably small. This reasoning implies that defective Ca^{2+} extrusion or sequestration is a critical factor leading to $[Ca^{2+}]_i$ overload during exposure to OFR. Individual Ca^{2+} extrusion and sequestering mechanisms are susceptible to influence by OFR (Fig. 11), including the SR (Hess *et al.*, 1983, 1984; Okabe *et al.*, 1988; Goldhaber and Liu, 1994), and ryanodine receptor (Boraso and Williams, 1994), the Na^+–Ca^{2+} exchanger (Reeves *et al.*, 1986; Dixon *et al.*, 1990; Coetzee *et al.*, 1994; Goldhaber, 1996), mitochondria (Harris *et al.*, 1982), and the sarcolemmal Ca^{2+} pump (Kaneko *et al.*, 1989). Interference with any of these $[Ca^{2+}]_i$ regulating processes could in theory result in cytosolic Ca^{2+} overload.

Inhibition of the Na^+ pump (Kim and Akera, 1987; Shattock and Matsu-ura, 1993) and increases in Na^+ window currents (Bhatnagar *et al.*, 1990)

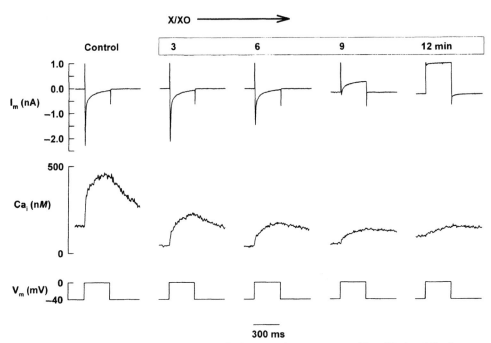

FIGURE 10   Effects of the free radical generating system xanthine (X, 1 mM) plus
xanthine oxidase (XO, 0.01 U/ml) on I_{Ca} and the $[Ca^{2+}]_i$ transient in whole cell patch-clamped
rabbit ventricular myocytes loaded with the $[Ca^{2+}]_i$ indicator fura-2. Control recordings of
membrane potential (I_m), $[Ca^{2+}]_i$, and membrane voltage (V_m) during 300-ms test pulses to
0 mV from a holding potential of -40 mV are displayed in the far left hand panel. Subsequent
panels contain tracings after 3, 6, 9, and 12 minutes exposure to X/XO. Note the rapid
decrease in the amplitude of I_{Ca}, which was associated with a decline in the amplitude of the
$[Ca^{2+}]_i$ transient.

by FRGS have also been described. Elevations of $[Na^+]_i$ in the setting of
stimulated Na^+–Ca^{2+} exchange is a likely cause of $[Ca^{2+}]_i$ overload after
prolonged ischemia. Many of the effects on $[Ca^{2+}]_i$ noted in the previous
studies may be complicated by the fact that OFR have been shown to
inhibit cardiac metabolism (McDonough *et al.*, 1987; Goldhaber *et al.*, 1989),
particularly glycolytic pathways (Goldhaber *et al.*, 1989; Corretti *et al.*, 1991;
Josephson *et al.*, 1991; Janero *et al.*, 1994), which also contributes to a
$[Ca^{2+}]_i$ overload state. Whether OFR-induced abnormalities in cellular Ca^{2+}
handling are primary or secondary to OFR-induced inhibition of metabolism is not clearly resolved.

2. Inflammatory Cytokines

In addition to oxygen radical species, neutrophils attracted to an ischemic
zone and coronary vascular endothelium may release other substances that

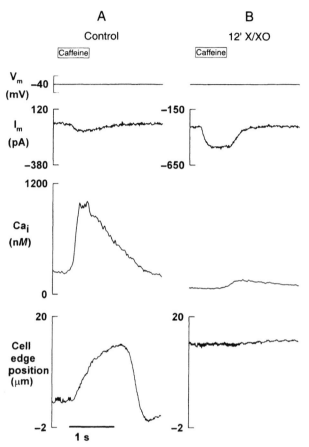

FIGURE 11 Effects of the free radical generating system xanthine (X, 1 mM) plus xanthine oxidase (XO, 0.01 U/ml) on I_m, $[Ca^{2+}]_i$, and cell edge position in a whole cell patch-clamped rabbit ventricular myocyte loaded with the $[Ca^{2+}]_i$ indicator fura-2. The cell was maintained at a constant membrane potential (V_m) of -40 mV during 650-ms exposure to 5 mM caffeine using a rapid extracellular solution exchange device. Under control conditions (panel A), application of caffeine releases SR Ca^{2+} stores, which induces a large $[Ca^{2+}]_i$ transient as well as an inward membrane current due to Na^+–Ca^{2+} exchange. After a 12-minute exposure to X/XO (panel B), the $[Ca^{2+}]_i$ transient induced by caffeine is much smaller, but the Na^+–Ca^{2+} exchange current is markedly increased, suggesting that Na^+–Ca^{2+} exchange was stimulated by the free radical generating system.

could have a deleterious effect on myocardial function. Proinflammatory cytokines, a class of secretory polypeptides that includes the interleukins (e.g., IL-1, IL-2, IL-6) and tumor necrosis factor-α (TNF-α), are produced by adhering leukocytes or endothelial cells in reperfused tissue (Zweier, 1988). Receptors for TNF-α are present in all somatic cells except erythrocytes (Beutler and Cerami, 1988). Cytokine levels are elevated in several clinical conditions associated with depressed contractility, including bacterial sepsis (Low-Friedrich *et al.,* 1992), myocardial infarction (Maury and Teppo, 1989), and postcardiopulmonary bypass (Finkel *et al.,* 1992). IL-2 receptors are elevated in patients with dilated cardiomyopathy (Limas *et al.,* 1995), and administration of IL-2 to patients for treatment of metastatic cancer is associated with reversible depression of cardiac function (Low-Friedrich *et al.,* 1992). Several authors have demonstrated direct inhibitory effects of inflammatory cytokines, white cell, and endothelial cell products on contractile function in cardiac muscle (Finkel *et al.,* 1992; Low-Friedrich *et al.,* 1992; Balligand *et al.,* 1993b; Kinugawa *et al.,* 1994; Shah *et al.,* 1994; Goldhaber *et al.,* 1996b). Finkel *et al.* (1992) showed that TNF-α, IL-2, and IL-6 caused a reversible decrease in contractility in isolated hamster papillary muscles. Other aspects of myocardial ischemia/reperfusion, particularly endothelial dysfunction (i.e., the loss of vascular relaxation in response to agents that release EDRF), can be mimicked by TNF-α. In one study, endothelial dysfunction could be prevented by the naturally occurring growth factor, transforming growth factor beta (TGF-β) when given before or immediately after ischemia (Lefer *et al.,* 1990a). TGF-β blocked the increase in circulating TNF-α activity and O_2^- production by the endothelium following ischemia/reperfusion. Because TGF-β is not a direct scavenger of oxygen radicals, it was suggested that preservation of function following reperfusion was probably related to a decrease in TNF-α with consequent decreases in leukocyte accumulation and O_2^- levels (Lefer *et al.,* 1990a).

The mechanism of cytokine-induced compromise of contractile activity is unknown. Cytokines induce "stress proteins" in cultured fetal mouse myocytes within 2 hours of exposure, but effects on contractility and development of arrhythmias occur much earlier after the onset of exposure to these agents (Low-Friedrich *et al.,* 1992). A more likely mechanism of acute action involves nitric oxide (NO). In the study by Finkel *et al.* (1992), the negative inotropic effects of the cytokines they used could be blocked by including an inhibitor of NO production (N^G-monomethyl-L-arginine, also known as L-NMMA) or enhanced by the addition of a promoter of NO synthesis—L-Arginine. Similar effects on contractility (also prevented by NO blockers) have been reported by other laboratories in isolated myocyte preparations exposed to TNF-α (Goldhaber *et al.,* 1996b) (Fig. 12), endotoxin, or macrophage-conditioned medium (Brady *et al.,* 1992; Balligand *et al.,* 1993b), the effects of which are believed to be mediated by cytokines (Tracey *et al.,* 1986; Beutler and Cerami, 1988; McKenna, 1990; Balligand *et al.,* 1993b). These results imply that induction of the NO synthesis pathways by cytokines may have a negative inotropic effect on the myocardium. In noncardiac tissues, it is fairly well established that cytokines increase NO by inducing transcription of NO synthase (Stuehr *et al.,* 1991). In vascular smooth muscle, the basal release of NO by a constitutive Ca^{2+}-

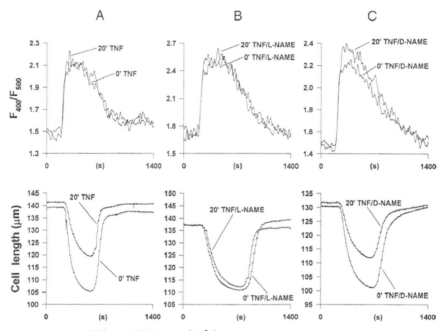

FIGURE 12   Effects of TNF-α on $[Ca^{2+}]_i$ and unloaded cell shortening in isolated rabbit ventricular myocytes loaded with the $[Ca^{2+}]_i$ indicator indo-1-AM and paced at 0.2 Hz using field stimulation. (A) 20-minute exposure to TNF-α (10,000 U/ml) caused a marked reduction in cell shortening amplitude and increased resting cell length without affecting the $[Ca^{2+}]_i$ transient. (B) The effect of TNF-α could be prevented by the nitric oxide synthesis inhibitor N^G-nitro-L-arginine methyl ester (L-NAME), but could not be prevented by L-NAME's inactive enantiomer D-NAME (C). These results suggest that the negative inotropic effect of TNF-α is mediated by nitric oxide. (Adapted with permission from Goldhaber *et al.*, 1996b.)

dependent NO synthase regulates vascular tone by activating guanylate cyclase, which increases cGMP and promotes smooth muscle relaxation (Lowenstein and Snyder, 1992). Bacterial endotoxin and cytokines can also activate a Ca^{2+}-*independent* NO synthase in endothelium (Radomski *et al.*, 1990) and vascular smooth muscle (Busse and Mulsch, 1990), which promotes smooth muscle relaxation by the same mechanism. A cytokine-induced Ca^{2+}-independent NO synthase (NOS2) has been identified in myocardial cells (Schulz *et al.*, 1992; Balligand *et al.*, 1994), and elevated cGMP has been shown to decrease twitch, dT/dt, and promote relaxation in ferret papillary muscles (Smith *et al.*, 1991). Cytokines decrease the contractile response of myocytes to catecholamines (Balligand *et al.*, 1993b), an effect that appears to be mediated by NO-induced increases in cGMP-dependent phosphodiesterase activity (Gulick *et al.*, 1989; Mery *et al.*, 1993). However, it takes several hours for cytokines to increase levels of NOS2 in cardiac cells (Tsujino *et al.*, 1994). This makes it difficult to explain the acute effects of cyotkines observed in some preparations. One possibility is that cytokines activate the constitutive, ordinarily Ca^{2+}-dependent NO synthase (NOS3), which is also present in cardiac cells (Balligand et al., 1993a). Acute NO-mediated negative inotropic effects of cytokines may be due to a decrease in myofilament Ca^{2+} responsiveness (Goldhaber et al., 1996b), although reductions in the $[Ca^{2+}]_i$ transient have also been observed acutely in response to TNF-α (Yokoyama et al., 1993).

3. Interaction of Cytokines and Free Radicals

An intriguing prospect is that some of the deleterious actions of cytokines may be mediated by OFR. For example, the cytotoxic effect of the combination of TNF-α, IL-1, and interferon gamma on pancreatic β-cells is associated with an increase in lipid peroxidation products and can be prevented by administration of an inhibitor of lipid peroxidation (Rabinovitch et al., 1992). TNF-α alone promotes superoxide anion generation by human neutrophils by a Ca^{2+}-dependent mechanism (Tsujimoto et al., 1986). TNF-α also elevates hydroxyl radical formation in a mouse tumorigenic fibroblast cell line (Yamauchi et al., 1989). In mouse hepatocytes, TNF-α induces oxidant stress (GSSG efflux) and ATP depletion. ATP depletion could be prevented by the addition of antioxidants or ruthenium red, which blocks ATP-dependent Ca^{2+} cycling across the inner mitochondrial membrane (Adamson and Billings, 1992). TNF-α may induce OFR production by damaging specific components of the mitochondrial electron transport chain, which directly increases OFR production by the mitochondria (Schulze-Osthoff et al., 1992). Another possibility is that TNF-α stimulates NO production, which raises hydroxyl radical levels via the peroxynitrite pathway (Beckman et al., 1990). Additional mechanisms of NO-induced injury have been described in bacterial cells. Although these studies suggest a role of OFR in the pathogenesis of cytokine injury, there is also evidence that OFR can participate in the synthesis of cytokines (IL-1) during hypoxia/reoxygenation in mononuclear phagocytes (Koga et al., 1992) and human umbilical vein endothelial cell monolayers (Ala et al., 1992). In some circumstances, cytokine exposure can have a beneficial effect by upregulating the levels of endogenous free radical scavengers. For example, conditioning of the myocardium with low-dose endotoxin or cytokines has been shown to increase myocardial catalase and Mn-superoxide dismutase levels and to reduce reperfusion injury (Wong and Goeddel, 1988; Brown et al., 1990, 1992; Eddy et al., 1992). Thus, the actions of cytokines and free radicals appear to be closely integrated.

4. Amphiphiles

These agents, which include lysophosphatidylcholine (LPC) and long chain acyl carnitines (LCAC), such as l-Palmitoylcarnitine (l-PC), accumulate in the myocardium within minutes during ischemia. They have been implicated in the pathogenesis of reperfusion injury, including contractile abnormalities and electrophysiological alterations leading to malignant arrhythmias as well as derangements of $[Ca^{2+}]_i$ homeostasis (Mock and Man, 1990; DaTorre et al., 1991). Agents that inhibit LPC production have been shown to limit reperfusion injury (Das et al., 1986). l-PC induces early and delayed afterdepolarizations and triggered activity consistent with a $[Ca^{2+}]_i$ overload mechanism (Meszaros and Pappano, 1990). The relationship of the electrophysiological effects of these agents to $[Ca^{2+}]_i$ overload is also suggested by the dependence of their effects on extracellular Ca^{2+} and studies showing that ryanodine can block the afterdepolarizations caused by l-PC (Meszaros and Pappano, 1990). The Ca^{2+}-sensitive dye indo-1 has been used in cultured neonatal ventricular myocytes to demonstrate that LPC increases $[Ca^{2+}]_i$ (Woodley et al., 1991). Because LPC can directly stimulate adenylate cyclase (Ahumada et al., 1979), an increase in I_{Ca} has been suggested as a mechanism of elevated $[Ca^{2+}]_i$. This was supported by studies that showed that increased Ca^{2+} accumulation caused by high doses of LPC (100 mM) could be blocked by verapamil if it was administered within 5 minutes of exposure (Sedlis et al., 1983). However, more recent studies have failed to confirm a benefit of Ca^{2+} blocking agents (Woodley et al., 1991) and LCAC caused a decrease in I_{Ca} when measured directly using the patch-clamp technique (Wu and Corr, 1992). LPC could lead to $[Ca^{2+}]_i$ overload by other means, for example by inducing a nonselective leak conductance, as has been described in guinea pig ventricular myocytes studied with the patch clamp technique and the $[Ca^{2+}]_i$ sensitive indicator fura-2 (Liu, et al., 1991a). Both LPC and LCAC have been shown to inhibit the Na^+ pump, which would favor Na^+ overload and reverse Na^+–Ca^{2+} exchange leading to cytosolic Ca^{2+} overload (Pitts and Okhuysen, 1984; Woodley et al., 1991; Tanaka et al., 1992). Both l-PC and LPC inhibit I_{Na} in guinea pig, and in the case of LPC this is an irreversible effect (Sato et al., 1992). However, LPC has been shown to induce a modified Na^+ current with unusual properties and with open channel activity at potentials ranging from -180 to $+30$ mV (Undrovinas et al., 1992), which could contribute to intracellular Na^+ accumulation and hence $[Ca^{2+}]_i$ overload by reverse Na^+–Ca^{2+} exchange. Similarly, LCAC induce a slow-inactivating current in rabbit ventricular myocytes that can be blocked by TTX, preventing $[Na^+]_i$ overload (Wu and Corr, 1994). LPC has been shown to inhibit Na^+–Ca^{2+} exchange in isolated canine sarcolemmal vesicles (Bersohn et al., 1991), an effect that would compromise the Ca^{2+} efflux (and influx) properties of Na^+–Ca^{2+} exchange. Finally, LPC has been reported to inhibit Ca^{2+} uptake by the SR (Brovkovitch et al., 1987; Ambudkar et al., 1988), whereas l-PC increased the open probability of SR Ca^{2+} release channels of rabbit and pig skeletal muscle incorporated into lipid bilayers (El-Hayek et al., 1993).

Thus, the effects of amphiphiles on $[Na^+]_i$ and $[Ca^{2+}]_i$ regulating mechanisms could lead to $[Ca^{2+}]_i$ overload and may contribute significantly to abnormalities of $[Ca^{2+}]_i$ regulation associated with ischemia and reperfusion that are not explained by metabolic inhibition alone.

5. Interactions of Free Radicals and Amphiphiles

Lipid peroxidation, a prominent effect of OFR, leads to oxidative deterioration and release of polyunsaturated lipids. The release of these unsaturated fatty acids may result in the production of lysophospholipids, especially under ischemic conditions when lysophospholipase-mediated catabolism is inhibited (Corr et al., 1984; Prasad and Das, 1989). This raises the interesting possibility that some of the effects attributed to direct actions of OFR may in fact be secondary to the generation of amphiphiles. Several amphiphiles (palmitoyl-CoA, l-PC or LPC) have been shown to potentiate OFR-induced lipid peroxidation of cell membranes and inhibition of the Na^+-K^+ pump (Mak et al., 1986). The effect of amphiphiles and OFR on the Na^+-K^+ pump was synergistic rather than additive. In a separate study by Sedlis et al. (1990), the depressant effects of LPC on contractile function were potentiated by superoxide and could be prevented using SOD. The mechanism leading to the potentiation of OFR effects by amphiphiles is unclear, but may involve insertion of amphiphilic monomers into the membrane lipid bilayer resulting in enhanced accessibility of OFR to the phospholipid unsaturated fatty acids, the preferred targets of OFR.

C. ISCHEMIC PRECONDITIONING

Ischemic preconditioning is a phenomenon whereby brief repetitive episodes of ischemia provide temporary protection against injury (Murry et al., 1986) and arrhythmias (Shiki and Hearse, 1987; Hagar et al., 1991) resulting from a subsequent more prolonged period of ischemia. Under experimental conditions, which often vary considerably from study to study, hearts that have been "preconditioned" exhibit less glycogenolysis, decreased accumulation of lactate and glycolytic intermediates, delayed ultrastructural damage, and a slowed rate of ATP loss (Murry et al., 1990), although the latter finding has not been found consistently (de Albuquerque et al., 1994). The mechanism of protection due to ischemic preconditioning is not clear. The protection afforded by preconditioning protocols may depend on an initial reduction in glycogen content (Wolfe et al., 1993). Preconditioned hearts subsequently exhibit less glycolysis (just enough to support cell survival) and reduced glycogen consumption during the prolonged ischemic episode. The decrease in glycolytic activity compared to nonpreconditioned hearts is sustained during postischemic reperfusion, although glucose oxidation rapidly returns to normal preischemic levels (Finegan et al., 1995). In this scheme, enhanced recovery of mechanical function is attributed to improved coupling between glycolysis and glucose oxidation. This would explain the reduction in proton production (de Albuquerque et al., 1994) and consequent protection from $[Na^+]_i$ and $[Ca^{2+}]_i$ overload observed in preconditioning (Steenbergen et al., 1993).

It is believed that the mechanism underlying preconditioning involves activation of adenosine A_1 receptors by the nucleotide breakdown product adenosine. Preconditioned dog hearts have increased activity of both ecto-solic and cytosolic forms of 5' nucleotidase (Kitakaze *et al.*, 1993), the enzyme responsible for breakdown of 5'-AMP to adenosine during isch-emia. Inhibition of ectosolic 5'-nucleotidase activity using α, β, methylene adenosine 5'-diphosphate attenuates enhanced release of adenosine by preconditioning protocols and blunts the infarct size-limiting effect of isch-emic preconditioning (Kitakaze *et al.*, 1994). In rabbit heart, adenosine receptor blockers can prevent the protection conferred by preconditioning, whereas intracoronary infusion of adenosine or an A_1-specific agonist is as protective as preconditioning itself (Liu *et al.*, 1991b). Similar results have been reported in dog (Auchampach and Gross, 1993) and pig (Yokota *et al.*, 1995), but the effect of A_1 receptor blockade is incomplete in rat (Liu and Downey, 1992). How adenosine receptor activation confers a benefit is uncertain. It is well known that adenosine receptors use G-proteins in their signal transduction (Stiles, 1992). Carbachol, a muscarinic receptor agonist whose actions are also coupled by G-proteins, protects the isolated rabbit heart as well as an A_1 receptor agonist, whereas pertussis toxin, a G-protein ribosylating agent, attenuates the preconditioning effect (Thorn-ton *et al.*, 1993a).

It has also been observed that protein kinase C (PKC) activation is an essential component of preconditioning. PKC antagonists inhibit the protection conferred by preconditioning and PKC activation with phorbol ester mimics preconditioning (Ytrehus *et al.*, 1994; Mitchell *et al.*, 1995). PKC may be activated by several different mechanisms, including the aden-osine A_1 receptor, muscarinic M_2 receptor, and α_1-adrenergic receptors, although the exact mechanism responsible for its activation during precon-ditioning is controversial. Translocation of PKC from the cytosol to the membrane is hypothesized to be an essential step in preconditioning (Liu *et al.*, 1994). According to this scheme, once PKC reaches the membrane it remains there as long as adenosine stimulation continues, and it is subse-quently available in its active form during the second ischemic episode to initiate kinase activity.

Activation of I_{KATP} is hypothesized to be the key effector that confers the preconditioning benefit. Blockade of I_{KATP} using the sulfonylurea gli-benclamide prevents the beneficial effects of preconditioning in a dog model (Gross and Auchampach, 1992) and in human atrial tissue (Speechly-Dick *et al.*, 1995), but not consistently in rabbit (Thornton *et al.*, 1993c) or rat (Liu and Downey, 1992). Interestingly, it has been shown that a PKC activator markedly shortens the time to activation of I_{KATP} during metabolic inhibition in isolated rabbit ventricular myocytes, but only if adenosine is also present (Liu *et al.*, 1996). The requirement for persistent exposure to adenosine is consistent with the following two observations: (1) the protection provided by direct PKC activators can be abolished by blocking adenosine receptors during ischemia (Liu *et al.*, 1994); and (2) the benefits of ischemic preconditioning or pretreatment with adenosine are abolished by adenosine receptor antagonists during ischemia (Toombs *et al.*, 1992; Thornton *et al.*, 1993b). At reperfusion, however, continued adenosine receptor stimulation is not necessary for the continued benefit of preconditioning (Thonrton *et al.*, 1993b).

A variety of interventions can confer benefits similar to those observed with ischemic preconditioning. These include brief exposures to CN (de Albuquerque *et al.*, 1994), hypoxia (Shizukuda *et al.*, 1992), α_1-adrenergic agents (Banerjee *et al.*, 1993), acetylcholine (Yao and Gross, 1993), stretch (Ovize *et al.*, 1994), and stress proteins (Yellon *et al.*, 1992). Whether the underlying mechanisms are similar to genuine ischemic preconditioning remains uncertain.

Taken together these findings suggest that adenosine receptor activation during the preconditioning and subsequent ischemic phases is an essential component of the preconditioning phenomenon. During the preconditioning phase, adenosine may stimulate PKC, resulting in translocation of cytosolic PKC to the SL membrane. In the continued presence of adenosine, PKC in the membrane is available to phosphorylate ATP-sensitive K^+ channels, either directly or perhaps indirectly through a G-protein mediated pathway, priming the channels so that they are more readily opened during the subsequent ischemic period. The consequent shortening of the action potential duration would result in a decreased Ca^{2+} current and, therefore, reduced contraction and cardiac work. Decreased work results in reduced ATP consumption and glycolytic flux, preservation of the intracellular adenine nucleotide pool, and maintenance of cellular ionic balance. Each of these effects would be expected to facilitate myocyte recovery at reperfusion.

In summary, reperfusion of cardiac tissue after short periods of ischemia results in impaired contractile function, known as myocardial stunning. Reperfusion after longer periods (>20–40 min) results in irreversible injury. Reperfusion arrhythmias are also common. Protection from these reperfusion abnormalities can be conferred by repetitive but brief ischemic episodes known as "preconditioning." Although ATP levels are depressed in reperfused ischemic myocardium, probably due to washout of nucleotide precursors because oxidative metabolism recovers rapidly, metabolic abnormalities are not directly responsible for contractile abnormalities. In the case of stunning, contractile abnormalities are caused by altered myofilament Ca^{2+} responsiveness caused by Ca^{2+}-activated protease activity. Irreversible injury is caused by massive Ca^{2+} overload that inhibits metabolism and activates proteolytic processes that interfere with normal cellular function.

The causes of $[Ca^{2+}]_i$ overload at reperfusion are not easily explained on the basis of metabolic inhibition alone. Several lines of evidence suggest that abnormalities in Ca^{2+}, Na^+, and K^+ handling during reperfusion are caused by oxygen-free radicals, amphiphiles, and cytokines elaborated during ischemia and reperfusion. Oxygen-free radicals have been shown to have direct effects on the SR, Na^+–Ca^{2+} exchanger, SL Ca^{2+} pump, Ca^{2+} leak channel, Na^+ channel, Na^+ pump, and inwardly rectifying K^+ channel. Cytokines induce the free radical NO and alter β-adrenergic responsiveness by activation of cGMP pathways. Amphiphiles inhibit the voltage-sensitive Na^+ channel and the Na^+ pump, while also activating a novel Na^+ channel at negative membrane potentials. Amphiphiles also disrupt normal SR function and SL integrity. Regardless of the mechanism of injury, reduced cardiac work during the ischemic period results in improved recovery at reperfusion.

REFERENCES

Adamson, G.M., and Billings, R.E. (1992). Tumor necrosis factor induced oxidative stress in isolated mouse hepatocytes. *Arch. Biochem. Biophys.* 294:223–229.

Ahumada, G.G., Bergmann, S.R., Carlson, E., Corr, P.B., and Sobel, B.E. (1979). Augmentation of cyclic AMP content induced by lysophosphatidyl choline in rabbit hearts. *Cardiovasc. Res.* 13:377–382.

Ala, Y., Palluy, O., Favero, J., Bonne, C., Modat, G., and Dornand, J. (1992). Hypoxia/reoxygenation stimulates endothelial cells to promote interleukin-1 and interleukin-6 production. Effects of free radical scavengers. *Agents Actions* 37:134–139.

Allen, D.G., Lee, J.A., and Smith, G.L. (1989). The consequences of simulated ischaemia on intracellular Ca^{++} and tension in isolated ferret ventricular muscle. *J. Physiol.* 410:297–323.

Allen, D.G., Morris, P.G., Orchard, C.H., and Pirolo, J.S. (1985). A nuclear magnetic resonance study of metabolism in the ferret heart during hypoxia and inhibition of glycolysis. *J. Physiol.* (London) 361:185–204.

Allen, D.G., and Orchard, C.H. (1983). Intracellular calcium concentration during hypoxia and metabolic inhibition in mammalian ventricular muscle. *J. Physiol.* 339:107–122.

Allen, D.G., and Orchard, C.H. (1987). Myocardial contractile function during ischaemia and hypoxia. *Circ. Res.* 60:153–168.

Ambrosio, G., Becker, L.C., Hutchins, G.M., Weisman, H.F., and Weisfeldt, M.L. (1986). Reduction in experimental infarct size by recombinant human superoxide dismutase: Insights into the pathophysiology of reperfusion injury. *Circulation* 74:1424–1433.

Ambrosio, G., Jacobus, W.E., Mitchell, M.C., Litt, M.R., and Becker, L.C. (1989). Effects of ATP precursors on ATP and free ADP content and functional recovery of postischemic hearts. *Am. J. Physiol.* 256:H560–566.

Ambudkar, I.S., Abdallah, E.S., and Shamoo, A.E. (1988). Lysophospholipid-mediated alterations in the calcium transport systems of skeletal and cardiac muscle sarcoplasmic reticulum. *Mol. Cell. Biochem.* 79:81–89.

Anderson, S.E., Murphy, E., Steenbergen, C., London, R.E., and Cala, P.M. (1990). Na-H exchange in myocardium: Effects of hypoxia and acidification on Na and Ca. *Am. J. Physiol.* 259:C940–C948.

Arai, A.E., Pantely, G.A., Thoma, W.J., Anselone, C.G., and Bristow, J.D. (1992). Energy metabolism and contractile function after 15 beats of moderate myocardial ischemia. *Circ. Res.* 70:1137–1145.

Ashcroft, S.J.H., and Ashcroft, F.M. (1990). Properties and functions of ATP-sensitive K-channels. *Cell. Signal.* 2:197–214.

Auchampach, J.A., and Gross, G.J. (1993). Adenosine-A(1) receptors, K(ATP) channels, and ischemic preconditioning in dogs. *Am. J. Physiol.* 264:H1327–H1336.

Auffermann, W., Wagner, S., Wu, S., Buser, P., Parmley, W.W., and Wikman-Coffelt, J. (1990). Calcium inhibition of glycolysis contributes to ischaemic injury. *Cardiovasc. Res.* 24:510–520.

Bailey, I.A., Radda, G.K., Seymour, A.M., and Williams, S.R. (1982). The effects of insulin on myocardial metabolism and acidosis in normoxia and ischaemia. A 31P-NMR study. *Biochim. Biophys. Acta* 720:17–27.

Bailey, I.A., Williams, S.R., Radda, G.K., and Gadian, D.G. (1981). Activity of phosphorylase in total global ischaemia in the rat heart. *Biochem. J.* 196:171–178.

Balaban, R.S., and Bader, J.P. (1984). Studies on the relationship between glycolysis and (Na$^+$ − K$^+$)-ATPase in cultured cells. *Biochim. Biophys. Acta* 804:419–426.

Balligand, J.L., Kelley, R.A., Marsden, P.A., Smith, T.W., and Michel, T. (1993a). Control of cardiac muscle cell function by an endogenous nitric oxide signaling system. *Proc. Natl. Acad. Sci. USA* 90:347–351.

Balligand, J.L., Ungureanu, D., Kelly, R.A., Kobzik, L., Pimental, D., Michel, T., and Smith, T.W. (1993b). Abnormal contractile function due to induction of nitric oxide synthesis in rat cardiac myocytes follows exposure to activated macrophage-conditioned medium. *J. Clin. Invest.* 91:2314–2319.

Balligand, J.L., Ungureanu-Longrois, D., Simmons, W.W., Pimental, D., Malinski, T.A., Kapturczak, M., Taha, Z., Lowenstein, C.J., Davidoff, A.J., and Kelly, R.A. (1994). Cytokine-inducible nitric oxide synthase (iNOS) expression in cardiac myocytes. Characterization and regulation of iNOS expression and detection of iNOS activity in single cardiac myocytes *in vitro*. *J. Biol. Chem.* 269:27580–27588.

Banerjee, A., Locke-Winter, C., Rogers, K.B., Mitchell, M.B., Brew, E.C., Cairns, C.B., Bensard, D.D., and Harken, A.H. (1993). Preconditioning against myocardial dysfunction after ischemia and reperfusion by an alpha 1-adrenergic mechanism. *Circ. Res.* 73:656–670.

Barrigon, S., Wang, S.Y., Ji, X.W., and Langer, G.A. (1996). Characterization of the calcium overload in cultured neonatal rat cardiomyocytes under metabolic inhibition. *J. Mol. Cell. Cardiol.* 28:1329–1337.

Barrington, P.L. (1994). Interactions of H$_2$O$_2$, EGTA and patch pipette recording methods in feline ventricular myocytes. *J. Mol. Cell. Cardiol.* 26:557–568.

Barrington, P.L., Meier, Jr., C.F., and Weglicki, W.B. (1988). Abnormal electrical activity induced by free radical generating systems in isolated cardiocytes. *J. Mol. Cell. Cardiol.* 20:1163–1178.

Barry, W.H., Peeters, G.A., Rasmussen, Jr., C.A.F., and Cunningham, M.J. (1987). Role of changes in [Ca^{++}]$_i$ in energy deprivation contracture. *Circ. Res.* 61:726–734.

Beckman, J.S., Beckman, T.W., Chen, J., Marshall, P.A., and Freeman, B.A. (1990). Apparent hydroxyl radical production by peroxynitrite: Implications for endothelial injury from nitric oxide and superoxide. *Proc. Natl. Acad. Sci. USA* 87:1620–1624.

Bekheit, S., Restivo, M., Boutjdir, M., Henkin, R., Gooyandeh, K., Assadi, M., Khatib, S., Gough, W.B., and El-Sherif, N. (1990). Effects of glyburide on ischemia-induced changes in extracellular potassium and local myocardial activation: A potential new approach to the management of ischemia-induced malignant ventricular arrhythmias. *Am. Heart J.* 119:1025–1033.

Berne, R.M., Knabb, R.M., Ely, S.W., and Rubio, R. (1983). Adenosine in the local regulation of blood flow: A brief overview. *Fed. Proc.* 42:3136–3142.

Bers, D.M., and Ellis, D. (1982). Intracellular calcium and sodium activity in sheep heart Purkinje fibres. Effect of changes of external sodium and intracellular pH. *Pflugers Arch.* 393:171–178.

Bersohn, M.M. (1995). Sodium pump inhibition in sarcolemma from ischemic hearts. *J. Mol. Cell. Cardiol.* 27:1483–1489.

Bersohn, M.M., Philipson, K.D., and Weiss, R.S. (1991). Lysophosphatidylcholine and sodium-calcium exchange in cardiac sarcolemma: Comparison with ischemia. *Am. J. Physiol.* 260:C433–C438.

Beutler, B., and Cerami, A. (1988). Tumor necrosis, cachexia, shock, and inflammation: A common mediator. *Annu. Rev. Biochem.* 57:505–518.

Bhatnagar, A., Srivastava, S.K., and Szabo, G. (1990). Oxidative stress alters specific membrane currents in isolated cardiac myocytes. *Circ. Res.* 67:535–549.

Bittle, J.A., Balschi, J.A., and Ingwall, J.S. (1987). Effects of norephinephrine infusion on myocardial high-energy phosphate content and turnover in the living rat. *J. Clin. Invest.* 79:1852–1859.

Bolli, R. (1992). Myocardial "stunning" in man. *Circulation* 86:1671–1691.

Boraso, A., and Williams, A.J. (1994). Modification of the gating of the cardiac sarcoplasmic reticulum Ca^{2+}-release channel by H_2O_2 and dithiothreitol. *Am. J. Physiol.* 267:H1010–H1016.

Brady, A.J., Poole-Wilson, P.A., Harding, S.E., and Warren, J.B. (1992). Nitric oxide production within cardiac myocytes reduces their contractility in endotoxemia. *Am. J. Physiol.* 263:H1963–H1966.

Braunwald, E., and Kloner, R.A. (1982). The stunned myocardium: Prolonged, postischemic ventricular dysfunction. *Circulation* 66:1146–1149.

Braunwald, E., and Rutherford, J.D. (1986). Reversible ischemic left ventricular dysfunction: Evidence for the "hibernating myocardium." *J. Am. Coll. Cardiol.* 8:1467–1470.

Bricknell, O.L., Daries, P.S., and Opie, L.H. (1981). A relationship between adenosine triphosphate, glycolysis and ischaemic contracture in the isolated rat heart. *J. Mol. Cell. Cardiol.* 13:941–945.

Bricknell, O.L., and Opie, L.H. (1978). Effects of substrates on tissue metabolic changes in the isolated rat heart during underperfusion and on release of lactate dehydrogenase and arrhythmias during reperfusion. *Circ. Res.* 43:102–115.

Bristow, J.D., Arai, A.E., Anselone, C.G., and Pantely, G.A. (1991). Response to myocardial ischemia as a regulated process. *Circulation* 84:2580–2587.

Brooks, W.M., and Willis, R.J. (1983). 31P nuclear magnetic resonance study of the recovery characteristics of high energy phosphate compounds and intracellular pH after global ischaemia in the perfused guinea-pig heart. *J. Mol. Cell. Cardiol.* 15:495–502.

Brovkovitch, V.M., Nikitchenko, Y.V., and Lemeshko, V.V. (1987). The damage of Ca^{++}-transporting function and lipid membrane component of the sarcoplasmic reticulum in total ischemia of the rat myocardium. *Byull. Eksp. Biol. Med.* 104:546–548.

Brown, J.M., Anderson, B.O., Repine, J.E., Shanley, P.F., White, C.W., Grosso, M.A., Banerjee, A., Bensard, D.D., and Harken, A.H. (1992). Neutrophils contribute to TNF induced myocardial tolerance to ischaemia. *J. Mol. Cell. Cardiol.* 24:485–495.

Brown, J.M., White, C.W., Terada, L.S., Grosso, M.A., Shanley, P.F., Mulvin, D.W., Banerjee, A., Whitman, G.J., Harken, A.H., and Repine, J.E. (1990). Interleukin 1 pretreatment decreases ischemia/reperfusion injury. *Proc. Natl. Acad. Sci. USA* 87:5026–5030.

Buja, L.M., Hagler, H.K., Parsons, D., Chien, K., Reynolds, R.C., and Willerson, J.T. (1985). Alterations of ultrastructure and elemental composition in cultured neonatal rat cardiac myocytes after metabolic inhibition with iodoacetic acid. *Lab. Invest.* 53:397–412.

Burnashev, N.A., Undrovinas, A.I., Fleidervish, I.A., and Rosenshtraukh, L.V. (1989). Ischemic poison lysophosphatidylcholine modifies heart sodium channels gating inducing long-lasting bursts of openings. *Pflugers Arch.* 415:125–126.

Buschmeier, B., Meyer, H.E., and Mayr, G.W. (1987). Characterization of the calmodulin-binding sites of muscle phosphofructokinase and comparison with known calmodulin-binding. *J. Biol. Chem.* 262:9454–9462.

Busse, R., and Mulsch, A. (1990). Induction of nitric oxide synthase by cytokines in vascular smooth. *FEBS Lett.* 275:87–90.

Butwell, N.B., Ramasamy, R., Lazar, I., Sherry, A.D., and Malloy, C.R. (1993). Effect of lidocaine on contracture, intracellular sodium, and pH in ischemic rat hearts. *Am. J. Physiol.* 264:H1884–H1889.

Chatham, J., Gilbert, H.F., and Radda, G.K. (1988). Inhibition of glucose phosphorylation by fatty acids in the perfused rat heart. *FEBS Lett.* 238:445–449.

Chien, K.R., Han, A., Sen, A. Buja, L.M., and Willerson, J.T. (1984). Accumulation of unesterified arachidonic acid in ischemic canine myocardium. *Circ. Res.* 54:313–322.

Clague, J.R., and Langer, G.A. (1994). The pathogenesis of free radical-induced calcium leak in cultured rat cardiomyocytes. *J. Mol. Cell. Cardiol.* 26:11–21.

Clarke, K., O'Connor, A.J., and Willis, R.J. (1987). Temporal relation between energy metabolism and myocardial function during ischemia and reperfusion. *Am. J. Physiol.* 253:H412–H421.

Cobbold, P.H., and Bourne, P.K. (1984). Aequorin measurements of free calcium in single heart cells. *Nature* 312:444–446.

Coetzee, W.A., Ichikawa, H., and Hearse, D.J. (1994). Oxidant stress inhibits Na-Ca-exchange current in cardiac myocytes: Mediation by sulfhydryl groups? *Am. J. Physiol.* 266:H909–H919.

Corr, P.B., Gross, R.W., and Sobel, B.E. (1984). Amphipathic metabolites and membrane dysfunction in ischemic myocardium. *Circ. Res.* 55:135–154.

Corretti, M.C., Korestsune, Y., Kusuoka, H., Chacko, V.P., Zweier, J.L., and Marban, E. (1991). Glycolytic inhibition and calcium overload as consequenes of exogenously-generated free radicals in rabbit hearts. *J. Clin. Invest.* 88:1014–1025.

Covell, J.W., Pool, P.E., and Braunwald, E. (1967). Effects of acutely induced ischemic heart failure on myocardial high energy phosphate stores. *Proc. Soc. Exp. Biol. Med.* 124:126–131.

Cross, H.R., Opie, L.H., Radda, G.K., and Clarke, K. (1996). Is a high glycogen content beneficial or detrimental to the ischemic rat heart? A controversy resolved. *Circ. Res.* 78:482–491.

Dahl, G., and Isenberg, G. (1980). Decoupling of heart muscle cells: Correlation with increased cytoplasmic calcium activity and with changes of nexus ultrastructure. *J. Membr. Biol.* 53:63–75.

Dale, W.E., Hale, C.C., Kim, H.D., and Rovetto, M.J. (1991). Myocardial glucose utilization. Failure of adenosine to alter it and inhibition by the adenosine analogue N6-(L-2-phenyliso-propyl)adenosine. *Circ. Res.* 69:791–799.

Das, D.K., Engelman, R.M., Rousou, J.A., Breyer, R.H., Otani, H., and Lemeshow, S. (1986). Role of membrane phospholipids in myocardial injury induced by ischemia and reperfusion. *Am. J. Physiol.* 251:H71–H79.

DaTorre, S.D., Creer, M.H., Pogwizd, S.M., and Corr, P.B. (1991). Amphipathic lipid metabolites and their relation to arrhythmogenesis in the ischemic heart. *J. Mol. Cell Cardiol.* 23 Suppl. 1:11–22.

de Albuquerque, C.P., Gerstenblith, G., and Weiss, R.G. (1994). Importance of metabolic inhibition and cellular pH in mediating preconditioning contractile and metabolic effects in rat hearts. *Circ. Res.* 74:139–150.

DeBoer, L.W., Ingwall, J.S., Kloner, R.A., and Braunwald, E. (1980). Prolonged derangements of canine myocardial purine metabolism after a brief coronary artery occlusion not associated with anatomic evidence of necrosis. *Proc. Natl. Acad. Sci. USA* 77:5471–5475.

Deitmer, J.W., and Ellis, D. (1980). Interactions between the regulation of the intracellular pH and sodium activity of sheep cardiac purkinje fibres. *J. Physiol.* 304:471–488.

Demaison, L., and Grynberg, A. (1994). Cellular and mitochondrial energy metabolism in the stunned myocardium. *Basic Res. Cardiol.* 89:293–307.

Dennis, S.C., Gevers, W., and Opie, L.H. (1991). Protons in ischemia: Where do they come from; where do they go to? *J. Mol. Cell Cardiol.* 23:1077–1086.

Deutsch, N., Alexander, L.D., Shang, P., and Weiss, J.N. (1994). Activation of ATP-sensitive K^+ channels and cellular K^+ loss. *Biophys. J.* 66:A427.

Deutsch, N., and Weiss, J.N. (1993). ATP-sensitive K channel modification by metabolic inhibition in isolated guinea-pig ventricular myocytes. *J. Physiol.* 465:163–179.

Deutsch, N., and Weiss, J.N. (1994). Effects of trypsin on cardiac ATP-sensitive K channels. *Am. J. Physiol.* 266:H613–H622.

Di Lisa, F., Blank, P.S., Colonna, R., Gambassi, G., Silverman, H.S., Stern, M.D., and Hansford, R.G. (1995). Mitochondrial membrane potential in single living adult rat cardiac myocytes exposed to anoxia or metabolic inhibition. *J. Physiol.* (London) 486:1–13.

Dixon, I.M., Kaneki, M., Hata, T., Panagia, V., and Dhalla, N.S. (1990). Alterations in membrane Ca^{2+} transport during oxidative stress. *Mol. Cell. Biochem.* 99:125–133.

Donoso, P., Mill, J.G., O'Neill, S.C., and Eisner, D.A. (1992). Fluorescence measurements of cytoplasmic and mitochondrial sodium concentration in rat ventricular myocytes. *J. Physiol.* 448:493–509.

Eberli, F.R., Weinberg, E.O., Grice, W.N., Horowitz, G.L., and Apstein, C.S. (1991). Protective effect of increased glycolytic substrate against systolic and diastolic dysfunction and increased coronary resistance from prolonged global underperfusion and reperfusion in isolated rabbit hearts perfused with erythrocyte suspensions. *Circ. Res.* 68:466–481.

Eddy, L.J., Goeddel, D.V., and Wong, G.H.W. (1992). Tumor necrosis factor-alpha pretreatment is protective in a rat model of myocardial ischemia-reperfusion injury. *Biochem. Biophys. Res. Commun.* 184:1056–1059.

Eisner, D.A., Nichols, C.G., O'Neill, S.C., Smith, G.L., and Valdeolmillos, M. (1989). The effects of metabolic inhibition on intracellular calcium and pH in isolated rat ventricular cells. *J. Physiol.* (London) 411:393–418.

Eisner, D.A., and Richards, D.E. (1982). Inhibition of the sodium pump by inorganic phosphate in resealed red ghosts. *J. Physiol.* (London) 326:1–10.

El-Hayek, R., Valdivia, C., Valdivia, H.H., Hogan, K., and Coronado, R. (1993). Activation of the Ca^{2+} release channel of skeletal muscle sarcoplasmic reticulum by palmitoyl carnitine. *Biophys. J.* 65:779–789.

Elliot, A.C., Smith, G.L., and Allen, D.G. (1989). Simultaneous measurements of action potential duration and intracellular ATP in isolated ferret hearts exposed to cyanide. *Circ. Res.* 64:583–591.

Ellis, D., and Noireaud, J. (1987). Intracellular pH in sheep purkinje fibres and ferret papillary muscles during hypoxia and recovery. *J. Physiol.* 383:125–141.

Ellis, S.G., Henschke, C.I., Sandor, T., Wynne, J., Braunwald, E., and Kloner, R.A. (1983). Time course of functional and biochemical recovery of myocardium by reperfusion. *J. Am. Coll. Cardiol.* 1:1047–1055.

Entman, M.L., Kaniike, K., Goldstein, M.A., Nelson, T.E., Bornet, E.P., Futch, T.W., and Schwartz, A. (1976). Association of glycogenolysis with cardiac sarcoplasmic reticulum. *J. Biol. Chem.* 251:3140–3146.

Fabiato, A., and Fabiato, F. (1978). Effects of pH on the myofilaments and the sarcoplasmic reticulum of skinned cells from cardiac and skeletal muscles. *J. Physiol.* (London) 276:233–255.

Ferrari, R., Ceconi, C., Curello, S., Guarnieri, C., Caldarera, C.M., Albertini, A., and Visioli, O. (1985). Oxygen-mediated myocardial damage during ischaemia and reperfusion: Role of the cellular defences against oxygen toxicity. *J. Mol. Cell Cardiol.* 17:937–945.

Figueredo, V.M., Brandes, R., Weiner, M.W., Massie, B.M., and Camacho, S.A. (1992). Cardiac contractile dysfunction during mild coronary flow reductions is due to an altered calcium-pressure relationship in rat hearts. *J. Clin. Invest.* 90:1794–1802.

Findlay, I. (1988a). ATP4- and ATPMg inhibit the ATP-sensitive K^+ channel of rat ventricular myocytes. *Pflugers Arch.-Eur. J. Physiol.* 412:37–41.

Findlay, I. (1988b). Effects of ADP upon the ATP-sensitive K^+ channel in rat ventricular myocytes. *J. Membr. Biol.* 101:83–92.

Finegan, B.A., Clanachan, A.S., Coulson, C.S., and Lopaschuk, G.D. (1992). Adenosine modification of energy substrate use in isolated hearts perfused with fatty acids. *Am. J. Physiol.* 262:H1501–H1507.

Finegan, B.A., Lopaschuk, G.D., Gandhi, M., and Clanachan, A.S. (1995). Ischemic preconditioning inhibits glycolysis and proton production in isolated working rat hearts. *Am. J. Physiol.* 269:H1767–H1775.

Finkel, M.S., Oddis, C.V., Jacob, T.D., Watkins, S.C., Hattler, B.G., and Simmons, R.L. (1992). Negative inotropic effects of cytokines on the heart mediated by nitric oxide. *Science* 257:387–389.

Flaherty, J.T., Pitt, B., Gruber, J.W., Heuser, R.R., Rothbaum, D.A., Burwell, L.R., George, B.S., Kereiakes, D.J., Deitchman, D., Gustafson, N., *et al.* (1994). Recombinant human superoxide dismutase (h-SOD) fails to improve recovery of ventricular function in patients undergoing coronary angioplasty for acute myocardial infarction. *Circulation* 89:1982–1991.

Gao, W.D., Liu, Y.G., Mellgren, R., and Marban, E. (1996). Intrinsic myofilament alterations underlying the decreased contractility of stunned myocardium. A consequence of Ca^{2+}-dependent proteolysis? *Circ. Res.* 78:455–465.

Gard, J.K., Kichura, G.M., Ackerman, J.J., Eisenberg, J.D., Billadello, J.J., Sobel, B.E., and Gross, R.W. (1985). Quantitative ^{31}P nuclear magnetic resonance analysis of metabolite concentrations in Langendorff-perfused rabbit hearts. *Biophys. J.* 48:803–813.

Garlick, P.B., Radda, G.K., and Seeley, P.J. (1979). Studies of acidosis in the ischaemic heart by phosphorus nuclear magnetic resonance. *Biochem. J.* 184:547–554.

Gaspardone, A., Shine, K.I., Seabrooke, S.R., and Poole-Wilson, P.A. (1986). Potassium loss from rabbit myocardium during hypoxia: Evidence for passive efflux linked to anion extrusion. *J. Mol. Cell Cardiol.* 18:389–399.

Gibbs, C.L., and Chapman, J.B. (1979). Cardiac energetics. *In* "Handbook of Physiology: The Cardiovascular System" (Berne, R.M., Sperelakis, N., and Geiger, S., eds.), pp. 775–804. American Physiological Society, Bethesda.

Glitsch, H.G., and Tappe, A. (1993). The Na^+/K^+ pump of cardiac Purkinje cells is preferentially fuelled by glycolytic ATP production. *Pflugers Arch.* 422:380–385.

Goldhaber, J.I. (1996). Free radicals enhance Na^+/Ca^{2+} exchange in ventricular myocytes. *Am. J. Physiol.* 271:823–833.

Goldhaber, J.I., Duong, T.K., and Weiss, J.N. (1996a). Activation of a non-selective cation current may drive net cellular K^+ loss during metabolic inhibition. *Biophys. J.* 70:A261.

Goldhaber, J.I., Ji, S., Lamp, S.T., and Weiss, J.N. (1989). Effects of exogenous free radicals on electromechanical function and metabolism in isolated rabbit and guinea pig ventricle: Implications for ischemia and reperfusion injury. *J. Clin. Invest.* 83:1800–1809.

Goldhaber, J.I., Kim, K.H., Natterson, P.D., Lawrence, T., Yang, P., and Weiss, J.N. (1996b). Effects of TNF-α on $[Ca^{2+}]_i$ and contractility in isolated adult rabbit ventricular myocytes. *Am. J. Physiol.* 271:1449–1455.

Goldhaber, J.I., and Liu, E. (1994). Excitation-contraction coupling in single guinea-pig ventricular myocytes exposed to hydrogen peroxide. *J. Physiol.* (London) 477:135–147.

Goldhaber, J.I., Parker, J.M., and Weiss, J.N. (1991). Mechanisms of excitation-contraction coupling failure during metabolic inhibition in guinea-pig ventricular myocytes. *J. Physiol.* (London) 443:371–386.

Goodwin, G.W., Arteaga, J.R., and Taegtmeyer, H. (1995). Glycogen turnover in the isolated working rat heart. *J. Biol. Chem.* 270:9234–9240.

Gorge, G., Chatelain, P., Schaper, J., and Lerch, R. (1991). Effect of increasing degrees of ischemic injury on myocardial oxidative metabolism early after reperfusion in isolated rat hearts. *Circ. Res.* 68:1681–1692.

Greenfield, R.A., and Swain, J.L. (1987). Disruption of myofibrillar energy use: Dual mechanisms that may contribute to postischemic dysfunction in stunned myocardium. *Circ. Res.* 60:283–289.

Grines, C.L., Browne, K.F., Marco, J., Rothbaum, D., Stone, G.W., O'Keefe, J., Overlie, P., Donohue, B., Chelliah, N., Timmis, G.C., et al. (1993). A comparison of immediate angioplasty with thrombolytic therapy for myocardial infarction. The Primary Angioplasty in Myocardial Infarction Study Group. *N. Engl. J. Med.* 328:673–679.

Grinwald, P.M. (1992). Sodium pump failure in hypoxia and reoxygenation. *J. Mol. Cell Cardiol.* 24:1393–1398.

Gross, G.J., and Auchampach, J.A. (1992). Blockade of ATP-sensitive potassium channels prevents myocardial preconditioning in dogs. *Circ. Res.* 70:223–233.

Guarnieri, T. (1987). Intracellular sodium-calcium dissociation in early contractile failure in hypoxic ferret papillary muscles. *J. Physiol.* 388:449–465.

Gulick, T., Chung, M.K., Pieper, S.J., Lange, L.G., and Schreiner, G.F. (1989). Interleukin 1 and tumor necrosis factor inhibit cardiac myocyte beta-adrenergic responsiveness. *Proc. Natl. Acad. Sci. USA* 86:6753–6757.

Hagar, J.M., Hale, S.L., and Kloner, R.A. (1991). Effect of preconditioning ischemia on reperfusion arrhythmias after coronary artery occlusion and reperfusion in the rat. *Circ. Res.* 68:61–68.

Haigney, M.C., Lakatta, E.G., Stern, M.D., and Silverman, H.S. (1994). Sodium channel blockade reduces hypoxic sodium loading and sodium-dependent calcium loading. *Circulation* 90:391–399.

Haigney, M.C.P., Miyata, H., Lakatta, E.G., Stern, M.D., and Silverman, H.S., (1992). Dependence of hypoxic cellular calcium loading on Na^+-Ca^{2+} exchange. *Circ. Res.* 71:547–557.

Han, J.W., Thieleczek, R., Varsanyi, M., and Heilmeyer, L.M., Jr. (1992). Compartmentalized ATP synthesis in skeletal muscle triads. *Biochem.* 31:377–384.

Harris, E.J., Booth, R., and Cooper, M.B. (1982). The effect of superoxide generation on the ability of mitochondria to take up and retain Ca^{2+}. *FEBS Lett.* 146:267–272.

Hasin, Y., and Barry, W.H. (1984). Myocardial metabolic inhibition and membrane potential, contraction, and potassium uptake. *Am. J. Physiol.* 247:H322–H329.

Haworth, R.A., Goknur, A.B., Hunter, D.R., Hegge, J.O., and Berkoff, H.A. (1987). Inhibition of calcium influx in isolated adult rat heart cells by ATP depletion. *Circ. Res.* 60:586–594.

Hearse, D.J. (1979). Oxygen deprivation and early myocardial contractile failure: Reassessment of the possible role of adenosine triphosphate. *Am. J. Cardiol.* 44:1115–1121.

Hearse, D.J., and Tosaki, A. (1987). Free radicals and reperfusion-induced arrhythmias: Protection by spin trap agent PBN in the rat heart. *Circ. Res.* 60:375–383.

Henning, S.L., Wambolt, R.B., Schonekess, B.O., Lopaschuk, G.D., and Allard, M.F. (1996). Contribution of glycogen to aerobic myocardial glucose utilization. *Circulation* 93:1549–1555.

Hess, M.L., Krause, S., and Kontos, H.A. (1983). Mediation of sarcoplasmic reticulum disruption in the ischemic myocardium: Proposed mechanism by the interaction of hydrogen ions and oxygen free radicals. *Adv. Exp. Med. Biol.* 161:377–389.

Hess, M.L., Okabe, E., Ash, P., and Kontos, H.A. (1984). Free radical mediation of the effects of acidosis on calcium transport by cardiac sarcoplasmic reticulum in whole heart homogenates. *Cardiovasc. Res.* 18:149–157.

Heytler, P.G., and Prichard, W.W. (1962). A new class of uncoupling agent. *Biochem. Biophys. Res. Commun.* 7:272–275.

Higgins, T.J.C., Allsopp, D., Bailey, P.J., and D'Souza, E.D.A. (1981). The relationship between glycolysis, fatty acid metabolism and membrane integrity in neonatal myocytes. *J. Mol. Cell Cardiol.* 13:599–615.

Hilgemann, D.W., Collins, A., and Matsuoka, S. (1992). Steady-state and dynamic properties of cardiac sodium-calcium exchange. Secondary modulation by cytoplasmic calcium and ATP. *J. Gen. Physiol.* 100:933–961.

Hill, J.L., and Gettes, L.S. (1980). Effect of acute coronary artery occlusion on local myocardial extracellular K^+ activity in swine. *Circulation* 64:768–777.

Hirche, H.J., Franz, C.H.R., Bos, L., Bissig, R., Lang, R., and Schramm, M. (1980). Myocardial extracellular K^+ and H^+ increase and noradrenaline release as possible cause of early arrhythmias following acute coronary artery occlusion in pigs. *J. Mol. Cell Cardiol.* 12:579–593.

Huang, W.H., and Askari, A. (1984). Regulation of $(Na^{++}K^+)$-ATPase by inorganic phosphate: pH dependence and physiological implications. *Biochem. Biophys. Res. Commun.* 123: 438–443.

Ichihara, K., and Abiko, Y. (1984). Rebound recovery of myocardial creatine phosphate with reperfusion after ischemia. *Am. Heart J.* 108: 1594–1597.

Ikenouchi, H., Kohmoto, O., McMillan, M., and Barry, W.H. (1991). Contributions of $[Ca^{++}]_i$, $[P]_i$, and pH_i to altered diastolic myocyte tone during partial metabolic inhibition. *J. Clin. Invest.* 88:55–61.

Ingwall, J.S. (1995). How high does intracellular sodium rise during acute myocardial ischaemia? A view from NMR spectroscopy. *Cardiovasc. Res.* 29:279.

Irisawa, H., and Kokubun, S. (1983). Modulation by intracellular ATP and cyclic AMP of the slow inward current in isolated single ventricular cells of the guinea-pig. *J. Physiol.* (London) 338:321–337.

Isenberg, G., Vereecke, J., van der Heyden, G., and Carmeliet, E. (1983). The shortening of the action potential by DNP in guinea-pig ventricular myocytes is mediated by an increase of a time-independent K conductance. *Pflugers Arch.* 397:251–259.

Ito, B.R. (1995). Gradual onset of myocardial ischemia results in reduced myocardial infarction. Association with reduced contractile function and metabolic downregulation. *Circulation* 91:2058–2070.

Ito, B.R., Tate, H. Kobayashi, M., and Schaper, W. (1987). Reversibly injured, postischemic canine myocardium retains normal contractile reserve. *Circ. Res.* 61:834–846.

Jacobus, W.E. (1985). Respiratory control and the integration of heart high-energy phosphate metabolism by mitochondrial creatine kinase. *Annu. Rev. Physiol.* 47:707–725.

Janero, D.R., Hreniuk, D., and Sharif, H.M. (1994). Hydroperoxide-induced oxidative stress impairs heart muscle cell carbohydrate metabolism. *Am. J. Physiol.* 266:C179–C188.

Janse, M.J., and Wit, A.L. (1989). Electrophysiological mechanisms of ventricular arrhythmias resulting from myocardial ischemia and infarction. *Physiol. Rev.* 69:1049–1168.

Jennings, R.B., and Ganote, C.E. (1974). Structural changes in myocardium during acute ischemia. *Circ. Res.* 34,35:III-156–III-172.

Jennings, R.B., Reimer, K.A., and Steenbergen, C. (1986). Myocardial ischemia revisited. The osmolar load, membrane damage, and reperfusion. *J. Mol. Cell Cardiol.* 18:769–780.

Jennings, R.B., Schaper, J., Hill, M.L., Steenbergen, C., and Reimer, K.A. (1985). Effects of reperfusion late in the phase of reversible ischemic injury. *Circ. Res.* 56:262–278.

Jeremy, R.W., Ambrosio, G., Pike, M.M., Jacobus, W.E., and Becker, L.C. (1993). The functional recovery of post-ischemic myocardium requires glycolysis during early reperfusion. *J. Mol. Cell Cardiol.* 25:261–276.

Jeremy, R.W., Koretsune, Y., Marban, E., and Becker, L.C. (1992). Relation between glycolysis and calcium homeostasis in postischemic myocardium. *Circ. Res.* 70:1180–1190.

Jones, D.P. (1986). Intracellular diffusion gradients of O_2 and ATP. *Am. J. Physiol.* 250:C663–C675.

Josephson, R.A., Silverman, H.S., Lakatta, E.G., Stern, M.D., and Zweier, J.L. (1991). Study of the mechanisms of hydrogen peroxide and hydroxyl free radical-induced cellular injury and calcium overload in cardiac myocytes. *J. Biol. Chem.* 266:2354–2361.

Kakar, S.S., Huang, W.H., and Askari, A. (1987). Control of cardiac sodium pump by long-chain acyl coenzymes A. *J. Biol. Chem.* 262:42–45.

Kameyama, M., Kakei, M., Sato, R., Shibasaki, T., Matsuda, H., and Irisawa, H. (1984). Intracellular Na^+ activates a K^+ channel in mammalian cardiac cells. *Nature* 309:354–356.

Kaminishi, T., and Kako, K.J. (1988). Mechanism of H_2O_2-induced acceleration of calcium influx into isolated rat heart myocytes. *In* "Biology of Isolated Adult Cardiac Myocytes" (Clark, W.A., Decker, R.S., and Bort, T.K., eds.), pp. 220–223. Elsevier, New York.

Kammermeier, H., Schmidt, P., and Jungling, E. (1982). Free energy change of ATP-hydrolysis: A causal factor of early hypoxic failure of the myocardium? *J. Mol. Cell Cardiol.* 14:267–277.

Kaneko, M., Beamish, R.E., and Dhalla, N.S. (1989). Depression of heart sarcolemmal Ca^{++}-pump activity by oxygen free radicals. *Am. J. Physiol.* 256:H368–H374.

Kantor, P.F., Coetzee, W.A., Carmeliet, E.E., Dennis, S.C., and Opie, L.H. (1990). Reduction of ischemic K^+ loss and arrhythmias in the rat hearts: Effect of glibenclamide, a sulfonylurea. *Circ. Res.* 66:478–485.

Karmazyn, M. (1988). Amiloride enhances postischemic ventricular recovery: Possible role of Na^+-H^+ exchange. *Am. J. Physiol.* 255:H608–H615.

Kentish, J.D. (1986). The effects of inorganic phosphate and creatine phosphate on force production in skinned muscles from rat ventricle. *J. Physiol.* 370:585–604.

Kihara, Y., Grossman, W., and Morgan, J.P. (1989). Direct measurement of changes in intracellular calcium transients during hypoxia, ischemia, and reperfusion of the intact mammalian heart. *Circ. Res.* 65:1029–1044.

Kim, D., and Clapham, D.E. (1989). Potassium channels in cardiac cells activated by arachidonic acid and phospholipids. *Science* 244:1174–1176.

Kim, D., and Smith, T.W. (1988). Cellular mechanisms underlying calcium-proton interactions in cultured chick ventricular cells. *J. Physiol.* 398:391–410.

Kim, M.S., and Akera, T. (1987). O_2 free radicals: Cause of ischemia-reperfusion injury to cardiac Na^+-K^+ ATPase. *Am. J. Physiol.* 252:H252–H257.

Kingsley, P.B., Sako, E.Y., Yang, M.Q., Zimmer, S.D., Ugurbil, K., Foker, J.E., and From, A.H.L. (1991). Ischemic contracture begins when anaerobic glycolysis stops: A 31P-NMR study of isolated rat hearts. *Am. J. Physiol.* 261:H469–H478.

Kingsley-Hickman, P.B., Sako, E.Y., Ugurbil, K., From, A.H., and Foker, J.E. (1990). 31P NMR measurement of mitochondrial uncoupling in isolated rat hearts. *J. Biol. Chem.* 265:1545–1550.

Kinugawa, K., Takahashi, T., Kohmoto, O., Yao, A., Aoyagi, T., Momomura, S., Hirata, Y., and Serizawa, T. (1994). Nitric oxide-mediated effects of interleukin-6 on $[Ca^{2+}]_i$ and cell contraction in cultured chick ventricular myocytes. *Circ. Res.* 75:285–295.

Kitakaze, M., Hori, M., Morioka, T., Minamino, T., Takashima, S., Sato, H., Shinozaki, Y., Chujo, M., Mori, H., Inoue, M., *et al.* (1994). Infarct size-limiting effect of ischemic preconditioning is blunted by inhibition of 5'-nucleotidase activity and attenuation of adenosine release. *Circulation* 89:1237–1246.

Kitakaze, M., Hori, M., Takashima, S., Sato, H., Inoue, M., and Kamada, T. (1993). Ischemic preconditioning increases adenosine release and 5'-nucleotidase activity during myocardial ischemia and reperfusion in dogs for myocardial salvage. *Circulation* 87:208–215.

Kitakaze, M., and Marban, E. (1989). Cellular mechanism of the modulation of contractile function by coronary perfusion pressure in ferret hearts. *J. Physiol.* 414:455–472.

Kitakaze, M., Weisfeldt, M.L., and Marban, E. (1988a). Acidosis during early reperfusion prevents myocardial stunning in perfused ferret hearts. *J. Clin. Invest.* 82:920–927.

Kitakaze, M., Weisman, H.F., and Marban, E. (1988b). Contractile dysfunction and ATP depletion after transient calcium overload in perfused ferret hearts. *Circulation* 77:685–695.

Kleber, A.G. (1983). Resting membrane potential, extracellular potassium activity, and intracellular sodium activity during acute global ischemia in isolated perfused guinea pig hearts. *Circ. Res.* 52:442–450.

Kleber, A.G. (1984). Extracellular potassium accumulation in acute myocardial ischaemia. *J. Mol. Cell Cardiol.* 16:389–394.

Kloner, R.A., DeBoer, L.W.V., Darsee, J.R., Ingwall, J.S., Hale, S., Tumas, J., and Braunwald, E. (1981). Prolonged abnormalities of myocardium salvaged by reperfusion. *Am. J. Physiol.* 241:H591–H599.

Kobayashi, K., and Neely, J.R. (1979). Control of maximum rates of glycolysis in rat cardiac muscle. *Circ. Res.* 44:166–175.

Koga, S., Ogawa, S., Kuwabara, K., Brett, J., Leavy, J.A., Ryan, J., Koga, Y., Plocinski, J., Benjamin, W., Burns, D.K., *et al.* (1992). Synthesis and release of interleukin 1 by reoxygenated human mononuclear phagocytes. *J. Clin. Invest.* 90:1007–1015.

Kohmoto, O., and Barry, W.H. (1989). Mechanism of protective effects of Ca^{++} channel blockers on energy deprivation contracture in cultured ventricular myocytes. *J. Pharmacol. Exp. Ther.* 248:871–878.

Koretsune, Y., Corretti, M.C., Kusuoka, H., and Marban, E. (1991). Mechanism of early contractile failure: Inexcitability, metabolic accumulation, or vascular collapse? *Circ. Res.* 68:255–262.

Koretsune, Y., and Marban, E. (1990). Mechanism of ischemic contracture in ferret hearts: Relative roles of $[Ca^{++}]_i$ elevation and ATP depletion. *Am. J. Physiol.* 258:H9–H16.

Kubler, W., and Katz, A.M. (1977). Mechanism of early "pump" failure of the ischemic heart: Possible role adenosine triphosphate depletion and inorganic phosphate accumulation. *Am. J. Cardiol.* 40:467–471.

Kusuoka, H., de Hurtado, M.C.C., and Marban, E. (1993). Role of sodium/calcium exchange in the mechanism of myocardial stunning. Protective effect of reperfusion with high sodium solution. *J. Am. Coll. Cardiol.* 21:240–248.

Kusuoka, H., and Marban, E. (1992). Cellular mechanisms of myocardial stunning. *Annu. Rev. Physiol.* 54:243–256.

Kusuoka, H., and Marban, E. (1994). Mechanism of the diastolic dysfunction induced by glycolytic inhibition. Does adenosine triphosphate derived from glycolysis play a favored role in cellular Ca^{2+} homeostasis in ferret myocardium? *J. Clin. Invest.* 93:1216–1223.

Kusuoka, H., Porterfield, J.K., Weisman, H.F., Weisfeldt, M.L., and Marban, E. (1987). Pathophysiology and pathogenesis of stunned myocardium. Depressed Ca^{2+} activation of contraction as a consequence of reperfusion-induced cellular calcium overload in ferret hearts. *J. Clin. Invest.* 79: 950–961.

Lamas, G.A., Flaker, G.C., Mitchell, G., Smith, S.C., Jr., Gersh, B.J., Wun, C.C., Moye, L., Rouleau, J.L., Rutherford, J.D., Pfeffer, M.A., *et al.* (1995). Effect of infarct artery patency on prognosis after acute myocardial infarction. The survival and ventricular enlargement investigators. *Circulation* 92:1101–1109.

Law, W.R., and McLane, M.P. (1991). Adenosine enhances myocardial glucose uptake only in the presence of insulin. *Metabolism* 40:947–952.

Laxson, D.D., Homans, D.C., Dai, X.Z., Sublett, E., and Bache, R.J. (1989). Oxygen consumption and coronary reactivity in postischemic myocardium. *Circ. Res.* 64:9–20.

Lederer, W.J., and Nichols, C.G. (1989). Nucleotide modulation of the activity of rat heart ATP-sensitive K^+ channels in isolated membrane patches. *J. Physiol.* 419:193–211.

Lederer, W.J., Nichols, C.G., and Smith, G.L. (1989). The mechanism of early contractile failure of isolated rat ventricular myocytes subjected to complete metabolic inhibition. *J. Physiol.* (London) 413:329–349.

Lee, J.A., and Allen, D.G. (1992). Changes in intracellular free calcium concentration during long exposures to simulated ischemia in isolated mammalian ventricular muscle. *Circ. Res.* 71:58–69.

Lefer, A.M., Tsao, P., Aoki, N., and Palladino, M.A., Jr. (1990a). Mediation of cardioprotection by transforming growth factor-beta. *Science* 249:61–64.

Lefer, D.J., Ma, X.L., Johnson, III, G., and Lefer, A.M. (1990b). Endothellum preservation as a major mechanism of cardioprotection by superoxide dismutase in myocardial ischemia-reperfusion injury. *Circulation* 82:III–70.

Lehninger, A.L., Nelson, D.L., and Cox, M.M. (1993). "Principles of Biochemistry." Worth Publishers, New York.

Liedtke, A.J., DeMaison, L., Eggleston, A.M., Cohen, L.M., and Nellis, S.H. (1988). Changes in substrate metabolism and effects of excess fatty acids in reperfused myocardium. *Circ. Res.* 62:535–542.

Limas, C.J., Goldenberg, I.F., and Limas, C. (1995). Soluble interleukin-2 receptor levels in patients with dilated cardiomyopathy. Correlation with disease severity and cardiac autoantibodies. *Circulation* 91:631–634.

Lipasti, J.A., Nevalainen, A., Alanen, K.A., and Tolvanen, M.A. (1984). Anaerobic glycolysis and the development of ischaemic contracture in isolated rat heart. *Cardiovasc. Res.* 18:145–148.

Lipton, P., and Robacker, K. (1983). Glycolysis and brain function: $[K^+]_o$ stimulation of protein synthesis K^+ uptake require glycolysis. *Fed. Proc.* 42:2875–2880.

Liu, E., Goldhaber, J.I., Weiss, J.N. (1991a). Effects of lysophosphatidylcholine on electrophysiological properties and excitation–contraction coupling in isolated guinea pig ventricular myocytes. *J. Clin. Invest.* 88:1819–1832.

Liu, G.S., Thornton, J., Van Winkle, D.M., Stanley, A.W.H., Olsson, R.A., and Downey, J.M. (1991b). Protection against infarction afforded by preconditioning is mediated by A1 adenosine receptors in rabbit heart. *Circulation* 84:350–356.

Liu, Y., Ytrehus, K., and Downey, J.M. (1994). Evidence that translocation of protein kinase C is a key event during ischemic preconditioning of rabbit myocardium. *J. Mol. Cell Cardiol.* 26:661–668.

Liu, Y.G., and Downey, J.M. (1992). Ischemic preconditioning protects against infarction in rat heart. *Am. J. Physiol.* 263:H1107–H1111.

Liu, Y.G., Gao, W.D., O'Rourke, B., and Marban, E. (1996). Synergistic modulation of ATP-sensitive K currents by protein kinase C and adenosine. Implications for ischemic preconditioning. *Circ. Res.* 78:443–454.

Lopaschuk, G.D., Spafford, M.A., Davies, N.J., and Wall, S.R. (1990). Glucose and palmitate oxidation in isolated working rat hearts reperfused after a period of transient global ischemia. *Circ. Res.* 66:546–553.

Lopaschuk, G.D., Wall, S.R., Olley, P.M., and Davies, N.J. (1988). Etomoxir, a carnitine palmitoyltransferase I inhibitor, protects hearts from fatty acid-induced ischemic injury independent of changes in long chain acylcarnitine. *Circ. Res.* 63:1036–1043.

Lowenstein, C.J., and Snyder, S.H. (1992). Nitric oxide, a novel biologic messenger. *Cell* 70:705–707.

Low-Friedrich, I., Weisensee, D., Mitrou, P., and Schoeppe, W. (1992). Cytokines induce stress protein formation in cultured cardiac myocytes. *Basic Res. Cardiol.* 87:12–18.

Lucchesi, B.R. (1990). Myocardial ischemia, reperfusion and free radical injury. *Am. J. Cardiol.* 65:14I–23I.

Lynch, R.M., and Balaban, R.S. (1986). The coupling of Na-K ATPase and glycolysis in the renal cell line MDCK. *Biophys. J.* 49:36a.

Lynch, R.M., and Paul, R.J. (1983). Compartmentation of glycolytic and glycogenolytic metabolism in vascular smooth muscle. *Science* 222:1344–1346.

MacLeod, K.T. (1989). Effects of hypoxia and metabolic inhibition on the intracellular sodium activity of mammalian ventricular muscle. *J. Physiol.* 416:455–468.

Mak, I.T., Kramer, J.H., and Weglicki, W.B. (1986). Potentiation of free radical-induced lipid peroxidative injury to sarcolemmal membranes by lipid amphiphiles. *J. Biol. Chem.* 261:1153–1157.

Mak, I.T., and Weglicki, W.B. (1988). Protection by á-blocking agents against free radical-medicated sarcolemmal lipid peroxidation. *Circ. Res.* 63:262–266.

Marban, E., Kitakaze, M., Chacko, V.P., and Pike, M.M. (1988). Ca^{++} transients in perfused hearts revealed by gated ^{19}F NMR spectroscopy. *Circ. Res.* 63:673–678.

Marban, E., Kitakaze, M., Koretsune, Y., Yue, D.T., Chacko, V.P., and Pike, M.M. (1990). Quantification of $[Ca^{++}]_i$ in perfused hearts. *Circ. Res.* 66:1255–1267.

Mathur, P.P., and Case, R.B. (1973). Phosphate loss during reversible myocardial ischaemia. *J. Mol. Cell Cardiol.* 5:375–393.

Matsuura, H., and Shattock, M.J. (1991). Membrane potential fluctuations and transient inward currents induced by reactive oxygen intermediates in isolated rabbit ventricular cells. *Circ. Res.* 68:319–329.

Maury, C.P., and Teppo, A.M. (1989). Circulating tumour necrosis factor-alpha (cachectin) in myocardial infarction. *J. Intern. Med.* 225:333–336.

Mayr, G.W. (1984). Interaction of calmodulin with muscle phosphofructokinase. Changes of aggregation state, conformation and catalytic activity of the enzyme. *Eur. J. Biochem.* 143:513–520.

McCord, J.M. (1974). Free radicals and inflammation: Protection of synovial fluid by superoxide dismutase. *Science* 185:529–531.

McDonald, T.F., and MacLeod, D.P. (1973). Metabolism and the electrical activity of anoxic ventricular muscle. *J. Physiol.* 229:559–583.

McDonough, K.H., Henry, J.J., and Spitzer, J.J. (1987). Effects of oxygen radicals on substrate oxidation by cardiac myocytes. *Biochim. Biophys. Acta* 926:127–131.

McKenna, T.M. (1990). Prolonged exposure of rat aorta to low levels of endotoxin *in vitro* results in impaired contractility. Association with vascular cytokine release. *J. Clin. Invest.* 86:160–168.

Medina, G., and Illingworth, J. (1980). Some factors affecting phosphate transport in a perfused rat heart preparation. *Biochem. J.* 188:297–311.

Medina, G., and Illingworth, J.A. (1984). Some hormonal effects on myocardial phosphate efflux. *Biochem. J.* 224:153–162.

Mejia-Alvarez, R., and Marban, E. (1992). Mechanism of the increase in intracellular sodium during metabolic inhibition: Direct evidence against mediation by voltage-dependent sodium channels. *J. Mol. Cell Cardiol.* 24:1307–1320.

Mercer, R.W., and Dunham, P.B. (1981). Membrane-bound ATP fuels the Na/K pump. Studies on membrane-bound glycolytic enzymes on inside-out vesicles from human red cell membranes. *J. Gen. Physiol.* 78:547–568.

Mery, P.F., Pavoine, C., Belhassen, L., Pecker, F., and Fischmeister, R. (1993). Nitric oxide regulates cardiac Ca^{2+} current. Involvement of cGMP-inhibited and cGMP-stimulated phosphodiesterases through guanylyl cyclase activation. *J. Biol. Chem.* 268:26286–26295.

Meszaros, J., and Pappano, A.J. (1990). Electrophysiological effects of L-palmitoylcarnitine in single ventricular myocytes. *Am. J. Physiol.* 258:H931–H938.

Mickle, D.A., del Nido, P.J., Wilson, G.J., Harding, R.D., and Romaschin, A.D. (1986). Exogenous substrate preference of the post-ischaemic myocardium. *Cardiovasc. Res.* 20:256–263.

Mitani, A., and Shattock, M.J. (1992). Role of Na-activated K channel, Na-K-Cl cotransport, and Na-K pump in $[K]_e$ changes during ischemia in rat heart. *Am. J. Physiol.* 263:H333–H340.

Mitchell, M.B., Meng, X.Z., Ao, L.H., Brown, J.M., Harken, A. H., and Banerjee, A. (1995). Preconditioning of isolated rat heart is mediated by protein kinase C. *Circ. Res.* 76:73–81.

Miyata, H., Lakatta, E.G., Stern, M.D., and Silverman, H.S. (1992). Relation of mitochondrial and cytosolic free calcium to cardiac myocyte recovery after exposure to anoxia. *Circ. Res.* 71:605–613.

Mock, T., and Man, R.Y.K. (1990). Mechanism of lysophosphatidylcholine accumulation in the ischemic canine heart. *Lipids* 25:357–362.

Mohabir, R., Lee, H.C., Kurz, R.W., and Clusin, W.T. (1991). Effects of ischemia and hypercarbic acidosis on myocyte calcium transients, contraction, and pH_i in perfused rabbit hearts. *Circ. Res.* 69:1525–1537.

Molaparast-Saless, F., Liedtke, A.J., and Nellis, S.H. (1987). Effects of the fatty acid blocking agents, oxfenicine and 4-bromocrotonic acid, on performance in aerobic and ischemic myocardium. *J. Mol. Cell Cardiol.* 19:509–520.

Murphy, E., LeFurgey, A., and Lieberman, M. (1987). Biochemical and structural changes in cultured heart cells induced by metabolic inhibition. *Am. J. Physiol.* 253:C700–C706.

Murphy, E., Perlman, M., London, R.E., and Steenbergen, C. (1991). Amiloride delays the ischemia-induced rise in cytosolic free calcium. *Circ. Res.* 68:1250–1258.

Murphy, J.G., Smith, T.W., and Marsh, J.D. (1988). Mechanisms of reoxygenation-induced calcium overload in cultured chick embryo heart cells. *Am. J. Physiol.* 254:H1133–H1141.

Murry, C.E., Jennings, R.B., and Reimer, K.A. (1986). Preconditioning with ischemia: A delay of lethal cell injury in ischemic myocardium. *Circulation* 74:1124–1136.

Murry, C.E., Richard, V.J., Reimer, K.A., and Jennings, R.B. (1990). Ischemic preconditioning slows energy metabolism and delays ultrastructural damage during a sustained ischemic episode. *Circ. Res.* 66:913–931.

Myears, D.W., Sobel, B.E., and Bergmann, S.R. (1987). Substrate use in ischemic and reperfused canine myocardium: Quantitative considerations. *Am. J. Physiol.* 253:H107–H114.

Nakamura, K., Kusuoka, H., Ambrosio, G., and Becker, L.C. (1993). Glycolysis is necessary to preserve myocardial Ca-2+ homeostasis during beta-adrenergic stimulation. *Am. J. Physiol.* 264:H670–H678.

Nayler, W.G., Poole-Wilson, P.A., and Williams, A. (1979). Hypoxia and calcium. *J. Mol. Cell Cardiol.* 11:683–706.

Neely, J.R., and Morgan, H.E. (1974). Relationship between carbohydrate and lipid metabolism and the energy balance of heart muscle. *Annu. Rev. Physiol.* 31:413–459.

Neely, J.R., Whitfield, C.F., and Morgan, H.E. (1970). Regulation of glycogenolysis in hearts: Effects of pressure development, glucose, and FFA. *Am. J. Physiol.* 219:1083–1088.

Neely, J.R., Whitmer, J.T., and Rovetto, M.J. (1975). Effect of coronary blood flow on glycolytic flux and intracellular pH in isolated rat hearts. *Circ. Res.* 37:733–741.

Neubauer, S., Newell, J.B., and Ingwall, J.S. (1992). Metabolic consequences and predictability of ventricular fibrillation in hypoxia A ^{31}P- and ^{23}Na-nuclear magnetic resonance study of the rat heart. *Circulation* 86:302–310.

Noma, A. (1983). ATP regulated K+ channels in cardiac muscle. *Nature* 305:147–148.

Noma, A., and Shibasaki, T. (1985). Membrane current through adenosine-triphosphate-regulated potassium channels in guinea-pig ventricular cells. *J. Physiol.* 363:463–480.

Okabe, E., Odajima, C., Taga, R., Kukreja, R.C., Hess, M.L., and Ito, H. (1988). The effect of oxygen free radicals on calcium permeability and calcium loading at steady state in cardiac sarcoplasmic reticulum. *Mol. Pharmacol.* 34:388–394.

Opie, L.H. (1983). *In* "Cardiac Metabolism" (Drake-Holland, A.J., and Nobel, M.I.M., eds.), pp. 279–307. John Wiley and Sons, New York.

Oram, J.F., Wenger, J.I., and Neely, J.R. (1975). Regulation of long chain fatty acid activation in heart muscle. *J. Biol. Chem.* 250:73–78.

Orchard, C.H., and Kentish, J.C. (1990). Effects of changes of pH on the contractile function of cardiac muscle. *Am. J. Physiol.* 258:C967–C981.

O'Rourke, B., Ramza, B.M., and Marban, E. (1994). Oscillations of membrane current and excitability driven by metabolic oscillations in heart cells. *Science* 265:962–966.

Ovize, M., Kloner, R.A., and Przyklenk, K. (1994). Stretch preconditions canine myocardium. *Am. J. Physiol.* 266:H137–H146.

Owen, P., Dennis, S., and Opie, L.H. (1990). Glucose flux rate regulates onset of ischemic contracture in globally underperfused rat hearts. *Circ. Res.* 66:344–354.

Parker, J.C., and Hoffman, J.F. (1967). The role of membrane phosphoglycerate kinase in the control of rate by active cation transport in human red blood cells. *J. Gen. Physiol.* 50:893–916.

Parr, D.R., Wimhurst, J.M., and Harris, E.J. (1975). Calcium-induced damage of rat heart mitochondria. *Cardiovasc. Res.* 9:366–372.

Pesaturo, J.A., and Gwathmey, J.K. (1990). The role of mitochondria and sarcoplasmic reticulum calcium handling upon reoxygenation of hypoxic myocardium. *Circ. Res.* 66:696–709.

Pierce, G.N., and Philipson, K.D. (1985). Binding of glycolytic enzymes to cardiac sarcolemmal and sarcoplasmic reticular membranes. *J. Biol. Chem.* 260:6862–6870.

Pike, M.M., Kitakaze, M., and Marban, E. (1990). ^{23}Na-NMR measurements of intracellular sodium in intact perfused ferret hearts during ischemia and reperfusion. *Am. J. Physiol.* 259:H1767–H1773.

Pirolo, J.S., and Allen, D.G. (1986). Assessment of techniques for preventing glycolysis in cardiac muscle. *Cardiovasc. Res.* 20:837–844.

Pitts, B.J., and Okhuysen, C.H. (1984). Effects of palmitoyl carnitine and LPC on cardiac sarcolemmal Na$^+$-K$^+$-ATPase. Am. J. Physiol. 247:H840–H846.

Poole-Wilson, P.A., Harding, D.P., Bourdillon, P.D.V., and Tones, M.A. (1984). Calcium out of control. *J. Mol. Cell Cardiol.* 16:175–187.

Prasad, M.R., and Das, D.K. (1989). Effect of oxygen-derived free radicals and oxidants on the degradation in vitro of membrane phospholipids. *Free. Radic. Res. Commun.* 7:381–388.

Prinzen, F.W., Arts, T., van der Vusse, G.J., Coumans, W.A., and Reneman, R.S. (1986). Gradients in fiber shortening and metabolism across ischemic left ventricular wall. *Am. J. Physiol.* 250:H255–H264.

Przyklenk, K., and Kloner, R.A. (1989). Effect of superoxide dismutase plus catalase, given at the time of reperfusion, on myocardial infarct size, contractile function, coronary microvasculature, and regional myocardial blood flow. *Circ. Res.* 64:86–96.

Puett, D.W., Forman, M.B., Cates, C.U., Wilson, B.H., Hande, K.R., Friesinger, G.C., and Virmani, R. (1987). Oxypurinol limits myocardial stunning but does not reduce infarct size after reperfusion. *Circulation* 76:678–686.

Rabinovitch, A., Suarez, W.L., Thomas, P.D., Strynadka, K., and Simpson, I. (1992). Cytotoxic effects of cytokines on rat islets: Evidence for involvement. *Diabetologia* 35:409–413.

Radomski, M.W., Palmer, R.M., and Moncada, S. (1990). Glucocorticoids inhibit the expression of an inducible, but not the constitutive, nitric oxide synthase in vascular endothelial cells. *Proc. Natl. Acad. Sci. USA* 87:10043–10047.

Rau, E.E., and Langer, G.A. (1978). Dissociation of energetic state and potassium loss from anoxic myocardium. *Am. J. Physiol.* 235:H537–H543.

Rau, E.E., Shine, K.I., and Langer, G.A. (1977). Potassium exchange and mechanical performance in anoxic mammalian myocardium. *Am. J. Physiol.* 232:H85–H94.

Reeves, J.P., Bailey, C.A., and Hale, C.C. (1986). Redox modification of sodium-calcium exchange activity in cardiac sarcolemmal vesicles. *J. Biol. Chem.* 261:4948–4955.

Reimer, K.A., Hill, M.L., and Jennings, R.B. (1981). Prolonged depletion of ATP and of the adenine nucleotide pool due to delayed resynthesis of adenine nucleotides following reversible myocardial ischemic injury in dogs. *J. Mol. Cell Cardiol.* 13:229–239.

Ribalet, B., Ciani, S., and Eddlestone, G.T. (1989). Modulation of ATP-sensitive K channels in RINm5F cells by phosphorylation and G proteins. *Biophys. J.* 55:587a.

Rovetto, M.J., Whitmer, J.T., and Neely, J.R. (1973). Comparison of the effects of anoxia and whole heart ischemia on carbohydrate utilization in isolated working rat hearts. *Circ. Res.* 32:699–711.

Rowe, G.T., Eaton, L.R., and Hess, M.L. (1984). Neutrophil-derived, oxygen free radical-mediated cardiovascular dysfunction. *J. Mol. Cell Cardiol.* 16:1075–1079.

Runnman, E.M., Lamp, S.T., and Weiss, J.N. (1990). Enhanced utilization of exogenous glucose improves cardiac function in hypoxic rabbit ventricle without increasing total glycolytic flux. *J. Clin. Invest.* 86:1222–1233.

Saddik, M., and Lopaschuk, G.D. (1991). Myocardial triglyceride turnover and contribution to energy substrate utilization in isolated working rat hearts. *J. Biol. Chem.* 266:8162–8170.

Sato, T., Kiyosue, T., and Arita, M. (1992). Inhibitory effects of palmitoylcarnitine and lysophosphatidylcholine on the sodium current of cardiac ventricular cells. *Pflugers Arch.* 420:94–100.

Satoh, H., Hayashi, H., Katoh, H., Terada, H., and Kobayashi, A. (1995). Na$^+$/H$^+$ and Na$^+$/Ca^{2+} exchange in regulation of [Na$^+$]$_i$ and [Ca^{2+}]$_i$ during metabolic inhibition. *Am. J. Physiol.* 268:H1239–1248.

Schaefer, S., Schwartz, G.G., Gober, J.R., Wong, A.K., Camacho, S.A., Massie, B., and Weiner, M.W. (1990). Relationship between myocardial metabolites and contractile abnormalities during graded regional ischemia. Phosphorus-31 nuclear magnetic studies of porcine myocardium *in vivo*. *J. Clin. Invest.* 85:706–713.

Scholz, W., Albus, U., Lang, H.J., Linz, W., Martorana, P.A., Englert, H.C., and Scholkens, B.A. (1993). Hoe 694, a new Na^+/H^+ exchange inhibitor and its effects in cardiac ischaemia. *Br. J. Pharmacol.* 109:562–568.

Schrader, J. (1990). Adenosine. A homeostatic metabolite in cardiac energy metabolism. *Circulation* 81:389–391.

Schulz, R., Nava, E., and Moncada, S. (1992). Induction and potential biological relevance of a Ca(2+)-independent nitric oxide synthase in the myocardium. *Br. J. Pharmacol.* 105:575–580.

Schulze-Osthoff, K., Bakker, A.C., Vanhaesebroeck, B., Beyaert, R., Jacob, W.A., and Fiers, W. (1992). Cytotoxic activity of tumor necrosis factor is mediated by early damage of mitochondrial functions. Evidence for the involvement of radical generation. *J. Biol. Chem.* 267:5317–5323.

Schuurmans Stekhoven, F., and Bonting, S.L. (1981). Transport adenosine triphosphatases: Properties and functions. *Physiol. Rev.* 61:1–76.

Schwaiger, M., Neese, R.A., Araujo, L., Wyns, W., Wisnecki, J.A., Sochor, H., Swank, S., Kulber, D., Selin, C., Phelps, M., *et al.* (1989). Sustained nonoxidative glucose utilization and depletion of glycogen in reperfused canine myocardium. *J. Am. Coll. Cardiol.* 13:745–754.

Schwaiger, M., Schelbert, H.R., Ellison, D., Hansen, H., Yeatman, L., Vinten-Johansen, J., Selin, C., Barrio, J., and Phelps, M.E. (1985a). Sustained regional abnormalities in cardiac metabolism after transient ischemia in the chronic dog model. *J. Am. Coll. Cardiol.* 6:336–347.

Schwaiger, M., Schelbert, H.R., Keen, R., Vinten-Johansen, J., Hansen, H., Selin, C., Barrio, J., Huang, S.C., and Phelps, M.E. (1985b). Retention and clearance of C-11 palmitic acid in ischemic and reperfused canine myocardium. *J. Am. Coll. Cardiol.* 6:311–320.

Schwartz, G.G., Schaefer, S., Meyerhoff, D.J., Gober, J., Fochler, P., Massie, B., and Weiner, M.W. (1990). Dynamic relation between myocardial contractility and energy metabolism during and following brief coronary occlusion in the pig. *Circ. Res.* 67:490–500.

Sedlis, S.P., Corr, P.B., Sobel, B.E., and Ahumada, G.G. (1983). Lysophosphatidyl choline potentiates Ca^{2+} accumulation in rat cardiac myocytes. *Am. J. Physiol.* 244:H32–H38.

Sedlis, S.P., Sequeira, J.M., and Altszuler, H.M. (1990). Potentiation of the depressant effects of lysophosphatidylcholine on contractile properties of cultured cardiac myocytes by acidosis and superoxide radical. *J. Lab. Clin. Med.* 115:203–216.

Seki, S., and MacLeod, K.T. (1995). Effects of anoxia on intracellular Ca^{2+} and contraction in isolated guinea pig cardiac myocytes. *Am. J. Physiol.* 268:H1045–1052.

Shah, A.M., Mebazaa, A., Wetzel, R.C., and Lakatta, E.G. (1994). Novel cardiac myofilament desensitizing factor released by endocardial and vascular endothelial cells. *Circulation* 89:2492–2497.

Shattock, M.J., and Matsuura, H. (1993). Measurement of Na^+-K^+ pump current in isolated rabbit ventricular myocytes using the whole-cell voltage-clamp technique. Inhibition of the pump by oxidant stress. *Circ. Res.* 72:91–101.

Shen, A.C., and Jennings, R.B. (1972). Myocardial calcium and magnesium in acute ischemic injury. *Am. J. Pathol.* 67:417–440.

Shieh, R.C., Goldhaber, J.I., Stuart, J.S., and Weiss, J.N. (1994). Lactate transport in mammalian ventricle. General properties and relation to K^+ fluxes. *Circ. Res.* 74:829–838.

Shigekawa, M., Dougherty, J.P., and Katz, A.M. (1978). Reaction mechanism of Ca^{2+}-dependent ATP hydrolysis by skeletal muscle sarcoplasmic reticulum in the absence of added alkali metal salts. I. Characterization of steady state ATP hydrolysis and comparison with that in the presence of KCl. *J. Biol. Chem.* 253:1442–1450.

Shiki, K., and Hearse, D.J. (1987). Preconditioning of ischemic myocardium: Reperfusion-induced arrhythmias. *Am. J. Physiol.* 253:H1470–H1476.

Shine, K.I., Douglas, A.M., and Ricchiuti, N. (1976). Ischemia in isolated interventricular septa: Mechanical events. *Am. J. Physiol.* 231:1225–1232.

Shizukuda, Y., Mallet, R.T., Lee, S.C., and Downey, H.F. (1992). Hypoxic preconditioning of ischaemic canine myocardium. *Cardiovasc. Res.* 26:534–542.

Shlafer, M., Myers, C.L., and Adkins, S. (1987). Mitochondrial hydrogen peroxide generation and activities of glutathione peroxidase and superoxide dismutase following global ischemia. *J. Mol. Cell Cardiol.* 19:1195–1206.

Silverman, H.S., Ninomiya, M., Blank, P.S., Hano, O., Miyata, H., Spurgeon, H.A., Lakatta, E.G., and Stern, M.D. (1991). A cellular mechanism for impaired posthypoxic relaxation in isolated cardiac myocytes. *Circ. Res.* 69:196–208.

Smith, G.L., Donoso, P., Bauer, C.J., and Eisner, D.A. (1993). Relationship between intracellular pH and metabolite concentrations during metabolic inhibition in isolated ferret heart. *J. Physiol.* (London) 472:11–22.

Smith, J.A., Shah, A.M., and Lewis, M.J. (1991). Factors released from endocardium of the ferret and pig modulate myocardial contraction. *J. Physiol.* (London) 439:1–14.

Snowdowne, K.W., Ertel, R.J., and Borle, A.B. (1985). Measurement of cytosolic calcium with aequorin in dispersed rat ventricular cells. *J. Mol. Cell. Cardiol.* 17:233–241.

Soei, L.K., Sassen, L.M., Fan, D.S., van Veen, T., Krams, R., and Verdouw, P.D. (1994). Myofibrillar Ca^{2+} sensitization predominantly enhances function and mechanical efficiency of stunned myocardium. *Circulation* 90:959–969.

Sparks, Jr., H.V., and Bardenheuer, H. (1986). Regulation of adenosine formation by the heart. *Circ. Res.* 58:193–201.

Speechly-Dick, M.E., Grover, G.J., and Yellon, D.M. (1995). Does ischemic preconditioning in the human involve protein kinase C and the ATP-dependent K^+ channel? Studies of contractile function after simulated ischemia in an atrial *in vitro* model. *Circ. Res.* 77:1030–1035.

Steenbergen, C., Murphy, E., Levy, L., and London, R.E. (1987). Elevation in cytosolic free calcium concentration early in myocardial ischemia in perfused rat heart. *Circ. Res.* 60:700–707.

Steenbergen, C., Murphy, E., Watts, J.A., and London, R.E. (1990). Correlation between cytosolic free calcium, contracture, ATP, and irreversible ischemic injury in perfused rat heart. *Circ. Res.* 66:135–146.

Steenbergen, C., Perlman, M.E., London, R.E., and Murphy, E. (1993). Mechanism of preconditioning. Ionic alterations. *Circ. Res.* 72:112–125.

Stern, M.D. 1992. Theory of excitation–contraction coupling in cardiac muscle. *Biophys. J.* 63:497–517.

Stern, M.D., Silverman, H.S., Houser, S.R., Josephson, R.A., Capogrossi, M.C., Nichols, C.G., Lederer, J.W., and Lakatta, E.G. (1988). Anoxic contractile failure in rat heart myocytes is caused by failure of intracellular calcium release due to alteration of the action potential. *Proc. Natl. Acad. Sci. USA* 85:6954–6958.

Stewart, L.C., Deslauriers, R., and Kupriyanov, V.V. (1994). Relationships between cytosolic [ATP], [ATP]/[ADP] and ionic fluxes in the perfused rat heart: A P-31, Na-23 and Rb-87 NMR study. *J. Mol. Cell Cardiol.* 26:1377–1392.

Stiles, G.L. 1992. Adenosine receptors. *J. Biol. Chem.* 267:6451–6454.

Stuehr, D.J., Cho, J.J., Kwon, N.S., Weise, M.F., and Nathan, C.F. (1991). Purification and characterization of the cytokine-induced macrophage nitric oxide synthase: An FAD- and FMN-containing flavoprotein. *Proc. Natl. Acad. Sci. USA* 88:7773–7777.

Tanaka, M., Gilbert, J., and Pappano, A.J. (1992). Inhibition of sodium pump by 1-palmitoyl-carnitine in single guinea-pig ventricular myocytes. *J. Mol. Cell Cardiol.* 24:711–719.

Tani, M., and Neely, J.R. (1989). Role of intracellular Na^+ in Ca^{2+} overload and depressed recovery of ventricular function of reperfused ischemic rat hearts. Possible involvement of H^+-Na^+ and Na^+-Ca^{2+} exchange. *Circ. Res.* 65:1045–1056.

Taniguchi, J., Noma, A., and Irisawa, H. (1983). Modification of the cardiac action potential by intracellular injection of adenosine triphosphate and related substances in guinea pig single ventricular cells. *Circ. Res.* 53:131–139.

Terzic, A., Jahangir, A., and Kurachi, Y. (1995). Cardiac ATP-sensitive K^+ channels: Regulation by intracellular nucleotides and K^+ channel-opening drugs. *Am. J. Physiol.* 38:C525–C545.

Thornton, J.D., Liu, G.S., and Downey, J.M. (1993a). Pretreatment with pertussis toxin blocks the protective effects of preconditioning. Evidence for a G-protein mechanism. *J. Mol. Cell Cardiol.* 25:311–320.

Thornton, J.D., Thornton, C.S., and Downey, J.M. (1993b). Effect of adenosine receptor blockade: Preventing protective preconditioning depends on time of initiation. *Am. J. Physiol.* 265:H504–H508.

Thornton, J.D., Thornton, C.S., Sterling, D.L., and Downey, J.M. (1993c). Blockade of ATP-sensitive potassium channels increases infarct size but does not prevent preconditioning in rabbit hearts. *Circ. Res.* 72:44–49.

Tillisch, J., Brunken, R., Marshall, R., Schwaiger, M., Mandelkern, M., Phelps, M., and Schelbert, H. (1986). Reversibility of cardiac wall-motion abnormalities predicted by positron tomography. *N. Engl. J. Med.* 314:884–888.

Toombs, C.F., McGee, S., Johnston, W.E., and Vinten-Johansen, J. (1992). Myocardial protective effects of adenosine. Infarct size reduction with pretreatment and continued receptor stimulation during ischemia. *Circulation* 86:986–994.

Tracey, K.J., Beutler, B., Lowry, S.F., Merryweather, J., Wolpe, S., Milsark, I.W., Hariri, R.J., Fahey, T.J., Zentella, A., and Albert, J.D. (1986). Shock and tissue injury induced by recombinant human cachectin. *Science* 234:470–474.

Tsujimoto, M., Yokota, S., Vilcek, J., and Weissmann, G. (1986). Tumor necrosis factor provokes superoxide anion generation from neutrophils. *Biochem. Biophys. Res. Commun.* 137:1094–1100.

Tsujino, M., Hirata, Y., Imai, T., Kanno, K., Eguchi, S., Ito, H., and Marumo, F. (1994). Induction of nitric oxide synthase gene by interleukin-1 beta in cultured rat cardiocytes. *Circulation* 90:375–383.

Undrovinas, A.I., Fleidervish, I.A., and Makielski, J.C. (1992). Inward sodium current at resting potentials in single cardiac myocytes induced by the ischemic metabolite lysophosphatidylcholine. *Circ. Res.* 71:1231–1241.

Van Bilsen, M., van der Vusse, G.J., Coumans, W.A., de Groot, M.J., Willemsen, P.H., and Reneman, R.S. (1989). Degradation of adenine nucleotides in ischemic and reperfused rat heart. *Am. J. Physiol.* 257:H47–H54.

van Bilsen, M., Van der Vusse, G.J., Willemsen, P.H.M., Coumans, W.A., Roemen, T.H.M., and Reneman, R.S. (1989). Lipid alterations in isolated, working rat hearts during ischemia and reperfusion: Its relation to myocardial damage. *Circ. Res.* 64:304–314.

Venkatesh, N., Lamp, S.T., and Weiss, J.N. (1991). Sulfonylureas, ATP-sensitive K^+ channels, and cellular K^+ loss during hypoxia, ischemia and metabolic inhibition in mammalian ventricle. *Circ. Res.* 69:623–637.

Vial, C., Font, B., Goldschmidt, D., Pearlman, A.S., and DeLaye, J. (1978). Regional myocardial energetics during brief periods of coronary occlusion and reperfusion: Comparison with S-T segment changes. *Cardiovasc. Res.* 12:470–647.

Vleugels, A., Vereecke, J., and Carmeliet, E. (1980). Ionic currents during hypoxia in voltage-clamped cat ventricular muscle. *Circ. Res.* 47:501–508.

Wang, S.Y., Clague, J.R., and Langer, G.A. (1995). Increase in calcium leak channel activity by metabolic inhibition or hydrogen peroxide in rat ventricular myocytes and its inhibition by polycation. *J. Mol. Cell Cardiol.* 27:211–222.

Webb, J.L. (1966). "Enzyme and Metabolic Inhibitors" pp. 1–3. Academic Press, New York.

Weiss, J., Couper, G.S., Hiltbrand, B., and Shine, K.I. (1984). Role of acidosis in early contractile dysfunction during ischemia: Evidence from pH_o measurements. *Am. J. Physiol.* 247:H760–H767.

Weiss, J., and Hiltbrand, B. (1985). Functional compartmentation of glycolytic versus oxidative metabolism in isolated rabbit heart. *J. Clin. Invest.* 75:436–447.

Weiss, J. N., and Lamp, S.T. (1987). Glycolysis preferentially inhibits ATP-sensitive K^+ channels in isolated guinea pig cardiac myocytes. *Science* 238:67–69.

Weiss, J.N., and Lamp, S.T. (1989). Cardiac ATP-sensitive K^+ channels: Evidence for preferential regulation by glycolysis. *J. Gen. Physiol.* 94:911–935.

Weiss, J.N., Lamp, S.T., and Shine, K.I. (1989). Cellular K^+ loss and anion efflux during myocardial ischemia and metabolic inhibition. *Am. J. Physiol.* 256:H1165–H1175.

Weiss, J.N., and Venkatesh, N. (1993). Metabolic regulation of cardiac ATP-sensitive K^+ channels. *Cardiovasc. Drugs Ther.* 7:499–505.

Weiss, J.N., Venkatesh, N., and Lamp, S.T. (1992). ATP-sensitive K^+ channels and cellular K^+ loss in hypoxic and ischaemic mammalian ventricle. *J. Physiol.* 447:649–73.

Werns, S.W., Shea, M.J., Driscoll, E.M., Cohen, C., Abrams, G.D., Pitt, B., and Lucchesi, B.R. (1985). The independent effects of oxygen radical scavengers on canine infarct size. *Circ. Res.* 56:895–898.

Wilde, A.A., and Aksnes, G. (1995). Myocardial potassium loss and cell depolarisation in ischaemia and hypoxia. *Cardiovasc. Res.* 29:1–15.

Wilde, A.A.M., Escande, D., Schumacher, C.A., Thuringer, D., Mestre, M., Fiolet, J.W.T., and Janse, M.J. (1990). Potassium accumulation in the globally ischemic mammalian heart: A role for the ATP-sensitive K^+ channel. *Circ. Res.* 67:835–843.

Wilde, A.A.M., and Kleber, A.G. (1986). The combined effects of hypoxia, high K^+, and acidosis on the intracellular sodium activity and resting potential in guinea pig papillary muscle. *Circ. Res.* 58:249–256.

Wolfe, C.L., Gilbert, H.F., Brindle, K.M., and Radda, G.K. (1988). Determination of buffering capacity of rat myocardium during ischemia. *Biochim. Biophys. Acta* 971:9–20.

Wolfe, C.L., Sievers, R.E., Visseren, F.L., and Donnelly, T.J. (1993). Loss of myocardial protection after preconditioning correlates with the time course of glycogen recovery within the preconditioned segment. *Circulation* 87:881–892.

Wong, G.H., and Goeddel, D.V. (1988). Induction of manganous superoxide dismutase by tumor necrosis factor: Possible protective mechanism. *Science* 242:941–944.

Woodley, S.L., Ikenouchi, H., and Barry, W.H. (1991). Lysophosphatidylcholine increases cytosolic calcium in ventricular myocytes by direct action on the sarcolemma. *J. Mol. Cell Cardiol.* 23:671–680.

Wu, J., and Corr, P.B. (1994). Palmitoyl carnitine modifies sodium currents and induces transient inward current in ventricular myocytes. *Am. J. Physiol.* 266:H1034–H1046.

Wu, J.Y., and Corr, P.B. (1992). Influence of long-chain acylcarnitines on voltage-dependent calcium current in adult ventricular myocytes. *Am. J. Physiol.* 263:H410–H417.

Wyatt, D.A., Edmunds, M.C., Rubio, R., Berne, R.M., Lasley, R.D., and Mentzer, Jr., R.M. (1989). Adenosine stimulates glycolytic flux in isolated perfused rat hearts by A1-adenosine receptors. *Am. J. Physiol.* 257:H1952–H1957.

Xu, K.Y., Zweier, J.L., and Becker, L.C. (1995). Functional coupling between glycolysis and sarcoplasmic reticulum Ca^{2+} transport. *Circ. Res.* 77:88–97.

Yamauchi, N., Kuriyama, H., Watanabe, N., Neda, H., Maeda, M., and Niitsu, Y. (1989). Intracellular hydroxyl radical production induced by recombinant human tumor necrosis factor and its implication in the killing of tumor cells *in vitro. Cancer Res.* 49:1671–1675.

Yan, G.X., Chen, J., Yamada, K.A., Kleber, A.G., and Corr, P.B. (1996). Contribution of extracellular K^+ accumulation in myocardial ischaemia of the rabbit. *J. Physiol.* (London) 490:215–228.

Yan, G.X., and Kleber, A.G. (1992). Changes in extracellular and intracellular pH in ischemic rabbit papillary muscle. *Circ. Res.* 71:460–470.

Yao, Z.H., and Gross, G.J. (1993). Acetylcholine mimics ischemic preconditioning via a glibenclamide-sensitive mechanism in dogs. *Am. J. Physiol.* 264:H2221–H2225.

Yellon, D.M., Alkhulaifi, A.M., Browne, E.E., and Pugsley, W.B. (1992). Ischaemic preconditioning limits infarct size in the rat heart. *Cardiovasc. Res.* 26:983–987.

Yokota, R., Fujiwara, H., Miyamae, M, Tanaka, M., Yamasaki, K., Itoh, S., Koga, K., Yabuuchi, Y., and Sasayama, S. (1995). Transient adenosine infusion before ischemia and reperfusion protects against metabolic damage in pig hearts. *Am. J. Physiol.* 268:H1149–H1157.

Yokoyama, T., Vaca, L., Rossen, R.D., Durante, W., Hazarika, P., and Mann, D.L. (1993). Cellular basis for the negative inotropic effects of tumor necrosis factor-alpha in the adult mammalian heart. *J. Clin. Invest.* 92:2303–2312.

Ytrehus, K., Liu, Y.G., and Downey, J.M. (1994). Preconditioning protects ischemic rabbit heart by protein kinase C activation. *Am. J. Physiol.* 266:H1145–H1152.

Zweier, J.L. (1988). Measurement of superoxide-derived free radicals in the reperfused heart. Evidence for a free radical mechanism of reperfusion injury. *J. Biol. Chem.* 263:1353–1357.

Zweier, J.L., Flaherty, J.T., and Weisfeldt, M.L. (1987). Direct measurement of free radical generation following reperfusion of ischemic myocardium. *Proc. Natl. Acad. Sci. USA* 84:1404–1407.

INDEX

Printed and bound by CPI Group (UK) Ltd, Croydon, CR0 4YY

08/05/2025

01864988-0001